Springer Proceedings in Business and Economics

More information about this series at http://www.springer.com/series/11960

Jocelyn Bellemare • Serge Carrier
Kjeld Nielsen • Frank T. Piller
Editors

Managing Complexity

Proceedings of the 8th World Conference on Mass Customization, Personalization, and Co-Creation (MCPC 2015), Montreal, Canada, October 20th-22th, 2015

Springer

Editors
Jocelyn Bellemare
Department of Management & Technology
University of Quebec at Montreal
Montreal, QC, Canada

Serge Carrier
Department of Management & Technology
University of Quebec at Montreal
Montreal, QC, Canada

Kjeld Nielsen
Department of Mechanical and
Manufacturing Engineering
Aalborg University
Aalborg, Denmark

Frank T. Piller
Technology and Innovation Management
RWTH Aachen University
Aachen, Germany

ISSN 2198-7246 ISSN 2198-7254 (electronic)
Springer Proceedings in Business and Economics
ISBN 978-3-319-29056-0 ISBN 978-3-319-29058-4 (eBook)
DOI 10.1007/978-3-319-29058-4

Printed on acid-free paper

This Springer imprint is published by SpringerNature
The registered company is Springer International Publishing AG Switzerland.

Preface

Twenty years ago mass customization was acknowledged as the "New Frontier in Business Competition." The first MCPC conference was hosted by Hong Kong University of Science and Technology in 2001. Since then the MCPC conference has grown to become the primary conference for presenting and discussing current issues and recent developments within the fields of mass customization, personalization, and customer co-creation. The 2015 MCPC conference, the eighth in the series, for which the contributions are presented in this book, was hosted by the School of Management at the University du Québec in Montréal, Canada. For this edition, the emphasis was placed on "managing complexity."

Research on management in general, and on mass customization more specifically, has evolved tremendously over the last few years. Often more focused on the link between theory and practice, it has allowed researchers and practitioners to present new viable business models. Yet, relying on the most recent technological advances, these new business models often increase the level of complexity in management. The creation of added value and market differentiation are a direct result of one's ability to manage this complexity. Hence, since operational excellence is strongly correlated to the ability to simplify complex entities, good managers must have an intuitive feel for this characteristic, understand it, and work hard to reduce it where possible. Unfortunately, the literature more often obscures the subject more than it reveals it.

The MCPC 2015 was a multitrack conference featuring a combination of high profile keynotes with expert talks, panel discussions, paper sessions, workshops, receptions, and much more. While it was devoted to sharing and discussing the latest research in the field, MCPC 2015 strongly emphasized real-life applications. The MCPC conference is truly unique among conferences in that, since its beginning, it has attracted an equal share of practitioners and academics/researchers. This year, more than 200 academics, entrepreneurs, and management experts presented the most recent developments in mass customization and co-creation. A total of 60 conferences (academic, projects, case studies, etc.) were offered to the participants.

Recognized for the quality of its practice-oriented education, the excellence of its applied research, and its international presence, the School of Management (ESG

UQAM) asserts its leadership with a bold vision for the future. In presenting the MCPC 2015 conference, it strived to engage academics, business leaders, and consultants in fundamental debates on managing complexity.

This book presents the latest research from the worldwide MCPC community bringing together the new thoughts and results from various disciplines within the topics of:

- Complexity management of knowledge-based systems in manufacturing design and production
- Sustainable mass customization
- Fashion, apparel, and footwear applications
- Manufacturing systems for MCPC
- Product modeling
- Choice navigation
- MCPC applications
- Solution space development
- Co-creation and open innovation

All papers have been peer reviewed to ensure the same high quality as seen on previous MCPC conferences.

The organizing committee would like to thank the MCPC community for the support for this conference, hoping that all participants, academic and industrial, will benefit from the presentations and discussions.

Montreal, Canada — Jocelyn Bellemare
Montreal, Canada — Serge Carrier
Aalborg, Denmark — Kjeld Nielsen
Aachen, Germany — Frank T. Piller

Contents

Contributors

Erlend Alfnes Department of Production and Quality Engineering, Norwegian University of Science and Technology, Trondheim, Norway

Ann-Louise Andersen Department of Mechanical and Manufacturing Engineering, Aalborg University, Aalborg, Denmark

Mads Bejlegaard Department of Mechanical and Manufacturing Engineering, Aalborg University, Aalborg, Denmark

Jocelyn Bellemare Department of Management and Technology, University of Quebec in Montreal (UQAM), Montreal, QC, Canada

Paul Blažek cyLEDGE Media GmbH, Vienna, Austria

Harold Boeck School of Management, ESG-UQAM, Montreal, QC, Canada

Martin Bonev Department of Engineering Management, Technical University of Denmark, Lyngby, Denmark

Jacob Bossen Department of Mechanical and Manufacturing Engineering, Aalborg University, Aalborg, Denmark

Thomas Ditlev Brunoe Department of Mechanical and Manufacturing Engineering, Aalborg University, Aalborg, Denmark

Carlo Carbone UQAM, Ecole de Design, Montreal, QC, Canada

Serge Carrier Department of Management and Technology, University of Quebec in Montreal (UQAM), Montreal, QC, Canada

Valérie Côté University of Montreal, Montreal, QC, Canada

François Courtemanche Tech3lab, HEC Montreal, Montréal, QC, Canada

Aldo Duchi BWI Center für Industrial Management, ETH Zürich, Zürich, Switzerland

Robert J. Freund Burgwald, Germany

Caroline Gagnon Laval University, Quebec, QC, Canada

Paul Christoph Gembarski Institute of Product Development, Welfengarten, Hannover, Germany

Valerie Gohier Imarklab, Montréal, QC, Canada

Philipp Gottwald Institute of Product Development, Welfengarten, Hannover, Germany

Leontin K. Grafmüller HHL Leipzig Graduate School of Management, Leipzig, Germany

Bill Grier AM4U Inc., Rancho Cucamonga, CA, USA

Hagen Habicht HHL Leipzig Graduate School of Management, Leipzig, Germany

Jan Willem Hoftijzer University of Technology Delft, Delft, The Netherlands

Simone Honetz cyLEDGE Media GmbH, Vienna, Austria

Lars Hvam Department of Engineering Management, Technical University of Denmark, Lyngby, Denmark

Kim Noergaard Jensen Department of Mechanical and Manufacturing Engineering, Aalborg University, Aalborg East, Denmark

Ron Journée Windesheim University of Applied Sciences, Zwolle, The Netherlands

Alexandre Joyce Concordia University, Montreal, QC, Canada

Thomas Kampschulte Aftermarket, thyssenkrupp Bilstein GmbH, Ennepetal, Germany

Peter Kilduff California State Polytechnic University, Pomona, CA, USA

Angella Kim University of Minnesota, St. Paul, MN, USA

Monika Kolb cyLEDGE Media GmbH, Vienna, Austria

Jayoung Koo University of Minnesota, St. Paul, MN, USA

Katrin Kristjansdottir Department of Engineering Management, Technical University of Denmark, Lyngby, Denmark

Matthias Kulcke Hamburg University of Technology, Hamburg, Germany

Roland Lachmayer Institute of Product Development, Welfengarten, Hannover, Germany

Nataly Levesque School of Management, ESG-UQAM, Montreal, QC, Canada

Haibing Li Institut für Produktentwicklung und Gerätebau, Leibniz Universität Hannover, Hannover, Germany

Søren Munch Lindhard Department of Mechanical and Manufacturing Engineering, Aalborg University, Aalborg East, Denmark

R. Bastian Lippert Institute of Product Development, Welfengarten, Hannover, Germany

Hans Lundberg School of Business and Economics, Linnaeus University, Växjö, Sweden

Omid Maghazei BWI Center für Industrial Management, ETH Zürich, Zürich, Switzerland

Aurelie Merles École de Management, Université de Grenoble, Grenoble, France

Basem Eid Mohamed Department of Architecture, Abu Dhabi University, Abu Dhabi, UAE

Anna Myrodia Department of Engineering Management, Technical University of Denmark, Lyngby, Denmark

Antonia Natsika School of Engineering, Demokritus University of Thrace, Xanthi, Greece

Kjeld Nielsen Department of Mechanical and Manufacturing Engineering, Aalborg University, Aalborg, Denmark

F. Xavier Olleros Department of Management and Technology, University of Quebec in Montreal (UQAM), Montreal, QC, Canada

Daniele Parisi Department of Management Engineering, Politecnico di Milano, Milan, Italy

Birgit Penzenstadler Department of Computer Engineering and Computer Science, California State University, Long Beach, CA, USA

Florian Pichonneau Sidlee, Montreal, QC, Canada

Sandrine Prom Tep Department of Marketing, ESG, University of Quebec in Montréal, Montreal, QC, Canada

Paul Schönsleben BWI Center für Industrial Management, ETH Zürich, Zürich, Switzerland

Muditha Senanayake California State Polytechnic University, Pomona, CA, USA

Sylvain Sénécal Department of Marketing, HEC Montreal, Montreal, QC, Canada

Sara Shafiee Department of Engineering Management, Technical University of Denmark, Lyngby, Denmark

Lars Skjelstad SINTEF Industrial Management, Trondheim, Norway

Anik St-Onge Department of Marketing, University of Quebec in Montreal (UQAM), Montreal, QC, Canada

Simon Haahr Storbjerg Department of Mechanical and Manufacturing Engineering, Aalborg University, Aalborg, Denmark

Clarissa Streichsbier cyLEDGE Media GmbH, Vienna, Austria

Ian Sutherland IEDC-Bled School of Management, Bled, Slovenia

Filippo Tamburini Department of Management Engineering, Politecnico di Milano, Milan, Italy

Stig B. Taps Department of Mechanical and Manufacturing Engineering, Aalborg University, Aalborg, Denmark

Maria K. Thomassen SINTEF Technology and Society, Industrial Management, Trondheim, Norway

Natasha Thoreson University of Minnesota, St. Paul, MN, USA

Alexander Tsigkas School of Engineering, Demokritus University of Thrace, Xanthi, Greece

Marcel Weber Windesheim University of Applied Sciences, Zwolle, The Netherlands

3CI Customer Co-Creation, Waalwijk, The Netherlands

Juanjuan Wu Department of Design, Housing, and Apparel, University of Minnesota, St. Paul, MN, USA

Chapter 1
Mass Customization in the Building and Construction Industry

Kim Noergaard Jensen, Kjeld Nielsen, Thomas Ditlev Brunoe, and Søren Munch Lindhard

1.1 Introduction

The productivity in the Danish construction industry has doubled since 1966, which is significantly less than other sectors in Denmark (Fig. 1.1). The construction industry employs approx. 25 % [4] of the private workforce in Denmark, and this industry is currently facing a number of challenges, including a lot of burden on costs that makes companies continuously searching for initiatives to reduce production costs to meet competition.

Increasing industrialization has achieved results in other industries in Denmark in terms of increasing productivity. We interpret increased industrialization as increasing utilization of new technologies for production, streamlining, and constant development of production processes and other correlated support processes. Productivity is here measured as output per performed working hour for the entire Danish economy [7].

One of the reasons that construction industry is having less degree of industrialization is that construction industry, opposite the standardized products that formed the basis for the industrial revolution, is very different and often one of a kind, and therefore, it may seem difficult or challenging to streamline and optimize processes as "assembly line production" [1, 3].

However, over the past decades, industrial production has gone through a process in which more and more companies are offering customized products [13] "at a price near Mass Production [1]" under the production strategy called mass customization [11].

K.N. Jensen (✉) • K. Nielsen • T.D. Brunoe • S.M. Lindhard
Department of Mechanical and Manufacturing Engineering, Aalborg University, Fibigerstraede 16, 9220 Aalborg East, Denmark
e-mail: knj@m-tech.aau.dk

J. Bellemare et al. (eds.), *Managing Complexity*, Springer Proceedings in Business and Economics, DOI 10.1007/978-3-319-29058-4_1

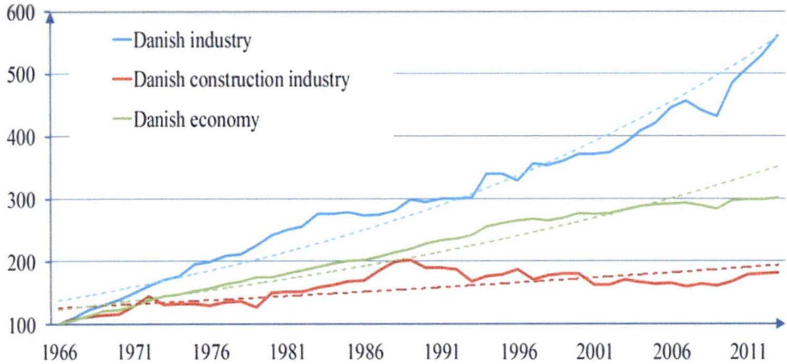

Fig. 1.1 Labor productivity by industry, unit cost, and time (Year (index 1966=100, 1966-price level chain figures) [Statistics Bank NATP23])

The construction industry's traditional demand for customization, e.g., distinctive architecture, function, quality, timeframe, and environment, is difficult to reconcile with traditional industrialization (standardization, mass production). However, mass customization characterizes the requirement of flexible products and processes.

In mass customization, applied IT tools like automated business processes, product configurators, flexible production processes, and product design allow a high degree of customization where the end customer can choose from millions of product variants and chose the flavor that just matches unique needs for a low price (cost minimization) [6, 11]. These principles are widely and with great success used in, e.g., automotive and computer industry [11, 12].

The applied principles behind mass customization enable industrial production of customized products [11], and for the construction industry, a great potential may result in applying these principles, as they face the challenge of producing products with high variety and often one-of-a-kind production [6].

Some segments of the building industry supplying the construction industry have already implemented parts of mass customization where some manufactures of windows, doors, kitchen, housing, and bath products offer customized products manufactured in a highly automated and flexible production [2, 5].

A current research project has as a goal to increase knowledge and utilization of mass customization in Danish construction industries. The objectives are to make Danish companies in the construction industry capable of implementing the principles of mass customization leading to increasing industrialization and productivity. Previous work in this project found that the participant companies all are planning to be more mass customization oriented (volume, variants) [8].

Recent research shows that companies that utilize mass customization must have three fundamental capabilities [12]:

1. "Solution Space Development: the ability to identify how customer requirements are different and develop products that can effectively adapt to these individual requirements through the product platforms or modularization."

2. "Choice Navigation: the ability to guide the customer to select or configure the product that matches his/she requirements."
3. "Robust Process Design: the ability to efficiently to produce a large batch of products at low cost that typically is achieved by using the flexible manufacturing systems."

Furthermore, the objectives of the project are to develop these three capabilities for companies in the construction industry in order to utilize mass customization. These companies will be able to realize a greater growth potential, as they will be able to meet the needs from their customers faster and at a lower cost [9]. This may also lead to an increase in the ratio of exports in those companies, because of the development of capabilities focusing on a greater share of the market [9]. In relation to the three capabilities, the project includes for each of the companies the following activities organized as conferences, networking, and workshops [12]:

- Solution Space Development
  - Screen of the current product structure to determine to which extent modularization is used and identify the variety in the product portfolio.
  - Choose areas, e.g., specific parts of products, where the company can benefit from using modularization in a short and a long term to gain competitive advantages.
- Choice Navigation
  - Identify opportunities for application of product configuration for sale and specification of products in a sale situation.
  - Develop prototypes of configurators.
- Robust Process Design
  - Analyze current production processes to identify the potential of automation (business processes and physical production processes) and the use of flexible manufacturing equipment.
- General
  - Together with the company, conduct workshops where the knowledge gained about the tools and methods is incorporated on selected specific areas.

The participating companies will, during the project, gain knowledge and skills in the tools and methods to be utilized, so that after the participation, they are able to properly select and apply the tools for developing mass customization. The innovation potential of the participating companies should be able to meet the customers demand for unique products cost effectively by:

- Developing new and more cost-effective business and production processes.
- Developing products that can be customized more efficiently than traditional building products.

Apart from the innovation objectives in the project, there are also formulated research objectives. The research objectives are to analyze how companies in the Danish construction industry can typically benefit from utilizing mass customization principles. Furthermore, the objective is to identify which specific challenges these companies typically face when implementing mass customization, since these challenges are expected to differentiate from those met in the general manufacturing industry. Finally, the objective is to adapt methods for enhancing the performance within mass customization, so that they are applicable in the construction industry.

This chapter addresses the research objectives, by discussing the potential in applying mass customization and some of the challenges mentioned initially. This chapter summarizes the results of a questionnaire done in cooperation with a number of companies that are a part of the supply chain of the construction industry.

This chapter answers how a number of companies within the construction industry see themselves, the competitors, and the customers. How do they see the market demands of customized products and, furthermore, how do they rate their capabilities and the value proposition, the cost perspectives, and the production technology available for supporting the advantages related to utilizing of mass customization?

1.2 Questionnaire Approach

The methodology includes a number of companies participating in a questionnaire concerning their considerations of themselves, the markets/customers, the competitors, and the technology available related to mass customization.

The focus approach between the involved actors (Fig. 1.2) is as follows:

- Capabilities of the focal companies of providing customized products.
- Market/customers' demands and value proposition of customized products.
- Competitiveness of customized products.
- Supplying of flexible manufacturing technologies for making customized products.

The purpose of the questionnaire is to increase the knowledge and understanding of how the involved companies consider the following:

1. The market demand for customized products to determine the future demand trend of customized products and to see if there is any relation between planned initiatives of the companies and market demand.
2. The value proposition by offering customized products and to see if there is any relation between the market demand of customized products and the value proposition by offering customized products.
3. Their relationship to their customer concerning efficiently communication and to see if there is any relation between communication and their ability to handle change management projects.
4. Themselves compared to competitor's position of making and delivering customized products and to see if there is any relation between the market demands,

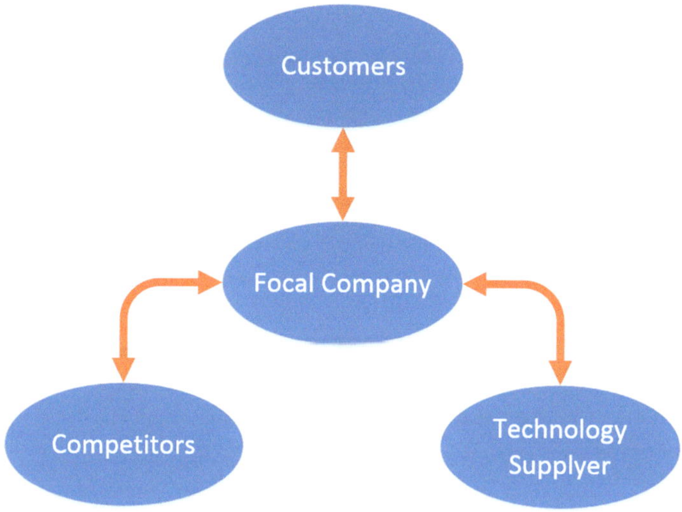

Fig. 1.2 Actors involved

the relative position of the competitors, and the future plans of the involved companies for offering customized products.

5. Their own cost perspectives related to offering customized products and to see if there is any relation between their cost perspectives and capabilities of delivering customized products and between their cost perspectives and the value proposition of delivering customized products.
6. Their future plans for delivering customized products and to see if there is any relation or gaps between their capabilities and future plans of offering customized products.
7. The industry's experience toward offering more flexibility in production technology and to determine to which extent production technology suppliers support the companies of making customized products to meet the market demands.

To address this, the following nine questions are created and asked to each of the involved companies:

1. To what extent do customers want customized products?
2. To what extent do customized products add value?
3. To what extent is customization of products costly?
4. To what extent do competitors selling customized products?
5. To which extent does the industry experience a trend toward more flexible production technology?
6. To what extent can customized products be distributed quickly and efficiently to customers?

7. To what extent does your company have efficient communication relations to their customers?
8. To what extent does your company have a desire to introduce customized solutions?
9. To what extent do companies manage change projects?

Each of the seven companies answered the above nine questions relative to the following classification:

1. No degree
2. To a lesser extent
3. To some extent
4. Great degree
5. A very high degree

The findings from the questionnaire are summarized in Sect. 1.4.

1.3 Questionnaire Participants

Nine companies have participated in the questionnaire, and the companies represent the following types of industries:

The company provides materials for all types of construction of single-family homes, townhouses, and industrial and commercial properties and institutions. Regardless of building type, the products contribute to the functional, individual, and profitable solutions. The products are flexible, and without the use of special tools, changes and adjustments can be very accurately made at the construction site. The concrete element is a fully breathable material that creates a flexible building process, and it is both environmentally friendly and fully recyclable.

The company is a large steel producer in Europe that develops and supplies a wide range of building components and solutions to the building and construction industry in some European countries. The components include trapezoidal profiles, sandwich panels, sinus profiles, wall cladding, architectural panels, and residential roof tiles and systems. The company provides customized products, focuses on introduction of new solutions, manufactures and distributes steel products, as well as provides services in design, technology, and consulting.

The company represented many places in the world and offers a wide range of products of titanium zinc for external roofing, roof drainage, façades, and internal designed solutions for companies, hotels, and public and private spaces. The services include illustrations, measures, dimensions, part numbers, etc., on products.

The company is a consulting firm that advises a number of companies in the construction industry and companies that supply the construction industry. In this context, this company represents a significant number of companies in the construction industry.

The company has many DIY stores in Denmark and Greenland and is the building industry's supplier of building materials and tools for craftsmen and private. It is crucial for the company that both the products and the advice customers receive in the shops are of the highest quality.

The company is a large and leading supplier of innovative products and systems based on stone wool. The company creates sustainable solutions to protect life, assets, and the environment based on one of nature's most abundant resources. Products are used for building insulation, industrial and technical insulation, and acoustic ceiling.

The company makes key elements for buildings serving acoustic comfort and a healthy indoor climate. The company manufactures acoustic panels from the natural material wood and cement and designs, develops, and manufactures a broad variety of panels in Denmark from local materials and delivers the products worldwide.

1.4 Results

The results of the questionnaire present relative to the research questions as mentioned in Chap. 2. A radar chart is used as this graphical method can display multivariate data in the form of a two-dimensional chart. The presented radar charts show the answers from each of the involved companies combined with the value of their answers to one or more questions.

Q1 in Fig. 1.3 shows a trend toward customers increasingly demanding customized products even though it is not significant for all the companies; likewise there is no relation between planned initiatives of the companies and market demand (Q1, Q8) even though it focuses on offering more customized products to their customers. There is a slight trend of the industry's experience toward offering more flexibility in production technology, so to some extent, the production technology suppliers support the companies of making customized.

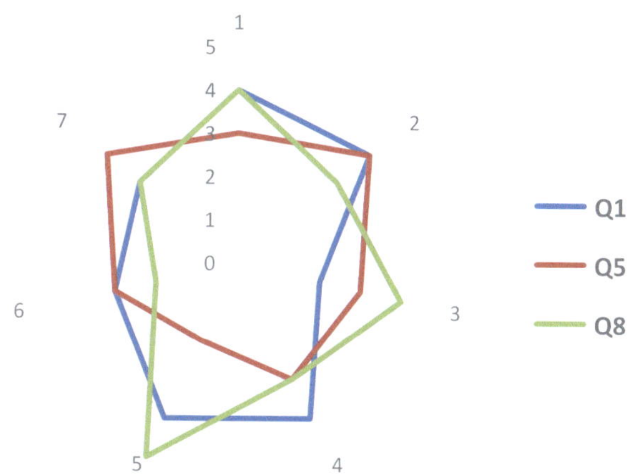

Fig. 1.3 Shows to what extent the customers want customized products (Q1), to which extent the industry experiences a trend toward more flexible production technology (Q5), and to what extent the companies have a desire to introduce customized solutions (Q8)

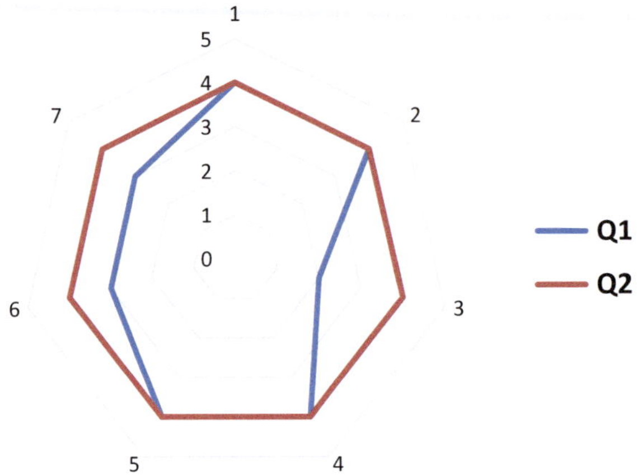

Fig. 1.4 Shows to what extent the customers want customized products (Q1) and to what extent the customized products add value to their customers (Q2)

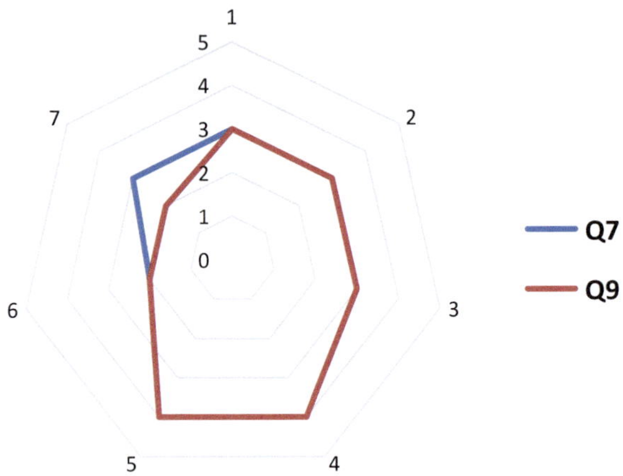

Fig. 1.5 Shows to what extent the company has efficient communication relations to their customers (Q7) and to what extent the companies manage change projects (Q9)

Figure 1.4 shows a significant trend toward adding value to the customer by offering customized products, which to some extent is in a balance with the market demand of customized products.

Figure 1.5 shows that the companies consider their relationship to their customer concerning efficient communication as high and almost at the same level as their ability to handle change management projects. This shows a significant relation between communication and their ability to handle change management projects, which can indicate a high custom oriented change readiness of the companies.

Figure 1.6 shows that some companies are considering themselves higher, equal, and lower compared to competitor's position of making and delivering customized products.

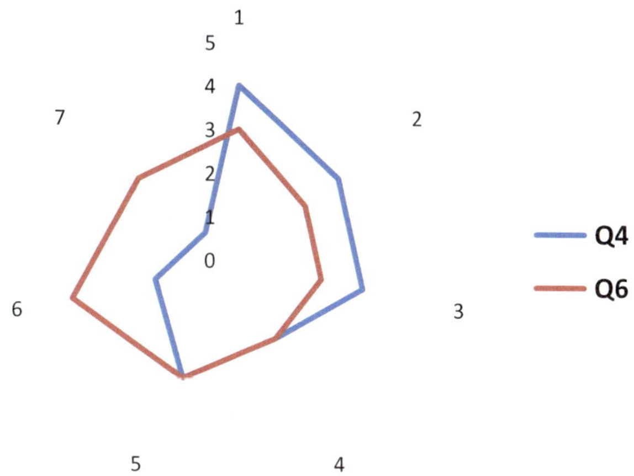

Fig. 1.6 Shows the companies quickly and efficiently distribute customized products (Q6)

Figure 1.7 determines if there is any relation between the market demands, the relative position of the competitors, and the plans of the involved companies for offering customized products, which seems not to be the case.

Figure 1.8 shows that the companies are considering own cost perspectives related to offering customized products as relatively significant, but most of all they consider the value adding significantly higher.

Figure 1.9 shows a slight trend telling that companies are considering their cost perspectives relatively high and high compared to their capabilities of delivering customized products, which can indicate room for improvement of their production and delivering flexibility.

Figure 1.10 shows that the companies have a relative high wish or plans for delivering customized products; but there does not seem to be any relation or gaps between their capabilities and plans of offering customized products.

1.5 Discussion

The results indicate that each company has different viewpoints in relation to the perspectives and utilization of mass customization even though all participants prior have gone through the same introduction to mass customization philosophy and introduction. This may be due to many reasons, e.g., individual skills, background, role within the company, knowledge about products, customers, competitors, and knowledge about internal business strategy.

The results from the questionnaire show that the involved companies' customers all want customized products (Fig. 1.4) even at a very high degree, and the result shows as well that customized products add value to their customers (Fig. 1.4). This

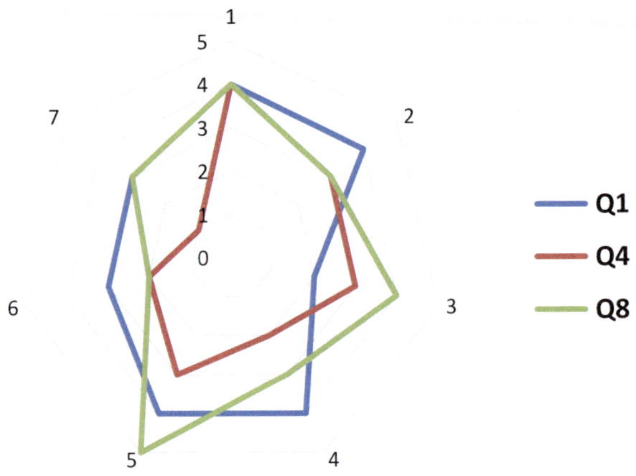

Fig. 1.7 Shows to what extent the customers want customized products (Q1), to what extent their competitors are selling customized products (Q4), and to what extent the companies have a desire to introduce customized solutions (Q8)

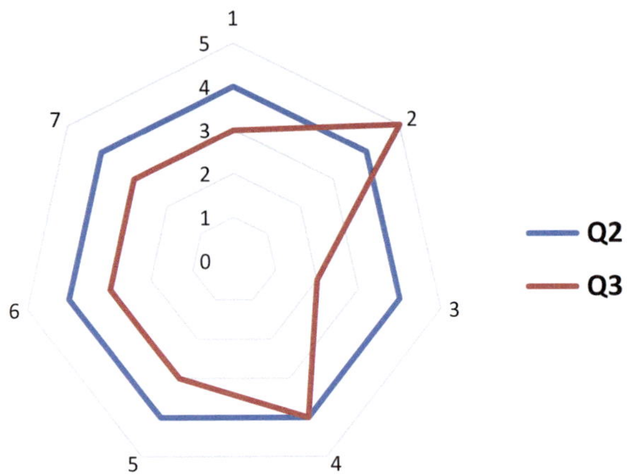

Fig. 1.8 Shows to what extent the customized products add value (Q2) and to what extent the customization of products is costly (Q3)

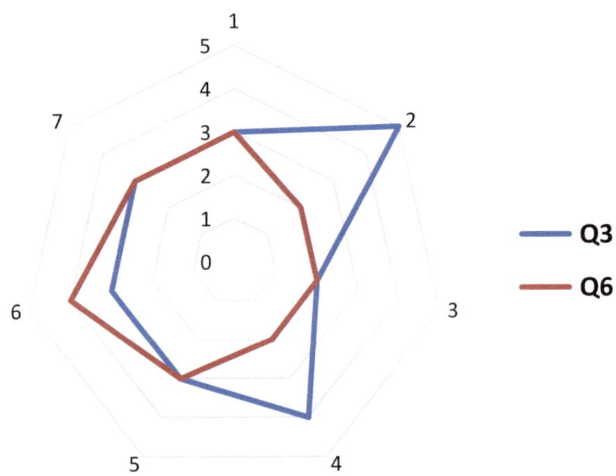

Fig. 1.9 Shows to what extent the customization of products is costly (Q3) and to what extent the customized products quickly and efficiently can distribute to customers (6)

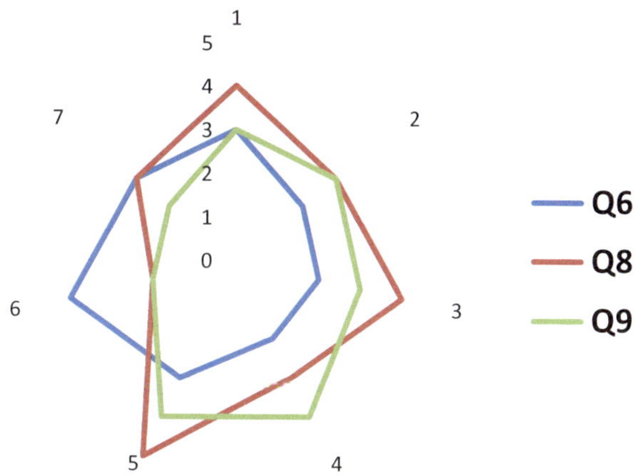

Fig. 1.10 Shows to what extent the companies quickly and efficiently distribute customized products (Q6), to what extent the companies have a desire to introduce customized solutions (Q8), and to what extent the companies manage change projects (Q9)

can indicate an increasing market demand of customized products, which are supported by the company's desire of offering customized solutions, as most of the involved companies plan to introduce customized solutions (Fig. 1.3). The result shows that the companies are considering the cost perspectives high related to making and delivering customized products to their customers (Fig. 1.9), even though many companies are considering an increasing approach from the industry toward offering more flexible production technology (Fig. 1.3). The results indicate that many of the involved companies are custom oriented as they have efficient communication relations to their customers and have relative strong change management capabilities (Fig. 1.5).

The innovative contribution for all of the involved companies is as a part of the collaboration during the project to increase the knowledge of tools and methods used for mass customization, which leads to increasing productivity. Therefore, by moving toward a higher degree of utilization of mass customization, it will improve the individual company and make small improvements to the Danish construction industry by implementing the principles of mass customization to increasing productivity.

The research contribution of the project is to clarify the situation of "where we are today" and to determine the development potential for the companies, and since it is early days in the project, the statistical material will improve when involving more companies in the workshops. The reliability of the data set was checked using the Cronbach's alpha test, and to be acceptable, values have to be of at least 0.7 [10]. Cronbach's alpha calculates to 0.387 (based on nine items), which is not acceptable as evidence-based research. However, this chapter is case based, and the results of the workshops done together with a limited number of companies give useful indications and serve to gain knowledge and experience of the involved companies for further use along with the improvement initiatives planned for the involved companies.

We experienced a need for intensive information and knowledge prior to any mass customization workshop or development process to ensure that everyone is in alignment and well informed by sufficient knowledge to contribute to the mass customization process. Therefore, we are recommending a systematic process to inform the parties with necessary information and knowledge.

References

1. Batchelor, R.: Henry ford, mass production, modernism, and design, vol. 1. Manchester University Press, Manchester (1994)
2. Benros, D., Duarte, J.: An integrated system for providing mass customized housing. Automat. Constr. **18**, 310–320 (2009)
3. Bohnstedt, K.D.: Enabling facilitation of mass customization via partnering in the construction industry. In: Brunoe, T.D., Nielsen, K., Joergensen, K.A., Taps, S.B. (eds.) Proceedings of the 7th World Conference on Mass Customization, Personalization, and Co-Creation (MCPC 2014), Aalborg, Denmark, February 4th–7th, 2014. Springer, Berlin (2014)
4. Boligministeriet, B., Force, B.T.: Byggeriets fremtid: Fra tradition til innovation (2000).
5. Dean, P., Tu, Y., Xue, D.: A framework for generating product production information for mass customization. Int. J. Adv. Manuf. Technol. **38**, 1244–1259 (2008)
6. Dean, P., Tu, Y., Xue, D.: An information system for One-of-a-kind production. Int. J. Prod. Res. **47**, 1071–1087 (2009)
7. dst.dk: Dansk Statistik. Www.Dst.Dk. 2013 (2013)
8. Jensen, K.N., Nielsen, K., Brunoe, T.D.: Application of mass customization in the construction industry. APMS **1**, 161–168 (2015)
9. Jiao, J., Ma, Q., Tseng, M.M.: Towards high value-added products and services: mass customization and beyond. Technovation **23**, 809–821 (2003)
10. Kline, P.: Handbook of psychological testing. Routledge, London (2013)
11. Pine, B.J.: Mass customization: the new frontier in business competition. Harvard Business School Press, Boston, MA (1993)
12. Salvador, F., De Holan, P.M., Piller, F.: Cracking the code of mass customization. MIT Sloan Manage. Rev. **50**, 71–78 (2009)
13. Walcher, D., Piller, F.T.: The customization 500, 1st edn. Lulu, Aachen (2011)

Chapter 2
Reconfiguring Variety, Profitability, and Postponement for Product Customization with Global Supply Chains

Martin Bonev, Anna Myrodia, and Lars Hvam

2.1 Introduction

With the emerging area of mass customization, researchers and practitioners alike have acknowledged a growing trend toward higher product variety and customization. Customizing a product can be described as the process of configuring a product variant by selecting predesigned components within a selected scope of offered variety [1]. Companies employ customization as a means to differentiate from their competitors by providing unique customer value [2]. Although many positive commercial advantages can be named from offering extensive customization [3], recently a stronger focus has been laid on the downside of the added supply chain complexity [4]. Higher product mixes created through diverse manufacturing strategies have been identified as major complexity drivers throughout value chains [5], often leading to reduced operational performances, such as longer lead times, poorer quality, and increased costs [6, 7]. Hence, integrating approaches to complexity management into the framework of supply chain management (SCM) has become compulsory [8].

A major concern in SCM is to systematically and strategically coordinate material flows across companies with the objective of reducing cost and achieving competitive advantages [9]. To account for the immanent complexity from customization, the scope of SCM needs to be aligned with aspects of variant management and postponement, i.e., the degree to which customization is provided throughout

M. Bonev (✉) • A. Myrodia • L. Hvam
Department of Engineering Management, Technical University of Denmark,
Produktionstorvet Building 426–030F, 2800 Kgs. Lyngby, Denmark
e-mail: mbon@dtu.dk; annamyr@dtu.dk; lahv@dtu.dk

J. Bellemare et al. (eds.), *Managing Complexity*, Springer Proceedings
in Business and Economics, DOI 10.1007/978-3-319-29058-4_2

the supply chain [10, 11]. This chapter adds to the existing knowledge of how supply chains dealing with varying degree of customization can handle the arising complexity. Based on a literature study on designing and managing supply chain networks for customization (Sect. 2.2), Sect. 2.3 introduces a suggested approach for the reconfiguration of the network design. Next, a case study is presented in Sect. 2.4, where empirical evidence is provided on how postponement and substation may positively reduce complexity and simultaneously increase companies' overall profitability and operational performance.

2.2 Literature Review

2.2.1 *Product Customization with Global Supply Chain Networks*

To compete on international markets, manufacturing companies are organizing their business processes around a global supply chain network [12]. Figure 2.1 displays a conceptual model of a hypothetical supply chain network design. From a high-level perspective, supply chains may typically include activities related to engineering and purchasing, manufacturing, assembly, distribution, and sales. To serve the needs of local markets, traditionally these activities have in their simplest form been established within the country of origin. With globalization firms have over time been moving toward international markets, for which some of the supply chain requires to be outsourced or physically displayed [13]. As indicated in Fig. 2.1, depending on the sales strategy, to secure lead times and product delivery, sales may, for example, be displaced to target markets, thereby establishing local sales

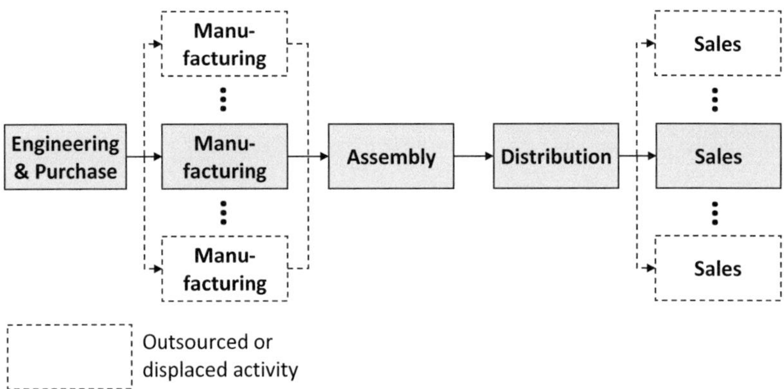

Fig. 2.1 Conceptual global supply chain network with outsourced or displayed manufacturing and sales

channels. To lower product costs or to focus on key competences, manufacturing on the other hand may be outsourced or displaced to low-cost countries, keeping the final assembly of components in the country of origin [14]. An example of this approach can be seen in the apparel industry, where products are designed in the country of origin, often manufactured in others, and sold locally within target markets [15]. In more general terms, the relative cost advantage of low-cost countries and the small value added to the final products is often named to be the main motivation for emphasizing this particular part of the supply chain, like manufacturing [16]. To this end, several studies have investigated the possible gains and motivation from reconfiguring supply chain networks. While major part of the research suggests an overall positive effect on the firm's performance, few studies also point out the potential risks with this strategy [17].

In addition to the network design of a particular supply chain, offering product customization requires consideration about the product design and production planning and control system. The degree to which customization is provided can vary across the entire product portfolio of a company and is often described through the relative involvement of customers with the companies' supply chain, i.e., to the customer order decoupling point (CODP) [18]. As displayed in Fig. 2.2, the more supply chain activities are directly related to a particular customer order, the higher is the degree of the offered variety and the early in the supply chain the CODP is placed. Literature names a few distinct product planning and control systems allowing for customization, depending on the relative placement of the CODP [19]. In an Engineer-to-Order (ETO) situation, components have to be engineered based on a specific request from customers, forcing all subsequent activities to be directly engaged in fulfilling the order. Due to the early customer involvement, typically ETO products obtain a large amount of variety, but their production volumes are low [20]. In a Make-to-Order (MTO) scenario, predesigned and available components are used for manufacturing and subsequent assembly of the product variants. In case both engineering and manufacturing activities are performed based on forecast, subassemblies from stock are used in the assembly process to Assemble-to-Order (ATO) the requested product variant. To account for a high

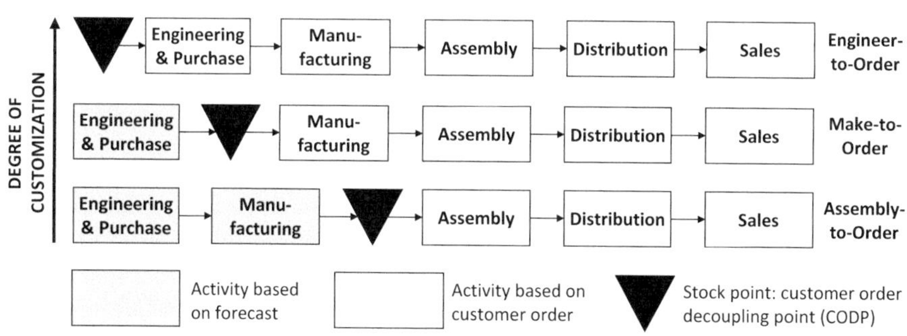

Fig. 2.2 Degree of customization and placement of the CODP

amount of final variety, a modular product design has been reported to facilitate the separation between manufacturing of components and (final) assembly [21]. With the so-called modular product architecture, components or modules can be produced or outsourced based on forecast and recombined according to the requirements of the customer [22]. This would allow the company to postpone the CODP closer toward the customer, i.e., to an MTO or ATO situation. The so-called Type III postponement strategy aims at capitalizing on standardization and modularity, thereby achieving economies of scale [23].

2.2.2 Supply Chain Performance and Reconfiguration

Despite the rather simplistic view on the production process, dividing the different production planning and control systems according to the placement of the CODP helps to define clear strategies for a particular supply chain network design. Decisions about a suitable configuration of the network may be related to key operational performance measures of a company, such as to cost and time [24]. From customers' perspective, higher degree of customization allows for more engagement in the supply chain and hence to more unique product designs. However, since more activities have to be performed after a specific order has been placed, there is a tradeoff between the uniqueness of the product design and the related delivery time and cost. In general, the higher the number of activities performed for a customer, the bigger the sum of the individual lead times of each process [2]. Moreover, unique designs with higher engineering engagement have often proved to be more costly and less quality assured [25]. Since a higher percentage of the supply chain is performed based on a distinctive customer requirement, processes are less standardized and may involve ad hoc and unproven tasks which require stronger coordination effort [20]. On the other hand, with an MTO and ATO strategy, the increased standardization of components and processes combined with reduced delivery times has shown to be particularly useful for products with moderate or limited variety and high volumes [18]. Therefore, setting the right strategy for the production planning and control system can have a wide-ranging impact on the profitability of the provided portfolio.

Traditionally, decisions about the placement of the CODP are made based on inventory management theories and may include aspects of inventory cost, lead time requirements toward the market, sales volume and order frequency, and scope of offered variety [26, 27]. Accordingly, items with low volumes and high variety should be organized around an early placement of the CODP and vice versa. Recent literature however emphasizes that more and diverse customization significantly increases supply chain complexity, making cost allocation and prices estimations less accurate [8]. Planning with higher product variety often leads to overestimated profits, where the complexity-induced cost of the supply chain is not taken appropriately into account by traditional accounting methods [28]. Schuh et al. (2008) discuss complexity from two forces [29]. External complexity occurs due to desired customer requirements. This defines the number of the offered product variety. Internal complexity describes the processes, parts, and product designs across supply chain needed to provide the demanded product variety. Reducing the internal complexity as much as possible by obtaining the necessary external complexity is seen as a guiding principle for managing the complexity across supply chains [1].

A common way to identify unnecessary external complexity is to investigate the realized contribution margins (CMs) for each variant according to the Pareto principle [30]. As studies have shown, in complex supply chains, a large amount of the sold variants do not contribute if at all to the turnover of firms. Instead, a major part of the turnover is generated from a small amount of the variety [31]. In order to classify which variants to keep and which to reduce or replace, a categorization into A, B, and C products is typically performed [32]. Once unprofitable variants are identified, various initiatives can be enforced to reduce the related complexity. Depending on the product design and the supply chain network, such initiatives may include the increase of modularity [33], postponement [11], or product standardization through increasing component commonality [34].

Yet, due to the rather sensitive operational data, empirical-based research considering both analysis on margins and the related initiatives is rare. Hence, the main focus of this research is to find empirical evidence on how to identify the most profitable product variety for product customization regarding production strategy and supply chain setup. In particular this research attempts to answer the following research question:

RQ1: How can the operational and financial performance of a supply chain network for customized products be improved?

This research question is answered based on the three subquestions:

RQ1.1: How can customized products be categorized relative to their degree of customization?

RQ1.2: How can the potential for a postponement of the CODP and a standardization strategy be identified?

RQ1.3: How can postponement and standardization effects on costs and contributions margins be quantified?

2.3 Suggested Approach

As stated in the previous sections, complexity creates uneven cost distribution across the different product variants. Based on the literature, moving the CODP toward the front-end is an effective approach to complexity cost reduction. However, in cases where the manufacturer produces not only ATO products but also MTO and ETO, the setup varies a lot among the different production strategies. On top of that, the profitability assessment may be calculated through several approaches. Recent literature suggests that in order to have a clear picture of the "high runners" and the "long tail," both CM and sales volume have to be taken into consideration in the profitability analysis.

In alignment with related contributions, this research suggests an approach for profitability analysis and complexity reduction, which can be applied to manufacturing companies with different production strategies. In order to analyze the profitability, an ABC product categorization is performed. Each product is grouped into A, B, or C based on its CM and net revenue (NR), which enables the consideration of the sales volume for each variant. To reduce the supply chain complexity, two coordinated methods are considered. The first one relates to postponement of the CODP and the resulting product standardization. Besides, complexity reduction theory suggests the development of modular products that consist of standard subassemblies. In that way when an order is placed by the customer, the final configuration of the product can takes place with an MTO or ATO approach. This strategy reduces lead time, complexity cost, and production cost. The second method discusses the provided variety of the product portfolio in terms of cannibalization and profitability. Related literature highlights that the increasing variety offered to the customers does not necessarily indicate that a wider range of application is covered. In order to ensure that variants with different production cost and sales volumes are not offered with similar properties and applications, product merging through substitution is suggested. This is done by analyzing the bill of materials (BOMs) and the CMs of these variants.

2.4 Case Study

2.4.1 Data Collection

The suggested methodology is applied on a case study of a Danish manufacturer of pumps. The company produces standardized as well as more specialized products with an ATO, MTO, or ETO strategy. The main market requirements for pumps are reliability, functionality, design, price, delivery performance, and solution flexibility. The product portfolio of the company includes pumps for chemical, environmental, heavy, and petrochemical duty and for general purpose. The data collection is performed through the company's internal database and includes BOMs, total cost, NR, sales volume, production strategy, and country of production and distribution, on finished good level. The sample size refers to sales within a 2-year period (2012, 2013). Semistructured interviews with project managers are performed, in order to verify the accuracy of the data acquisition.

As suggested in literature, since part of the supply chain is based on forecast, the ATO products have relatively shorter lead times and better delivery performances. MTO products are produced based on an order received from the distribution center (DC). They consist of standard parts, which additionally require special treatment, and are produced in low runs. Before their components can be produced, BOM and prices have to be verified, which results in longer lead times compared to the ATO variants. Special customer requirements are treated as ETO products and hence obtain longer lead times and higher cost in comparison to the ATO and MTO products. A significant difference between an MTO and an ETO product is that for the latter, a dedicated production setup is required, which involves alternative processes and tooling. Moreover, the R&D department is also involved in the enquiry and quotation process, to verify the feasibility of the customer's requirements and to ensure the supply chain capabilities.

The company acquires two production sites, one in Denmark and one in China, and three DCs, one in each of the following countries: Denmark, China, and the USA. The DCs in China and Denmark deliver products produced to the respective site; the North America market is supplied by either China or Denmark. However, the products distributed in Denmark are produced in two ways; either they are entirely produced in Denmark (local) or they are produced as standard semifinished units (SFU) in China, and then the final configuration and testing is performed in Denmark (Figs. 2.3 and 2.4).

The sample size focuses on one representative product family consisting of 299 variants, the heavy duty (HD) pumps consisting of a modular product architecture.

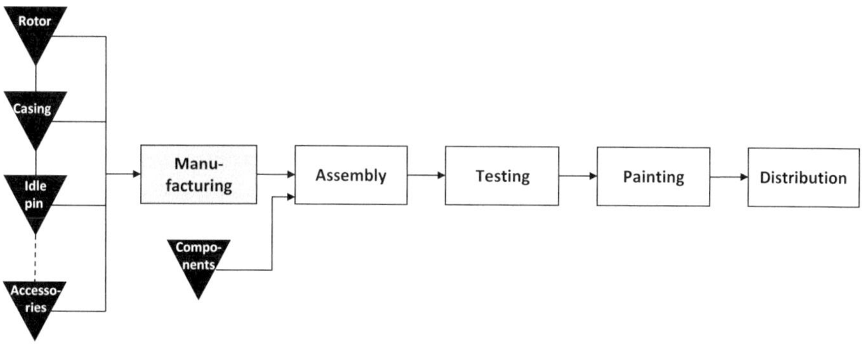

Fig. 2.3 Local production in Denmark

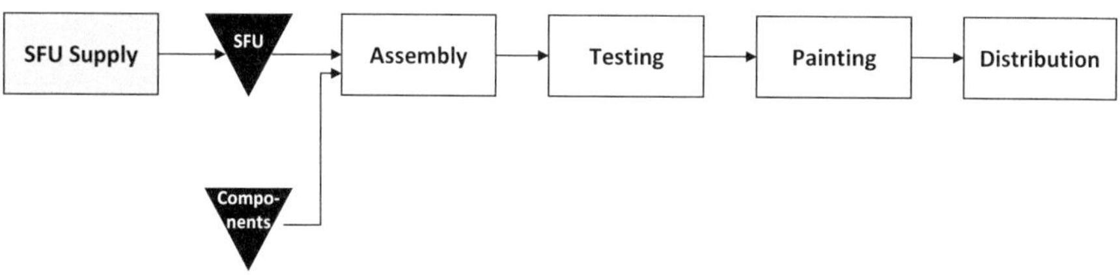

Fig. 2.4 SFU production in China and final configuration in Denmark

The particular product family is selected due to its significant share of the total sales, which accounts for 60.61 % of the total revenue. Moreover, HD pumps are offered based on all three production strategies with a distribution of 32, 33, and 34 % between ATO, MTO, and ETO accordingly. To limit the scope of analysis, the sample size refers to products being sold from the DC in Denmark.

2.4.2 Analysis and Results

Currently, the company categorizes the products as A, B, and C based on their inventory turnover and their picking frequency. The results from this internal ABC analysis are presented in the following table (Table 2.1).

The ABC categorization is based on internal experience. Products are categorized as A if they have inventory turnover higher than or equal to three and picking frequency higher than or equal to 20. B products are indicated by inventory turnover equal to two and picking frequency between three and 20. Finally, C products have inventory turnover less or equal to one and picking frequency less or equal to three. All the data refers to a 12-month period.

Both parameters, inventory turnover and picking frequency, are related to the sales volume of the products. However, with this internal categorization approach, none of the measures accounts for the CM of the products. Yet according to the literature, in order to draw conclusions regarding the profitability of a product, the NR and production cost have to be taken into consideration. This results in questioning the accuracy of the internal ABC product categorization.

By implementing the suggested methodology, an ABC analysis is performed, which categorizes the products based on the NR and CM instead. The CM is calculated as the difference of the NR from the direct production cost, where direct production cost includes the cost of material and labor. The following table presents the results of the ABC analysis (Table 2.2).

When comparing the results from the two ABC analyses, it can be concluded that in the company's perspective, many C products are kept in stock (81.6 %), which leads to increasing inventory costs and consequently complexity costs. From the suggested ABC analysis, the ratio of C products is relatively lower (77.3 %). Yet the distribution of products varies between the two analyses, indicating that further research is required to identify the cause of this divergence.

Table 2.1 Internal ABC analysis

Inventory turnover	Picking frequency			
	Category	A (>20)	B (4–20)	C (0–3)
	A (≥3)	18	2	0
	B (2)	11	24	5
	C (0–1)	3	46	190

Table 2.2 ABC product categorization based on CM and NR

NR	CM			
	Category	A	B	C
	A	38	23	11
	B	0	7	88
	C	0	0	132

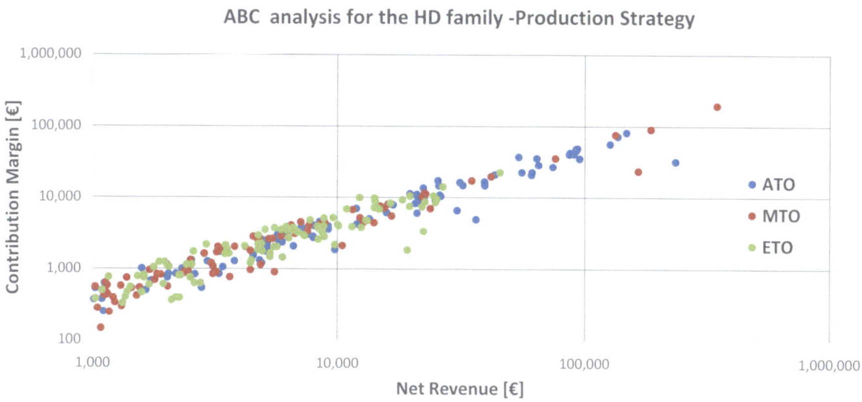

Fig. 2.5 ABC product categorization by production strategy

To gain better understanding of how postponement may be applied, the results are displayed in relation to the three production strategies (ATO, MTO, ETO). In other words, the products are categorized into A, B, or C, based on their NR and CM, revealing a significant difference between how the type of products are included under each production strategy.

As displayed in Fig. 2.5 above, 60 % of the ATO products are categorized as C products. 29 % of the ATO variants are categorized as A and the remaining 11 % as B products. However, this result highly contradicts to the internal categorization of a product ATO. ATO products are standardized, are produced in large batches, and are high runners. That implies that ATO products have lower production cost and higher revenue, which would result in higher CM and, consequently, in an A product. Less contradictory, only 8 % of the MTO belong to A and 87 % to C products. Finally, as expected only 2 % of the ETO products are A and 88 % C.

In detail, the following table presents the total cost, NR, CM, number of variants, and sales volume per production strategy.

The results from Fig. 2.6 indicate that the ATO products are more profitable, contribute far more to the company's profitability, and are sold in higher volume. However, this again does not conform with the result from the internal ABC analysis (see Table 2.1), which shows that 60 % of the ATO products are C. Based on the

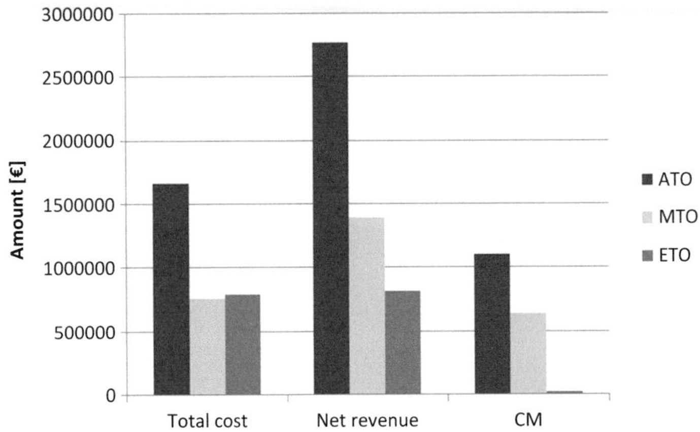

Fig. 2.6 Comparison of the financial data from the three production strategies

above, a re-categorization of the products under the three production strategies is recommended.

By following the suggested research method, two approaches are implemented. The first one aims at increasing the standardization of the ATO products. The company, as discussed above, uses SFU manufactured in China as preassemblies for the ATO products. The products including these SFU have significantly lower production cost. However, out of the 97 ATO variants, only in 8 % of the cases outsourcing through SFUs is used. The following Table 2.3 gathers the relevant financial data for the products produced in China and in Denmark.

To identify the potential for outsourcing, products with similar properties and sizes produced in Denmark and China are investigated. By increasing the number of SFUs used in the final assemblies, the overall number of variants produced is significantly reduced, thereby decreasing the complexity of the supply chain. The following Table 2.4 illustrates the results of those calculations.

For further product standardization, a re-categorization of the products among the three production strategies (ATO, MTO, ETO) is examined. Products with same sizes are analyzed based on their production strategy with the intention to move as many products as possible to the ATO category. Decisions are made after comparing the BOM and the functional properties of the products. This analysis results in increasing the standardization of 36 products, or 12 % of the portfolio. In detail, 18 MTO and 18 ETO products are moved to ATO category. The financial impact is illustrated in the following figure (Fig. 2.7).

Summarizing the results from the two standardization methods discussed above, it can be seen that the total cost of the HD family is decreased by 4.3 %. The impact of the implementation on the NR is not significant, due to the lower sales price the standardized products have compared to the customized ones. Yet, the increase in

Table 2.3 ATO products

Production country		Cost (€)	NR (€)	CM (€)	# of variants	Sales volume
CH	Sum	8.826	14.269	5.444	8	273
	Aver	1.103	1.784	680	–	–
DK	Sum	109.347	194.853	85.505	89	1264
	Aver	1.229	2.189	961	–	–

Table 2.4 Financial data after implementing the SFU standardization

	Before (€)	After (€)	Difference (€)
CM	3.370.800	3.388.987	18.187
Revenue	6.436.071	6.076.030	–360.041
Cost	3.065.271	2.687.043	–378.228

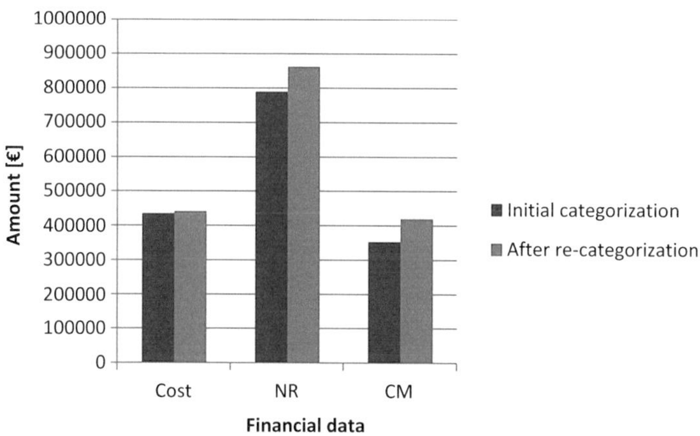

Fig. 2.7 Comparison of financial analysis of the production strategy categorization

the CM by 18 % (from 354.299€ to 419.314€) indicates that the profitability of the new product portfolio has been positively affected (Table 2.5).

Next, the potential for substitution is being investigated. The analysis is made in ten groups of products that have the same size. In particular 98 product variants are merged into 44, where 20 out of them are merged into 13 products that have SFUs produced in China as preassemblies. By merging the products, 54 variants can be eliminated, which additionally reinforces the standardization of the product family.

In order to estimate the total effect on the company's profitability after implementing the suggested method of both product standardization and variant

Table 2.5 Total impact on the HD family

	Before (€)	After (€)	Total impact (%)
Total revenue	4.977.942	4.996.389	0.4
Total cost	3.212.839	3.074.773	–4.30
Total CM	1.765.103	1.921.616	8.9

Table 2.6 Sensitivity analysis with four scenarios

	A (%)	B (%)	C (%)	D (%)
Cost	–20	–20	–20	–30
Sales price	0	–5	–5	–10
Sales volume	5	10	0	20

Table 2.7 Impact of the four scenarios

	1 (%)	2 (%)	3 (%)	4 (%)
Cost	–3	–2	–4.1	–0.8
NR	1.8	1.7	–1.2	1.5
CM	10.5	8.3	9.9	5.1

substitution, a sensitivity analysis is performed. The following table describes the four combinations that are used in order to gain a better understanding of the impact of the approach on the CM of the product family (Table 2.6).

For each of the above scenarios the cost, NR and CM are calculated. The results are as follows (Table 2.7):

The negative percentages indicate that there is a reduction after the implementation of the suggested approaches. The results demonstrate that the CM is increased in every case. It is worth mentioning that even in scenario 4, where there is no increase in the sales volume, the CM is increased considerably. As a result, the outcome of the sensitivity analysis indicates that the application of the suggested methods for product standardization and variant elimination has an impact on reduction of complexity costs and increase profitability.

2.5 Conclusion

This research examined the effect of postponement and product substitution on profitability and complexity reduction in the manufacturing industry. The suggested methodology was developed based on recent research studies and is further supported by empirical evidence. A particular pump manufacturer considered being highly representative for this research was used as a case study, due to its diverse production strategy with different degrees of customization and a global supply chain network. The case study investigated variants profitability and identified the realized degree of customization of a selected product range.

The results indicated that there is a significant improvement of the product's profitability once the standardization and substitution method is applied. By managing the existing variety of the product portfolio, eliminating the variants that add no value and/or no additional properties, and postponing the CODP, the operation performance in terms of profitability and lead time was improved. An 18 % increase in the CM of the ATO products was achieved by standardizing 12 % of the variants. Furthermore, additional effects were estimated from a subsequent variant substitution.

Despite being one of the rare empirical-based studies within this research field, since the results are supported by a single case study, the main limitation to this research is the generalizability. This provides opportunity for further research which would help to investigate the impact of the suggested approach on the different cost elements and complexity costs across a number of cases. Likewise, the distribution of complexity costs over the product range and the effect of the portfolio standardization and substitution are to be further examined. Here, additional case studies may to allow the generalization of the suggested method and further enhance the external validity of the results.

References

1. ElMaraghy, H., Schuh, G., ElMaraghy, W., Piller, F., Schönsleben, P., Tseng, M., Bernard, A.: Product variety management. CIRP Ann. Manuf. Technol. **62**, 629–652 (2013)
2. Piller, F.T., Moeslein, K., Stotko, C.M.: Does mass customization pay? An economic approach to evaluate customer integration. Prod. Plan. Control **15**(4), 435–444 (2004)
3. Salvador, F., Forza, C.: Configuring products to address the customization-responsiveness squeeze: a survey of management issues and opportunities. Int. J. Prod. Econ. **91**, 273–291 (2004)
4. Hoole, R.: Five ways to simplify your supply chain. Supply Chain Manag. Int. J. **10**, 3–6 (2005)
5. Serdarasan, S.: A review of supply chain complexity drivers. Comput. Ind. Eng. **66**(3), 533–540 (2013)
6. Åhlström, P., Westbrook, R.: Implications of mass customization for operations management: an exploratory survey. Int. J. Oper. Prod. Manag. **19**, 262–275 (1999)
7. Blecker, T., Abdelkafi, N., Kaluza, B., Friedrich, G.: Controlling variety-induced complexity in mass customisation: a key metrics-based approach. Int. J. Mass Custom. **1**(2–3), 272–298 (2006)
8. Bozarth, C.C., Warsing, D.P., Flynn, B.B., Flynn, E.J.: The impact of supply chain complexity on manufacturing plant performance. J. Oper. Manag. **27**(1), 78–93 (2009)
9. Mentzer, J.T., Dewitt, W., Keebler, J.S., Min, S., Nix, N.W., Smith, C.D., Zacharia, Z.G.: Defining supply chain management. J. Bus. Logist. **22**(2), 1–25 (2001)
10. Lamothe, J., Hadj-Hamou, K., Aldanondo, M.: An optimization model for selecting a product family and designing its supply chain. Eur. J. Oper. Res. **169**(3), 1030–1047 (2006)
11. Trentin, A., Salvador, F., Forza, C., Rungtusanatham, M.J.: Operationalising form postponement from a decision-making perspective. Int. J. Prod. Res. **49**(7), 1977–1999 (2011)
12. Makhija, M.V., Kim, K., Williamson, S.D.: Measuring globalization of industries using a national industry approach: empirical evidence across five countries and over time. J. Int. Bus. Stud. **28**, 679–710 (1997)

13. Kumar, N., Andersson, D., Rehme, J.: Logistics of low cost country sourcing. Int. J. Logist. Res. Appl. **13**, 143–160 (2010)
14. Kusaba, K., Moser, R., Rodrigues, A.M.: Low-cost country sourcing competence: a conceptual framework and empirical analysis. J. Supply Chain Manag. **47**, 73–93 (2011)
15. Kumar, S., Arbi, A.S.: Outsourcing strategies for apparel manufacture: a case study. J. Manuf. Technol. Manag. **19**, 73–91 (2008)
16. Fredriksson, A., Jonsson, P.: Assessing consequences of low-cost sourcing in China. Int. J. Phys. Distrib. Logist. Manag. **39**, 227–249 (2009)
17. Horn, P., Schiele, H., Werner, W.: The 'ugly twins': failed low-wage-country sourcing projects and their expensive replacements. J. Purch. Supply Manag. **19**(1), 27–38 (2013)
18. Duray, R.: Mass customization origins: mass or custom manufacturing? Int. J. Oper. Prod. Manag. **22**(3), 314–328 (2002)
19. Rudberg, M., Wikner, J.: Mass customization in terms of the customer order decoupling point. Prod. Plan. Control **15**(4), 445–458 (2004)
20. Caron, F., Fiore, A.: 'Engineer to order' companies: how to integrate manufacturing and innovative processes. Int. J. Proj. Manag. **13**(5), 313–319 (1995)
21. Kusiak, A.: Integrated product and process design: a modularity perspective. J. Eng. Des. **13**(3), 223–231 (2002)
22. Mikkola, J.H.: Management of product architecture modularity for mass customization: modeling and theoretical considerations. IEEE Trans. Eng. Manag. **54**(1), 57–69 (2007)
23. Forza, C., Salvador, F., Trentin, A.: Form postponement effects on operational performance: a typological theory. Int. J. Oper. Prod. Manag. **28**(11), 1067–1094 (2008)
24. Neely, A., Gregory, M., Platts, K.: Performance measurement system design: a literature review and research agenda. Int. J. Oper. Prod. Manag. **25**(12), 1228–1263 (2005)
25. Ulrikkeholm, J.B., Hvam, L.: The cost of customising: assessing the performance of a modular product programme. Int. J. Prod. Dev. **19**(4), 214–230 (2014)
26. De Toni, A., Caputo, M., Vinelli, A.: Production management techniques: push-pull classification and application conditions. Int. J. Oper. Prod. Manag. **8**(2), 35–51 (1988)
27. de Koster, R., Le-Duc, T., Roodbergen, K.J.: Design and control of warehouse order picking: a literature review. Eur. J. Oper. Res. **182**, 481–501 (2007)
28. Cooper, J.C.: Logistics strategies for global businesses. Int. J. Phys. Distrib. Logist. Manag. **23**, 12–23 (1993)
29. Schuh, G., Sauer, A., Doering, S.: Managing complexity in industrial collaborations. Int. J. Prod. Res. **46**(9), 2485–2498 (2008)
30. Pareto, V., Schwier, A.S.: Manual of political economy. Augustus M. Kelley, New York (1971). p. 504
31. Brynjolfsson, E., Hu, Y.J., Simester, D.: Goodbye Pareto principle, hello long tail: the effect of search costs on the concentration of product sales. Manag. Sci. **57**(8), 1373–1386 (2011)
32. Yücel, E., Karaesmen, F., Salman, F.S., Türkay, M.: Optimizing product assortment under customer-driven demand substitution. Eur. J. Oper. Res. **199**(3), 759–768 (2009)
33. Gualandris, J., Kalchschmidt, M.: Product and process modularity: improving flexibility and reducing supplier failure risk. Int. J. Prod. Res. **51**(19), 5757–5770 (2013)
34. Labro, E.: The cost effects of component commonality: a literature review through a management-accounting lens. Manuf. Serv. Oper. Manag. **6**(4), 358–367 (2004)

Chapter 3
Mass Customization Challenges of Engineer-to-Order Manufacturing

Maria K. Thomassen and Erlend Alfnes

3.1 Introduction

Mass customization (MC) is the capability to offer individually tailored products on a large scale [1]. Moreover, it is about developing, producing, marketing, and delivering affordable goods and services with enough variety and customization possibilities that nearly everyone finds exactly what they want [2]. The concept offers new opportunities to companies combining a mass production tradition with a high level of customization, maintaining high efficiency while offering highly customized products. MC is considered a dominant form of production in business-to-business and business-to-consumer, high-end, and major consumer markets [3]. MC has got great attention in several industries during the last two decades, but its adoption in practice has been slow seen in terms of the increasing interest and major potential [4–6].

Engineer-to-order (ETO) manufacturing environments are typically characterized by high levels of product and process variation, high product complexity and deep product structures, and low production volumes. Each new order involves product design and development based upon customer specifications, and products are typically highly customized. Moreover, design, delivery speed, and flexibility are typical order winners, and the customer order decoupling point (CODP) is typically positioned at the very start of production [7].

MC literature has traditionally focused on the transition of mass producers, defining strategies to increase customization without any loss of efficiency, while there are few MC studies taking the perspective of custom producers such as ETO

M.K. Thomassen (✉)
SINTEF Technology and Society, Industrial Management, Trondheim, Norway
e-mail: maria.thomassen@sintef.no

E. Alfnes
Department of Production and Quality Engineering, Norwegian University of Science and Technology, Trondheim, Norway
e-mail: erlend.alfnes@ntnu.no

J. Bellemare et al. (eds.), *Managing Complexity*, Springer Proceedings in Business and Economics, DOI 10.1007/978-3-319-29058-4_3

companies that seek to increase efficiency while maintaining a high customization level [8]. MC and ETO are two different production strategies, but these can be combined into a hybrid MC-ETO strategy [9] or standardized customization strategy [8]. The motivation of ETO companies to move toward MC includes benefits such as reduced delivery times, more precise cost calculations, and reduced specification costs, by increasing standardization of customized products, i.e., limiting product variety [8].

However, compared to mass producers, the MC movement of ETO companies seems more complex [8] as ETO companies meet major challenges when moving toward further standardization, i.e., seeking efficiency in customization of products. ETO manufacturing does not necessarily involve high volumes such as in mass, but often imply low volume production. Current knowledge on the adoption of MC principles is developed with primary focus on mass producers and provides only limited guidance on ETO settings. The problem is that since most knowledge on MC is typically developed for mass producers, its relevance in ETO settings may be questioned. Current studies on MC in ETO focus on product design and configurator issues, and there is a research gap related to major manufacturing challenges.

In this paper, general MC principles are tested in a case company to identify major implementation challenges. The aim is to provide further empirical insights to issues in the intersection between MC and ETO manufacturing that are critical for the development of MC principles that are better suited for ETO manufacturing.

3.2 Methodology

This study is based upon a literature review and a case study of an ETO company. The purpose of the literature was to investigate major challenges of MC in ETO settings. Literature searches in academic databases and reviews of identified articles were carried out in several iterations.

A framework of critical areas for MC manufacturing [10] was chosen for structuring and analyzing the empirical data. This framework was chosen because it addresses several relevant MC areas in manufacturing and takes both mass production and handcraft production into consideration.

An empirical case study approach was chosen since there was a need to develop further detailed insights to issues of implementing MC principles in ETO. A single case was necessary to ensure enough detail and in-depth insights to major issues of a typical ETO situation. Case company selection criteria included that their operations were characterized as ETO, and they had long tradition of efficiency improvement work in production. They also had put a lot of effort into this work, and they experienced major challenges attempting to increase efficiency in operations while maintaining high product customization and experienced customer value.

The company is characterized as ETO. The products are complex and heavy and are produced in low volume and in high variety. Operations include both parts fabrication such as cutting, welding, grinding, and machining and assembly processes.

There is high flexibility in the replenishment of parts since these can be fabricated in-house and purchased from suppliers.

Case company data was collected in several iterations over a 3-year period. Interviews and discussions with key personnel including plant manager and logistics manager, planning managers, and planners were carried out combined with plant visits. Most of the data was collected in all-day workshops with case company representatives. Both qualitative and quantitative data were collected; data extracts from the ERP system were mainly used to verify data collected from interviews and discussions. Data was used to describe key issues of the case company related to the general MC implementation principles.

3.3 Literature Review

3.3.1 *MC Challenges in ETO Manufacturing*

While there is a significant amount of MC research in view of mass production, MC has got limited attention in research of manufacturing settings with high customization [9]. The review identified few studies only that deal with issues related to the MC transition of ETO companies. This chapter briefly presents major challenges identified in the literature review.

A general feature of the MC transition is that it typically implies an increased standardization of engineering work [11]. Major issues are therefore related to the design stages including new product development and order-specific engineering phases [9]. The decision to offer less product variety may compromise the entire business foundation of an ETO company [12]. A common challenge is to find the right balance between flexibility and standardization, i.e., to ensure an appropriate level of flexibility to meet customer demands relative to a rational level of commonality between product designs [8].

Definition of a predefined solution space is a key MC capability [13]. However, defining boundaries of a stable product solution space may turn out to be a highly complex task in ETO companies [9]. There is a risk that the solution space is not adequately large to satisfy all customers' requirements [8]. Since ETO products are often based upon a knowledge-based design, they are difficult to standardize to a degree that allows configuration [8, 9]. Concerning procurement, achieving a reciprocal understanding of needs and interdependencies in the supply chain related to the definition of the solution space is seen as a challenge [9]. Another related challenge is to organize and structure product lines into families, platforms, and modular structures and make knowledge more explicit [9].

Simplification of product designs offered may have unfortunate consequences since it may lead to loss of innovative capabilities, greater risk of product imitations, and organizational resistance to simplifying the engineering work [8].

The rigidity of traditional ICT systems is a major challenge for configuring customized products and manufacturing processes [9]. To ensure flexible manufacturing operations, the support of increased information richness of products and processes is further necessary [9].

Furthermore, issues are related to the standardization of knowledge of repetitive design tasks for automation requiring high technical and social competence of engineering team leaders in the development of customer specifications [9]. The required amount of know-how and skills is also challenging especially in the design and use of product configurators [9].

A tight integration between NPD, sales, and engineering support is also necessary to ensure efficient matching of customer needs with defined product variety [9]. With regard to the supply chain, issues are related to managing relationships with more suppliers, spending more time on sourcing market research, and investing in SCM systems integration that are necessary to ensure efficient sourcing and shipping of small quantities of highly differentiated products [9].

3.3.2 MC Manufacturing Principles

A set of implementation guidelines was selected as the starting point for the development of an adjusted MC manufacturing strategy approach for ETO companies. The guidelines are structured into eight main decision areas including market interaction, product, ICT, manufacturing technology, processes, manufacturing planning and control, supply chain integration, and work organization [10].

Some guidelines are only valid for mass producers or handcraft producers that aim to implement the mass customization strategy. It is assumed that ETO production resembles most to the situation of handcraft producers, and thereby, these guidelines are prioritized over the guidelines for mass producers. The guidelines are summarized in Table 3.1 below. Guidelines specifically valid for mass producers are marked with (a) and handcraft producers with (b).

In the following chapter, these principles are tested in a case company to reveal major concerns of implementation in an ETO setting.

3.4 Test of General MC Principles in an ETO Case Company

3.4.1 Market Interaction

The general guideline suggests that the market interaction strategy should be changed into MTO or ATO. Also, mass producers should position the CODP upstream, while it should be positioned downstream for handcraft producers.

In the case company, production orders are based upon customer orders, and engineering is needed to specify a new customer order, i.e., ETO. Since engineering is a major competitive advantage in this market, the underlying ETO

Table 3.1 MC manufacturing strategy implementation guidelines [10]

Decision area	Guidelines
1. Market interaction	Change the market interaction strategy to MTO or ATO
	(a) Aim to position CODP upstream in the value chain
	(b) Aim to position CODP downstream in the value chain
2. Products	Offer high level of customization on components/modules that represent the highest added value to customers
	Make a product program based on similar design elements for all product families
	(a) Modularize components to enhance the variability for the customers
	(b) Standardize components to reduce the complexity for the manufacturing
3. ICT	Establish online order registration
	Establish a product configurator
	Guide the customer through the order process and visualize the choices
	Strive for seamless integration of all information system (CAD/CAM, product configurator, ERP, order tracking, etc.)
4. Manufacturing technology	Strive for automation in manufacturing, but balance it toward the flexibility obtained by human resources
	Utilize efficient technology in processes upstream of CODP
	Utilize responsive and flexible technology (FMS) in customer-specific processes
5. Processes	Establish a product-oriented material flow
	Design a layout that reduces nonvalue added processes
	Manufacturing processes should perform operations based on digitally transferred information about customer specifications
6. Manufacturing planning and control	Introduce demand-driven replenishment of standard components and modules
	Define and prioritize criteria for sequencing of orders in customer-specific processes
	Aim to introduce push-pull principle in processes upstream of CODP
	Aim to introduce push-pull principle (FIFO) downstream of CODP
7. Supply chain integration	Establish JIT partnership with suppliers of standard components/modules
	Allow key suppliers of customer-specific components online access to the order system
	Establish rapid distribution channels to all the markets areas
8. Work organization	Train operators to be multiskilled
	Educate operators in multiple tasks
	Develop a flexible job rotation and job allocation system

strategy is an essential business foundation. The process of specifying new customer orders typically starts well in advance, between 6 months up to several years. There is thus often a high certainty in the long-term delivery plan. However, it is common that change orders, changed times of delivery, and engineering specification adjustments are defined after the start of production. Forecasts of expected new orders are used to initialize production and purchasing to deal with long lead times and ensure high process efficiency in production.

Spare parts are produced to stock due to the criticality of the delivery time of such parts. For new products, the customer delivery lead time is typically significantly longer than the required production lead time. In practice however, due to late change orders and order specifications, the actual time between that orders is completely specified until delivery is often significantly shorter than the production lead time. This means that the company has decided to start production before the order has been fully specified. Some years ago, the CODP was placed at the very start of operations. However, in order to keep a high level of resource utilization, the company has moved the CODP further downstream. Today, the primary CODP is therefore located at the parts inventory.

Even though there is a unique drawing for each new product, most parts are produced to stock long in advance of start of assembly operations since few parts are customer unique, and parts are often interchangeable. At the same time, there is limited degree of parts commonality as the product variety is high with respect to material and size leading to high inventory levels.

The principle of moving the CODP further downstream for highly standardized products may be further investigated in the case company to systematically achieve additional efficiency gains in production including lower inventory costs and WIP levels. In order to define CODP location that permits further differentiated control of product flows, it is critical to more systematically distinguish between products based upon level of standardization or customization.

3.4.2 Products

It is suggested that high level of customization should be offered on components or modules that represent the highest added value to customers. It is further proposed to form a product program based on similar design elements for all product families. Mass producers are recommended to modularize components to enhance the variability for the customers, while handcraft producers should standardize components to reduce the complexity for the manufacturing.

In the case company, about 80 % of a product's parts are delivered as standard parts. Some components are customized more often than others. However, the company has not defined any specific limitations regarding what components that may or may not be customized. The products may therefore in theory be entirely customized to meet specific needs of each unique customer. A new drawing is created for each new product.

The company has an overall product program consisting of four main product families that include products with similar design elements. Since each product family in turn consists of a wide range of variants and models, the product variability is high.

Even though only a small part of the total number of components of a final product are actually customized, the high number and variety of components still implies high complexity in the company's production processes. Further standardization of components may of course help the company to reduce complexity. However, there is also a risk that increased standardization will have consequences for the company's ability to deliver customer-specific products and thereby its competitive position.

3.4.3 *ICT*

ICT-related principles include the establishment of online order registration and a product configurator. It is also recommended that customers are guided through the order process and choices are visualized. Moreover, all information systems should be seamlessly integrated.

The case company does not have a product configurator but utilizes CAD/CAM software to visually support the interaction with customers during the sales and order specification process. Drawings and engineering specifications are available via the ERP system. These are also used for generating work orders for parts fabrication. Engineering changes are frequent throughout the production process, and it is critical that changes are taken into consideration as early as possible to avoid rework or build up inventory. Increased ICT integration with regard to engineering change information in the company could improve current practices by rapid communication of changes from engineering to production so that these can be taken into consideration in the production process without delay.

3.4.4 *Manufacturing Technology*

It is suggested to strive for automation in manufacturing, but balance it toward the flexibility obtained by human resources. This implies the use of efficient technology in processes upstream of CODP and of more responsive and flexible technology (FMS) in customer-specific processes.

The case company has a long tradition of automation in flexible machine resources used for parts fabrication and has several ongoing initiatives related to welding and grinding process automation. Highly efficient and at the same time flexible technology is typically applied in upstream production processes with focus on parts fabrication. For example, flexible machine resources are used to produce both customer-specific parts and standard parts. However, there is a major potential to also automate the physical handling of materials and products in the plant as well as consider flexible robot technology in assembly operations.

3.4.5 Processes

With regard to manufacturing processes, companies should establish a product-oriented material flow with a layout that reduces nonvalue added processes. Also, operations should be performed based on digitally transferred information about customer specifications.

The company experiences major challenges with long lead times in production, and there is lack of flow and product focus in the plant. This can be explained by a long tradition of high resource efficiency and strong focus on machine capacity utilization. Value added time may be improved by a new layout. However, most machine equipment in the plant is heavy and large and therefore difficult to move. These resources are also shared as they are used for parts fabrication to all product families. Information about customer specifications are directly transferred to machine operators, and software programs are uploaded to machine resources used for automated fabrication of parts. Flow orientation of processes has high priority in the company to reduce production lead time and increase value added time relative to nonvalue added time. Focus in this work is on the interface between machine resources and assembly operations.

3.4.6 Manufacturing Planning and Control

The recommended design of planning and control processes is to large extent determined by the position of the CODP. Demand-driven (just-in-time) replenishment should be established for standard components and modules. Typically these are produced upstream of the CODP, but also downstream customer-specific processes will contain some standard components that can be replenished. Sequencing rules that takes delivery dates, capacity constraints, and setup times into account should be introduced downstream of the CODP in order to synchronize the production of different components of a customer order and to roughly keep the pace of the bottleneck. The flow upstream of the CODP should be based on supermarkets and pull, while the downstream flow should be based on first-in-first-out (FIFO) lanes.

The company has a traditional forecast-driven replenishment of materials. The supply of components is controlled through material requirement planning (MRP). The MRP calculates planned work orders and purchase orders based on the company's customer order backlog. They are now introducing a standard pull system for the supply of standard inexpensive short lead time items. However, the majority of parts are either customized, capital intensive, or long lead time items that will be ordered based on MRP calculations. Work orders, drawings, and work instructions for machining, welding, subassembly, final assembly, etc., are released to the different departments of the factory. The flow between operations is to some extent controlled by the due dates on work orders, but the flow is not synchronized, and delays due to missing parts are common. Most of the production is customer specific, and the company is now establishing sequencing rules in order establish takt and a synchronized flow from the CODP. Customization, deep product structures, a large product mix, and high variation work content make this synchronization challenging. A push-pull planning model is developed to ensure that all customized and standard parts for a complete product are delivered just-in-time to the assembly operations and that all operations in each value stream are producing to the same takt.

3.4.7 Supply Chain Integration

The supply chain for mass customized products needs to be streamlined and integrated from end-customers to suppliers. The recommended design for effective supply is to replenish key standard components just-in-time based on partnerships with the most important suppliers. Customized components need to be made to a specific customer order, and suppliers of customized components should be allowed online access to the order system in order to build what the customer want. Mass customization requires fast deliveries of products right after they are built, and the establishment of direct distribution channels to customers is recommended.

The company is manufacturing capital-intensive goods. Most products are built for new construction projects and are ordered months ahead. Delivery precision is key performance objective for products to new construction projects. However, they also deliver spare parts for the service market, and fast delivery time is crucial for these deliverables. The product delivery ratio is high in the service market. The current strategy to meet the delivery requirement in the service market is therefore to customize and convert a similar product with more slack in delivery time. All customized components are made in-house. Materials are purchased on forecasts and stored in sufficient quantities to meet any change in demand. Standard components are ordered weeks before delivery date in order to ensure that all components are available in time for final assembly.

3.4.8 Work Organization

To build customized products efficiently and with short delivery time requires a flexible workforce. Delivery times should be kept short even if mix and volumes fluctuate. A recommended strategy to cut lead times is to train operators and engineers to be multiskilled and able to handle a larger share of the order cycle. Labor efficiency should be high even when demand fluctuates, and the need for different types of jobs varies. Operators should be educated in multiple tasks, and it is recommended to develop a flexible job rotation and job allocation system that can adapt to fluctuations.

The case company engineers and manufactures advanced and complex products where some tasks require specialized skills and knowledge. For most products, one engineer has the total responsibility for a customer order and provides a single point of contact for the customer. The need for specialized knowledge hampers a fully developed job rotation system at the shop floor. Operators are organized in teams for each function and can do various tasks within machining, welding, assembly, testing, etc., but the rotation between different disciplines is limited.

3.5 Discussion

The literature review revealed several relevant issues of MC in ETO settings. A set of general MC principles were also tested in an ETO company to identify concerns of the specific company setting. Based on the literature review and the case study, major concerns related to decision areas are shown in Table 3.2.

Literature suggests that most issues are related to design phases and that operations, logistics, and procurement rely upon improvements in upstream engineering and design processes [9]. The interdependent relationship between MC capabilities including solution space, robust processes, and choice navigation [13] however proposes that MC capabilities should be developed coherently. The case study shows that issues in the early engineering and design phases are important, but that they are not isolated to these areas. Rather, concerns seem to be related to multiple and interdependent areas. This means that MC capabilities involving several areas are to be developed in parallel rather than in a sequential mode starting with product design. To ensure coherency between changes of both products and production processes in ETO companies, further considerations are needed with regard to achieving synergies between MC capabilities.

3.6 Conclusions

Research on the adoption of the MC strategy in manufacturing companies is dominated by studies on the transition of industrial mass producers to become mass customizers. Consequently, the knowledge base of the application of MC in companies with high degree of customization and crafting is still limited.

An underlying assumption of this work is that even though ETO companies may benefit from applying MC principles, these principles have different implications for such settings compared to when MC is applied in mass production. A literature review was carried out to identify major challenges in applying MC in ETO settings. This was followed by an in-depth case study of an ETO company with focus on testing a set of general MC principles.

The study revealed that major issues of MC in ETO are interdependent across several decision areas and involve manufacturing as well as engineering and design phases. There is limited knowledge of challenges for manufacturing compared to

Table 3.2 Major MC concerns of ETO companies

Decision area	Guidelines	Major concerns for ETO
1. Market interaction	Change the market interaction strategy to MTO or ATO	Production lead time is often long compared to delivery lead time. Difficulties in moving CODP downstream in high-mix, low volume manufacturing without increasing inventory cost
	(a) Aim to position CODP upstream in the value chain	
	(b) Aim to position CODP downstream in the value chain	
2. Products	Offer high level of customization on components/modules that represent the highest added value to customers	High variation in what features are customized in different orders
	Make a product program based on similar design elements for all product families	Engineered products often have unique structures that are difficult to standardize
	(a) Modularize components to enhance the variability for the customers	
	(b) Standardize components to reduce the complexity for the manufacturing	
3. ICT	Establish online order registration	Orders often include engineering changes that cannot easily be specified or communicated via online order registration and product configuration. Requires standardized components and modularization
	Establish a product configurator	Engineering work is highly manual and requires expert knowledge
	Guide the customer through the order process and visualize the choices	
	Strive for seamless integration of all information system (CAD/CAM, product configurator, ERP, order tracking, etc.)	
4. Manufacturing technology	Strive for automation in manufacturing, but balance it toward the flexibility obtained by human resources	Low volumes and high levels of customization make automation challenging
	Utilize efficient technology in processes upstream of CODP	
	Utilize responsive and flexible technology (FMS) in customer-specific processes	

(continued)

Table 3.2 (continued)

Decision area	Guidelines	Major concerns for ETO
5. Processes	Establish a product-oriented material flow	Job shop layout with high degree of shared resources makes flow orientation in processes and layout difficult
	Design a layout that reduces nonvalue added processes	Machine programming tends to be time-consuming as new CAD drawings are required for each new product Manual control might be more suitable
	Manufacturing processes should perform operations based on digitally transferred information about customer specifications	
6. Manufacturing planning and control	Introduce demand-driven replenishment of standard components and modules	Pure pull-based replenishment is less suitable for manufacturing characterized by low volumes and high customization
	Define and prioritize criteria for sequencing of orders in customer-specific processes	Push-pull principles need to be combined to ensure lean flow
	Aim to introduce pull-principle in processes upstream of CODP	
	Aim to introduce push-principle (FIFO) downstream of CODP	
7. Supply chain integration	Establish JIT partnership with suppliers of standard components/modules	Low volumes of standard components, which makes it difficult to establish JIT partnerships
	Allow key suppliers of customer-specific components online access to the order system	Due to frequent engineering changes and complex flows, there is high uncertainty in shared plans
	Establish rapid distribution channels to all the markets areas	Total delivery time depends upon the level of product complexity and engineering changes
8. Work organization	Train operators to be multiskilled	The need for specialized knowledge hampers a fully developed job rotation system at the shop floor. Rotation between different disciplines is challenging
	Educate operators in multiple tasks	
	Develop a flexible job rotation and job allocation system	

product design and engineering issues. This paper provides more in-depth insights to practical consequences of implementing MC in ETO manufacturing. Several directions for further research are suggested. The general MC implementation framework could be further adjusted to ETO settings. The challenges presented in this work may also be tested in additional cases to add even more details to current issues. The study includes one single case of an ETO company. The concerns addressed here should be investigated in other ETO manufacturing settings with different product and production characteristics in a multiple case study. In order to contribute to further implementation of MC in ETO, there is a need to develop new methods and tools for successful development and deployment of MC-based solutions in ETO companies. This also includes new approaches to MC implementation that consider a closer integration between several areas, i.e., manufacturing, NPD and engineering, and so on.

Acknowledgements This research has been supported by the Research Council of Norway. We are grateful for the contribution from the case company representatives involved in this research. We would also like to thank the reviewers for their valuable feedback on an earlier manuscript version of this paper.

References

1. Zipkin, P.: The limits of mass customization. Sloan Manag. Rev. **42**, 81–87 (2001)
2. Pine II, J.B.: Mass customization; The New frontier in business competition. Harvard Business School Press, Boston (1993)
3. Fogliatto, F.S., da Silveira, G.J., Borenstein, D.: The mass customization decade: an updated review of the literature. Int. J. Prod. Econ. **138**(1), 14–25 (2012)
4. Comstock, M., Johansen, K., Winroth, M.: From mass production to mass customization: enabling perspectives from the Swedish mobile telephone industry. Prod. Plan. Control **15**(4), 362–372 (2004)
5. Alfnes, E., Skjelstad, L.: How to implement the mass customisation strategy: guidelines for manufacturing companies. In: MCPC 2007, Boston (2007).
6. Skjelstad, L.: Modell og retningslinjer for masseprodusert skreddersøm, p. 246. Norges teknisk-naturvitenskapelige universitet, Trondheim (2008).
7. Olhager, J.: Strategic positioning of the order penetration point. Int. J. Prod. Econ. **85**(3), 319–329 (2003)
8. Haug, A., Ladeby, K., Edwards, K.: From engineer-to-order to mass customization. Manag. Res. News **32**(7), 633–644 (2009)
9. Duchi, A., et al.: Motivations and challenges for engineer-to-order companies moving toward mass customization. In: Advances in production management systems. Innovative and knowledge-based production management in a global-local world, pp. 320–327. Springer (2014).
10. Alfnes, E., Skjelstad L.: 1.2 How to implement a mass customization strategy: guidelines for manufacturing companies. In: Research in mass customization and personalization: strategies and concepts-applications and cases, p. 44. (2009).
11. Rudberg, M., Wikner, J.: Mass customization in terms of the customer order decoupling point. Prod. Plan. Control **15**(4), 445–458 (2004)
12. Bossen, J., et al.: An engineer-to-order mass customization development framework. In: Advances in production management systems. Innovative and knowledge-based production management in a global-local world, pp. 116–123. Springer (2014).
13. Salvador, F., De Holan, P.M., Piller, F.: Cracking the code of mass customization. MIT Sloan Manag. Rev. **50**(3), 71–78 (2009)

Chapter 4
An Open-Source Model of Collaboration and Customization in Architecture

Carlo Carbone and Basem Eid Mohamed

4.1 Architecture, Customs, Industry, and Customization

Customization is a central theme in architecture. Architecture and building construction are typically singular undertakings expressing individuality both in terms of character and customs. Based on tradition, social context, site specificity, and human relations, the production of architecture is defined by one-off prototypes seeking creative uniqueness tailored to users' specific needs. On a primary level, the idea of custom architecture is connected to characterizing one's boundaries and outlining a framework for social interaction.

Mass customization in architecture relates less with primary needs as it does to the commercialized methods of production generated by the industrial revolution. This type of made-to-order personalization designates adaptable and flexible models of production. This adaptability encompasses the capacity to oblige individuals' desires in a mass manufacturing process. Within the field of architecture, mass customization relates predominantly to industrialized building systems as these systems imply a business model of mass production.

Industrialized building systems, prefabrication of architecture or off-site fabrication of sub-assemblies, are not new strategies. Some have described prefabrication as the oldest new idea in architecture [1]. This prefabrication model in architecture is based on the experiments of many generations of builders. From Roman military engineers to medieval master guilds and to Great Britain's early industrialists, all prepared components off-site (precut stones, precut or notched wooden beams, iron beams) to facilitate on-site construction.

C. Carbone (✉)
UQAM, Ecole de Design, Montreal, Canada
e-mail: carbone.carlo@uqam.ca

B.E. Mohamed
Department of Architecture, Abu Dhabi University, Abu Dhabi, UAE
e-mail: bassem.eid@mail.mcgill.ca

J. Bellemare et al. (eds.), *Managing Complexity*, Springer Proceedings in Business and Economics, DOI 10.1007/978-3-319-29058-4_4

The building industry today is highly industrialized. Architects and builders pick and assemble continuously produced components (doors, windows, beams, finishes, etc.), repeating a highly inefficient design and construction process for each building. This high level of custom building implies waste at almost every level of a building's production but affirms a perceived uniqueness. Our examination of the industry intended to elucidate connections and potentials between mass customization and industrialized building systems with the intention of elevating architecture in terms of efficiency, quality, and personalization.

The desire for an industrialized building process that optimizes construction efficiency, costs, and mass production has spanned eras, customs, cultures, and even public policies [2]. The history of architecture and prefabricated construction recounts this sometimes confluent but often divergent tale. The early twentieth-century economic crises, social turmoil, and industrial development shaped icons of prefabricated architecture. Projects such as Lustron[1] in the United States, AIROH (Aircraft Industries Research Organisation on Housing)[2] in Great Britain, government-owned and government-operated precast concrete panel plants[3] in the USSR, and Sekisui Heim M1 by Sekisui Chemical in Japan[4] all convey the modernist twentieth-century fantasy of factory-produced architecture [3]. Often supported by the transfer of military knowledge and processes to civilian industries, many manufactured architecture experiments were also supported by mega-housing programs in their respective countries [3].

Architectural projects spawned by new industrial materials and methods sustained the founding principles of modernity. From Konrad Wachsmann to Jean Prouvé and Buckminster Fuller [3], the goal of an industrialized, quality, and low-cost architecture for the many was a recurrent obsession for the modern architect.

Since modernity's union of architecture and industry, both fields (architecture and prefabricated construction) have outlined divergent trajectories. Architecture established an idealized representation of prefabrication, while the prefabricated construction industry has largely remained in a mass production paradigm[5] [2]; early debatable construction methods and repetitive design contributed to the negative connotation that the industry is still trying to relinquish. The evolution from "mobile home" to "modular houses" and to "manufactured homes" suggests a long but stigmatized history.

---

[1] www.lustron.org

[2] For a project description, see Carbone, C. (2014)—*Prefabrication experiments (10) Aircraft Industries Research Organisation for Housing—the A.I.R.O.H. house* retrieved from http://prefabricate.blogspot.ca

[3] Carbone, C. (2014)—*Prefabrication experiments (22) Precast concrete (pieces, panels and boxes) in postwar U.S.S.R.* retrieved from http://prefabricate.blogspot.ca

[4] For a project description, see Carbone, C. (2015)—*Prefabrication experiments (62) Sekisui Chemical's Sekisui Heim M1* retrieved from http://prefabricate.blogspot.ca

[5] "since the 1950s architects have retreated from this position, distancing themselves from the factory…factory produced has become a style," Davies C., *The Prefabricated Home*, Reaktion Books, 2005, p51

Mass customization in architecture, although not identified as such, has been the core dispute in the tumultuous relationship between architecture and industrialized building systems. Sigfried Gideon[6] [4], when analyzing the work of Walter Gropius on the relationship between architecture and industry, spoke of the superficial uniqueness of everyday architecture and how the need for this uniqueness hindered the development of industrially produced systems for architecture.

A little over 100 years after Gropius' manifesto on industrialized building systems[7] [5] for housing, architecture and industrialization are converging once more, this time with regards to new information technology and its potential to induce mass customization strategies within architecture. Big data is changing the way architects collaborate [6] and is generating a new paradigm of collaboration and customization within the industry. The theory that data management will encourage prefabrication was highlighted by producers and architects surveyed in the McGraw-Hill's report, *Prefabrication and Modularization: Increasing Productivity in the Construction Industry* [7]. Notwithstanding this trend, we suggest that a lack of inventiveness and ancient connotations still stifle innovation potential in this industry sector.

The development of information management software augmented by territorial, demographic, and environmental issues is leading a transformation of our design criteria and lends itself to new production and construction methods. Informed data management is central to this revolution in design and construction methods. However construction as a whole remains relatively distant from these contemporary tools' overall potential. A revolution in design, construction, and management methods articulated to data management will induce a shift toward information-based collaboration provoking an environment conducive to an open exchange of ideas. Autodesk Seek[8] seems to point in the direction of information sharing but not specifically for industrialized building systems.

Our study, financed in part by the Société d'Habitation du Québec, set out to map and characterize the prefabricated building industry in North America, particularly

[6] "Gropius' and Wachsmann's Packaged House system, with its carefully worked out designs of standardized building components, is in the direct line of future development, especially in its concentration upon the production of easily transportable and easily assembled multi-purpose unit parts and not upon the production of complete standardized house types. Nevertheless it had no financial success. Why is this ? .... These difficulties, in the last resort, lie within the present attitude of the house purchaser. No matter how identical in plan and appearance his house may be to all its neighbors in its suburban setting, the man building his own home still likes to believe that he is getting an individual, personal, handmade product." Giedion, S., *Walter Gropius, Work and Team Work*, Reinhold, 1954, New York, p76

[7] Gropius submitted his "program zur Gründung einer allgemeine Hausbaugellscahaft auf künsterlich einheitlicher grundlage", m.b.H. (Program for the Founding of a General Housing-Construction Company Following Artistically Uniform Principles) to Rathenau of AEG in April, 1910"; see Herbert G., *The dream of the factory-made house: Walter Gropius and Konrad Wachsmann*, MIT press, 1984, Cambridge, p33.

[8] http://seek.autodesk.com/search.htm

existing systems and their customization strategies to examine the potential for cross-pollination between prefab producers.

4.1.1 Customizable Architecture and Its Relationship to Industrialized Building Systems

Modular building, industrialized building systems, manufactured housing, prefabricated architecture, and the mobile home all share the genetics of early twentieth-century Fordism as applied to building construction. The advantages of a climate-controlled environment, standardization, waste reduction, labor efficiency, and bulk material procurement all contributed to the development of the desire for a factory-produced architecture. Advocated as a necessary change in housing production to serve the rapid urbanization that accompanied industrialization, the mass production of architecture in a factory echoed the mass production of other commodities. The convergence of industrial production, architecture, and urbanization was particularly fertile for the design of industrially minded customizable architectural prototypes [3].

The open plan (*plan libre*), proposed by Le Corbusier in 1909 under the name DOM-INO (domicile—innovation), was a structural system emblematic of the union of architecture and industrial production[9] [8]. The free or open plan combined new materials and methods, and reinforced concrete, toward an open post and slab structure that allowed planning flexibility and customization. A grid of small posts or columns defined space horizontally and vertically. This grid replaced preindustrial load-bearing walls and allowed for freedom in planning and three-dimensional organizations. The column/slab system is used today in the construction of most commercial buildings for flexible arrangements. This open plan "plan libre" was a revolution in architecture.

In addition to Le Corbusier's DOM-INO, many architects explored industrialized building systems for housing and pursued tactics for flexibility and adaptability. The Weissenhof neighborhood project orchestrated by Mies van der Rohe at the request of the city of Stuttgart, Germany, in 1927 encompassed 21 proposals by 16 architects. This exhibition of modern placemaking included proposals from Bruno Taut, Le Corbusier, and Walter Gropius and portrayed a potential for the industrialization of architecture.

In America, California more specifically, the Case Study House Program fused industry, architects, and the quest for an industrial but individualized architecture. Implemented by *Arts &Architecture* magazine with the support of its editor John Entenza, the Case Study House Program was based on modern values of innovation, scalability, reproducibility, affordability, and personalization. Thirteen out of

[9] "*Architecture ou révolution., he touches on the idea of revolution, both technical and political. By the former, he clearly meant the industrial revolution, already achieved through the mass production of automobiles; by the latter, he presumably intended revolutionary socialism fermenting beneath the surface of society and due primarily, in his view, to the fact that the working class was ill-housed.*" Frampton K, *Le Corbusier*, Thames and Hudson, 2001, New York, p31

the 36 residential prototypes were built on the conviction that architecture could be both mass produced and fitted to owners' personalities. In 1949, fed by European avant-garde influences and the transfer of knowledge acquired in military service, Charles Eames designed the Case Study House 8 and collaborated on the Case Study House 9 [9]. Eames explored an open frame structure, a clear span space, structured by a steel skeleton leaving considerable flexibility to potential occupants and users. This variability similar to what Le Corbusier had developed was based on ready-made industrialized components.

In continuing, developing, studying, and probing modern architecture's strategies, N.J. Habraken published *Supports: An Alternative to Mass Housing* in 1972 [10]. This progressive publication was the foundation of the "open building" theory [11], which aims to increase personalization, adaptability, and flexibility of architecture over time. Habraken proposed the separation of common infrastructure (supports) and personal systems (infill) to inform customizable building planning based on a shared substructure. Kendall and Teicher [11] reiterated and continue to sustain these ideas within the "open building" theoretical framework.

The establishment of "open building theory" was influenced by collaborative and customizable building systems that were examined or explored during the twentieth century as patterns for client-based personalization in design and production. Timber Structures Inc.'s Mobilcore provides one such example combining the strengths of on- and off-site construction within a larger made-to-measure framework. Published in the April 15, 1946 edition of *Life* magazine, the 8×24 ft. (2.4×7.2 m) Mobilcore[10] included all fixtures and appliances. The box-unit service core was divided into bath, mechanical room, and kitchen. For US$2700 (approximately 40 % of a total house price of the era), one could purchase a unit, have it delivered on-site, and then build a custom-made house around it. The organizational variability was articulated to a stable nucleus that optimized factory production for the complex parts of a building.

This type of box-unit construction for mass customization can also be seen in an even more systemic level in Sekisui Chemical's[11] first experiment into the housing market. Sekisui Chemical produced its first modular light steel frame box-unit in 1971: the Sekisui Heim M1.[12] The box-unit's commercial success contributed to lowering its construction costs and increased production capacity and illustrated the then attainable factory-produced adaptable house. The basic module unit was a rectangular prism composed of light-gauge steel-framed edges, which included walls, floors, ceiling, and service cabinets. Multiple cabinet organizations were available and this user-defined element exemplified the beginnings of mass customization strategies within the industry. Each box-unit could be juxtaposed or stacked with complete box-units or a 2/3 fragment of a unit. The stitching of adjacent units was simplified by the juxtaposition of structural edge members.

---

[10] *Life* magazine April 15, 1946—*Wyatt will use all kinds of building to get the job done*, p34

[11] http://www.sekisuichemical.com/about/division/housing/index.html#h_01

[12] http://www.sekisuiheimm1.com/index_english.html#

The system's variability, in plan and in section, challenged the mass production paradigm that defined most industrialized construction systems. The 2.4 m×4.8 m box-units were based on a familiar 2:1 tatami mat proportion. Each house included a distinct tatami room relating to traditional Japanese housing. This combination of industrialization, variability, and tradition established a new era for prefabricated architecture.

4.2 Industrial Cross-Pollination Toward Innovation

The customizable prefabrication that seemed evident to Timber Structures Inc. or to Sekisui Chemical appears to be permeating into the building industry today. Combining the flexibility of frame construction with factory-produced cores or modules creates a formidable, open customizable industrialized building system.

Kieran and Timberlake's Loblolly house[13] or Alastair Parvin's Wiki-house[14] make a case for an open-source approach to the mass customization of architecture. The theories imbedded within these prototypical projects support our pursuit for a comprehensive strategy for open collaboration toward quality and sustainable architecture.

The manufactured building industry developed from the application of new technologies to the ongoing urbanization of cities. Demographic and economic changes caused by the industrial society pressured government, which placed the burden directly on private industry to solve the growing housing crises. The postindustrial nuclear house and its privatization were the main constituents of the rapid suburbanization of North America, which established the single-family home constructed on-site by a wood frame builder as the nucleus of North American housing and building culture. The manufactured housing industry could not compete with the prevalence of the on-site builder and, as noted by Gideon [4], the superficial customization offered by traditional homebuilder; the purchaser's aspired uniqueness was however offset by an overwhelming homogeneity.

Today's social heterogeneity combined with environmental priorities, progressive design tools, and information management software is federating a fertile environment for the greater use of prefabricated building systems and their customization. Our research focused on customization as it relates to mapping potential collaboration and links between prefab manufacturers. Our research strategy aimed to organize this potential and start working to offer the means and tools for interactive online interaction. These tools would create an environment for choosing, composing, and assembling components and sub-assemblies for buildings, an "open" language for architecture.

[13] "Loblolly House"—American Institute of Architects case study retreived from http://www.aia.org/aiaucmp/groups/aia/documents/pdf/aiab081572.pdf

[14] http://www.wikihouse.cc

4.2.1 *Research Strategy and Its Evolution*

In an evolving attempt to engage an open-source type collaboration in architecture and to reflect on the industry's potential, we established two complementary data structures: a catalogue of building systems and an annotated list of companies as a basis for a larger industry analysis and as knowledge incubators. We envision these tools as the starting point of an online reference point for prefabrication strategies and their crossbreeding (see Fig. 4.1 for a list fragment and Fig. 4.2 for a sample catalogue page).

Our growing annotated list of 800 companies was undertaken in 2014 and continues to be compiled by cross-referencing literature, trade associations, modular building groups, and a comprehensive keyword search on the internet. This annotated list along with proposed catalogue of case studies is the catalyst for a growing research project that shares information about the industry and more importantly strives to involve manufactures and stakeholders in an agenda of collaborative construction of knowledge.

The list of producers also allowed us to triangulate existing data and to generate a point of view in terms of how the industry works and how it could evolve. The share quantity of "box-unit or module" type producers depicts an industry still dependent on one type of prefabrication. This modular sector represented 72 % of our list. The assembled data also presented the archaic "pattern-book of house types" as the significant model for customization within the industry.

CANADA

COLOMBIE-BRITANNIQUE

ENTREPRISE	Adresse	Ville	Code Postal	AFFILIATION	TYPE	SUR SITE	MODULES	MATÉRIAUX	PERSONNALISATION	SITE WEB
Minaean International Corporation	1055 W Georgia St	Vancouver	V6E 3P3	X	Résidentiel (multi-familiale), kit de	Assemblage	Sectionnel	Métal, panneaux	Système de panneaux adaptable, maison résidentiel, multi-familiale, mass	http://www.minaean.com/
Moduline	1175 Railway St,	Penticton	V2A 5X5	Champion Homes	Résidentiel+Commercial		Sectionnel + Mobile	Bois	Choix limité à des plans pré-définis. Possibilité de modifications de ceux-ci selon	http://www.moduline.ca
Chaparral Industries	3075 Sexsmith	Kelowna			Résidentiel (modulaire,	Installation et Assemblage	Sectionnel + Mobile	Bois	Non définie	www.chaparralhomes.com
Shelter Modular	3294 262 St	Aldergrove	V4W 2X2		Résidentiel temporaire (workforce	Non définie	Sectionnel + Mobile	Bois et métal (conteneurs	Possibilité de créer son propre plan.	www.sheltermodular.com
Britco	1825 Tower Rd	Agassiz	V0M 1A2		Résidentiel collectif		Sectionnel			www.britco.com
Freeport Industries	3522 Red Cloud Way	Westbank	V4T 2G9		Résidentiel (Individuel et collectif),	Assemblage	Sectionnel	Bois	Non définie	www.freeportindustries.ca
Harmony Homes	833 Finns Rd	Kelowna	V1X 5B8	Indiana MHA (MHI	Résidentiel (Multi-familial)	Assemblage	Sectionnel	Bois	Projet custom	http://www.harmonyhomes.net/
Northern Trailer	3355 Sugarloaf	Kamloops	V2C 6B7		Résidences de travail (camp)		Sectionnel			http://www.horizonnorth.ca/
Cratex Container Sales	7664 80 St	Delta	V4G 1C1		Industriel et commercial		Sectionnel Conteneur	Conteneurs		http://www.cratexcontainer.com/
Atco Structures & Logistics	982 Boundary	Prince George	V2N 5T2		Logement de travailleur (camp) et		Sectionnel			http://www.atcosl.com/fr-ca/
Outbuildings	2128 Front St.	North Vancouver	V7H 1A5		Résidentiel (tiny) et agricole		Sectionnel	Bois		http://www.outbuildings.ca/
A-Mobile Mini		Vancouver			Commercial et industriel		Sectionnel + Mobile	Métal		https://www.mobilemini.com/
Eagles Homes	1292 Trans Canada	Salmon Arm	V1E 2T3		Résidentiel et promoteur résidentiel -		Sectionnel + Mobile	Bois	Système sectionnel	www.eaglehomes.ca/
Amco Modular Homes	7401 Industrial	Lantzville	V0R 2H0	Moduline	Résidentiel		Sectionnel + Mobile	Bois		http://www.amcohomes.ca/
Best Buy Homes	1433 Velocity St	Kelowna	V1V 3C2		Résidentiel		Sectionnel + Mobile	Bois		http://www.bestbuyhousing.com/default.asp?LocationSelection=10
Bob Paterson Homes inc.	1200 South Mackenzie	Williams Lake	v2g 3y1				Structural			http://www.bobpatersonhomes.com/
BROOKSWOOD HOMES LTD	3229 - 200th	Langley	V3A 4W4				Sectionnel			http://www.brookswoodhomes.com/
Chaparral Industries Inc.	3075 Sexsmith	Kelowna,	V1X 7T1		résidentiel, coimmercial,		Sectionnel			http://www.chaparralhomes.com/
Columbia River Homes	195 West Airport	Castlegar	V1N 4M5			Offre l'assemblage	Sectionnel			http://www.columbiariverhomes.ca/
Countryside Home Sales	2425 Cranbrook	Cranbrook	V1C 3T3	Winfield Home Systems, a	Résdientiel (neuf ou réutilisé)	Offre le transport	Sectionnel		Offre commercial : paiement mensuel pour acheter.	http://www.countrysidehomesales.ca/
Countryside Manufactured Homes Ltd.	401 Cree Dr	Kamloops	V2H 1G7		Résidentiel		Sectionnel + Mobile			http://csmh.ca/

Fig. 4.1 Excerpt from the list, to see the complete list go to https://drive.google.com/file/d/0B5Te_qsSnKzpWG9KTjdMS0EzZWc/view?usp-sharing

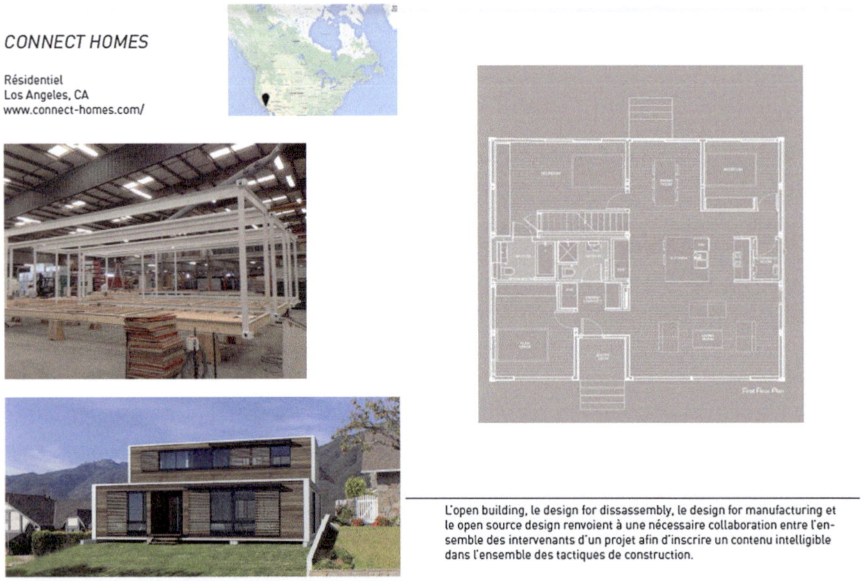

Fig. 4.2 Sample systems catalogue page—images are screen shots from Connect Homes' web site: sample page from report available at https://drive.google.com/file/d/0B5Te_qsSnKzpWG9K-TjdMS0EzZWc/view?usp=sharing (*Source*: Author)

The prefab directory was organized by location: the United States and Canada (province or state and precise address and contact information), by production approaches (modules, panels, components/kits, hybrids) [12] and by customization strategies. Other complementary characteristics such as materials or construction details helped measure dissimilar systems within the same prefabrication model. For example, a module could be framed in wood, steel, or concrete. The data fields are also evolving as we inform our characterization process, as there is no currently accepted theoretical model for grouping industrialized building methods; there are many [13].

Our methodological positions for the data's organization, type (modules, panels and pieces), content (sub-assemblies), and context (United States and Canada) were framed by our objective to elevate our local industry with regard to its undervalued potential and its current production. This localized point of view was also supported by a complementary objective of addressing a market where industrialized building has not taken a foothold.

This specificity is important as Asia, Europe, and Australia have a different industrialized building legacy. The North American market carries a vision of prefabrication and its potential customization influenced by the American dream of the single-family dwelling. In order to stimulate a paradigm shift toward open collaboration and customization, the traditional box-unit modular prefabrication model in the United States and Canada will have to be rerouted toward other building types and strategies. Our preliminary work has allowed us to compare our research with other industry characterizations and has revealed a great potential for innovation within a somewhat conceptually suppressed industry.

4.2.2 Current Approaches to Mass Customization

Our classification positioned modular (room, sections, or house boxes) at 72 %, panel (walls or floors) construction at 21 %, and kit (pieces or components) construction at 13 %. We observed a substantial overlap between modular and panel construction both in terms of business model and market share. The modular segment at 72 % showcased a limited potential for customization as each module is either juxtaposed or stacked. This does not impede customization practices at more internal levels of production such as finishes, interior systems, materials, and fabrication methods (cutting and assembling). The modular and panel segment is largely based on similar construction and customization strategies. A number of manufacturers have begun integrating building information modeling (BIM) [14] and have established a potential for a new type of custom prefab. BIM is changing the way architects and industry collaborate and is creating a fertile environment in which design and production could merge. This is the case of Premier Building Systems[15] from Washington, USA, which articulate their sales pitch to a capacity to tailor fit the home within a system of standard structural insulated panels. This innovation is occurring at a sluggish pace and mostly by experimental projects that are not being mass produced.

The lack of innovation is largely forged by archaic views of building construction and mass production. Our simple cataloguing system of boxes, panels, and pieces, although not the industry standard, illustrated this lack of innovation as most companies share similar business models. The list's secondary objective was to foster a potential cross-utilization of systems: boxes for service cores, panels for building envelope, and pieces for open and adaptable frames. These potential relationships between manufactures and builders could stimulate the industry allowing stakeholders to understand how systems work, their agility, and how they can be employed together toward quality and singular architecture.

The work of Kieran Timberlake for Loblolly house,[16] the work of Bensonwood Homes for corewall,[17] or Project Frog's[18] language of components point out the conclusive capability of industrialized building components to accelerate innovation and cooperation. A collaborative model based on an informed pedigree of interrelated systems could initiate a new era for prefabricated building systems.

Although we did not find large-scale examples of this type of "open" customization, we did find examples of information technology and computer modeling technology driving mass customization. This pattern will continue to drive architecture and industrial collaboration [7]. The outdated conceptual limits between design, fabrication, and construction are collapsing under powerful information management tools for construction [14]. We found that the companies that are

[15] http://www.premiersips.com

[16] "Loblolly House"—American Institute of Architects case study retrieved from http://www.aia.org/aiaucmp/groups/aia/documents/pdf/aiab081572.pdf

[17] http://www.openprototype.com/press/corewall.pdf

[18] http://projectfrog.com

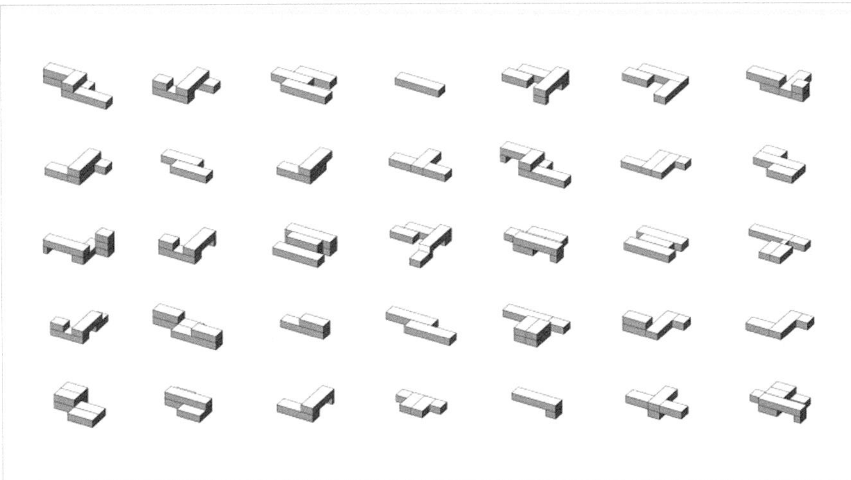

Fig. 4.3 Screen shot of modular configurations from Resolution: 4 Architecture's web site (*Source*: http://re4a.com/the-modern-modular/)

employing technology toward mass customization strategies are at the forefront of innovation, but are fairly marginal in relation to the industry as a whole. It is important to note again that this model applies to what we found within the North American market; Japan, Scandinavia, and Australia would most certainly have given us a totally different data structure as their industrialized building industries have advanced within a different social and contextual framework.

In conjunction with the flexibility and responsiveness of systems to meet various contemporary realities, the housing market is moving toward a customization pattern. The proliferation of lifestyle types is increasing demand for choice and is shifting the marketplace. The diversity of multiple family structures, behavioral individualization, and aging population structures underline the need for new design criteria with variability as its benchmark. The following examples highlight some of the efforts to implement customization in the housing industry either by architects or manufactures. The focus was on design strategies, as well as tools for customization.

(a) The case of Resolution: 4 Architecture: *The Modern Modular*[19] (see Fig. 4.3)
Developed in 2006 by a New York firm, Resolution: 4 Architects, this approach showcases standardization of prefabricated box modules and their potential aggregation. Variable in both vertical and horizontal juxtapositions, the box-unit configurations vary in H, I, L, T, and Z shapes. Each volume is completed off-site and then stitched to other volumes on-site. This adaptable and modular design process responds to an assortment of choices and lifestyles.

[19] http://re4a.com/the-modern-modular/

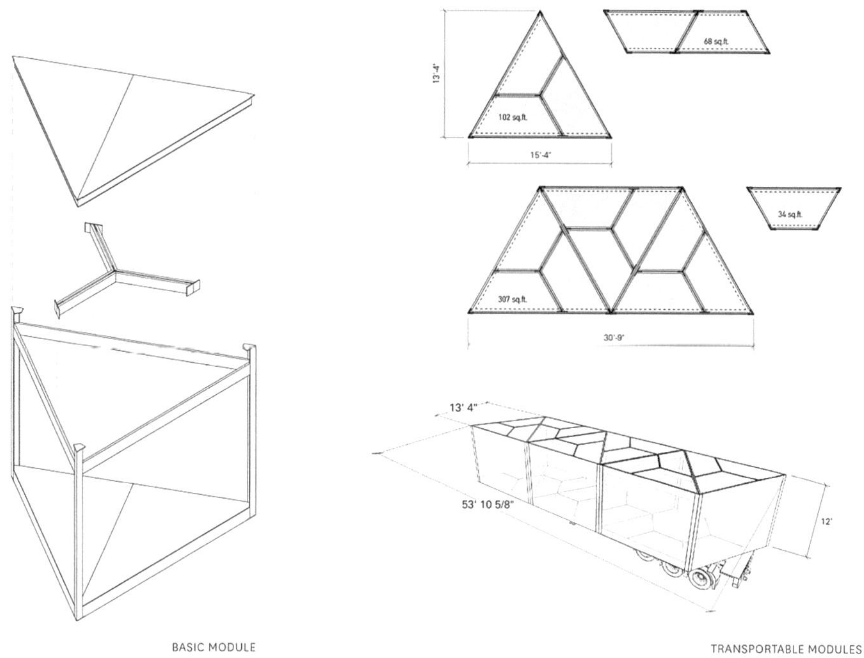

Fig. 4.4 Screen shot of the HOMB basic planning unit from Method Homes' web site (*Source*: http://skylabarchitecture.com/work/taft-residence/#slide1)

Similar to Sekisui Heim M1 in Japan, the volumetric variability is conceived as a response to individual requirements. Each unit is defined by its use: (bathroom, kitchen, office), lifestyle (home office) or living areas. This method of exploring architectural uniqueness within a system of standard components interprets a mass customization based on variable juxtapositions of a kit of space types and functions.

(b) The case of Method Homes: HOMB Modular Prefab[20] (see Fig. 4.4)

The "HOMB," an inhabitable honeycomb, was co-developed by Skylab Architects and Portland Oregon's Method Homes. The system is founded on the adaptability, strength, and compositional agility of triangles. Articulated to architecture's geometric heritage, the system reveals the unlimited flexibility of geometric compositions. Similar to Swiss architect Justus Dahinden's Trigon 65,[21] putting 100 ft two triangles together in multiple geometries or architectural compositions generates infinitely adjustable plans. Allowing users to choose window sizes, finishes, and materials further enhances the made-to-measure capacity of this geometric planning grid.

[20] http://skylabarchitecture.com/work/taft-residence/#slide1

[21] For a project description, see Carbone, C. (2014)—*Prefabrication experiments (40) TRIGON 65* retrieved from http://prefabricate.blogspot.ca

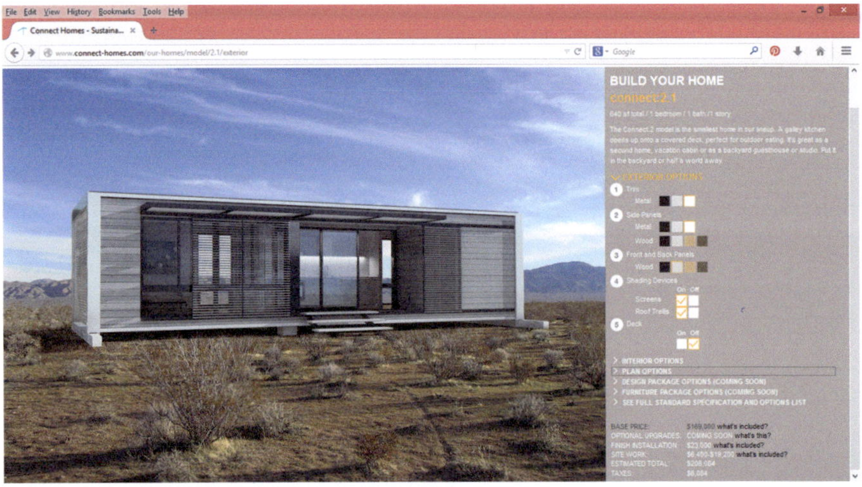

Fig. 4.5 Screen shot of configurator from Connect Homes' web site (*Source*: http://www.connect-homes.com)

This collaboration between architects and manufactures reveals a mass customization system characterized by its ordered variability for planning. HOMB modular prefab also includes choices for augmenting energy efficiency and reducing the buildings environmental footprint.

(c) The case of Connect Homes[22] (see Fig. 4.5)

Founded by two architects, Jared Levy and Scott Gordon, recognized for their contribution to Marmol Radziner Architects, this patent-pending modular system relates a simple modern aesthetic to sustainable design values. Ninety percent of the process is articulated to off-site production. Each design achieves a baseline sliver LEED (Leadership in Energy and Environment Design) and further design parameters achieve "gold" or "platinum" certification levels and even "net zero" energy use.

Aiming for an advanced level of personalization, the company has proposed a web interface where potential buyers can choose a predefined model and refine it with multiple options. The interface allows the user to choose spatial configuration, finishes, energy systems, and a myriad of elements to add to the basic design and tailor the design to preset individualized options. Analogous to the automobile industry, each selected option adds and modifies the design's cost in real time. This web interface typifies mass customization based on an option-controlled standardization.

(d) The case of Project Frog[23] (see Fig. 4.6)

Project Frog exemplifies the use of technology in generating their own "open-ended" architectural language of components. The variability of industrialized

[22] http://www.connect-homes.com/

[23] http://projectfrog.com/performance/technology/

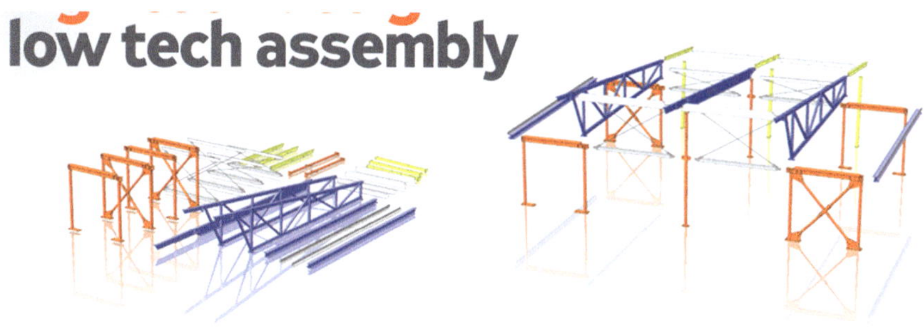

Fig. 4.6 Screen shot of Project Frog's Kit-of-Parts from the company web site (*Source*: http://projectfrog.com/performance/systemized)

components conjures images of Ikea's business model for furniture or the general panel house designed by Wachsmann and Gropius [5]. Connecting uniqueness, standardization, and production efficiency, Project Frog is based on the manufacturing of high-performance technically advanced standard components predetermined for systemic adaptability and agility. This system employs consistent assemblies toward a diversity of building types and organizations.

An information management system and the encoded components enable a wealth of interaction possibilities. Parametric information modeling monitors material criteria, life cycle criteria, energy-saving criteria, and building performance. This precise and integrated design leads to a "lean" [15] production process that reduces waste at all levels of design and manufacturing. This mass customization based on computer modeling from design to production imbeds performance monitoring and control at all stages of the project's production.

(e) The case of Honka Canada[24] (see Fig. 4.7)

Honka has been producing timber houses from massive planks or logs since 1958. Informed by Finnish building culture and its link to timber and forestry, the company's production articulates traditional wood-working knowledge with contemporary design and fabrication tools and elucidates a state of the art streamlined relationship between conceptual design and manufacturing.

Each house is a unique design and an assembly of digitally controlled cut pine logs. Astute profiles provide stability, strength, and weather tightness. Each computer model represents a specific project and is transferred to digital fabrication once the design is approved. Machinery translates the design. This mass customization manufacturing method can reduce waste and epitomizes a just-in-time prefab tailored to a specific user.

[24] http://east.honka.ca/en/why-log-home

Fig. 4.7 Honka Canada's digitally cut log profiles—screen shots from web site (*Source*: http://east.honka.ca/en/node/219)

4.2.3 An Untapped Potential

The preceding examples demonstrate an unexploited potential. Current customization applies to different levels either to increase choice or to achieve made-to-order designs. Customization is not limited to the design. Contemporary design and manufacturing tools enable an efficient flow of information containing the parameters to modify, tweak, and intervene at different stages of production.

These mass customization strategies and levels of customization are permeating and will continue to transform the industry. The flexible aggregation of standardized components, geometric modular adaptability, programmed design variables, encoded components, and digital fabrication are a few strategies we have observed.

In each case technology is establishing a potential to redefine industrialized building systems toward architectural singularity.

The five previous examples also represent an informed collaboration between architectural design and industry enriching innovation and products. This collaboration could challenge the systemic lack of interaction between architecture and the prefab housing industry. Working from two diverse, distinct, and complementary perspectives, these two fields must be connected. A knowledge incubator could be a point of connection between these diverse stakeholders, as a confluence of factors seems to point to prefabrication as an important strategy for efficient and streamlined resource management. Still, only a small percentage of single-family housing starts to employ prefabrication, about 12 % in Canada, and even a more marginal amount employ technologically advanced customization strategies, with many still only using the age-old "plan pattern-book" for customization. A large proportion of companies are just producing houses in a factory as they would on-site.

Our study has led us to imagine and conceive of a knowledge incubator, a collaborative online tool that could hypothetically increase both industry/architecture connections and prefab use within the building sector. We are currently establishing an online "wiki-prefab" platform that would engage producers, stakeholders, and technology toward a network of potential hybrids and toward a library of potential ready to use informed components for architectural design. This library of components is not a new idea. Autodesk "Seek" already employs online networking for collaborating and sharing building information. Our proposal, similar in strategy, addresses not only the need for informed components for architectural modeling but also for the value of combining industrial production and architectural design.

We believe this industrial cross-pollination is an ingredient for accelerating change.

4.3 Sixty-Four Years Later, Accelerating Change

In his 1951 *The Prefabrication of Houses* [16], Burnham noted that in the turbulent era of the early twentieth-century America, even with the encouraged growth, prefabricated building systems never truly permeated American building culture. Though highly subsidized, factory production of houses never achieved its potential to provide a lower-cost and higher-quality alternative to traditionally built housing. The extremely competitive, low-cost, low-overhead, and entrenched building culture reinforced on-site wood frame construction and relegated the factory-built house to a market share that stabilized at no more than one out of eight or ten dwellings produced.

Sixty-four years after the work of the Albert Farwell Bemis Foundation, our corresponding and evolving project draws on similar values of a better product (sustainable, efficient) for a larger part of the population. Our analysis shows that the prefabricated building industry has developed in a parallel, somewhat divergent model to the housing industry and even more so to the practice of architecture.

This divergence already recognized at the beginning of industrialization increased during the twentieth century and has impeded a fertile cross-contamination between architecture and industrialized building systems.

Today, factory-produced building systems are evolving into sustainable, resource-responsible, and customizable options for housing but still only garner a fraction of production in North America.

Digital fabrication, automation, mass customization, and lean construction are becoming typical in factories around the world and can contribute to a renewed, durable, and ecological building culture. Within the contemporary convergence of a renewed production process, an appetite for sustainable housing options, and a demographic shift toward heterogeneity, the manufactured housing industry can be an important player in establishing creative building and housing concepts to serve the market's ever-evolving lifestyles and family structures.

Prefabrication and Modularization: Increasing Productivity in the Construction Industry [7] discusses the different conditions of increased competitiveness for prefabricated construction systems. Articulated to a variety of topics, such as the lack of skilled labor, waste reduction, increased productivity, and reduced construction time, a larger environmental awareness is driving increased attention to off-site fabrication. The contemporary building culture defined by an integrated design process and digital conceptualization is also more conducive to factory fabrication. Furthermore resource-optimized factory production is also accepted as a valid and superior alternative to complex and resource-intensive on-site building. Harnessing this potential of factory-based construction hinges on a new creativity.

Our current research established the need to share knowledge within the architectural profession and throughout the building industry to both recharge prefab's potential and erase age-old connotations. Ancient mass production models no longer limit architecture's long-lasting objective of creative uniqueness. Today's tools inform a creative process that allows us to understand that the perceived uniqueness of the architectural process is being overtaken by uniqueness imbedded in variable processes that leverage technology toward holistic approaches.

Encouraged by our preliminary mapping of the North American prefab industry, our aim of breeding knowledge exchange within the industry has led us to define and imagine an online collaborative "wiki" as an open-source model for collaboration and customization in architecture. We are currently collaborating with and calling upon trade associations, academics, and manufacturers to establish a test version of this information management tool to assess our premise. We suggest that we are at the cusp of a new and "open" era for the oldest new idea in architecture.

References

1. Stevens, K.: Can prefab deliver? http://archpaper.com/news/articles.asp?id=1877#.VRmZdbqI8pw (2008). Accessed 18 Aug 2015
2. Davies, C.: The prefabricated home. Reaktion books, London (2005)

3. Bergdoll, B., Christensen, P.: Home delivery: fabricating the modern dwelling. Birkhauser, Basel (2008)
4. Giedion, S.: Walter Gropius, work and team work. Reinhold, New York (1954)
5. Herbert, G.: The dream of the factory-made house: Walter Gropius and konrad wachsmann. MIT Press, Cambridge (1984)
6. Phillips, A.: How big data is transforming the construction industry (January 27, 2015). http://www.constructionglobal.com/equipmentit/399/How-big-data-is-transforming-the-construction-industry. Accessed 18 Aug 2015
7. SmartMarket Report, Prefabrication and modularization: increasing productivity in the construction industry. McGraw-Hill Construction. http://www.modular.org/htmlPage.aspx?name=McGrawHill_Prefabrication. Accessed 18 Aug 2015
8. Frampton, K.: Le Corbusier. Thames and Hudson, New York (2001)
9. McCoy, E.: Case study houses 1945-1962, 2nd edn. Hennessey & Ingalis, Los Angeles (1977)
10. Habraken, N.J.: Supports: an alternative to mass housing. The Architectural Press, Praeger, New York (1972)
11. Kendall, S., Teicher, J.: Residential open building. Spon Press, London (2000)
12. Dietz, A.G.H., Cutler, L.S.: Industrialized building systems for housing. MIT Press, Cambridge (1971)
13. Kamar, K.A.M., Hamid, Z.A., Azman, A., Ahamad, M.S.S.: Industrialized building systems (IBS), revisiting issues of classification. Int. J. Emerg. Sci. **1**(2), 120–132 (2011)
14. Eastman, C., Teicholz, P., Sacks, R., Liston, K.: BIM handbook: a guide to building information modeling for owners, managers, designers, engineers, and contractors. N.J. Wiley, Hoboken (2008)
15. Rubrich, L.: An introduction to lean construction. WCM and associates LLC (2012)
16. Burnham, K.: The Albert Farwell Bemis foundation: the prefabrication of houses. MIT Press, Cambridge (1951)

Chapter 5
Information-Driven Customization: A Profile-Matching Model

Basem Eid Mohamed and Carlo Carbone

5.1 Introduction

During the early twentieth century, as a result of automobile manufacturer Henry Ford's "serial production" model, architects began questioning the idea of technology transfer from automotive industry to buildings. Ford's method was based on the manufacturing of large standardized components and the systematization of production processes, to improve on existing methods of sequential production. Accordingly, systematic repetition of processes and components lowered costs through economies of scale. The most notable outcome of such an effort was the continuous flow of the mass production of Ford's product, making the Model T a remarkably affordable car [1].

As many sectors of the market at that time initiated the shift toward mass production models, architects and builders sought to understand why factory production has revolutionized the creation of formerly hand-crafted objects, as well as modern mobility, such as automobiles, while the building industry has been largely resistant to such transformation. This enquiry informed the work of many architects: Le Corbusier, Walter Gropius, Frank Lloyd Wright, Buckminster Fuller, and Jean Prouvé, resulting in a wide field of notable experimentations in mass production techniques and their implementation [2].

During the post-World War II housing boom, mass production became a necessity, being viewed as a logical solution in response to high demand for housing. Builders aimed for cost and quality control and so production of mass housing. Later on, the viability of mass production model rose and fell in other economic

B.E. Mohamed (✉)
Department of Architecture, Abu Dhabi University, Abu Dhabi, UAE
e-mail: basem.mohamed@adu.ac.ae

C. Carbone
UQAM, Ecole de Design, Montreal, QC, Canada
e-mail: carbone.carlo@uqam.ca

J. Bellemare et al. (eds.), *Managing Complexity*, Springer Proceedings in Business and Economics, DOI 10.1007/978-3-319-29058-4_5

sectors over time [2]. By the 1970s, significant increase in demand for personalized goods and products, coupled with further decline and obsolescence of the standard mass production paradigm, a new model was introduced primarily by Toyota in the automotive industry. First known as the Toyota Production System (TPS), this model became broadly referred to as lean production in early 1990s [2]. Along with further advancements in manufacturing technologies, such a paradigm was received enthusiastically, leading to a new definition referred to as *mass customization*[1] [3].

Many segments of diverse industries including investment and consumer goods, such as machinery, telecommunication systems, cars, furniture, personal computers, and watches, have been switching toward customization in response to consumer demand. Given the principle importance of customer satisfaction, the adoption of such production strategy has proven attractive to companies seeking to remain competitive. Pertaining to the building industry, mass customization would seem to hold great potential as buildings can become superbly unique and highly customized products. Employing design and fabrication tools such as computer-aided design (CAD) and computer-aided manufacturing (CAM), modern digitally integrated production processes have offered a paradigmatic shift in construction ideology. Individual building components can now be mass customized in ways previously considered impossible. This method permits optimal variance in response to differing contextual conditions, such as uniquely shaped and sized structural components or variable openings [4].

Over the last two decades, many architects, researchers, and prefab companies have demonstrated a surge of interest in emergent digital design and fabrication techniques toward shifting traditional production models to new ones of customization [5]. Accordingly, a valuable amount of research and industry efforts have been developed in various domains, specifically during the last decade, exploring how to efficiently adopt mass customization in the housing realm. Nevertheless, while research endeavors resulted in pragmatic solutions, the industry has been continuously facing challenging limitations that have inhibited the complete adoption of such an approach, due to several reasons, primarily as mass customization implies a new business model.

This chapter presents a framework for constructing an advanced configuration system for mass customization of prefabricated housing, one that can be considered a milestone in the housing industry. The focus is on information management subsystem, denoted with devising a profile-matching model to enable assigning a housing prototype to specific homebuyers' profile. Accordingly, the chapter is designed to the following structure: first, an overview of current applications of mass customization in the realm of architecture, with focus on practices in the housing industry, leading to problem definition, and second, an overview of the proposed advanced configuration system and its rational. Finally, the profile-matching model as a core element within the proposed advanced configuration system.

[1] Pine (1993) defines mass customization as a production strategy that integrates mass production principles with the process of producing custom products.

5.2 Mass Customization in Architecture

Erecting a building is often achieved through the assembly of various configured components, leading to the construction of a unique structure. Most buildings fabricate elements on-site by the direct processing of materials such as concrete and masonry. In parallel with off-site fabrication, specialized manufacturers create customized products and components to be integrated with on-site activities, thus further improving productivity. This mode of production employs speculative digital design and manufacturing techniques within CAD/CAM environments, to model and then fabricate distinctive building elements. On the one hand, CAD tools in the form of parametric modeling platforms offer the development of comprehensive 3D models of complex shapes comprising detailed data of building objects. On the other hand, CAM techniques offer significant capabilities for transforming digital models into physical objects through Computer Numerical Control (CNC) machinery, in unprecedented precision [6].

Kieran and Timberlake (2004) argued that mass customization has increasingly influenced construction processes and products over the past few decades. Nevertheless, there are two challenges in applying such a production strategy to architecture. First is evaluating the efficacy of a product, being customizable in concordance with design codes and regulations, and accurate manufacturing. Commonly, consumer products are modularized to limit customization through high levels of standardization, something that might be challenging to achieve in the realm of architecture. Second is establishing an integrated information management model to enable effective levels of communication between designers and manufacturers. Such a model would also work on the consumers' end, to build profiles and then transform it into set of quantifiable requirements.

5.2.1 The Case of Housing

There have been several successful research explorations and production attempts that seek to use advancements in design and manufacturing technologies to go beyond a conventional custom-build practices in the housing industry. These explorations tackled the concept of mass customization from two angles: design and production. While some of these attempts can be traced back to the early 1970s, for instance, the work of John Habraken[2] [7], the following section highlights a number of prior attempts and evaluates their successes and limitations with focus on

[2] Habraken (1972) developed the "Theory of Supports," which was both a significant contribution to the field of mass housing and also based upon the principle of user participation. The theory distinguished between two fundamental components: "supports," regarded as a physical entity or the rigid part of the building, and "infills," the flexible part that could be adjusted on different levels – social, industrial, economic, and organizational. The system was designed to facilitate variations to floor plans over time while also accommodating the design of dwellings to meet the diverse standards of housing in any particular society.

technological applications, in the form of information transfer and management tools, as primary enabler for customization. The review of research and industry applications aims at rationalizing the development of the proposed advanced configuration system.

Perhaps one of the leading investigations on how computational technologies, materials, and strategies for design can allow for customization in the housing industry is the work by House_n, a former digital media and housing research group at MIT Department of Architecture. In an article titled "A New Epoch: Automated Design Tools for the Mass Customization of Housing" (2001) [8], the authors explored how automated design tools may help architects to develop better solutions for mass housing projects, facilitating a shift from mass production toward mass customization. The authors defined three necessary elements for the mass customization of housing:

- Preference engine: A framework aimed at building a user profile by collecting and refining these responses.
- Design engine: A computational-based design system that encodes data, collected by the preference engine, into a shape grammar that defines the architectural strategy.
- Production system: A digitally controlled production system that can extract information, including geometric data, from the digital design model.

Duarte [9–11] capitalized on the same approach through developing a comprehensive model for the mass customization of housing, built around an interactive computer program that would generate housing designs following a given language. The design system used description and shape grammar as technical mediums for coding design rules. Later on, many other efforts followed the same models, by looking specifically at computational tools as effective enablers for mass customization in the housing realm. On the other hand, further endeavors tackled new territories in digital fabrication, exploring how CNC machinery could contribute to the process of mass customization of housing.

Along with a rising awareness of the internet's potential as an interactive medium that might be applied toward the customization of prefabricated housing in the North American market, Huang [12] developed a model to support homebuyers' participation in the design of their dwellings, based on a decision support system. This model employed an interactive questionnaire that guided users in a sequential process toward finding the appropriate solution, relying on a catalogue of prefabricated modular housing systems. Housing prototypes were built within a Building Information Modeling (BIM) platform, allowing for ease in interchange of components within a digital environment, as well as data transfer for fabrication.

The interest in mass customization was stimulated by economic potential, the opportunity for solving certain social problems, and the technical challenge involved. Such an interest has taken various levels in architecture, engineering, and construction in general and in design computation, in particular, becoming as a mainstream research topic in recent years. For other approaches to the mass customization of housing, see Benros, D. and Duarte [13], Botha and Sass [14], Noguchi [15, 16], and Matcha and Quasten [17].

Fig. 5.1 Screenshots from configuration system by LivingHomes and Blu Homes, consecutively. The process has high visualization qualities in both cases, where LivingHomes utilize static images that changes in response to user selection, while Blu Homes employ an interactive 3D configurator plug-in (Source: http://www.livinghomes.net/configure.html?model, http://www.bluhomes.com/homeconfigurator/)

Pertaining to the housing industry, such an approach of internet-based customization platforms, online configurators, has been explored by a number of prefabricated housing companies in Europe, Japan, and the United States, following similar applications in the automotive, clothing, and computer industries. Commonly, this model of networked interface engages customers in the design of their homes through a sequence of decision-making processes that ultimately lead to customization. Housing companies engaged in this field must invest in building a large database of varying housing prototypes, which are searchable by type, area, average cost, and number of bedrooms. Once a housing model is selected, homebuyers are prompted to access a visually interactive configuration tool that offers selections of different exterior/interior finishes, roof styles, and systems. Figure 5.1 demonstrates some of the applications implemented by two companies operating in the North America: LivingHomes and Blu Homes.

To reflect on the appraisal of both research efforts and industry applications, it can be said that there is a disparity between intended research goals and current applications within the housing market, with regard to the level of customization, and applied technologies. Accordingly, this chapter proposes a comprehensive approach that aims at bridging the gap between what has been proposed in research and current industry practices. The advanced configuration system offers homebuyers greater control over the design of their dwellings, toward high level of customization. Regarded as a core component, the profile-matching model represents a methodology to match homebuyers' profiles to housing prototypes beyond standard industry practices, one that aims at contemplating their lifestyle.

5.3 An Advanced Configuration System

Blecker et al. (2005) defined product configuration systems as information tools through which the order-taking process is automated, thereby recording customer requirements without the need for external human intermediaries [18]. In his 2005 book Democratizing Innovation, Von Hippel described product configurators in terms of "tool kits," whereby customers are provided with required tools to configure a product based on their needs. Configuration systems are commonly implemented in the online interface between a producer and its customers. These systems are aimed at supporting customers throughout the configuration process, so as to produce a product in accordance with their particular and individual requirements. Von Hippel (2005) further stated that the main technical constituent of a configuration system is a knowledge-based component, built around two subcomponents: its database and its configuration logic [19]. While the database comprises the whole set of component types, variants, and their instances, the configuration logic identifies the constraints existing between different components, which ensure valid product variants.

Borrowing from successful approaches in the North American prefab housing market, the system described in this chapter is built around a browser-based configuration tool to enable the functionality of a multilevel interactive process. The system tends to offer homebuyers categorized levels of intervention, supported by two primary functions embedded within the configuration logic. Firstly, a recommendation agent that seeks the most suitable housing models in correspondence to a homebuyers' profiling procedure. Secondly, a hierarchically structured product configurator allows homebuyers to customize their selected housing unit. As a result, this configuration process is broken down into three consecutive levels:

(a) Room block modifications: Configures room blocks, particularly those of the kitchen and bathroom.
(b) Appearance: Configures surface materials, color, and texture selections of exterior and interior components.
(c) Appliances/systems: Configures kitchen appliances, laundry, air conditioning, and heating/cooling systems.

Figure 5.2 represents an overview of the proposed configuration system and its various levels. Each phase links to the next one via automated information transfer, to improve efficiency of the system. The end of the process is in the form of a loop, controlled by approval, or return to configuration, a decision made by homebuyer and backed by a decision support subsystem, in the form of a pricing, and performance modules.

5.3.1 *The Recommendation Agent: Profile Matching*

The notion of building homebuyer's profile, including generating and then maintaining it, is regarded as a crucial issue in customization. Diverse research has focused on recommendation agents that utilize user profile data and information filtering techniques to generate and then maintain a user profile. Montaner et al. (2003) [20] presented taxonomy of techniques for profile generation and maintenance and profile exploitation. With regard to profile generation and maintenance, the authors proposed five dimensions; the system will require a method for profile representation, generation of initial profile, source of relevance feedback, profile

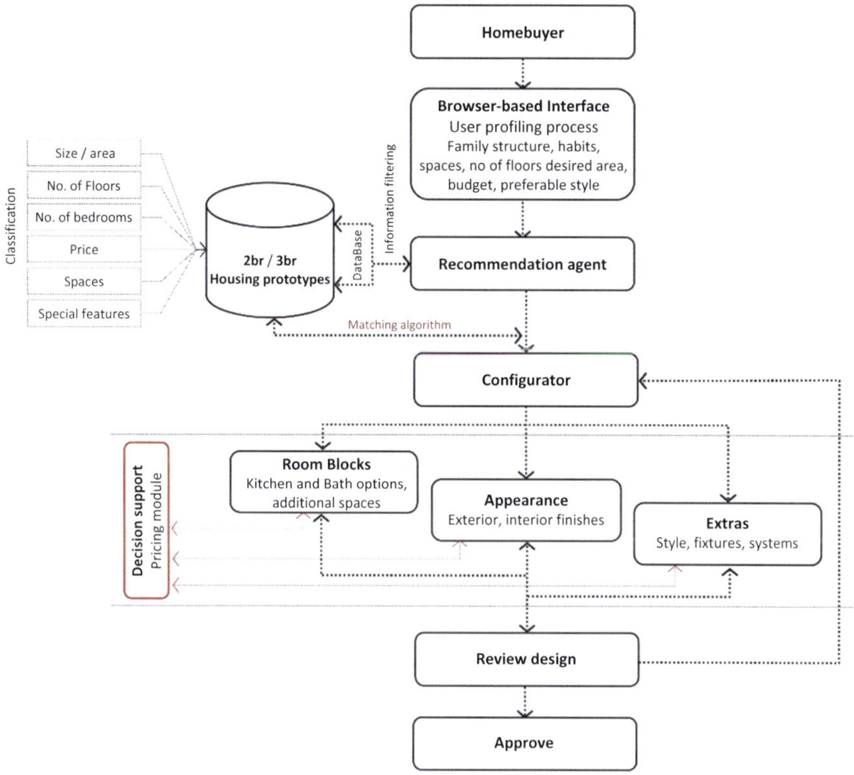

Fig. 5.2 Proposed structure of the configuration system, detailing the procedures of each phase (*Source*: The author)

learning technique, and profile adaptation technique. One of the interesting representation methods, vector space model, utilizes vectors to represent items by associating them with a value, which can be a Boolean or a real number. On the other hand, profile exploitation aims filtering information to make recommendations. Three main dimensions were defined for providing accurate data through information filtering: content-based filtering, collaborative-based filtering, and a hybrid approach that merges between both. While content-based filtering uses detailed description of products normally in the form of vectors or item matrices, collaborative-based filtering relies more on matching users with similar interests and then makes recommendations on this basis. Finally, hybrid approach merges, which would be employed in the search process, features of content based and collaborative based as they prove to be complementary [20, 21].

A hybrid approach can be employed toward collecting data and creating user profile to assist in generating and matching solutions within the proposed advanced configuration model. Such an approach targets building an understanding of demographic/psychographic qualities and household activities. The priority of each activity should be represented as a vector of values, acting as inputs driving force of the design.

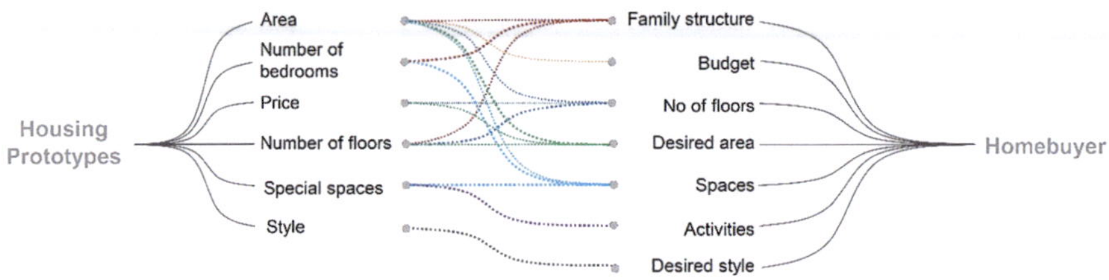

Fig. 5.3 Proposed logic behind the recommendation agent (*Source*: The author)

Users may also differentiate the relative importance of each requirement, based on their preferred lifestyle. Figure 5.3 represents a taxonomy of homebuyer's profile variables and its relationship to housing prototype classification. Such variables and classifications represent some of the most common qualities a company might offer, though they can be adjusted to suit the needs of each individual manufacturer.

The outcome of the profile generation process would feed the search engine to appraise various housing prototypes, thus matching homebuyer's requirements with the most suitable models. The main function embedded within the recommendation agent is the matching algorithm, one that matches a user's profile to a set of housing alternatives from a pre-designed database. This process can be described as the synthesis phase of the problem formulation elements. The algorithm derives its mechanism from conditional programming in the form of if/then/else structures – basic constructs of an expert system. The algorithm matches related attributes and isolates irrelevant housing profiles. A simple pseudocode example of the matching algorithm is as follows:

Start
Get household variables (i.e.: family structure, area, number of floors)
IF household variable is less than or equal to three
    THEN eliminate three-bedroom houses
    or ELSE keep list
IF budget variable is less than or equal to xxx
    THEN eliminate models with area greater than xxx
  IF number of floors variable is equal to one
THEN eliminate two floor models
Display remaining models
Repeat until a model is selected
End

Within the matching process, some relationships are simple and direct, while others may attain greater levels of complexity. Consequently, in order for the recommendation agent to perform efficiently, user and solution profiles must deconstruct into sublevels, with a matching method instituted between them. In some cases, homebuyers can prioritize certain requirements, assisting in narrowing their choices by eliminating redundant prototypes. This process is regulated through a relationship matrix that translates various connections into a vector of values.

5.3.2 Implementation Model

While the processes of implementing the proposed customization system can be devised on various levels and phases, a collaborative approach is necessary. These levels would address each of the configuration system components, with the aim of creating a coherent implementation mechanism. The goals of this process and the research that informs its objectives are as follows:

- *Develop a comprehensive product platform*: Creating a product platform aims at providing the necessary taxonomy for positioning different products and structuring their interrelationships. Further advances in mass customization processes require the improved functional and technical variety of products within the platform. Such diversification would bring new product functionalities, in addition to diverse technologies, design methods, manufacturing processes, components, and assemblies.
- *Develop parametric, three-dimensional BIM models for product platform components*: Parametric modeling, in addition to BIM, creates the means by which architects, engineers, planners, manufacturers, and component fabricators can communicate together in a fully integrated environment. BIM forms the basis for design and, when supported by digital prototyping tools, would contain all the information required for the fabrication and assembly of components and subcomponents in a dwelling unit. Such integration would produce multiple advantages, including efficient structural and technical coordination, maintenance of information and design model integrity, collaboration in design and production through the better data management, schedules of procurement, and greater control over fabrication, assembly, and construction.
- *Develop an appropriate method of representation*: In most cases, homebuyers lack the experience and training to fully interpret architectural drawings and technical features in a house's design. Traditionally, a salesperson would assist homebuyers in overcoming any difficulties or confusion. In the case of a browser-based configuration process, the salesperson is generally absent. Consequently, the system's visualization technique becomes ever more important. Current models of cloud-based interactive 3D environments can offer a practical tool for building the configuration system's representation model. Based on 3D virtual environment, such platform would present homebuyers with dynamic, real-time configurations of spaces and surfaces. Finally, the system would be supported by a pricing module to reflect the effects of each selection on the total cost of the housing unit.
- *Develop the profile matching model*: Implementing a search algorithm is considered to be an easy programming exercise. The challenging part would be to efficiently translate the relationship between the homebuyer's profile and housing prototypes classification. The system proposes a matching methodology based on an expert system construct.
- *Develop browser-based interface*: Given that the configuration process would operate online through the company's website, the quality, structure, and design of the interface play an important role in the process, as it represents the medium

of a virtual dialogue between homebuyer and designer. Signifying the company's brand and the homebuyer's aspirations, the interface must offer visual cues for quality and success.

The implementation of the configuration system is considered as a collaborative multidisciplinary process, aimed at engaging various forms of expertise from different fields, in order to deliver high-quality, customizable prefabricated housing prototypes that are fundamentally responsive to market demand.

5.4 Reflections

Recent research efforts on mass customization of housing reveal three main areas of interest: first, design systems that would enable the generation of customized housing designs according to customers' needs and requirements and, second, managing data from collection to design and production, including interactive platforms and BIM environments and digital prototyping tools toward more efficient data management models. Finally, the role digital fabrication techniques can play to handle complexities associated with the process of customization. This chapter focuses primarily on managing management throughout the customization process, with emphasison profile-matching model. Such a model is implemented within an advanced configuration system, one that aims at pushing industry boundaries toward higher levels of customization. The proposed system positions itself at an intermediate level, between published research efforts and industry application.

The principal function of the proposed system is a recommendation agent, initialized by input, then filtering users' profile data. The search logic employs if/then/else functions to search through a database of housing prototypes, classified according to a specific criteria: area, price, number of floors, and spatial arrangements. To match a housing model to a specific homebuyer's profile, the search follows a decision tree. Once a model is selected, further configuration options are offered: room blocks, exterior and interior finishes, and systems, till the homebuyer reviews the final design and becomes ready to finalize the buying process.

The system presents potential applicability within prefabricated housing industry, being technologically accessible, advocating a multidisciplinary approach. Nevertheless, one of the challenges that face configuration systems is that when building a product platform, the number of alternatives has to be limited to three or four options in order to avoid additional overhead cost. Additionally, mass customization implies adopting flexible, technology-oriented business models, where in some cases companies are not ready for such a change.

This chapter presents a framework that can be pursued when designing a design system to customize housing. The proposed framework is generic and flexible for further developments. Currently, researchers are working on validating such framework through two case studies: the first within a prefabricated single-family housing system, while the second focuses on multi-family housing. The reason behind selecting such case studies is due to challenges associated with design and construction of those two housing typologies. Prefabricated housing required setting various constraints regarding design and production. On the other hand, multi-family housing involves managing large amount of homebuyer's information which, in some cases, difficult to handle. Once the framework is validated, it will open future opportunities to mass customize housing more efficiently.

References

1. Davies, C.: The Prefabricated Home. Reaktion Books, London (2005)
2. Kieran, S., Timberlake, J.: Refabricating Architecture: How Manufacturing Methodologies Are Poised to Transform Building Construction. McGraw-Hill, New York (2004)
3. Pine, B.J.: Mass Customization: The New Frontier in Business Competition. Harvard Business School Press, Boston, Massachusetts (1993)
4. Kolarevic, B. (ed.): Architecture in the Digital Age: Design and Manufacturing. Spon Press, New York, NY (2003)
5. Bergdoll, B., Christensen, P., Broadhurst, R.: Home Delivery: Fabricating the Modern Dwelling. Museum of Modern Art, New York (2010)
6. Smith, R.E.: Prefab Architecture: A Guide to Modular Design and Construction. Wiley, Hoboken (2010)
7. Habraken, J.: Supports: An Alternative to Mass Housing. Praeger, New York (1972)
8. Larson, K., Tapia, M.A., Duarte, J.P.: A New Epoch: Automated Design Tools for the Mass Customization of Housing. A + U, vol. 366, pp. 116–121 (2001)
9. Duarte, J.P.: Customizing mass housing: a discursive grammar for Siza's Malagueira houses. Ph.D. Thesis, Massachusetts Institute of Technology, Cambridge (2001)
10. Duarte, J.P.: A discursive grammar for customizing mass housing: the case of Siza's houses at Malagueira. Autom. Constr. **14**, 265–275 (2005)
11. Duarte, J.P.: Towards the mass customization of housing: the grammar of Siza's houses at Malagueira. Environ. Plan B: Plan. Des. **32**, 347–380 (2005)
12. Huang, C.-H.: Using internet and query approach of customizing prefabricated houses. Ph.D., Illinois Institute of Technology, Chicago (2008)
13. Benros, D., Duarte, J.P.: An integrated system for providing mass customized housing. Autom. Constr. **18**(3), 310–320 (2009)
14. Botha, M., Sass, L.: The instant house: design and digital fabrication of housing for developing environments. Int. J. Archit. Comput. **4**(4), 109–123 (2006)
15. Noguchi, M.: The effect of the quality-oriented production approach on the delivery of prefabricated homes in Japan. J. Housing Built Environ. **18**(4), 353–364 (2003)
16. Noguchi, M.: A choice model for mass customization of lower—cost and higher—performance housing in sustainable development. Ph.D. Dissertation, McGill University, Montreal (2004)
17. Matcha, H., Quasten, G.: A parametric-typological tool: more diversity for mass produced single family homes through parameterized design and customized mass production. Paper presented at the eCAADe Computation: the New Realm of Architectural Design, Istanbul (2009)
18. Blecker, T., Friedrich, G.: Mass Customization: Challenges and Solutions. Springer, New York (2006)
19. Von Hippel, E.: Democratizing Innovation. MIT Press, Cambridge (2005)
20. Montaner, M., Lopez, B., Rosa, J.: A taxonomy of recommender agents on the internet. Artif. Intell. Rev. **19**, 285–330 (2005)
21. Larson, K., Smithwick, D.: Beyond the configurator: collecting accurate data for an architectural design recommendation engine. http://cp.media.mit.edu/research/papers (2010)

Chapter 6
The Potential of Product Customization Using Technologies of Additive Manufacturing

Roland Lachmayer, Paul Christoph Gembarski, Philipp Gottwald, and R. Bastian Lippert

6.1 Introduction

In the course of current technological possibilities, product development gains new ways of thinking by adapting innovative aspects in the product development process. In addition, to approach in the context of "industry 4.0," similar to intelligent factories, a trend toward additive manufacturing processes can be identified [2, 20, 21].

The significance of additive manufacturing technologies becomes more important for the development of modern products. As a result, development times can be shortened and the degree of customization can be increased. On the one hand, this depends on the different additive manufacturing technologies, which can be considered in the product development process. Thereby process steps can be parallelized. An overview of the various possibilities is depicted in Fig. 6.1. It can be seen that the three groups (according to [22]) include different development stages and thereby consist of certain model levels.

On the other hand, the customer has the option to specify characteristics for the end part by using additive manufacturing. In order to do this, the provider defines the relevant interfaces to set up the design space for the customer. This procedure enables the reduction of iteration loops.

Considering additive manufacturing, differences toward conventional manufacturing techniques can be identified. These restrict the application from an economic and technical point of view. The comparison between the traditional mass production and the additive manufacturing results in an area of conflict. In this context, the advantages and disadvantages between additive manufacturing and conventional

R. Lachmayer • P.C. Gembarski • P. Gottwald • R.B. Lippert (✉)
Institute of Product Development, Welfengarten 1A, 30167 Hannover, Germany
e-mail: Lippert@ipeg.uni-hannover.de

J. Bellemare et al. (eds.), *Managing Complexity*, Springer Proceedings in Business and Economics, DOI 10.1007/978-3-319-29058-4_6

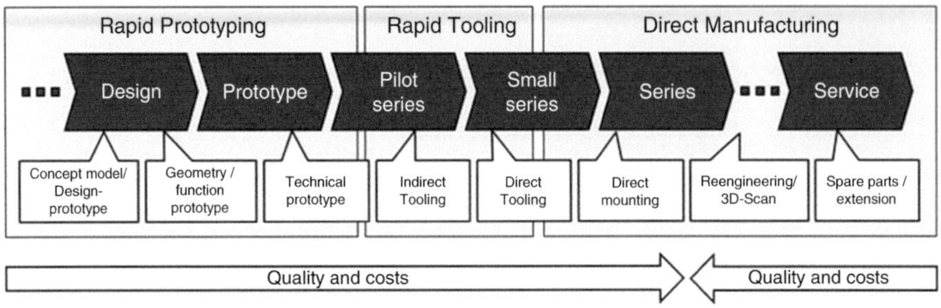

Fig. 6.1 Potential for additive manufacturing technologies in a product development process

production depend on the requirements of the product and its application [1]. Especially for product customization and the production of small lot sizes, the additive manufacturing shows its large potential [19]. Therefore, the influence of these both characteristics toward a useful application of AM is analyzed.

It is difficult to define for which exact parameters the additive manufacturing is better suited than a conventional technology. In literature there exist some approaches for this purpose, which deal with the description of subjective characteristics.

To illustrate the potential of the additive manufacturing and to promote its use in practice, *Buchmayr* demonstrates a SWOT analysis (S = strength, W = weaknesses, O = opportunities, T = threats) for SLM components. He establishes a relation matrix by defining criteria for each of the three sectors. As an example, the high material efficiency or the suitability for small lot sizes is defined as strengths. By contrast, the possibility for customization was classified as an opportunity. Overall, 54 criteria are defined for this relation matrix. Using this SWOT relationship matrix, S/O, W/O, W/T, and S/T strategies can be generated. For example, the combination of the small lot sizes (S) and the possibility for customization (O) offers the application of AM for a product [6].

A further approach for the implementation is given by *Mellor*. Within an analysis for a framework of implementation, a comparison between the SLM technology and the high-pressure die casting (HPDC) is accomplished. As a result, a break-even point at a lot size of about 42 pieces is described. Thus, a smaller lot size is well suited for additive manufacturing. The framework depends only to lot size effects, component geometry, and design, and the degree of customization is disregarded [16].

Berman goes a step further and describes a comparison of additive manufacturing and mass customization (MC) for end parts. Therefore, the four characteristics (1) manufacturing technology, (2) supply chain integration requirements, (3) economic benefits, and (4) range of products are defined. Based on these characteristics, suitable product types are described for AM and MC. A comparison of the described fields of application results in intersections. Therefore, the products which are described within this intersection are customizable by AM [3].

Further approaches are based on similar principles. By the comparison of the advantages and disadvantages for specific component, an analysis is carried out. Using these basic principles, an approach for properly positioning components has to be described, in which component characteristics can be compared analytically.

6.2 Method

There are three groups, in which additive manufacturing technologies can be classified. With rapid prototyping (RP), sample parts and prototypes are created, the rapid tooling (RT) is used to manufacture tools such as casting molds, and with the direct manufacturing (DM), end parts with a (near-) net shape can be produced. Thereby, the RP technology is mainly applied at the early stages of the product development process, RT for preparing series production and DM for end-part production itself [22].

The AM technologies allow to interrupt the sequential procedure of product development and then parallelize process steps. Especially with the application of RP, the customer can be involved in early design phase and affect the final product according to his wishes. When thinking about DM, this idea can be continued. On the one hand, the manufacturer can print a customized product. Thanks to the tool-free layer structure, already small lot sizes may be produced economically. On the other hand, the production step can be outsourced toward the customer, so that the product responsibility is passed to the user. Taking these two aspects into account, the possibility of customization has to be examined.

Though the DM is strongly component dependent, a case distinction is needed to be analyzed. Therefore, a matrix is used, which applied the lot size compared to the degree of customization. These two properties can be identified as major determinants [7]. As depicted in Fig. 6.2, the quadrants for generalized mass and single production as well as for customized mass and single production arise.

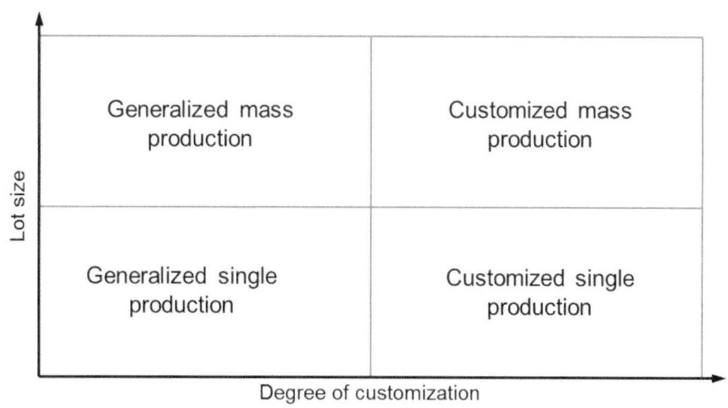

Fig. 6.2 Customization compared to the lot size

	Impact on in-house engineering	Impact on in-house data management	Impact on in-house manufacturing	Customer Integration Level
Tuning Customization	very low	very low	very low	middle
Cosmetic Customization	very low	low	very low	low
Set-Up Customization	low	middle	very low	very low
Composition Customization	middle	middle	low	middle
Aesthetic co-design	high	high	high	high
Function co-design	very high	very high	very high	very high

Fig. 6.3 Degrees of customization [8]

According to the "Product Process Change Matrix" from [4], the resulting groups coincided with existing classifications. Thereby, the axis *lot size* represents the number of produced goods. The axis *degree of customization* points the possibility of involving customers. The classification can be specified using a scale. *Gembarski and Lachmayer* defined six degrees for customization, which are depicted in Fig. 6.3 [8].

The first degree is defined as *tuning customization*. A standard product becomes refined by another supplier in the supply chain in order to adapt the standard product to either special applications or to markets with only few customers in general [8].

The second degree, named *cosmetic customization*, can be defined as a standard product, which can be presented differently to different customers [9]. It allows changing the outer appearance of the product itself to a defined degree, which usually can, e.g., be realized through another paining.

The *setup customization* stands for mechatronic devices, where the mechanical part is kept the same, but its behavior is controlled differently by the software component [12]. An example is the iPhone-iTunes ecosystem. The process of manufacturing is not affected [10].

The third degree of customization is called *composition customization*. It corresponds to the common assemble-to-order strategy, where different subassemblies are assembled together to a product using standardized interfaces [17].

At the *aesthetic codesign*, the customer defines his own outer appearance of a product. This is not only limited to modifications of color and texture; also the shape can be influenced according to the given manufacturing processes [8].

Gembarski and Lachmayer defined the most far-reaching degree of customization as *function codesign*. Here also the functional building blocks are determined by the customer [8, 18].

Referring to lot size and degree of customization, the potential of additive manufacturing for each of the four quadrants as mentioned before was analyzed by the use of demonstrator components. These demonstrators show different characteristics and are representatives of different fields of application. In addition, the number of customers as well as their possible influences on the products' design, shape variation, or change of materials, for instance, is discussed further.

6.3 Potential Analysis

For determining the potential of additive manufacturing, four different demonstrators are identified. All components are chosen out of automotive applications. Two of these demonstrators, a shaft and a wheel carrier, descended from the mechanical field. The third demonstrator, a car key fob, is mainly from a mechatronic application. The last one is an optomechatronics system, presented as a reflector. These four demonstrators are depicted in Fig. 6.4.

The first analyzed demonstrator represents the *shaft* of a gear. In the automotive sector, gears are used for an optimal reinforcement of torques. In context of transmission technologies, the components of the assemblies include a high degree of standardization. The used shafts are severely limited in their design freedom by the coupled standard components of the gear. Thereby, the design parameters are restricted regarding their functionality as well as their design space [5, 11]. Shafts are manufactured in mass production for different series. Concomitantly the standardization of applied components limits the degree of customization that this characteristic can be declared to tuning customization. An application in the additive manufacturing is rather unsuitable.

The wheel carrier from a wheel suspension of a race car represents the second demonstrator. The application is identified for motor racing, wherein the design space of components is quite high. Design parameters are just depending from the application points of the other components of the chassis. Every racing team develops their own design of this component regarding lightweight design and reliability [5, 13]. This leads to a small series, close to the lot size 1. With the consideration of the high freedom for the design, the degree of customization can be identified as

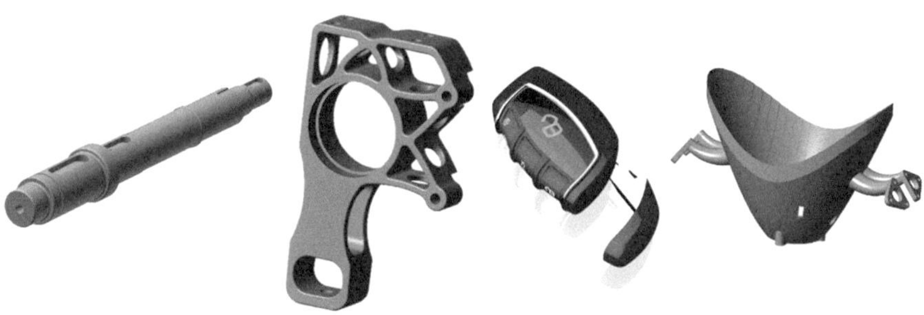

Fig. 6.4 Overview of the demonstrators (from *left* to *right*: shaft, wheel carrier, car key fob, and reflector)

function codesign. For this area, a production in the additive manufacturing is quite suitable, because complex geometries as well as a high level of functional integration can be achieved.

As a third component, a car key fob is chosen. Because this component is necessary for each car, the production of a high lot size is necessary. Nevertheless, the car key fob should present an impression of the associated car so the design space is high and the design parameters can be varied. On the one hand, restrictions for possible variations are limited by the mechatronic content. On the other hand, a limitation based on the challenges of design impact and technology development occurs [14]. Derived from these facts, the degree of customization is high and can be assigned to the aesthetic codesign class. For the adjustment of the large component variations, the additive manufacturing is a suitable technology.

The fourth component is an additional reflector for an automotive spotlight. Particularly, a headlight becomes a design feature at modern cars. Thus, its importance has grown, especially for luxury vehicles. In order to create different light spectra, the coating of the reflector interface is decisive [15]. Moreover, the lot size for such a component is quite low because an additional light source is not regularly installed in headlights of cars. The adjustment of the reflector is mainly caused through the coating since the shape determines the light distribution which may not be altered. That degree of customization fits with the cosmetic customization, in which the color and texture can be affected by the customer. The application of additive manufacturing technologies dependent on the component design, e.g., a high functional integration, like integrated cooling channels, can be easily achieved by additive manufacturing.

After the analysis of the characteristics regarding lot size and degree of customization, the four demonstrators are classified in the matrix (Fig. 6.5). In addition to Fig. 6.2, the six degrees of customization are added to the matrix.

It can be recognized that all the four demonstrators are classified to different degrees of customization considering their lot size. Moreover, the different characteristics point out new potentials for the application of AM processes in the context of the DM.

With regard to the demonstrators, especially the customized single production offers a great opportunity for additive manufacturing. The application of the generalized single production as well as the customized mass production depends on further characteristics of the component. However, the generalized mass production is almost entirely unsuitable.

For visualization, Fig. 6.6 depicts a transition for a suitable and unsuitable application of additive manufacturing. The exact determination of this transition region depends on variety drivers, such as the design of the component.

By referring the suitable area to the different degrees of customization, the functional codesign and the aesthetic codesign are qualified for AM. By outsourcing the function design as well as the freedom of design toward the customer, the technology can be applied. For the composition customization, the usage of AM has to be weighed. If there are defined interfaces which have to be connected, parts can be

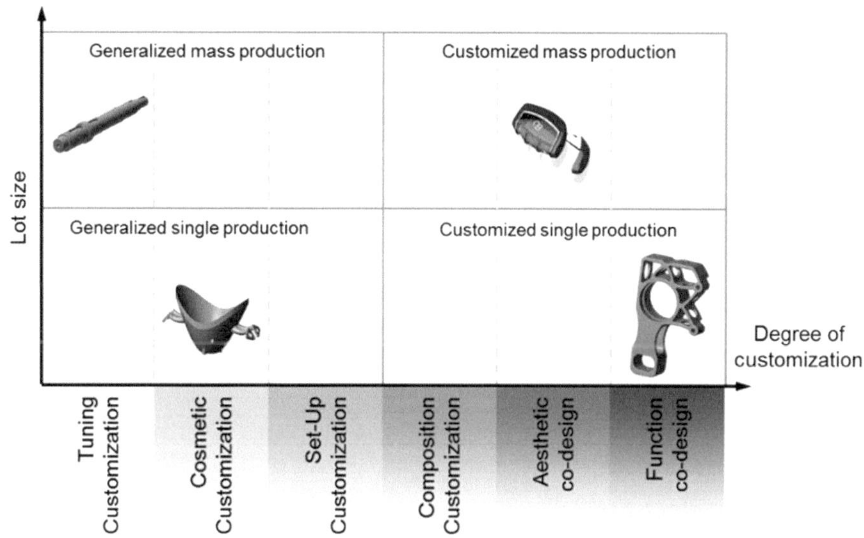

Fig. 6.5 Classification of demonstrators

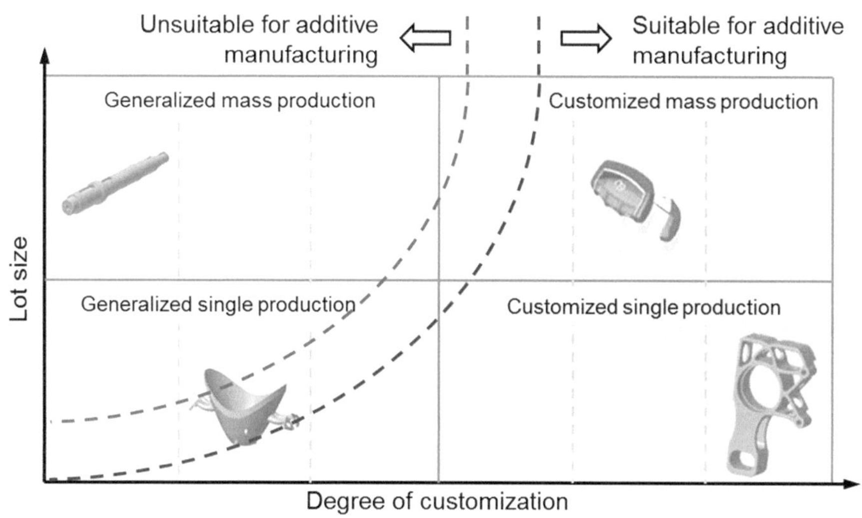

Fig. 6.6 Potential applications for additive manufacturing comparing the degree of customization and lot size

manufactured in AM. However, this decision has to be made case specific, because component geometries and application areas have to be considered.

6.4 Application Example

In this section, the design of parts for a tea brewing machine is presented as application example. As a particular feature of this tea brewing machine, the adaptability to the kitchen or room furniture is provided. This is achievable by exchangeable covers, depicted in Fig. 6.7. Since the functionalities and the basic design of the machine remain the same, this degree of customization can be defined as design co-creation.

Marketing identified two key customer groups: The first is hoteliers who want to distinguish themselves from competitors by integrating also electrical devices into the room concept for the single categories they provide. The second group of key customers is consumers who are willing to pay a premium price for a customized tea brewing machine.

Fig. 6.7 Tea brewing machine with exchangeable covers

For the first, a constant demand and lot sizes with up to 500 pieces are estimated, the latter has an inconstant demand, and lot sizes in the range of 1–5 pieces are predicted. The matrix above indicates that additive manufacturing is suitable and has to be considered in product development and value chain.

In order to define the principle solution space for design exploration, the production centers are defined first. The coverings will be manufactured in ABS plastics on a laser sintering machine, so no additional support structures have to be taken into account. Furthermore, the process restrictions, for example, the material thickness or the dimensions of the process chamber, have to be considered for the design process.

After the design interfaces and connection points to the main body of the tea brewing machine have been defined, a design configurator is developed so that the user can create his own covers (Fig. 6.8). Additional to the shape, the color can be chosen from a given list since the processed parts are dipcoated.

According to the maximum dimensions of the coverings, which are restricted due to the limitations of the process chamber, a maximum count of 60 pieces can be manufactured in one job. The build time is approximately 30 h including cooling, cleaning and dip-coating. Switching to an SLS machine with a bigger process chamber would allow a parallel production of 320 pieces in one job at duration of 90 h. An example of different cover configurations is depicted in Fig. 6.9.

Next step in the development of the design configurator is the addition of a case base so that custom designs can be saved, altered, and rated by other users in order to address the configuration experience and competition. Basis for this is indeed additive manufacturing as enabler to (a partial) MC of the tea brewing machine in this example.

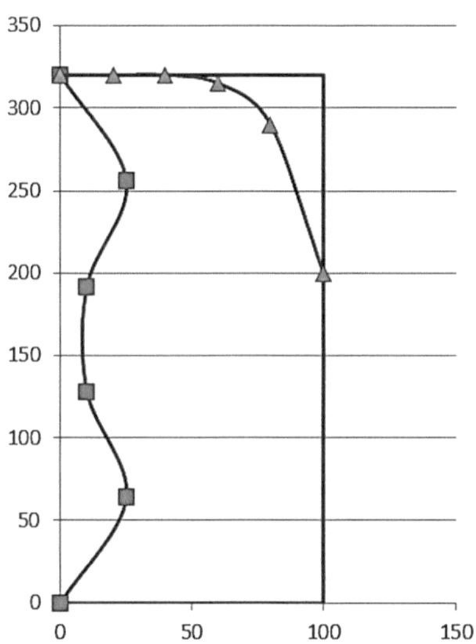

Configuration		
Design Space	Height	Width
	320	100
	max. 320	max. 100
	x	y
Left Cut	0	0
	25	64
	10	128
	10	192
	25	256
	0	320
Upper Cut	0	320
	20	320
	40	320
	60	315
	80	290
	100	200
Embossed Text	IPeG	
Color	yellow	

Fig. 6.8 Configuration dialog for custom covers

Fig. 6.9 Different cover configurations

6.5 Conclusion

It has been shown that the DM (in the course of additive manufacturing) can be suitable, regarding lot size effects as well as the degree of customization. By comparing these two aspects, a potential matrix was set up. This matrix includes four classifications: the generalized mass production, the generalized single production, the customized mass production, and the customized single production. Thereby the degree of customization depends on a six-divided scale.

By analyzing four different demonstrators, the quadrants were described. Based on the consideration of the application, which depends on the lot size and degree of customization, a boundary for the potential of additive manufacturing is defined. This boundary is intended to facilitate the selection of the manufacturing process of a component.

Based on a case study, an exemplary development process is demonstrated. Hereby the aesthetic codesign is in focus of the investigation. By the determination of a design space, the customer is enabled to formulate his specific needs according to the product configuration process.

The implementation of additive manufacturing for customized products requires an accurate definition of interfaces. These are digital interfaces for the CAD model, geometric interfaces of the produced component, and informative interfaces for the design space for the customer. In addition to the standardization of digital models, especially guidelines for the design have to be integrated. For a practicability application, beside of information storages, the integration of this knowledge in the development process is necessary.

References

1. Bauwens, M., Iacomella, F., Mendoza, N., Burke, J., Pinchen, C., Léonard, A., Mootoosamy, E.: Synthetic overview of the collaborative economy. Orange Labs und P2P Foundation. http://p2p.coop/files/reports/collaborative-economy-2012.pdf (2014). Accessed 21 May 2014
2. Benkler, Y.: The wealth of networks. How social production transforms markets and freedom. Yale University Press, New Haven. http://www.benkler.org/Benkler_Wealth_Of_Networks.pdf (2013). Accessed 19 Dec 2013
3. Berman, B.: 3-D printing: the new industrial revolution. Kelley School of Business, Hempstead (2011)
4. Boynton, A.C., Victor, B., Pine, B. J.: New competetive strategies: Challanges to organizations and information technology, IBM Systems Journal Volume 32, Number 1, Page 40 (1993)
5. Braess, H.-H. (Hrsg.), Seiffert, U. (Hrsg.): Vieweg Handbuch Kraftfahrzeugtechnik, Verlag: Springer Vieweg; Auflag: 7., aktual. Aufl. (2013)
6. Buchmayr, B.: Eine SWOT-Analyse zum Einsatz der Additiven fertigung für metallische Bauteile. BHM, Springer (2014)
7. Gebhardt, A.: Generative Fertigungsverfahren: Additive Manufacturing und 3D Drucken für Prototyping—Tooling—Produktion. Hanser, München (2013)
8. Gembarski, P. C., Lachmayer, R.: Degrees of Customization and Sales Support Systems - Enablers to Sustainibility in Mass Customization. In: Proceedings of the 20th International Conference on Engineering Design (ICED15), Volume 1, Page 193, Milan, Italy (2015)
9. Gilmore, J., Pine, J: The four faces of mass customization; BHM, Springer. Harv. Bus. Rev. **73**(1), 1–8 (1997)
10. Gräßler, I.: Kundenindividuelle Massenproduktion: Entwicklung, Vorbereitung der Herstellung, Veränderungsmanagement. Springer, Heidelberg (2014)
11. Haberhauer, H., Bodenstein, F.: Maschinenelemente. Gestaltung, Berechnung, Anwendung. Springer-Lehrbuch (2014)
12. Jørgensen, K.A.: Customization and customer-product learning. Chesbrough (2011)
13. Lachmayer, R., Mozgova, I., Deiters, A., Sauthoff, B., Gottwald, P.: Design evolution by integration of life cycle information in knowledge repositories, consilience and innovation in design. In: 5th International Congress of International Association of Societies of Design Research, Tokyo, Japan, pp. 309–318 (2013)
14. Lachmayer, R., Weiss, F., Deiters, A., Lippert, R.B.: Education in cross enterprise engineering design. In: dmi: 19th Academic Design Management Conference 2014, London, pp. 2885–2901, 2–4 Sept 2014
15. Lachmayer, R., Wolf, A., Kloppenburg, G.: Reflektorbasierte Laser-Lichtmodule als Zusatzfernlicht für die Fahrzeugbeleuchtung. In: Tagungsband 21. Gemeinschaftstagung, LICHT 2014, 21.-24.09.2014, Den Haag
16. Mellor, S., Zhang, D., Hao, L.: Additive manufacturing: a framework for implementation. Int. J. Prod. Econ. United Kingdom (2013)
17. Pahl, G., Beitz, W., Feldhusen, J., Grote, K.H.: Pahl/Beitz Konstruktionslehre. Springer, Heidelberg (2013)
18. Reichwald, R., Piller, F.: Interaktive Wertschöpfung. Gabler, Wiesbaden (2009)
19. Schmid, M.: AM auf dem Weg in die Produktion. KunststoffXTRA, St. Gallen (2014)
20. Strangler, D., Maxwell, K.: DIY Producer Society, Innovations/Ausgabe 7, Nummer 3 (2012)
21. The Economist: a third industrial revolution. Special Report Manufacturing and Innovation. Online verfügbar unter: http://www.economist.com/node/21552901/print (2013). Accessed 10 June 2013
22. Verein Deutscher Ingenieure: VDI 3405: Additive manufacturing processes, rapid manufacturing—basics, definitions, processes. VDI-Gesellschaft Produktion und Logistik

Chapter 7
Conceptual Model for Developing Platform-Centric Production Architectures

Jacob Bossen, Thomas Ditlev Brunoe, Mads Bejlegaard, and Kjeld Nielsen

7.1 Introduction

The potentials of using platforms for product design have been exploited for decades by developing modular components that can be recombined into new products [2]. This fundamental principle of reusability is also one of the very basic mechanisms for increasing the robust process design capability as a mass customizer [26]. Although focus on creating modular products has had increasing attention since Pine [23] popularised mass customization, the robust process design capability has more widespread impact than product modularity. In fact, the original contribution of the robust process design capability states that such capability covers all organisational and value-chain resources [26] and hence motivates the idea of also having a modular production set-up. This is likely motivated by the shifting cost driver when comparing different product domains. For example, in the software industry, the main cost driver is the product design phase (which motivates product modularity), whereas in the manufacturing industry, the main cost driver lies within the production phase (motivating production modularity). Hence, reusability of, for example, production equipment becomes essential for increasing the robust process design capability for a mass customizer in the manufacturing industry.

The research on product modularity for robust process design is documented intensively and is referred to as—but not limited to—product modularization, product family design and product platform design. Although with different scope, several specific methods have emerged from this research with the purpose of designing product families and platforms [8, 11, 15]. One of the key outcomes of using such methods is awareness of a product architecture expressed through an architecture description, i.e. the structured mapping of functional elements in

J. Bossen (✉) • T.D. Brunoe • M. Bejlegaard • K. Nielsen
Department of Mechanical and Manufacturing Engineering, Aalborg University, Fibigerstræde 16, 9000 Aalborg, Denmark
e-mail: jbo@m-tech.aau.dk

J. Bellemare et al. (eds.), *Managing Complexity*, Springer Proceedings in Business and Economics, DOI 10.1007/978-3-319-29058-4_7

the product, the mapping of functional elements to physical components and the specification of interfaces among product elements [27]. The common elements of such product architecture are potential constitutional elements of a product platform [11], and hence the platform becomes a subpart of the architecture. Other definitions of a platform indirectly support this by including structure and interfaces descriptions [17].

Recent attempts have been made to motivate the use of platforms for production development by framing it as one of the two aspects in platform-based co-development and co-evolution of product and production systems [20]. Platform-based co-development can be categorised as integrated product development by means of both product and production platforms and hence something serving as a holistic approach to cope with competitive market conditions. However, not much attention has been given to this research area, and more specifically, it lacks attention to production platforms [4].

Some attempts have though been done to use concepts and constructs known in the research area of software architectures and software platforms for information systems [3, 13] for production architectures. The concept of viewpoint is particularly interesting, as it provides a way of dealing with different kind of stakeholders and their concerns in the system. In this perspective, ISO/IEC/IEEE 42010 [12] provides a generic conceptual model for architecture descriptions of software-intensive system that can be applicable for describing production architectures [13].

This paper contributes to research on production platforms by presenting a conceptual model that sets the context of concepts involved in defining a platform architecture. In other words, instantiations of the conceptual model in the industry will result in a domain-specific platform architecture model. To support the intention of the paper, the following research questions are used: How can a conceptual model be described for expressing the context of production platforms? How can the conceptual model be applied and instantiated to create a platform architecture model?

Above questions are answered by first presenting the method used for creating the conceptual model following a description with industrial examples. Finally, recommendations are given for applying the model and thereby obtaining a domain-specific production platform architecture.

7.2 Method

The conceptual model has been created in a design science research approach as one of more components in a design theory on platform-based co-development. Two case companies are involved in the research that experience induced complexity, from offering high product variance and increasing frequency of new product introductions (NPIs), when developing new production system elements. The conceptual model presented in this paper is a part of the third component in the design theory framework depicted in Fig. 7.1. This component is called principle of form and function, which represents the abstract "blueprint" that describes an artefact [10]. Thus,

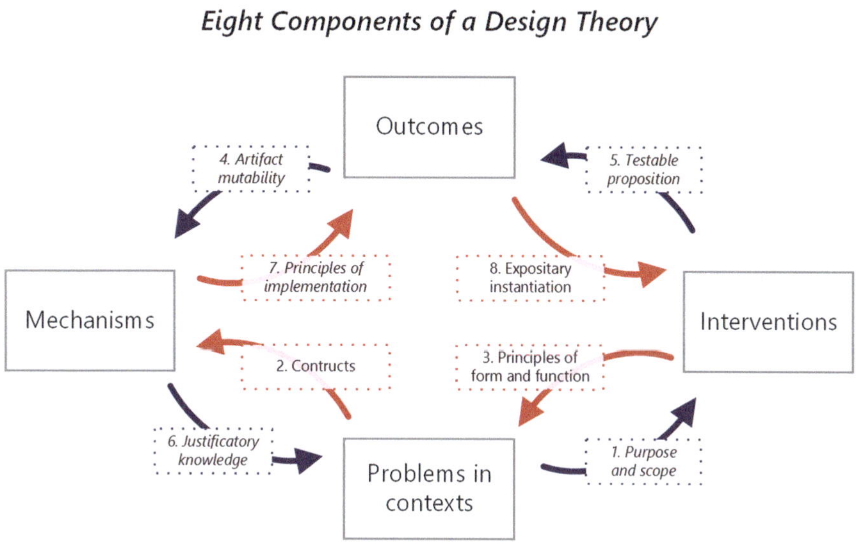

Fig. 7.1 Eight components of design theory [9]. Redrawn from [11]

the conceptual model can be considered a metamodel from which actual domain-specific production platform architecture models can be derived when instantiated.

The other components are not in focus of this paper, but for clarification, the context of the theory can shortly be summarised as:

- Problem in context: The case companies involved in the research experience induced complexity in development of new production system elements as a result of offering high product variance and increasing frequency of NPI. This is deflected in delayed projects and use of immature production technology when managers demand faster-time-to-market, faster-time-to-profit and global production systems with similar structures.
- Intervention: The use of production platforms in a platform-based co-development approach.
- Outcome: Project transparency through increased communication capability reduced investment in production equipment, risk management and risk reduction.
- Mechanism: Systematic reuse of equipment base and increased reusability through awareness of architecture thinking across production system elements.

Hence, the purpose and scope of the design theory can be considered as—but not limited to—(1) communicating production capabilities to enable efficient integrated product development, (2) reducing investment in NPI, (3) increasing awareness of necessity of architecture thinking and (4) enabling reuse of production system elements.

The artefacts involved in the theory cover both constructs, models and method, and one of these artefacts is presented in this paper, i.e. the conceptual model.

7.2.1 Method Used for Creating the Conceptual Model

The conceptual model is created by conducting action research and case studies with input from input from literature and evaluations. More specifically, the sequence of activities used to create the conceptual model can be summarised to the following:

1. Obtain theoretical knowledge on production platforms and related research areas by conducting a literature review.
2. Obtain empirical knowledge on best practice and challenges by interviewing relevant personal case companies with experience in platform projects.
3. Create first draft of a conceptual model with base in internal case-company personnel recommendations and ISO/IEC/IEEE 42010 [12].
4. Identify possible stakeholders to platform descriptions.
5. Apply the conceptual model by interviewing stakeholders. The purpose of the interview is to identify:
    (a) Title and department belonging to determine stakeholder type.
    (b) Examples of concerns in NPI projects for later design of viewpoints.
    (c) The purposes related to concerns in order to identify concern relations.
    (d) A recommendation of other relevant stakeholders in order to adjust the scope of platform architecture.
6. Evaluate conceptual model with input from stakeholders and update accordingly.
7. Repeat step 4–6 until a sufficient number of stakeholders have been interviewed.

Thus, the methodology used can be considered to involve several runs of iterative activities involving literature review, industry observations and interviews and evaluation of the model. The model is illustrated with a UML class diagram with conventions defined in ISO/IEC 19501 and based on the ISO/IEC/IEEE 42010 [12] standard for architecture descriptions. The reasoning for the latter is elaborated in the following section.

7.3 A Conceptual Model for Describing the Context of Production Platforms

As a product platform is a subset of a product architecture [11], a description of a platform must consequently be a part of an architecture description. By acknowledging that product and production are coequal systems [20] and that software architectures for information systems do not diverge significantly in basic concepts [13], it makes sense to create the conceptual model for production platforms with base in literature related to architecture descriptions such as [12, 13]. An essential notion in this perspective is that ISO/IEC/IEEE 42010 [12] differentiates between an architecture and an architecture description. Architecture is defined as the fundamental concepts or properties of a system in its environment embodied in its elements, relationships and the principle of its design and evolution, whereas the architecture description is the work product that expresses the architecture. Such distinction is not common in the literature identified on product and production architecture (such as [11, 24]). In fact, only one study took this standpoint [13].

Despite the weak penetration of this matter in product and production architecture research, the argumentation is proven reasonable for software architecture development and will be adapted to the conceptual model. Consequently, the platform description becomes a part of an architecture description in the model. This is elaborated in the Sect. 7.3.1. An overview of the conceptual model is shown in Fig. 7.2, but due to size constraints and entailed limited readability, the details are illustrated in separate figures as a part of the following description of the model.

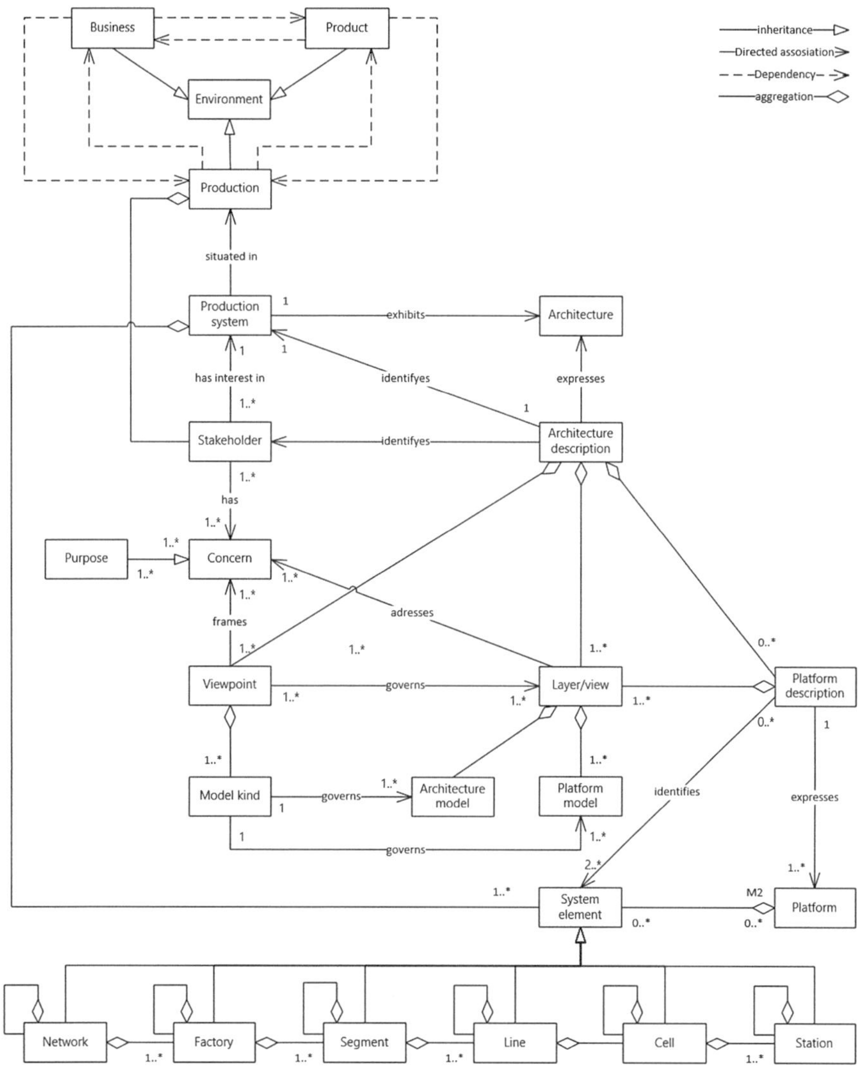

Fig. 7.2 A conceptual model for creating a production platform architecture

7.3.1 Platform Description

This paper focuses on the context for describing production platforms and advocates for the idea of framing it as a distinct part of an architecture description embodied in its own description. Naturally, such platform description will address aspects related to reusability concerns, which in principle could be handled by the architecture description itself [11, 12].

The complexity of developing new elements to global production systems with increasing attention to reuse motivates though a platform description that intuitively can communicate and address reusability-related concerns from several stakeholders. In this sense, the conceptual model for a platform description does not diverge significantly from a conceptual model for architecture descriptions [12] since the main purpose is still to address stakeholder concerns through the concept of viewpoints (see Sect. 7.3.2). The difference lies in the concerns treated: the platform does only address concerns that are caused by or related to a reusability concern, and we advocate having it as the centre of developing a production architecture. We refer to this as platform-centric production architecture development and consider it as a means contributing to the realisation of platform-based co-development of product and production system [20].

Figure 7.3 illustrates essential elements of the model that links to above argumentation. In the figure, levels of the Stuttgart Enterprise Model [29] and later related research of the same model [28] are used to show examples of relevant

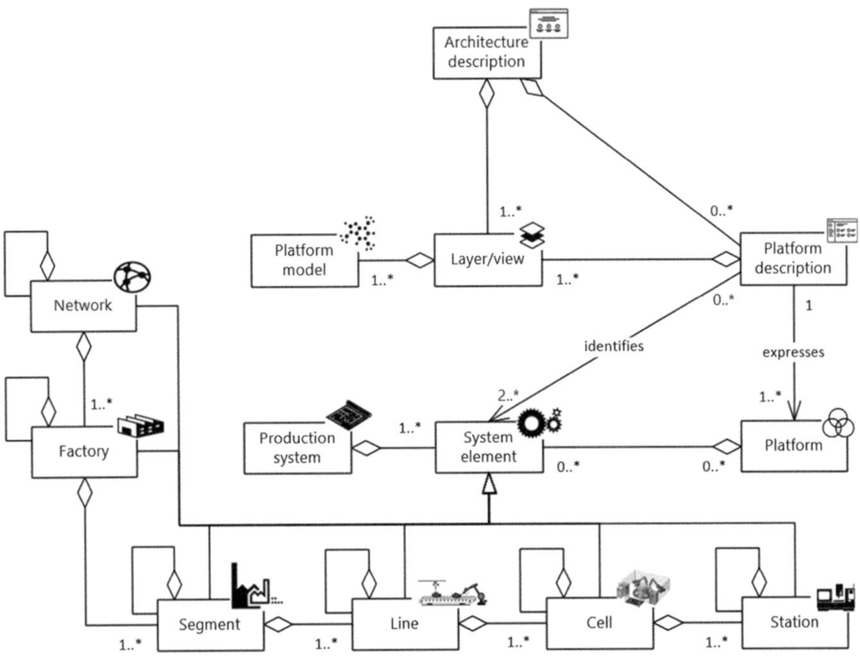

Fig. 7.3 The context of a platform description

system elements that can frame the scope of production platforms. Some larger companies can benefit from scoping production platforms from the factory level or segment level, while SMEs will scope the platforms from the line and cell level simply because they might only have one factory or one segment in a factory.

7.3.1.1 Platform and Co-evolution Concerns

Since production platforms must be updated and maintained throughout their life cycle [21], development is not the only major concern when talking platforms. The updates are most likely actions with cause in concerns from stakeholders belonging to a production environment, having internal and external concerns relating to a product or business environment. See illustration in Fig. 7.4. Some researchers refer to this aspect as co-evolution [18] or continuous platform development [22].

The conceptual model acknowledges the importance of co-evolution by NOT referring to a platform as common elements only. Instead, we define the platform as a value-adding collection of shared elements in an existing production system combined with low-risk future common elements. The latter is considered potential future elements of a production system that has been tested to comply with a

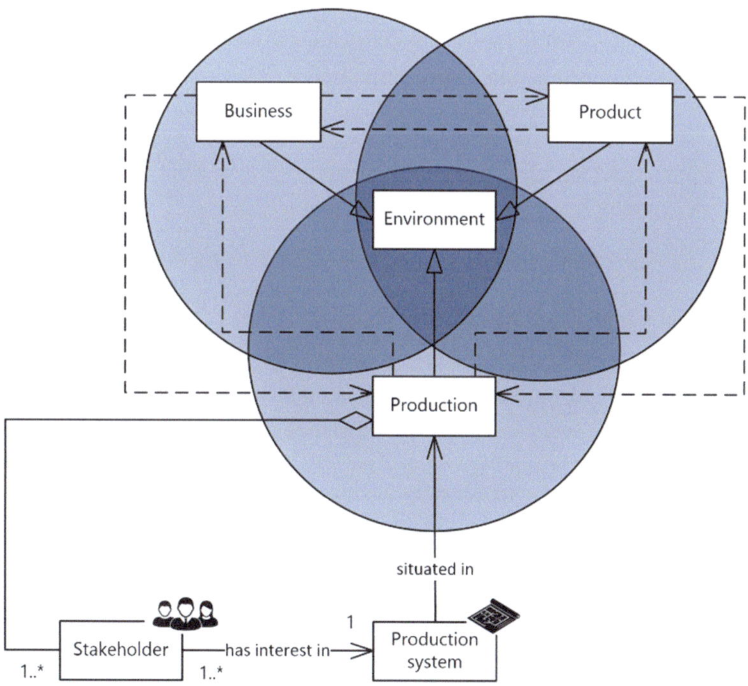

Fig. 7.4 Illustration of the three different environments that affect each other and thereby cause a dynamic environment for each of them

platform description and hence something only causing minor risks of system errors when implemented. Two identified industry examples that motivate co-evolution of production platforms are given in the following:

1. Planned production change: A new manufacturing execution system (MES) is planned to be implemented. The new MES is able to communicate with existing equipment through older protocols, but in later updates of the MES system, some of the oldest protocols are likely phased out. Hence, new production equipment must have support for newer communication protocols supported by the MES.
2. Planned product change: The product development department plans introduce a component with strict requirements for traceability in the production system such that the required product lifetime can both be secured and documented. As a consequence, future equipment must be able to give detailed process data to the MES—a capability not yet present for elements in the current production system.

The two evolution examples above show that a static platform based only on communality in existing equipment would likely limit the platform lifetime to the shortest lifetime of the products targeted. The diverging life cycles of a factory and its elements with cause in increasing frequency of NPI [7] are thus not treated. This instead motivates a platform complemented with future elements tested reliable in laboratory conditions. Besides a planned product change or a planned production change, several other motivational factors (so-called drivers) are identified in literature [5, 7, 16]. With an analogy in research on software product line engineering, the above approach to platform development can be referred to as a combination of extractive and reactive approach [1]: extractive approach because existing standard elements form the basis for transition towards platform development and reactive because the platform is prepared to be changed incrementally to meet new requirements.

7.3.2 *The Viewpoint Concept*

As indicated before, this model builds upon the way of expressing architecture from an industry standard, and therefore Fig. 7.2 shows a model similar to ISO/IEC/IEE 42010 [12]. Thus, the following explanation of the contents of the model does not diverge from the standard, but refers to examples that relate to specifying a production platform description.

7.3.2.1 Stakeholders of the Environment

Stakeholders are any persons, groups and organisations that have interest in the system considered [12], i.e. stakeholders with interest in the production platform in this case. Ideally, these stakeholders come from a production environment, but as strong dependencies to other environments exist—i.e. the business and product environment in the company—stakeholders most likely have responsibilities beyond internal operation activities.

Figure 7.4 illustrates the relations between a product, production and business environment of a company with inspiration from Landherr and Westkämper [16] who found the root cause of such dynamic environment in the compromises between cost, time and quality. Table 7.1 shows a collection of stakeholders of a potential platform which are identified in the case company. Jepsen [13] provides though also a larger list of production architecture stakeholders that can be used as an inspiration.

7.3.2.2 Concerns

A concern is the specific interest that comes from the stakeholder. For instance, in order to secure the product quality, an operation manager will have a reliability concern to the production system built with basis in platform elements. Thus, in order to use platforms and thereby address the reusability concern, the platform architecture must be able to communicate and document that the reliability of the platform elements is within the requirements from the operation manager. Table 7.1 shows more examples of concerns identified in the case company having direct or indirect relation to a reusability concern. A good concern is though both quantifiable and measurable [25], and hence the concerns identified can only be considered a first draft.

7.3.2.3 Viewpoints and Views

ISO/IEC/IEEE 42010 [12] states that a viewpoint is a way of looking at a system, whereas the view is the result of applying the viewpoint. The goal of a view is to address a set of stakeholder concerns through platform models (referred to as models of the platform theory in Sect. 7.2).

The viewpoint concept should be considered as guidance to constructing the actual view and may contain templates, methods, patterns and conventions (model kinds) to do so, as the concept origins from research on software architectures, patterns and templates are meant as design guides for creating code that solves a specific problem. In relation to production platforms, such constructs can be considered as principles for addressing concern and thereby solving a problem. This could, for instance, be a principle contained in a production development viewpoint stating that platform elements should always be designed with support for reconfigurable manufacturing system layouts defined by Koren and Shpitalni [14]. The purpose is here to address a strategic concern related to capacity scalability and variant flexibility.

A related example of this has been identified in a segment of the case company covering several lines that are producing similar components. Eighty percent of the considered lines had a linear line layout with similar process sequence and a

Table 7.1 Identified concerns in the case companies having relation to a reusability concern in development of new production system elements

Organisation	Stakeholder	Concern	Related concern(s)	Context examples	Purpose
Operation	Operation managers	Reliability	Maturity, process stability	Secure that engineering department only designs new lines where equipment and process technology have a high degree of maturity. Communicate possible issues on process stability such that proper process technology and equipment are used prospectively	Quality compliance
		Profitability	Efficiency, operating cost, resource utilisation	Give feedback to production engineering department on line, cell and station efficiency, resource utilisation and standard unit cost such that they can be improved in future projects. Secure compliance to the standard unit cost specified	Profitability compliance
		Usability	Learnability, operability	Keep track on available staff competences, and communicate possible gaps to engineering department. Secure sufficient training of staff. Hire and fire of staff	Competence compliance
		Maintainability	Stability	Secure that production system elements are maintained in order to keep stable production processes and reliable throughput. Communicate possible issues on maintainability to production engineering department	Production throughput compliance

Production development	Production engineers	Regulation compliance	Ergonomics, equipment safety, sustainability	The designed and procured equipment must be ergonomic, environmental friendly and safe to operate for the staff	Ensure good working conditions
		Usability	Learnability, operability	Ensure that designed equipment can be operated by conducting training sessions and designing an intuitive HMI. Provide process data in interactive HMIs such that staff can master processes issues and be less dependent of technology specialists	Ensure skilled staff
		Functionality	Interoperability, process stability	Process data analysis, requirement compliance sessions, process equipment specification, supplier audit, fixture design, tolerance calculation, line buildup, cycle time and flow analysis (LEAN), ramp-up/deployment, test equipment design	Ensure systematic design of production system elements
	Test engineers	Maintainability	Testability, traceability, version control, repeatability, accuracy, reduce scrap	Design reliable and reusable test equipment (accurate and repeatable) to detect process faults; possible faults must be traceable such that equipment can be adjusted and scrap minimised. Application software versions must be kept updated	Profitability and product quality compliance
	ICT Engineers	Functionality	Interoperability, traceability	Software update, version control, traceability, MES integration, protocol design/specification, data structure, protocol analysis, database design, application design	Ensure communication capabilities for production equipment
	Production engineering manager	Profitability	Project cost, reusability	NPI project planning must be planned and resources allocated. Hiring process specialists. Keeping deadlines for deliverables such that the NPI project is not delayed	Time-to-market and time-to-profit compliance

(continued)

Table 7.1 (continued)

Organisation	Stakeholder	Concern	Related concern(s)	Context examples	Purpose
Product development	Product engineers	Profitability	Development time, development cost, manufacturability	Stage-gate model deadlines must be kept in order to not delay the NPI project. Meetings with production development must be held in early stages to secure manufacturability and compliance to standard unit cost requirements	Time-to-market and time-to-profit compliance
		Regulation compliance	EuP directive compliance	Product must be designed according to regulations	Provide sustainable and safe products
		Functionality	Customer experience, interoperability, reliability, performance	Future products can comply with functionality and interfaces from older products	Business requirement compliance

modular-like separation of stations. The last 20 % had a compact U-shaped layout with different process sequences and fairly integrated structure of customised machines designed for the specific purpose. By comparing the lines, a generic process flow could be established for the majority of the lines with linear layout, whereas the U-shaped lines were labelled unique due to spatial and material interfaces. This witnesses that the elements of the lines with linear layout could likely be reused, whereas the U shaped was too optimised for the specific job with regard to a reusability concern. The internal production engineering personally expressed that a principle of having a noncompact linear layout would benefit a reusability concern in this specific situation. Furthermore, it was suggested to adjust future layouts according to a RMS layout principle in order to incorporate ability to scale capacity and product variance.

As the research on production architectures and production platforms is limited, it is difficult to find already defined generic viewpoints for this domain—the so-called library viewpoints [25]. However, several library viewpoints for software development exist though, and these can to some extent be applicable within the production domain [3]. For example, the functional viewpoint from [25] can quite easily be adapted and provided with model kinds and analytic techniques relevant for production platforms. Another example could be an evolution viewpoint based on the concepts of creating a technology radar for assessing process technology maturity [9].

The following examples are models that to some extent have been used in the case company with regard to production platforms:

- Process flow: Used to document the process flow AS-IS and TO-BE.
- Precedence diagrams: Used to document both limitation and change opportunities in the process flow.
- Function mean models: Used to document how functions of the line is realised in specific processes.
- Organ diagram: Used to give more specific details on interfaces.
- Generic organ diagram: Used to document which organ are shared across lines.

Above examples are also contained in models and methods having a larger scope than only functional viewpoint. Some examples of this are given below:

- The configurable component framework [6, 19]:
  Enables detailed modelling of large set of interfaces, functional requirements, design solution, design rationale and internal relationships. Such model contains information suitable for addressing engineering concerns but less suitable for addressing concerns from stakeholders with nonengineering background.
- The product family master plan [11]:
  Enables static modelling of both business, product and production information through composition and classification models. Suitable to illustrate the current AS-IS product for stakeholders with both engineering and nonengineering background and enables discussion on what to change in the future. The model lacks support for production system design though.

- The product family master plan[2] [15]:
  An updated version of the previously mentioned model. It contains more details on production aspects AS-IS but lacks support for production system design [20].

Above individual examples contain more than one viewpoint, and to some degree, they can be considered as containing viewpoints as well. During the literature review, only one full-defined viewpoint was though identified having being (1) described in accordance with the viewpoint concept defined in ISO/IEC/IEEE 42010 [12] and (2) intended for use in a production architecture or platform architecture. The viewpoint identified is named the production capability viewpoint [13] and contains a collection of five different models that enable communicating production capabilities to different stakeholders. This viewpoint is asset applicable as a viewpoint for a platform description, but more investigations should be done to support this statement.

7.4 Conclusion and Discussion

The contents presented in this paper are considered a first step towards creating a production platform theory that contains artefacts such as constructs, models and methods. In the conceptual model, these artefacts are put into context by, for example, presenting the models as platform models with the purpose of addressing specific stakeholder concerns related to reusability through the concept of viewpoints. With basis in ISO/IEC/IEEE 42010, the methods of the platform theory can be explained as contained in viewpoints, which purpose is to construct the actual views of the platform architecture. Furthermore, such viewpoints can contain templates, patterns, principles and conventions, which will provide the constructs for the production platform theory.

The created conceptual model hence puts words on how production platforms should be grasped in the manufacturing industry and provides examples of considerations that must be made with basis in research on software architectures. The examples include, e.g. empirical identified stakeholder concerns from a case company and library viewpoints identified in the literature.

In order to instantiate the conceptual model and thereby create a specific production platform architecture model fitting specific types of manufacturing industry domains, more stakeholders and concerns must though be identified (step 4–7 in Sect. 7.2.1). Those concerns will be the key to establish efficient views and viewpoints, which constitute a production platform architecture. In order to make operational concerns, they must though be both quantifiable and measurable in later versions.

References

1. Apel, S., Batory, D., Kästner, C., et al.: Feature-oriented software product lines. Springer, Berlin (2013)
2. Baldwin, C.Y., Woodard, C.J.: The architecture of platforms: a unified view. In: Gawer, A. (ed.) Platforms, markets and innovation, p. 19. EE Publishing, Cheltenham (2009)
3. Benkamoun, N., ElMaraghy, W., Huyet, A., et al.: Architecture framework for manufacturing system design. Procedia CIRP **17**, 88–93 (2014)
4. Bossen, J., Brunoe, T.D., Nielsen, K.: Platform-Based Production Development. IFIP Advances in Information and Communication Technology 459, 53–61 (2015)
5. Brunoe, T.D., Bossen, J., Nielsen, K.: Identification of drivers for modular production. IFIP Advances in Information and Communication Technology 459, 235–242 (2015)
6. Claesson, A.: A configurable component framework supporting platform-based product development. Ph.D. thesis, Chalmers University of Technology, Gothenburg (2006)
7. ElMaraghy, H.A., Wiendahl, H.P.: Changeability—an introduction. In: ElMaraghy, H.A. (ed.) Changeable and reconfigurable manufacturing systems, pp. 3–24. Springer, London (2009)
8. Ericsson, A., Erixon, G.: Controlling design variants: modular product platforms. ASME Press, New York (1999)
9. Ford, N.: Build your own technology radar. Whitepaper. Online. ThoughtWorks (2013)
10. Gregor, S., Jones, D.: The anatomy of a design theory. J. Assoc. Inf. Syst. **8**, 312–335 (2007)
11. Harlou, U.: Developing product families based on architectures: contribution to a theory of product families. ORBIT (2006)
12. ISO/IEC/IEEE 42010: Systems and software engineering—architecture description (ISO/IEC/IEEE 42010:2011). (2011)
13. Jepsen, A.D.: Architecture descriptions—a contribution to modeling of production system architecture. Ph.D. Thesis—Annual Report, DTU Mechanical Engineering (2015)
14. Koren, Y., Shpitalni, M.: Design of reconfigurable manufacturing systems. J. Manuf. Syst. **29**, 130–141 (2010)
15. Kvist, M.: Product family assessment. Ph.D. Thesis, DTU Management Engineering (2010)
16. Landherr, M., Westkämper, E.: Integrated product and assembly configuration using systematic modularization and flexible integration. Procedia CIRP **17**, 260–265 (2014)
17. Meyer, M.H., Lehnerd, A.P.: The power of product platforms: building value and cost leadership. Free Press, New York (1997)
18. Michaelis, M.T., Johannesson, H.: Platform approaches in manufacturing: considering integration with product platforms. In: Proceedings of the ASME 2011 International Design Engineering Technical Conferences & Computers and Information in Engineering Conference, Washington, DC, pp. 1115–1124 (2011)
19. Michaelis, M.T., Johannesson, H., ElMaraghy, H.A.: Function and process modeling for integrated product and manufacturing system platforms. Journal of Manufacturing Systems, Rev. **36**, 203–215 (2015)
20. Michaelis, M.T.: Co-development of products and manufacturing systems using integrated platform models. Chalmers University of Technology, Gothenburg (2013)
21. Michaelis, M.T.: Co-development of products and manufacturing systems-applying a configurable platform approach. Licentiate thesis, Chalmers University of Technology, Gothenburg (2011)
22. Nielsen, O.F.: Continuous platform development. DTU Management Engineering, Department of Management Engineering, Technical University of Denmark (2010)
23. Pine, I., Joseph, B., Victor, B.: Making mass customization work. Harv. Bus. Rev. **71**, 108–117 (1993)
24. Robertson, D., Ulrich, K.: Planning for product platforms. Sloan Manag. Rev. **39**, 19–32 (1998)
25. Rozanski, N., Woods, E.: Software systems architecture: working with stakeholders using viewpoints and perspectives. Addison-Wesley, Boston (2011)

26. Salvador, F., De Holan, P.M., Piller, F.: Cracking the code of mass customization. MIT Sloan Manag. Rev. **50**, 71–78 (2009)
27. Ulrich, K., Eppinger, S.D.: Product design and development. McGraw-Hill, New York (2003)
28. Wiendahl, H., ElMaraghy, H.A., Nyhuis, P., et al.: Changeable manufacturing-classification, design and operation. CIRP Ann. Manuf. Technol. **56**, 783–809 (2007)
29. Zahn, E.: Wandlungsfähige produktionsunternehmen: Das stuttgarter unternehmensmodell. Springer, Berlin (2009)

Chapter 8
From ETO to Mass Customization: A Two-Horizon ETO Enabling Process

Aldo Duchi, Filippo Tamburini, Daniele Parisi, Omid Maghazei, and Paul Schönsleben

8.1 Introduction

Engineer to order (ETO) is referred to a manufacturing strategy for highly customized products which are required to be designed and engineered in detail, based on the customer's order specifications [1–3]. On the one hand, highly specialized customer requirements pose various challenges in such systems, such as difficulties in accurate estimation of lead time and delivery dates, late changes and expensive reworks, poor product quality, and material waste [2]. On the other hand, globalization, margin shrink, increased competition, and dramatic technological advances raise crucial issues concerning the ETO firms to retain a competitive edge [2, 4, 5].

Mass customization (MC) is defined as "customer co-design process of products and services, which meets the needs of each individual customer with regard to certain product features. All operations of customization are performed within a fixed solution space, characterized by stable but still flexible and responsive processes" [6]. Solution space (SS) refers to the possible permutations of design parameters in order to fulfill future customer needs [7]. Salvador et al. introduce solution space development as a basic competence for MC firms [8].

MC, which aims at producing customized goods and services to satisfy individual needs of the customers [9], has several similarities with the ETO environment. Along the same lines, tackling a combination of ETO and MC settings, as a hybrid strategy toward ETO firms, is likely to provide room for improvement. As a consequence, a transition toward MC calls for fresh approaches and effective mechanisms to smooth this process and, in turn, to achieve superior results of formulating such a hybrid strategy.

A. Duchi (✉) • O. Maghazei • P. Schönsleben
BWI Center für Industrial Management, ETH Zürich, Zürich, Switzerland
e-mail: aduchi@ethz.ch

F. Tamburini • D. Parisi
Department of Management Engineering, Politecnico di Milano, Milan, Italy

J. Bellemare et al. (eds.), *Managing Complexity*, Springer Proceedings in Business and Economics, DOI 10.1007/978-3-319-29058-4_8

Interestingly, empirical evidences show a paradigm shift for ETO companies to formulate innovative strategies in order to employ MC in the ETO systems [10]. Past studies in the context of ETO environment have not entirely addressed the interplay between MC and ETO settings. Although at the strategic level, ETO-MC has been articulated by Haug as a hybrid strategy for ETO firms which gathers MC and ETO characteristics together, aiming at optimizing the internal processes [10]. However, the impact of solution space management in such setting, both in the short- and long-term outlook, needs further investigation.

Thus, this paper is outlined as follows. Having examined the extant literature on the solution space allows us to establish a conceptual framework to deal with the ETO-MC setting. Then, we would empirically examine our proposition through the context of three Swiss ETO companies. Finally, we underline the investigated success practices that are likely to enhance effective management of customization processes.

8.2 Solution Space Definition, Research Gap, and Questions

The aim of the following section is to provide a definition of "solution space" in pursuit of the ETO-MC settings and in turn to frame the research questions of this study.

8.2.1 The Solution Space

The definition of the solution space is a fundamental issue concerning the ETO-MC companies. Typically, Pine defines the solution space as a hyperplane shaped by all the possible permutation of the predefined design parameters [11]. This definition mainly derives from the idea of configurators, implemented in the mass customization environment.

The basic idea of the ETO systems is that the customer order decoupling point is prior to the engineering phase [12]. Therefore, it is not always a priori reason to reduce and enumerate the possible design parameters.

The spread of the configurators and the success of mass customizers have made complicated the definition of solution space. Henceforward, we divide the solution space into two classes: standard solution space (SSS) and nonstandard solution space (NSSS).

To define the SSS, we follow the definition of Hadzic that defines a product through a mathematical definition of three objects: variables, domain for the variables, and formulae [13]. The standard product belongs to this domain. It is worth emphasizing that, as long as the variables can be defined by a continuous domain, the SSS can be infinite accordingly. The standard products can be managed by the company as make-to-order (MTO) products [14]. This definition can be suitable for the commercial offer for the mass customizer, whereas in the ETO-MC context, the customer needs to choose over a wider set of variables and attributes compared to the one offered as standard products [10].

Similarly, it is necessary to define NSSS. A nonstandard product, which belongs to the NSSS, is the outcome of the violation of variables, domains of the variables, or formulae. The NSSS encompasses the hypothetic domain of technically feasible solution requested by the customer. This domain is infinite by definition. The sum of SSS and NSSS defines the solution space.

The advanced ETO companies are capable of representing their standard space by utilizing the configurators. The advantages pertaining to the implementation of the configurator are out of the scope of this study, although they have been entirely articulated in past studies [14–18]. Notice that a technical, computer-aided configurator is just the representation of the SSS and an efficient but costly implementation strategy. Nonetheless, firms that do not invest in a configurator are capable of shaping their standard offers through standard processes. This projection of a standard product at the operational level can be defined as "standard space of action."

8.2.2 Gap and Research Questions

Past studies adequately address the issues relating to companies switching processes from mass production to mass customization. For those companies, therefore, the solution space is standard. To date, the studies related to the transition from ETO to MC are mainly focused on the technological aspects [10]. However, the characteristics of the ETO-MC companies need further investigation. Their typical solution space can gather some additional relevant information from the clients within their operational processes. In particular, the design management of the nonstandard products is a significant process. The information and knowledge which are generated in the process of developing tailored solutions can lead to a modification of the SSS [19]. To date, the dynamics of this emerging challenge and the evolution of the solution space have not entirely explored. Therefore, this study aims at addressing the following research questions.

RQ.1: *How can the evolution of the SSS be described in an ETO-MC setting?*
RQ.2: *How can an ETO-MC setting operatively deal with the modifications of the SSS, in order to gain a sustainable competitive advantage?*

The first research question defines the base framework for ETO-MC companies and their solution space evolution of the order-specific engineering and its developed knowledge. The second research question underlines the operational management of this evolution. Two types of enabling processes are represented, with different levels of planning.

8.2.3 Methodology

This study is based on an extended literature review and multiple case studies. Firstly, the topics related to mass customization, ETO, knowledge, and information management have been considered.

Secondly, a series of interviews were conducted with the engineering managers and product line managers of three ETO companies in order to understand the real effectiveness of the research agenda. After a cross analysis, the resultant topic has been narrowed and defined. For the following 2 years, the authors and three Swiss leading companies closely collaborated together in order to address the identified issues. All of the practices and the tools presented below were identified, selected or developed, and implemented in the selected companies.

8.3 The ETO Enabling Process

The hybrid business model has some issues to be addressed. Firstly, the interaction between standard and nonstandard products is a specification of the ETO-MC environment. Secondly, the operational management and updating the solution space are the other major issues in such setting. A proxy of the competitive value of a firm in the ETO-MC sector can be identified pertaining to the success rate of the winning tenders. In a pure ETO sector, this value is less than 30 %, and, therefore, fast and cost-efficient processes are essential attributes in the engineering phase [20]. In this ETO-MC extension, the management of order-specific engineering with regard to the modules and subsystems of the final product is a crucial task. The relationship between knowledge and information reuse and successful performances is relevant to the ETO environment [21], and the focus of this study is on inclusion of relevant knowledge into their standard space of action.

In order to confront this challenge, two complementary processes are represented. The ETO enabling process and the effective practices related to the implementation process are represented by Schönsleben [14] and Duchi et al. [5]. This study extends the primary model that presents every nonstandard product as a triggering mechanism of exchanging know-how in the ETO value chain.

The extension includes two classes named "short-term ETO enabling process" and "long-term ETO enabling process." The former relates to the operational management of the specification fulfillment for the nonstandard products. The latter refers to the creation of a sustainable competitive advantage. Moreover, there is a cross-interaction between these two dynamics that will be further investigated.

8.3.1 Short-Term ETO Enabling Process

To have an in-depth understanding of the short-term ETO enabling process, it is necessary to deal with the order-specific engineering (Fig. 8.1).

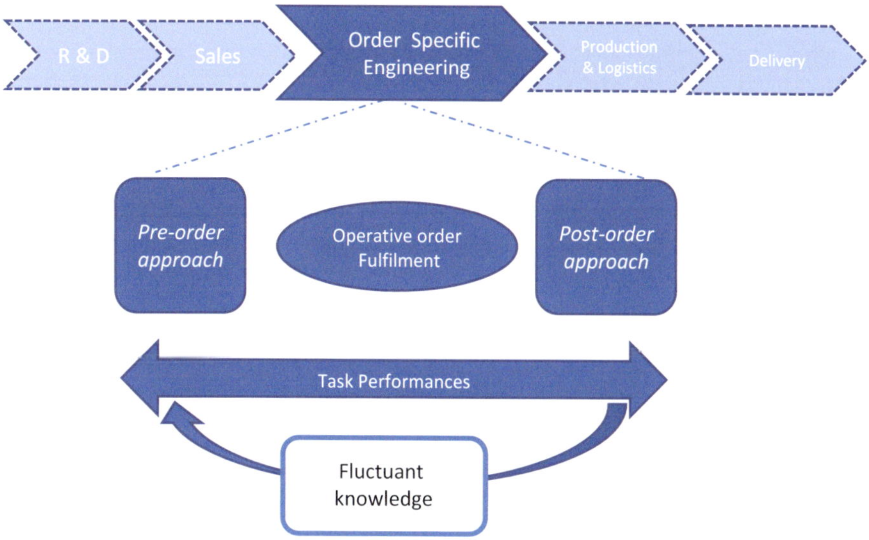

Fig. 8.1 Order-specific engineering

The order-specific engineering is a process activated by those requests that are submitted by the customers and requires the direct engineer work to develop a tailored product. Furthermore, this process is divided into the three sequential phases, such as preorder approach, operative order fulfillment, and post-order approach. The focus of this study will be on the design phase, and the other relevant aspects, e.g., cost estimation and pricing, are out of the scope of this study [22].

The first stage of the short-term (ST) ETO enabling process is the approach to tackle the development of a new required technical solution. It can include the correct receipt of the specification, the analysis (cost and design feasibility), the study of the issues, the communication with other functions, and the research of the previously created similar solutions.

The operative order fulfillment phase refers to the creation of the specific solution, the production of the new technical drafts, the selection of the right subcomponents, and the comprehensive assessment of the solution feasibility.

The last stage of ST ETO enabling process is the post-order process, which includes the communication of the created solution, the intra- and intercompany idea sharing, and the completion of the project. The efficient and effective capabilities to successfully carrying out these three phases will eventually lead to shorter lead times, cost reduction, and improvements in the quality of the offers [22].

Considering the company processes as a dynamic nature, it seems reasonable to assume that several issues have been already addressed out of the SSS boundaries. These processes are likely to lead to the creation of both explicit and implicit knowledge. Theoretically, the bodies of information and knowledge inside of the company are continuously changing and expanding [21]. It represents that the firm's and employee's capabilities have been developed during the past problem-solving issues, as well as all the past produced documentation. We coin this as "fluctuant knowledge." This knowledge represents the key enabler for the short-term ETO enabling process.

The ST ETO enabling process starts from the firm's awareness of the potential impact of this order-specific engineering process on the competitiveness outcomes. It aims at determining the process of acquisition, assimilation, transformation, and exploitation with regard to the order-specific engineering. Moreover, it establishes a framework to follow the actors in every phase [23]. Therefore, the ST ETO enabling process affects the operational level. It is worth mentioning that not only does this process focus on the reuse of the previously created knowledge but also it affects all the sharing, elaborating, and solution-renewing processes. Effective ETO-MC firms are able to implement a series of organizational and IT-based practices in order to determine the ST ETO enabling process.

8.3.2 Long-Term ETO Enabling Process

The generation of a sustainable competitive advantage is considered as a main goal of the knowledge management systems [24, 25]. This issue relates to the ETO-MC companies, and, therefore, the long-term ETO enabling process will operationally address it.

The main risks and the challenges of the knowledge management in such firms are linked to the owners of the knowledge, the so-called knowledge workers [26]. The knowledge is tied up to the human experiences and interactions [27]. In the previously discussed knowledge generation environment (order-specific engineering), human resources play a relevant key role. The tasks are performed to fulfill the delivery of a product belonging to the NSSS, and therefore they request a not automatable outcome. With the ST enabling process, the solutions are rationally stored, but their impact on the process performances will retain to the individuals.

ETO-MC settings can institutionalize the existing information and knowledge [28] toward the standard space of action, as a reliable basis to gain competitive advantage. The possibilities of standardization and formally institutionalization are the main drivers that lead the firms to compete with this hybrid business model.

The long-term ETO enabling process primarily gives the awareness of the long-term potential improvements inside of the firms in pursuit of fluctuant knowledge.

Needless to say, the whole created knowledge is not necessarily embodied in the standard space of action. The modifications call for different processes and IT tools. The improvement of the latter is a topic discussed by several scholars [29–31]. In the ETO-MC context, the concretization of these changes is visible through the solution space. Simply put, the solution space performs with the functionality of a knowledge repository. For instance, in reshaping of the SSS, some product variants are considered as MTO by the company. This means that the associated knowledge, belonging to the fluctuant knowledge, needs to be embodied in a relatively structured way. This argument raises a serious challenge since the nature of the fluctuant knowledge consists both tacit and explicit knowledge, and it is argued that the tacit knowledge cannot be easily incorporated in the IT systems [32, 33]. However, the knowledge which is subject of institutionalization is mainly related to the product knowledge and is less likely to be defined as a tacit knowledge [28].

The trade-off between the inclusion and exclusion of the information is loosely related to the trade-off between cost and benefit [10].

It is noteworthy that the renewal of the offer in terms of eliminating the obsoleted commercial portfolios that are not requested by the market anymore is a fundamental task. Therefore, not only the SSS will be increased in terms of its dimension but also it will be changed into different shapes.

The long-term ETO enabling process concretizes itself in the IT tools and in the organizational practices that lead the company to capture the relevant information inside the standard space of action. The reshaped commercial offer, represented in the SSS, is therefore the reflection of the new organizational and IT systems in order to have an accurate understanding and forecast of the market needs. Some successful cases are represented in the Sect. 8.4.

8.3.3 *Interaction Between Short- and Long-Term ETO Enabling Process*

The interaction between the short- and long-term ETO enabling process is outlined as follows. It is argued that these two processes should be seen as complementary.

Figure 8.2 illustrates the interaction between these two processes and the impact on the SSS and NSSS. A generic customer could either ask for a standard product, so this product belongs to the SSS and will be dealt with by the company as an MTO product, or ask for a product which belongs to the NSSS. The performances of the process of creating an NSSS are primarily connected to the effectiveness of the ST ETO enabling process, in which on the one hand defines and creates the fluctuant knowledge and on the other hand allows an effective knowledge retrieval. The companies attempt to fulfill the requests which are outside of the SSS. As a consequence, the available fluctuant knowledge will be further advanced. At a certain point when the fluctuant knowledge has more consistent patterns, it is likely to trigger the long-term ETO enabling process. Thus, this stimulates capturing such expertise, awareness, information, and newly created solutions into the SSS. As a consequence, the solution space is likely to be reshaped. This is in light of the fact that some information is added and some information is deleted accordingly. A relevant issue concerns with the definition of the relationship between these two processes.

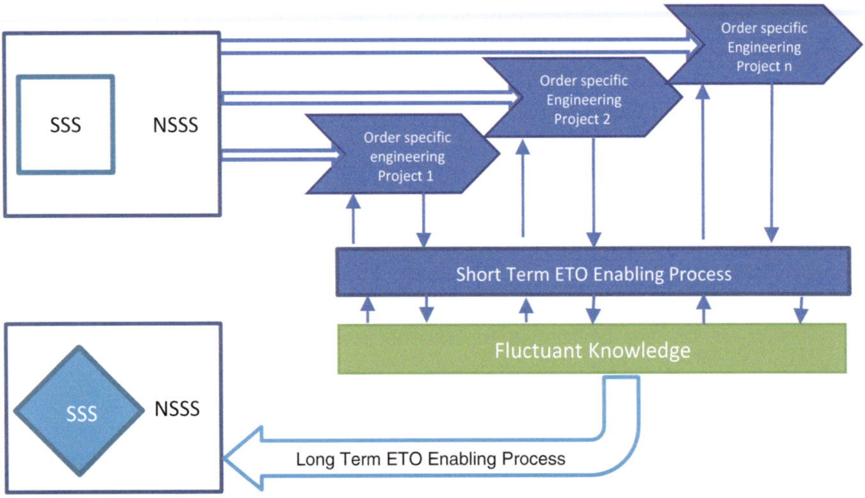

Fig. 8.2 The interaction between the short- and long-term ETO enabling processes

Firstly, it is believed that the LT ETO enabling process is less likely to operate without replication of the ST ETO enabling process. The LT ETO enabling process is triggered by the fluctuant knowledge, and its creation is connected to the effectiveness of the ST enabling process. Therefore, the more a firm is capable of developing the tools and practices to structure this knowledge, the more effective it can acquire relevant knowledge and information to eventually improve changes in the SSS.

Secondly, we believe that the ST ETO enabling process can be implemented without the existence of the LT ETO enabling process. However, this is an unsustainable approach which leads the firms toward inevitable loss of relevant knowledge and in turn degrades competitiveness.

8.4 Implementations

This chapter provides an overview of some practical implementations of the ST and LT ETO enabling processes. The practices employed by the three companies are classified according to two features: one of them relates to the temporal horizon of the enabling process, and the other examines whether the actors involved in the practice are internal or external in accordance with the boundary of the firm.

The first category distinguishes between practices contributing either to the ST or LT ETO enabling process. This distinction is useful to realize whether a practice is part of a firm's set routines or it has been triggered during the SSS renewal process. It is worth to emphasize that the same practice can be useful for both the enabling processes, although with diverse purposes and approaches.

The second category deals with the practices as *internal* that is implemented and operatively managed by internal actors and *external* that actors out of the boundary of the firm are expected to perform. According to external actors, we take the entire value chain into consideration; suppliers and customers are two notable instances. The subsidiaries that actively operate at the engineering phase are considered as internal actors in this study [9]. The motivation of the introduction of this distinction is that the relationship between the firm and the external actors is peculiar in ETO context, because every order fulfillment presents some unique dimensions.

8.4.1 The Companies

Table 8.1 represents the characteristics of the three case study companies relevant to this study.

8.4.2 Successful Implementations

8.4.2.1 Company A

Company A has developed the *naming convention*, a language protocol to univocally define product components and subsystems. Information that is related to the component variance and configuration is contained in the naming codification, and unnecessary searches in product data management are avoided. Substantial benefits arise from time savings and cost reductions that spread both internally and externally. For instance, in the case of strategic relationships with suppliers, this practice is useful to avoid recycles and to facilitate quick exchange of structured information. This practice therefore belongs to the ST enabling process and the actors are both internal and external.

A second practice is the *documentation sharing platform*. Company A has managed to create a single point of sharing and communication with the customers and the suppliers. Such IT tool enables company A to define the most convenient channel of communication for each document exchanged. Furthermore, it plays a vital role in capitalizing past documents exchanged in order to facilitate and speed up the information retrieving process. This practice is classified in the ST enabling process and with both internal and external actors.

Component information manager is the third practice that can drastically reduce the order-specific engineering lead time by leveraging past developed solutions. This tool helps the engineers in handling both technical- and process-related information. The database can be easily accessed through the web-based interface, and, by filling in some basic parameters, it could give back a past solution already developed with similar features.

Table 8.1 Case study companies

Company	A	B	C
Industry	Industrial steam turbines	Elevators	Concrete mixing and mineral processing plants
Product	Extremely diverse, ad hoc technologies for different customers	Slightly diverse, high degree of commonality among the product families	Highly diverse, different core technologies
Market (units/year)	Less than 50	More than 1000	300–500
Supporting IT systems/tools	Sales configurator	Design configurator	Design configurator
Company engineering knowledge	Engineering knowledge is based on the expertise on advanced precision engineering	The company B has to face with critical issues related to the use of the meta-knowledge. The European site retains most of the advanced technologies and know-how	The company relies on a deep and strong base of engineering meta-knowledge
Organizational practices	Engineer project is separated into subtasks, each of them refers to a specific component development that belongs to NSSS	One of the highest levels of efficiency among the main competitors in dealing with engineer tasks, thanks to the deployment of both a technical and a commercial configurator	Considerable system's complexity does not allow the firm to achieve high level of standardization in the short run
ETO enabling process	Profit margin increases with more customized tools	The high volume of specifications customized per year allows to identify communalities in the customer's requests and also to feed the company's SSS	The ETO enabling process shows several improvement margins due to the high-customized products
Maturity level	Low	High	Low

Although it is mostly used during the phase of pre-order approach, it has a relevant impact on the post-order approach, as every new drawing needs to be uploaded in order to update the meta-knowledge tool. It belongs to the ST ETO enabling process practices, with internal actors.

8.4.2.2 Company B

The *SE-AE process* is a process between the sales engineers and the application engineers. The sales engineers are responsible for the commercial offer of the standard products. They prepare a tender with the support of a technical configurator. Every time an order-specific engineering is required, the sales engineer calls for the assistance of the application engineer who develops a tailored solution as a feedback. An interesting attribute of the process is that the list of variable violations is automatically provided to AE toward an extension of the configurator, which resulted in reducing the error rates and their associate loops. The impact of this internal practice is relevant mostly on the ST ETO enabling process. This tool is also an enabler for the LT ETO enabling process because the electronically provided data contributes to the creation of the *fluctuant information*.

The *AE-PLM process* is between the *application engineers* and the *product line manager*. The product line managers are responsible for the commercial offers. Every 6 months, these two functions are called to evaluate the future commercial offer of the firm, reshaping the SSS. The value of this process is both on the creating and unlearning solutions. This practice belongs to the LT ETO enabling process with internal actors.

The former practices are supported by the data gathered toward the *Frequently Asked Request* (*FAR*) *Tool*. This IT tool handles the fluctuant knowledge created during the order-specific engineering. It allows the application engineers to evaluate the nonstandard task required by customers. It is possible to find some pattern and evaluate the most frequently violation of parameters (frequently in terms of range). This tool is the core of LT ETO enabling process that allows the company to gather relevant information from the non-SSS and evaluates its introduction into the standard boundaries. Furthermore, the internal actors are involved in this process.

8.4.2.3 Company C

The first practice that has been implemented in company C is *issue manager*. Whenever a problem is being reported during the life cycle of a product, it will be communicated to the engineering department, and the designs will be fixed accordingly. This mechanism aims at avoiding recycles and waste of resources. Functioning like a knowledge-based repository, on the one hand, it stimulates the possibility of increasing the level of efficiency of the product development phase and on the other hand represents a long-term arrangement in light of expanding the company's

Table 8.2 Summary and categorization of success practices

Success implementation	Company	Temporal horizon of the ETO enabling process	Range of action
Naming convention	A	Short term	Internal/external
Documentation sharing platform	A	Short term	Internal/external
Component information manager	A	Short term	Internal
AE-PLM process	B	Long term	Internal
SE-AE process	B	Short/long term	Internal
FAR tool	B	Long term	Internal
Issue manager	C	Long term	Internal
Sharing platform	C	Short/long term	Internal

meta-knowledge. This practice belongs thus to the long-term ETO enabling process with internal actors.

Serving different markets, the company C represents three different engineering locations spread out in different countries that often have to solve similar engineering tasks. The *sharing platform* has been developed in order to avoid repetition of engineer tasks. It is an IT tool that allows different functions to share information in a fast and structured way. The centralization of information represents not only a mean to save time and costs, but it mainly enables the transformation of special requests into valuable knowledge for the organization. This practice supports both the short- and long-term enabling process, and just internal actors are involved during this process.

In the following table, the success practices are summarized and categorized (Table 8.2).

8.5 Conclusions

Mass customization as a business reality and ETO setting as a manufacturing environment for highly customized products are making a customization-efficiency trade-off from different perspectives. Furthermore, they are facing serious challenges in pursuit of the hybrid strategy of ETO-MC. One of the challenges is linked to the definition of the commercial offer. ETO-MC companies decide for the rationalization of a batch of products (the ones represented in the SSS), but at the same time, they allow customers to customize them in several dimensions. This strategic decision needs to be sustained by an operative and continuous approach. A conceptual framework for the evolution of the SS has been presented in this study, based on the information and knowledge management in the design phase. The proposed conceptual framework not only suits the companies that can afford the investment in configurators but also the ones that can deal with standard offers with less advanced tools.

Several companies are likely to develop such framework and, in turn, to take it as a reliable basis for an understanding and rationalization of their internal processes. Furthermore, some practical cases have been presented that allow for having an in-depth understanding about the practical implementation. This study attempts to posit the research gap into the real ETO-MC environment. Therefore, we confronted the real challenges that our case study companies were facing with, and, indeed, we introduced and promoted success practices, leading to a more efficient ST enabling process and a sustainable readaptation of the solution space in the long term.

This study presents some limitations. The first one is driven by the different core competences and market of each ETO Company. Consequently, it is essential to strongly consider the relevance of the sector and the context of the firm to the identified success practices. We were not able thus to define generic operative practices. The second limitation relates to the focus of this study, which mostly deals with the engineering phase. It is necessary to examine related aspects of the downstream phases in the ETO value chain. The third limitation is concerned with the impact of the maturity of the company. This dimension is likely to affect the development of the SSS and the company's suitability for some practices instead of others.

The primary goal of this study is to define the different engineering capabilities that are essential for the sustainable development of the SSS. This suggests directions for future research. In concrete terms, it is essential to draw a meaningful distinction between diverse ETO settings in terms of, for instance, the degree of engineering complexity and customization. This is likely to lead to a deep understanding of the information and knowledge management requirements for the short- and long-term ETO enabling processes.

Acknowledgment This work is funded by the research project through the Swiss Commission for Technology and Innovation CTI (CTI no.: 15021.1 PFES-ES). The authors would like to thank all participating organizations for sharing their insights in the ETO industry.

References

1. Wikner, J., Rudberg, M.: Integrating production and engineering perspectives on the customer order decoupling point. Int. J. Oper. Prod. Manag. **25**(7), 623–641 (2005)
2. Pandit, A., Zhu, Y.: An ontology-based approach to support decision-making for the design of ETO (Engineer-To-Order) products. Autom. Constr. **16**(6), 759–770 (2007)
3. Schönsleben, P.: Integral logistics management—operations and supply chain management within and across companies. Taylor & Francis, Auerbach Publications, Boca Raton (2011)
4. Kumar, S., Wellbrock, J.: Improved new product development through enhanced design architecture for engineer-to-order companies. Int. J. Prod. Res. **47**(15), 4235–4254 (2009)
5. Duchi, A., Maghazei, O., Sili, D., Bassan, M., Schönsleben, P.: Advances in Production Management Systems: Innovative Production Management Towards Sustainable Growth: IFIP WG 5.7 International Conference, APMS 2015, Tokyo, Japan, September 7–9, 2015, Proceedings, Part II, S. Umeda, M. Nakano, H. Mizuyama, H. Hibino, D. Kiritsis, and G. von Cieminski, (Eds). Cham: Springer International Publishing, 223–231 (2015)
6. Piller, F.T.: Mass customization: reflections on the state of the concept. Int. J. Flex. Manuf. Syst. **16**(4), 313–334 (2004)

7. Zhang, L.L.: Product configuration: a review of the state-of-the-art and future research. Int. J. Prod. Res. **52**(21), 6381–6398 (2014)
8. Salvador, F., Forza, C.: Configuring products to address the customization-responsiveness squeeze: a survey of management issues and opportunities. Int. J. Prod. Econ. **91**(3), 273–291 (2004)
9. Willner, O., Powell, D., Duchi, A., Schönsleben, P.: Globally distributed engineering processes: making the distinction between engineer-to-order and make-to-order. Procedia CIRP **17**, 663–668 (2014)
10. Haug, A., Ladeby, K., Edwards, K.: From engineer-to-order to mass customization. Manag. Res. News **32**(7), 633–644 (2009)
11. Pine, I.B.J.: Mass customization: the new frontier in business competition. Harvard Business Press, Boston (1999)
12. Wortmann, J.C.: A classification scheme for master production schedule. In: Berg, C., French, D., Wilson, B. (eds.) Efficiency of manufacturing systems, vol. 11. Plenum Press, New York (1983)
13. Hadzic, T., Subbarayan, S., Jensen, R.M., Andersen, H.R.: Fast backtrack-free product configuration using a precompiled solution space representation. In: PETO, pp. 131–138 (2004)
14. Schönsleben, P.: Methods and tools that support a fast and efficient design-to-order process for parameterized product families. CIRP Ann. Manuf. Technol. **61**(1), 179–182 (2012)
15. Hvam, L., Haug, A., Mortensen, N.: Assessment of benefits from product configuration systems. In: 13th Workshop on Configuration, ECAI …, Lisboa, Portugal (2010)
16. Trentin, A., Perin, E., Forza, C.: Product configurator impact on product quality. Int. J. Prod. Econ. **135**(2), 850–859 (2012)
17. Trentin, A., Perin, E., Forza, C.: Sales configurator capabilities to avoid the product variety paradox: construct development and validation. Comput. Ind. **64**(4), 436–447 (2013)
18. Forza, C., Salvador, F.: Managing for variety in the order acquisition and fulfilment process: the contribution of product configuration systems. Int. J. Prod. Econ. **76**(1), 87–98 (2002)
19. Elgh, F.: Modeling and management of product knowledge in an engineer-to-order business model. In: ICED 11—18th International Conference on Engineering Design—Impacting Society Through Engineering Design, vol. 6, pp. 86–95 (2011)
20. Konijnendijk, P.A.: Coordinating marketing and manufacturing in ETO companies. Int. J. Prod. Econ. **37**, 19–26 (1994)
21. Silventoinen, A., Denger, A., Lampela, H., Papinniemi, J.: Challenges of information reuse in customer-oriented engineering networks. Int. J. Inf. Manage. **34**(6), 720–732 (2014)
22. Kingsman, B.G., De Souza, A.: A knowledge-based decision support system for cost estimation and pricing decisions in versatile manufacturing companies. Int. J. Prod. Econ. **53**(2), 119–139 (1997)
23. Zahra, S.A., George, G.: Absorptive capacity: a review, reconceptualization, and extension. Acad. Manage. Rev. **27**(2), 185–203 (2002)
24. Corso, M., Martini, A., Paolucci, E., Pellegrini, L.: Knowledge management in product innovation: an interpretative review. Int. J. Manag. Rev. **3**(4), 341 (2001)
25. Teece, D.J.: Strategies for managing knowledge assets: the role of firm structure and industrial context. Long Range Plan. **33**(1), 35–54 (2000)
26. Drucker, P.F.: Knowledge-worker productivity: the biggest challenge. IEEE Eng. Manag. Rev. **34**(2), 29 (2006)
27. McDermott, R.: Why information technology inspired but cannot deliver knowledge management. Calif. Manag. Rev. **41**(4), 103–117 (1999)
28. Crossan, M.M., Lane, H.W., White, R.E.: An organizational learning framework: from intuition to institution. Acad. Manag. Rev. **24**(3), 522–537 (1999)
29. Veldman, J., Klingenberg, W.: Applicability of the capability maturity model for engineer-to-order firms. Int. J. Technol. Manag. **48**(2), 219 (2009)
30. Forsman, S., Björngrim, N., Bystedt, A., Laitila, L., Bomark, P., Öhman, M.: Need for innovation in supplying engineer-to-order joinery products to construction: a case study in Sweden. Constr. Innov. Inf. Process. Manag. **12**(4), 464–491 (2012)

31. Roberts, S., Delaney, P.J.: Rapid reengineering teams: case study on maximizing information technology. In: Annual International Conference Proceedings—American Production and Inventory Control Society, pp. 70–75 (1996)
32. Johannessen, J.A., Olaisen, J., Olsen, B.: Mismanagement of tacit knowledge: the importance of tacit knowledge, the danger of information technology, and what to do about it. Int. J. Inf. Manag. **21**(1), 3–20 (2001)
33. Walsham, G.: Knowledge management: the benefits and limitations of computer systems. Eur. Manag. J. **19**(6), 599–608 (2001)

Chapter 9
Utilization of Mass Customization in Construction and Building Industry

Kjeld Nielsen, Thomas Ditlev Brunoe, Kim Noergaard Jensen, and Ann-Louise Andersen

9.1 Introduction

Mass customization is a widely adopted competitive strategy for delivering individually customized products at a cost near mass production, thereby meeting the increasing customer demand for inexpensive customized products [13, 19]. The concept of mass customization has already proved successful in many manufacturing companies, where modular product architectures, product configuration systems, and flexible and reconfigurable production setups are some of the key enablers [17, 18]. Nevertheless, shifting strategy to mass customization implies various challenges and depends largely on three fundamental capabilities: solution space development, robust process design, and choice navigation [19]. Thus, realizing mass customization requires that companies succeed in identifying the product attributes along which customer needs diverge, are able to share resources across product offerings, and succeed in supporting the customer effectively in configuring individual products. Prominent examples of the implementation of these mass customization principles are customized laptops, shoes tailored for individual customer needs, and cars that can be configured from hundreds of different options [19]. However, even though the concept of mass customization originally was coined in the late 1980s and expertise in this area continues to grow [6, 7], mass customization is still only an emerging concept in the construction and building industry [11, 12]. A number of aspects are noteworthy in regard to this. First of all, the construction and building industry is generally characterized by highly customized building projects. Building projects such as building residences, houses, and flats represent highly customizable products, where it has been proved that their successful customization is closely related to the level of

K. Nielsen (✉) • T.D. Brunoe • K.N. Jensen • A.-L. Andersen
Department of Mechanical and Manufacturing Engineering, Aalborg University, Aalborg, Denmark
e-mail: kni@m-tech.aau.dk

J. Bellemare et al. (eds.), *Managing Complexity*, Springer Proceedings in Business and Economics, DOI 10.1007/978-3-319-29058-4_9

satisfaction of the customer [11]. Secondly, recent studies conclude that productivity improvements in the construction industry are significantly lower than improvements in other sectors. For instance, the Danish construction industry has only doubled its productivity in terms of units cost and time since 1966, whereas Danish industry in general has gone through improvements of much greater magnitude [8]. Moreover, Bashford et al. [2] argue that despite the fact that many building projects, such as home building projects, possess characteristics that are similar to manufacturing processes and that applying management principles from manufacturing processes could be beneficial in terms of avoiding increasing delivery times for customers, increasing capital cost and work in progress, construction companies still tend to regard projects as small individual construction projects. In addition to this, Nahmens and Bindroo [15] found in a large survey among US homebuilders that operational performance generally deteriorates with an increase in customization, which means that ideal mass customization has not yet been reached.

Collectively, these considerations indicate that mass customization has great potential in the construction and building industry, as it is characterized by high variety and customizable deliverables, where individualization is closely linked to customer satisfaction, while management methods and productivity improvements seem to be lacking behind.

9.1.1 Mass Customization in Construction Industry

Currently, mass customization has not been widely explored in the construction and building research area. Therefore, only limited theoretical background for the implementation of mass customization in construction is currently present. A literature search in Thomson Reuters Web of Science revealed only 12 relevant papers addressing a combination of mass customization and construction industry or building industry. Furthermore, only approximately 25 relevant papers were found on the topic, by searching for mass customization literature in the categories of architecture, civil engineering, and construction building industry. However, it should be noted that additional literature concerning modular building and prefabrication exists, which indeed is related to mass customization, however, not explicitly stating the concept in the papers, e.g., [14, 16].

From an initial review of the identified papers on mass customization in the construction industry, it is evident that the main part focuses on issues related to the capability of choice navigation, whereas robust process design and solution space development are represented to lesser extent. For instance, da Rocha et al. [4] developed a method for analyzing and improving the configuration process of customized house building, considering problems such as the burden of choices. Similarly, multiple papers deal with design of configurations systems to support customization of buildings and housings, e.g., Friedmann et al. [10] and Duarte [9]. Frutos and Borenstein [11] developed an object-oriented model for an integrated process of exchanging information between the customer and the construction company, which enables the building of mass-customized houses. Benros and Duarte [3] describe an integrated system for providing mass-customized housings, containing a design system, a building system, and a computer system, that collectively make up a system for composing customized housing design and how to produce them using computer-aided design.

Besides the issue of configuring buildings, da Rocha and Kemmer [5] propose a method for delaying differentiation in the building of apartments, which includes defining the scope of the customization, which is the customizable attributes and the standard options, which is similar to defining the solution space. Hereafter, work packages in the building project that are influenced by the customization are identified and postponed as much as possible. Barlow et al. [1] also considered the location of the customer order decoupling point in the house building and argue that mass customization could be supported by several different supply chain models, which allows delivering houses with the appropriate degrees of customization to different market segments.

With this sparse amount of literature dealing with mass customization and the construction industry, it is evident that further research should be made in terms of realizing ideal mass customization in construction and improving productivity. In this regard, all three capabilities required for successful mass customization should be explored further, where issues related to solution space development and robust process design are currently least explored.

9.1.2 Research Q

Research Question: "How does the construction industry utilize mass customization, and which major challenges arise in doing so?"

To be able to address the research question, we have the outset in well-known concept of mass customization presented by Salvador et al., where the three fundamental capabilities in mass customization were introduced [19]:

- "Solution space development: the ability to identify how customer requirements are different and develop products that can effectively adapt to these individual requirements through the product platforms or modularization."
- "Choice navigation: the ability to guide the customer to select or configure the product that matches customers requirements."
- "Robust process design: the ability to efficiently produce a large batch of products at low cost that typically is achieved by using the flexible manufacturing systems."

As illustrated in Fig. 9.1, many stakeholders/partners are involved in the processes performed by the construction industries, indirectly or directly. In the project delivering the main data for the research, we have made a delimitation of the observation group to the building industry, specifically the suppliers of materials to the construction industry.

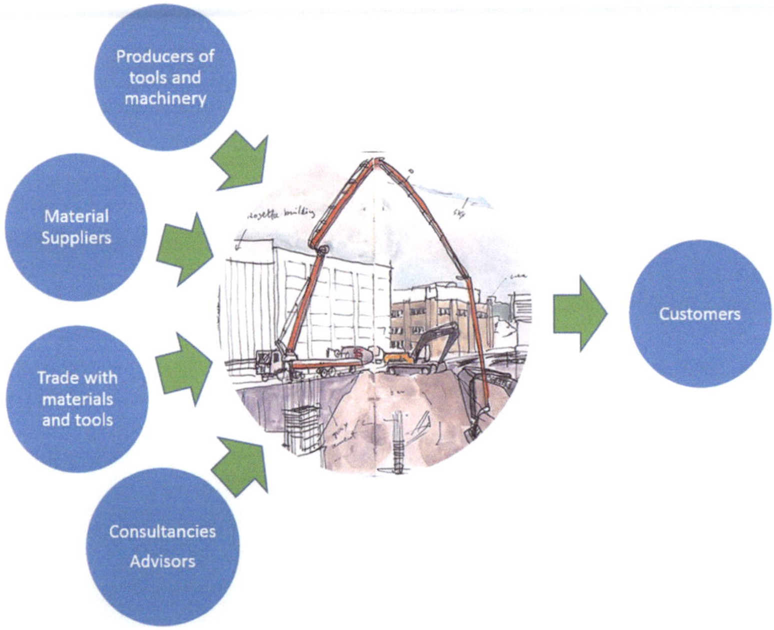

Fig. 9.1 Parties of the building and construction industry

9.2 Method

The empirical foundation for this research is a collection of observations, analyses, and results in literature reviews, studies, and cases and supported by observation in workshops. Data for the study has been collected during a period of 12 months, in a project carried out by three Danish research institutions, Aalborg University, DTU, and Alexandra Institute. The main focus in the project was introduction of mass customization for 20+ companies, all material suppliers to the construction industry. A series of workshops with presentation of mass customization and identification of potential companies for further cooperation has been performed. Each workshop had a standard agenda, with three focus areas to cover: (1) presentation of mass customization and the three fundamental capabilities [19]; (2) identification of the participating companies' understanding of mass customization including (2a) identifying whether or not they offer customizable products, (2b) how they offer customizable products (MTS, MTO, ATO, ETO), and (2c) into which degree they have organized the tasks and workload to adapt mass customization; and (3) performing a study about mass customization.

A few of these companies were selected for further workshops: firstly, with an introduction at company level of mass customization and a workshop to identify the largest potential in relation to become a (better) mass customizer in relation to construction industry and, secondly, followed two workshops with presentation of tools to improve the performance and implementation of the tools at company level. This was done as a sprint over a few months to be able to indicate the potential for improvement for the company and factual to be able to see (some) improvements—the companies have been introduced to the tools and potentials by the researchers involved, and the rest of the work has been done based on company resources.

Data for study one was collected at a workshop in phase 2 of the workshop after a short introduction to mass customization. Participating companies were asked to map themselves in relation to a coordinate system with mass production and mass produced products with indicating volume at y-axis and x-axis indication variants and closer to engineer-to-order business (see Fig. 9.2). Secondly, the companies were asked to map the future state of the company in relation to development of mass customization, in the same map. Thirdly, they were asked to map where their main competitors are and lastly map their customers' current expectations in relation to mass customization or more specifically customers' request for customized products.

Data for study 2 was collected at a company-specific workshop. After a short introduction to basic mass customization, the employees were asked to (1) map how they individually feel where the company is in a map as illustrated in Fig. 9.2 and (2) indicate which direction the company from their perspective should move to fulfill future expectations.

One of the two case companies has been part of the previously mentioned project, and the other is a case company we have used in several projects over the last 5 years.

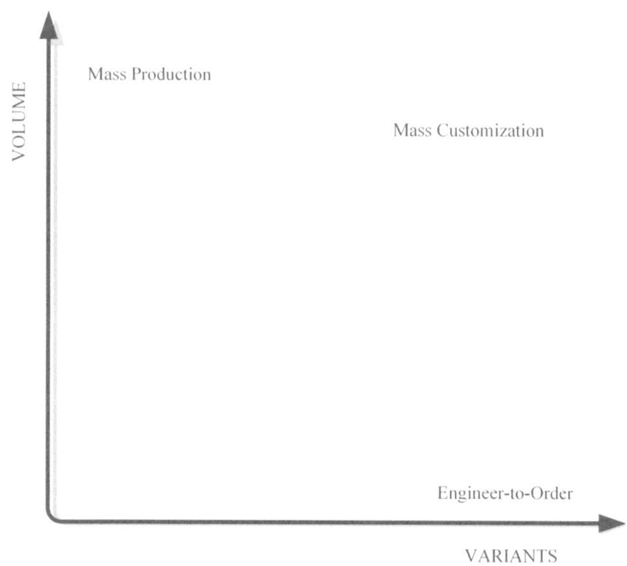

Fig. 9.2 Raw map for mass customization mapping

9.3 Results

9.3.1 Study 1

The results of study one are presented in Fig. 9.3. Based on a basic mass customization introduction, companies were asked to map themselves accordingly in a map as presented in Fig. 9.2. The current position (the box) mapping indicates that these four companies are customizers and the expectations of future mass customization state (the arrow) show an expected transition toward becoming more (better) mass customizer. The self-mapping of customers' (trolley) expectation indicates that these companies, besides one having request for more customized products than they have in the solutions space, the blue, orange, and red trolleys, are positioned on the right side of the current position of the companies. Lastly, the result of the study indicates that these companies map the competitors with less customization.

9.3.2 Study 2

This study was done in one company with six representatives from this company, representing management, sales, manufacturing, planning, and engineering. The outset was a short introduction to mass customization followed by the two questions specified in Chap. 2. The answers are illustrated in Fig. 9.4. The result of the current

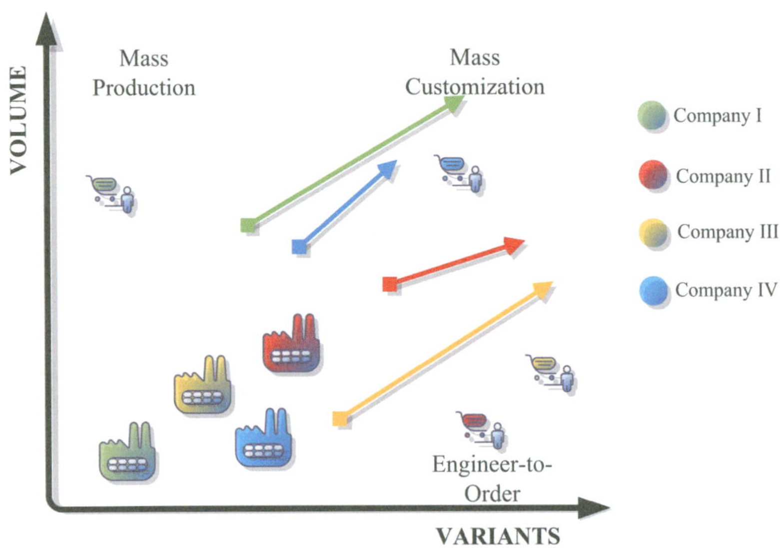

Fig. 9.3 Mass customization self-mapping of four material suppliers to construction industries (*box*=current position, *arrow*=future position, trolley=customers expectation, factory=competitors position)

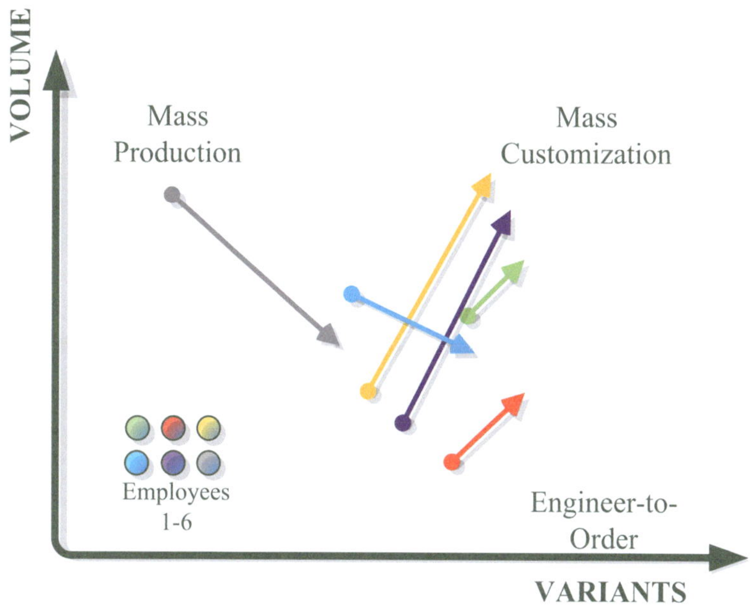

Fig. 9.4 Mass customization mapping based on six employees from the same company (*dot* = statement of current position of mass customization, *arrow* = expectation of future mass customization position)

position was except of one answer an even current mass customization position of the company. The expectation of the future mass customization state was less even, both when it comes to direction of transition and how wide.

9.3.3 Case A

Case A is a large Danish manufacturer within the doors and windows industry. The company is part of a global corporation manufacturing different components for constructing houses. Sales are done through various sales channels; however, no direct sales are made to end users. Sales are done through lumberyards, DIY retailers, and builders merchants, i.e., never directly to either contractors or homeowners. No standard products exist in the product portfolio, and every product is thus customized to fit the individual customer's requirements. The customer can choose from a range of different "models," which have a fixed structure. Each model can be customized in terms of dimensions and colors as well as different accessories like handles and locks. Requests for products that are outside the predefined solution space are rejected, as the company only wishes to produce products at their current level of automation. This means that inventories are very low; however, it also implies that meeting requirements for short lead times can be a challenge. This is particularly the case during the springtime and the summer, where the demand peaks in this particular market. This, on the other hand, also implies that if the company can do better than the competitors in reducing lead times, this can become a major competitive advantage. Company A has had a strong focus on developing robust processes, and a very large part of the operations in the manufacturing processes, which have traditionally been carried out as craft work, is now automated. This includes all wood processing, which is the major part of the manufacturing process, including cutting, planning, routing, sanding, and painting. The assembly processes have not been found profitable to automate and are thus currently carried out by hand. There is a close integration between the order system and the control system for the individual manufacturing processes, which implies that little re-typing of information is needed once the order is registered in the order system.

As described above, company A has been successful in establishing very robust processes and has had these processes for several years. Choice navigation, however, is something in which company A has had less success. Company A has had difficulties in developing a product configurator that can be used externally by, e.g., DIY retailers. One of the reasons for this is apparently conservatism in the industry, but apart from that, the retailers selling the products typically sell product from competitors also as well as other product types with high variety. This means that the retailers must know how to use and be willing to use a high number of different configurators. So far, the company has not been very successful in pushing a configurator solution to external sales channels, which means that orders are typically given by sending an e-mail with text specifications or a fax. This however means that a lot of re-typing is necessary, which is usually not value adding. Furthermore, every time a specification is re-typed, a risk of errors is introduced. This ultimately implies that the company frequently delivers a product, which is not corresponding to the customer's expectations. This, however, is a common issue in the construction industry. One other issue that company A faces is that the information path is somewhat long between the end user and company A, making it difficult to gather information about customer experience, which would usually be simple for a mass customizer using a product configurator.

Finally, company A has experienced some challenges in relation to solution space development. The company currently has no systems for monitoring which variety is being utilized and which variety is not. This may imply that items are being stocked unnecessarily, manufacturing flexibility is too high, and too many variants and options are presented to the customers.

9.3.4 Case B

Case B is a medium-sized manufacturer also within the doors and windows industry. This company's products are considered aimed at the high-end market and are aimed at a niche where certain technical certifications are necessary, which is outside the scope of most low-cost competitors. The sales channels for company B are very similar to those described for company A; however, since company B serves a niche market, company B is often in direct contact with architects and can thus influence the sales process this way. The products of company B are always completely customized, and thus, no stock of standard products is kept. Much like company A, the products of company B are customized by selecting a model, which has a predefined structure, and then dimensions, colors, and accessories are added. Contrary to company A, however, company B is not as reluctant toward selling products, which are outside the predefined solution space. Being a medium-sized manufacturer operating in a niche market, it is expected that customizations beyond the predefined solution space can be accommodated, however, at a price premium. The company states that this is one of the major reasons for the company being competitive in the market—being able to deliver almost anything although it is not in the catalogues. The company however acknowledges that they occasionally sell products, which are too far off their usual solution space, thus rendering these orders unprofitable, since they need too much manual processing and special solutions. Case B has a semiautomated production. All assembly is done manually, but some preceding processes are automated. For instance, CNC routers are used to shape holes for hinges, locks, and handles. The information needed for the automated processes is to some extent fed directly from the order system.

In terms of choice navigation, the company has not yet introduced a product configuration system, but is currently planning to do so in relation to the introduction of a new ERP system. The first step in introducing product configuration is to establish an internal product configurator, which can be used by internal sales people and sales supporters, making it easier to enter orders into the ERP system and do validity checks. There are no current plans of making a product configurator available for the direct customers. A major issue in relation to developing the product configurator is to decide which products should be included as "standard" configurable products and which should not be included, and instead be handled as "special orders." The reason for this being an issue is that it takes resources to include a product or an option in the product configurator, but subsequently the sales and order process becomes more efficient and resources are saved. However, if only a low volume of a particular product type is sold, the resources used for implementing this product type cannot be justified in saving in the business processes. Addressing this tradeoff has proven difficult for company B.

Finally, in relation to solution space development, the company has recently acknowledged that the product portfolio is characterized by little reuse of designs and parts, and little effort has been put into modularizing and standardizing. Currently, the company is going through a process of consolidating their product variety as seen from a manufacturing point of view.

9.4 Discussion and Conclusion

In introduction, the literature review indicates that there is a limited research within the field of interest. The literature indicates that it seems most focus in research so far has been on choice navigation and very little focus on solution space development and robust process design. Most literature is focused on modularity as a driver for customization and productivity.

Study 1 indicates that companies have customizable products, but their customers request products with a higher degree of customization, which may imply that these companies suffer with lower income, because engineering, planning, production, and manufacturing processes constantly are pushed to the limit and often beyond the solution space which the production is optimized for. Secondly, the study indicates that these companies seek to become (better) mass customizers.

Study 2 indicates that more knowledge about mass customization in relation to construction and building industries is needed. Within this company, the study indicates an inhomogeneous conception of mass customization which may imply lack of knowledge. A workshop series held specifically for this company in study 2 made it clear that the conception of mass customization among the employees was diverse, mainly because of the lack of common knowledge about mass customization and lack of common company strategy for implementing mass customization.

Case A shows that this company focused solely on robust process design with highly automated processes, and the company has since made the strategy about this in one of Denmark's most successful window and door manufacturers. However, they have difficulties in the development of their solution space, mainly because they did not consider that initially when they approached mass customization and secondly because they have not been able to develop choice navigation as a useful tool, which results in heavy use of resources when processing orders. This case indicates that knowledge and tools about solution space development and choice navigation have to be further developed. Furthermore, it indicates a general problem in the construction industry; manufacturers of building materials sell through a retail tier, which implies that sales people need to be able to use multiple different configurators, which is a huge barrier for the adoption of IT-supported choice navigation.

Case B shows that this company with focus on fulfilling all customers' needs as long as it is close to a standard program requires resources in order processing, engineering resources for specification, and hand-carried documentation in manufacturing, all very time-consuming and thus costly. The company does not have predefined and organized specification processes, which makes it impossible to process unique customer needs as standard products, with simple processes as in a product configurator. A workshop series carried out with the company showed that their approach to solution space development and choice navigation was somewhat ad hoc. The case also revealed that an introduction of few and simple tools to organize the specification process and product architecture has made the company able to simplify processes and reduce resource consumption.

Generally, this research indicates that there is a lack of industry-specific research for mass customization in the construction industry. Based on the different studies included in this paper, it is also clear that there is a widespread intention about utilizing mass customization in the construction industry. In order for academia to assist the construction industry in making this transition, various research gaps must be addressed. Future research must focus on identifying why construction industry is different from other industries where mass customization has been successful, for instance, in terms of structural issues, multiple sales channels, and multiple parallel vendors, traditions and specification process spanning multiple legal entities, as well as a tendency to design "one-off" solutions for every new building making standardization difficult.

References

1. Barlow, J., Childerhouse, P., Gann, D., et al.: Choice and delivery in housebuilding: lessons from Japan for UK housebuilders. Build. Res. Inf. **31**, 134–145 (2003)
2. Bashford, H., Walsh, K., Sawhney, A.: Production system loading-cycle time relationship in residential construction. J. Constr. Eng. Manag.-Asce **131**, 15–22 (2005)
3. Benros, D., Duarte, J.: An integrated system for providing mass customized housing. Autom. Constr. **18**, 310–320 (2009)
4. da Rocha, C.G., Formoso, C.T.: Configuring product variants in customisation strategies for house-building. Open House Int. **38**, 48–56 (2013)
5. da Rocha, C.G., Kemmer, S.L.: Method to implement delayed product differentiation in construction of high-rise apartment building projects. J. Constr. Eng. Manage. **139**, (2013) UNSP 05013001
6. Davis, S.M.: Future perfect. Addison-Wesley Publishing co, Reading, MA (1987)
7. Davis, S.: From future perfect: mass customizing. Plan. Rev. **17**, 16–21 (1989)
8. dst.dk: Dansk Statistik - Www.Dst.Dk. 2013 (2013)
9. Duarte, J.: A discursive grammar for customizing mass housing: the case of Siza's houses at malagueira. Autom. Constr. **14**, 265–275 (2005)
10. Friedman, A., Sprecher, A., Mohamed, B.E.: A computer-based system for mass customization of prefabricated housing. Open House Int. **38**, 20–30 (2013)
11. Frutos, J.D., Borenstein, D.: Object-oriented model for customer-building company interaction in mass customization environment. J. Constr. Eng. Manag. **129**, 302–313 (2003)
12. Frutos, J.D., Santos, E.R., Borenstein, D.: Decision support system for product configuration in mass customization environments. Concurr. Eng. **12**, 131–144 (2004)
13. Gilmore, J.H., Pine, B.J.: Markets of one: creating customer-unique value through mass customization. Harvard Business School, Boston, Mass (2000)
14. Lawson, R.M., Ogden, R.G., Bergin, R.: Application of modular construction in high-rise buildings. J. Archit. Eng. (2011)
15. Nahmens, I., Bindroo, V.: Is customization fruitful in industrialized homebuilding industry? J. Constr. Eng. Manag.-Asce **137**, 1027–1035 (2011)
16. Otreba, M., & Menzel, K.: A modular facade design approach in buildings renovations. Ework and Ebusiness in Architecture, Engineering and Construction pp. 609–616 (2012).
17. Pine, B.J.: Mass customization: the new frontier in business competition. Harvard Business School Press, Boston, Mass (1999)
18. Pine, I., Joseph, B., Victor, B.: Making mass customization work. Harv. Bus. Rev. **71**, 108–117 (1993)
19. Salvador, F., De Holan, P.M., Piller, F.: Cracking the code of mass customization. MIT Sloan Manag. Rev. **50**, 71–78 (2009)

Chapter 10
Challenges in Choice Navigation for SMEs

Kjeld Nielsen, Thomas D. Brunoe, Lars Skjelstad, and Maria Thomassen

10.1 Introduction

Since mass customization was introduced by Davis [1] and later popularized and described more in detail by Pine in 1993 [6], it has been introduced and implemented in a large number of industries in different countries.

Much of the literature found on mass customization focuses on cases of large enterprises. The majority of cases which are presented in research and which are highlighted in literature as examples of successful mass customizers are usually from manufacturers oriented toward consumer markets such as consumer electronics or automobiles, which are typically dominated by very large players.

However, although mass customization was originally described as a way for mass producers to increase variety to increase sales and prices, other types of companies have also utilized the principles of mass customization. This includes cases of engineer-to-order companies, which have originally had a very high variety, but can utilize the principles of mass customization to become more efficient, by applying, e.g., modularization, product configuration, etc. This tendency has become more common as mass customization has gained recognition in many different industries and as the methods and technology for implementing mass customization have become more mature and more broadly known. This has also implied that not only large companies pursue mass customization, but more small and medium enterprises (SMEs) are also beginning to implement mass customization, and even startups are founded where the initial idea is providing mass-customized products.

K. Nielsen (✉) • T.D. Brunoe
Department of Mechanical and Manufacturing Engineering,
Aalborg University, Aalborg, Denmark
e-mail: kni@m-tech.aau.dk

L. Skjelstad • M. Thomassen
SINTEF Industrial Management, Trondheim, Norway

J. Bellemare et al. (eds.), *Managing Complexity*, Springer Proceedings
in Business and Economics, DOI 10.1007/978-3-319-29058-4_10

However, referring to previous literature reviews on general mass customization, very little literature has been published specifically on mass customization in SMEs [2, 8]. A literature search performed on Thomson Reuters Web of Science and Elsevier Scopus revealed only 39 contributions directly or indirectly focused on mass customization in SMEs, whereas almost 3000 contributions could be identified for mass customization in general. Hence, a research focus on mass customization in SMEs is almost absent. However, it is reported by multiple contributions that mass customization holds a great potential for SMEs in terms of growth and profitability [3, 9].

From the literature search, only a few case studies were found which document the advantages and challenges when SMEs implement mass customization. One case study focused narrowly on how automation of welding processes could be utilized in a metal furniture manufacturer [4]. By Ismail et al. [3], two cases are presented, involving companies manufacturing children's playground equipment and luxury domestic showers illustrating how tools from mass customization can benefit SMEs. Finally by Orsila and Aho [5], a case study was presented of how an e-commerce system could improve business processes in a manufacturer of custom semiconductor products. The authors however believe that it would be beneficial for both researchers and practitioners to understand the mechanisms of mass customization in SMEs leading to benefits and challenges, which is the focus of this paper.

Salvador et al. [7] introduced the three fundamental capabilities of mass customization: (1) solution space development, "identify the product attributes along which customer needs diverge"; (2) robust process design, "reuse or recombine existing organizational and value-chain resources to fulfill a stream of differentiated customers needs"; and (3) choice navigation, "support customers in identifying their own solutions while minimizing complexity and the burden of choice." These capabilities are generic and must thus be possessed by large enterprises as well as SMEs. However, as it has long been recognized that SMEs are fundamentally different in terms of strategy, operation, etc. [11], it is expected that the approach to implement these capabilities will differ between large enterprises and SMEs.

In this paper, the primary focus will be on choice navigation and how SMEs can utilize choice navigation to improve business. The research question of the paper is:

What are the benefits which SMEs can gain from utilizing choice navigation, and which challenges may occur in doing so?

In this study, we apply the definition of SMEs from the European Union, i.e., maximum of 250 employees combined with a limit of maximum EUR 50 million annual turnover or a maximum of EUR 43 million annual balance sheet.

10.1.1 Forms of Choice Navigation

Choice navigation is basically the capability to help customers define which product matches their requirements and preferences and help them purchase this product. Choice navigation thus involves presenting the customers with which options they can choose from, allowing them to review combinations of these options and obtain information about the products in terms of different properties such as performance, appearance, price, delivery times, etc. What is commonly associated with choice navigation is the type of software, which is called product configurators. A product configurator is a software tool, which presents the user, who may be a customer or a sales person, with different options to choose from, and the user can by selecting different options configure a product. Many different types of product configurators exist; some are intended for back-end configuration and thus focus primarily on obtaining all the information needed for creating a quotation or providing information of the subsequent manufacturing, while front-end configurators are intended for usage by the end customers, and thus, the focus is more on communicating the different options to choose from and also frequently to visualize the appearance of the products.

However, choice navigation covers more than just product configurators. Another approach is what is called "assortment matching" or "product selectors." This concept differs from product configurators, since product selectors aid the customer in choosing from predefined variants. Product configurators on the other hand allow users to choose from predefined options within a certain solution space, and it is thus not necessary to choose from a range of predefined variants. Assortment matching or product selectors are also software tools, which can be used by either end customers or sales people.

An even simpler form of choice navigation however also exists. Arguably, the simplest form of choice navigation is simply presenting customers or sales people with the products and options that can be chosen from using lists or catalogues. This may be software supported but may also be paper based. This form of choice navigation has no constraints on whether products are predefined or configurable. While this approach "supports customers in identifying their own solutions," which is part of the choice navigation capability, it does not to the same extent as product configurators and product selectors "minimize complexity and burden of choice" which is also part of the choice navigation capability.

Another form of choice navigation, which is presented by Salvador et al. [7], is "embedded configuration." This is defined as "Products that understand how they should adapt to the customer and then reconfigure themselves accordingly" [7]. In this case, the user does not do anything explicitly to configure the product. Examples of this include cars that change the driving dynamics according to different situations, e.g., driving on the freeway vs. parking, or a cell phone, which detects a low battery level and reconfigures to minimize power consumption and maximize remaining battery life. In this type of choice navigation, there is no distinction between front-end and back-end configuration, as no users do explicit configuration.

10.2 Methodology

To answer the research question, a multi-case study was selected. The case study approach was selected to be able to do an explorative investigation of choice navigation in industry. The choice of a multi-case study was made to increase generalizability, which can, in some cases, be a challenge for single-case studies.

The cases, which were included in the study, were chosen among participants in three different projects focusing on development of mass customization capabilities in industry, which also included SMEs. The projects were run by SINTEF in Norway and Aalborg University in Denmark (AAU). The criteria for selecting case companies were that they must (1) be SMEs, (2) do mass customization, and (3) have some experience with choice navigation. Company names are not mentioned to ensure anonymity.

Throughout the projects, SINTEF and AAU have been in close collaboration with the case companies and have significant insight in their approaches to the three mass customization capabilities including choice navigation. The case studies are based partly on this insight and partly on interviews conducted specifically for this research.

Based on this knowledge, each case is described below, where the characteristics of each company are outlined; the form of choice navigation applied is described together with benefits gained from this as well as challenges.

10.3 Case Studies

10.3.1 Case 1

Case company 1 is a manufacturer of high-end building components used for private homes and public buildings such as museums, offices, etc. The products are sold for new building projects as well as for renovations of existing buildings. The core of the business is to manufacture highly customized products of very high quality, and the company does thus not compete on the low-end market with high price sensitivity. The company serves different types of customers; however, products are never sold directly to the end user but rather through a retailer. Usually the sales process will go through an architect or a contractor; however in rare cases, end users will have direct contact to the case company. The company has approximately 30 employees.

Addressing their approach to choice navigation, the case company currently does not apply product configurators or product selectors; however, a list of product types and options (although not complete) is available to end customers, which to some extent helps customers in choosing the right product. Internally, choice navigation is handled by using Spreadsheet templates, which state exactly which information is needed before a product can be produced. These spreadsheets however do not have any data validation ensuring that the options selected or the dimensions specified are within what can actually be manufactured. The benefits however are that no product is sold, without sufficient product information that is present for the manufacturing processes.

The company has a wish to establish a product configurator, which they expect customers to be able to use. This will enable customers to configure products themselves, potentially increasing revenue and reducing the load on internal sales people, sales supporters, and the technical department. The main challenges in doing this however are however partly to chose an appropriate product configurator software and partly to define the variety to be offered in the configurator. The reason for the latter being a challenge is that the company wishes to offer a range of configurable "standard" products accounting for the majority of the revenue, but still keep the door open for "special" products which are outside the configurable scope, but attractive to sell due to much higher markups. The distinction between "standard" and "special" products has proven very difficult for this company.

10.3.2 Case 2

Case company 2 manufactures simple components used for equipment and machinery installed in buildings, primarily used in the manufacturing industry. The products are assembled from a number of very simple components. The products are included as parts of products delivered by the customers of the case company. A sales organization of agents and specialized B2B sub-suppliers are in charge of selling the company's products.

With the use of catalogs and option lists and supported by external sales representatives, product configurations are made to match customer's equipment and machinery; the configured products will have a unique part number. In regard to planning, scheduling, and supply chain, the part number has no references internally, which in busy periods often cause huge delays. Engineering department will process new customer part numbers (new assembly/new configuration) before the order can be processed by manufacturing. Scheduling (estimating) a delivery time of each new or recurring order is based on manually counting of availability of individual parts included in the customer part number.

The company has decided that future development of new products or options should be applicable to an online product configurator, making the customers able to select their product options themselves, fulfilling their specific needs, and making the case company able to decompose the part number into individual parts, for scheduling, planning, and supply chain purposes. An analysis has shown that the existing product lines and product families' product architecture cannot support these requirements. Developing an initial product family (a forerunner) and introducing the online configurator indicates large potential; not only a major reduction in involvement of engineering department in all orders but also the potential of decomposing the customer product into individual parts for scheduling, planning, and supply chain purposes has been very promising. One major challenge for the case company is to convince customers to change their unique product to a new product—same product from a functionally point of view, but with slight physical changes and with a new part number.

10.3.3 Case 3

Company 3 is a manufacturer of heavy high-quality machinery equipment that facilitates bulk material handling and transport for the agricultural, industrial, and waste sectors. The company has built up a strong market position with focus on high-quality and customer service, despite a price-focused market. Their products are offered in different models, which are further adjusted to specific market requirements. Products may also be customized to the needs of individual customers. The company is represented on all continents via a network of distributors and agents. Customers are typically companies in the agricultural sector that sell and distribute the company's products to individual farmers. However, end users have direct contact with the company's service support, and end customers may also buy the products directly from the company. The company has 75 employees.

The company offers about 30 customer choices that are available via a list of product models and options. A salesperson is often involved in the product selection process helping the customer to specify the needed product features options and individually customized features. The production process typically starts before the final order is confirmed. The company purchases and produces parts and subassemblies based on forecast, while products are assembled based upon customer orders. Products mainly consist of standard parts, and the majority of options are introduced in the final assembly process.

The company has recently introduced a new product range, with low production volumes in an initial phase. As product volumes increase, the company is aware of the need for a product configurator to support the sales process. Today, most sales are carried out via distributors. A configurator may support the expansion of the internal sales organization of the company, with more sales directly to end customers.

Most challenges are related to complexity, competence, and resources. The company has limited resources to invest in a new complex and often expensive software and implementation project. Necessary internal IT competence and resources to maintain and use the solution must be in place. Since the sales process is mainly carried out by external partners, training is needed to utilize the tool. In addition, distributors may have to deal with a wide range of different product configurators depending on the number of suppliers. With regard to end users, a configurator might be challenging to use, for instance, by farmers with limited IT familiarity. The company also sells its products to customers all over the world, so the configurator must be adapted to users in a wide array of geographical markets, risking to increase complexity further.

10.3.4 Case 4

Case company 4 has approx. 25 employees and manufactures and sells kitchen equipment. Mostly their own products and designs but also imported brands from foreign manufacturers are kept on stock and sold in Norway. Customers are primarily Norwegian kitchen manufacturers, but also private persons that renovate their homes or cabins can order new products directly. Increasingly, entrepreneurs offer customers to specify this kind of equipment as part of the contract when signing for a new house or flat.

First, customers can choose between many models, and on all domestic manufactured types, customers can specify dimensions and colors to fit their needs optimally. To some extent, the functionality can be customized too. For custom products, the company charges a premium price, exploiting the business potential in mass customization.

Sales are primarily done through contracts with kitchen manufacturers. Buying a new kitchen is a big investment, and customers often go into discussions with more than one supplier to get different prospects and offers. In these design and decision processes, also case company 2's products are discussed and decided upon as an integrated part of the kitchen. For this purpose, the case company has developed brochures and a website as their front-end solution, to be used both by the kitchen sales personnel in meetings and end customers at home.

When orders are placed, they are sent to the case company on either mail or fax. The company then registers orders manually into their business system before printing necessary manufacturing documents. On these documents, customer-specific solutions are written as text into designated fields. For several reasons, the company wishes to implement an electronic configurator solution. As they say, "…this would save us resources, reduce risk of human errors, and speed up processes." The market for customized products, and hence the volume of communication with customers, is increasing, and the company is starting to look into configurators suitable to them.

The main reason for not having an electronic configurator in place already is that technology costs and that they lack knowledge in this field of software. Also, to implement a front- and back-end solution is a time- and resource-consuming process that has so far been put on hold. There is probably an optimal time for changing over to using a configurator, somewhere along the increasing demand for these products among customers. In the beginning, the volume of specific orders is manageable manually, but at one time the amount of unique orders calls for an automated system. It might be that the company has over passed this time now.

10.3.5 Case 5

Case 5 is a winter sports equipment manufacturer. With its approx. 80 employees, the company is a major player within winter sports globally, being the second largest manufacturer of these high-quality products in the world. Sales are through sport shops, but the company considers end users their customers too.

It is of great importance that every customer gets a product that is suited to his/her body measures and qualifications to be able to perform on top level. At the same time, for capacity and employee reasons, products are made year round, and not only in the winter season when the products are sold and used. The company therefore has to manufacture products during summer months without knowing the exact need of every customer. Hence, the strategy is to actively differentiate product characteristics continuously within defined limits during summer production, in order to have a range of slightly different products ready for the high season. This way there will most likely be a suitable product for you available in the shop. In winter, products can be tailor-made to some extent for single customers in parallel with ordinary deliveries to retailers. Direct customer orders can be fulfilled in a week or two, which is sufficient for dedicated athletes. They are normally well prepared and plan for achieving new products way ahead.

The company keeps track of all products unique characteristics using embedded RFID tags. The products then are matched to each single customer during the new sales process in the shops. To solve this, an online tablet connected to the product database is used to enter customer-specific data and find a perfect match every time.

The system is best described as a matchmaking system, or a front-end system. But, sales data are recorded and can be used to replace products at the retailers, forming a sort of back-end functionality.

Previously products were sold by more or less skilled sales personnel at retailers and sport shops. This had some disadvantages: first, sales took much longer time per customer, because several products had to be fetched, measured, and tested with the customer. Now the optimal product is identified electronically based on customer input and just collected from the shelf. Second, the accuracy was lower leading to some sales ending with customers getting the wrong product. This of course leads to disappointed customers, complaints, and possibly lost future sales. Another advantage with the new system is the mentioned sales-data back-loop to the manufacturer.

The new system required a lot of resources and time to develop, but has ended as the most "high-tech" solution in the market and hence contributes to branding among young people.

10.3.6 Case 6

Case 6 is small-sized workwear company, with approximately 12 employees. The company setup is similar to the majority of textile companies, organized with design department, operations department, and sales department, and with major manufacturing in Far East and minor manufacturing in Poland. The case company sells its workwear products primarily to the service sector both private and public customers, half of it as direct sales supported by own sales force and the other half sold through service providers to the service industry, all B2B sales. Most direct sales are based on a standard product catalogue, with minor customization as logo prints or embroidery, and sales to service providers are often customized standard products with minor design changes as well as prints and embroidery. The case company does the prints, and locally based sup-suppliers do the embroidery for direct sales standard products mainly, because of the volume in small-medium size, whereas sales through service providers often are significantly higher volume which makes it possible to have all products fully customized by the Far East manufactures. The case company had some year ago designed a product family specifically for medium-sized customers in a specific sector in the service market, where the level of customization was raised significantly. The company developed a product that was customizable both in design, in combination of parts, and pattern of textile and additional design features as pockets, piping, zippers, and buttons. The business strategy was aiming for a market which the case company has difficulties to reach, mainly because the volume was low and customization was high and the sensitivity on price was low. The business setup of this customizable product family was direct sales supported by a selector—originally it was the idea to have a tablet product configurator, but due to cost, they decided to make a manual selector, presenting the solutions space with a selector tool like you choose the color of paint or children playing with cutout dolls. Further, because the expectation was low volume, they made a supply chain and operations setup with Polish sup-suppliers, with direct delivery to end customers.

The case company closed the product family after a year, mainly because they did not have orders; minor reason was difficulties in the operations setup.

The case company has based on its experiences chosen not to open for further choice navigation direct to customers, neither with selectors nor product configurators in a similar setup. They have decided that product configuration as an internal tool can be valuable in several relations, and there is a potential to reduce design cost and operation cost with the use of product configurator. Based on this, they are now considering how they will organize the future design approach and are considering how a product architecture approach can assist such strategy.

10.4 Results

Comparing all of the six cases included in this study, all of them are mass customizers and all of them apply choice navigation in some form. One company applied assortment matching, which has been very successful, in reducing load on sales organization, branding the company and ensuring a better fit between customer needs and the sold product. The remaining five have considered implementing a product configurator. The reasons for wanting a product configuration system are quite similar across the five cases and include:

- Increase sales by allowing direct sales to customers rather than through traditional sales channels
- Reduce resource load on sales organization by "outsourcing" the choice navigation process to customers
- Reduce load on the technical department by automating the sales delivery process using information from the configurator

- Support expansion in sales volume, sales organization, and geographical sales area
- Reduce the risk of human errors by validating input and by eliminating manual type in of orders
- Reduce design cost

It should be noted that these reasons are expected benefits of implementing a product configuration systems, and this case study thus cannot report if these expected benefits are realized. However, the expected benefits are very similar to what has been reported from larger companies and are thus considered feasible to achieve.

Although great benefits may be expected from implementing a more sophisticated system for choice navigation, such as product configurators, significant challenges can be expected in relation to this. As with the expected benefits, the challenges reported are very similar across the case companies. The challenges highlighted the most by the case companies are as follows:

- Selecting the right software for the specific application.
- Defining the solution space to be introduced in the choice navigation system.
- Resistance toward or lack of skills from customers in relation to using the product configuration system.
- Large software and implementation investment is a barrier especially in SMEs.
- Lack of internal IT competences to implement systems.
- Adaption to different geographical markets.

As with the advantages highlighted above, the challenges related to choice navigation are mostly expected challenges, as they relate primarily to, e.g., product configuration systems, which are not implemented yet. However, these expected challenges can be interpreted as barriers toward implementing more sophisticated systems for choice navigation.

10.5 Conclusion

The objective of this paper was to investigate the benefits and challenges related to choice navigation in small and medium enterprises. This was addressed by performing multiple case studies in six different SMEs in Norway and Denmark. It was found that all of the SMEs were applying choice navigation, although most of them in a very simple form, using catalogues, list of product families and options, and manual order entry forms. One SME had great success with implementing an assortment matching system, whereas a second system failed in implementing an "analogue configurator." The expected benefits of implementing a product configurator, which five of the six companies were considering, were quite similar across the cases, related primarily to saving resources, increasing sales through direct sales, and reducing errors. The challenges and expected challenges if implementing product configuration systems almost all related to investment, resources, and skills and to some extent also resistance toward adopting product configuration systems and configured products.

When interpreting the results, it seems that many SMEs have potential in utilizing more advanced forms of choice navigation, such as product configuration, to increase sales and reduce resources per sold product. However, many SMEs are reluctant to implement these systems for various reasons, which may limit their growth potential. Reasons why the SMEs are reluctant may be related to the fact that an investment in such systems is rather large compared to their profits and thus involves some risk. Furthermore, the resources needed to run an implementation project may be significant compared to the total available resources in the organizations, limiting the free resources for, e.g., product development projects. These risks are less significant in large enterprises, which is why implementation in SMEs must be handled differently than in large enterprises. As reported above, very little literature exists on choice navigation in SMEs, and it is thus recommended by the authors that more research be done within this specific area.

Acknowledgement We are grateful for the support of the Research Council of Norway to this research.

References

1. Davis, S.: From Future Perfect: Mass Customizing. Planning Review (1989)
2. Fogliatto, F.S., da Silveira, G.J.C., Borenstein, D.: The mass customization decade: an updated review of the literature. Int. J. Prod. Econ. **138**(1), 14–25 (2012)
3. Ismail, H., Reid, I., Mooney, J., et al.: How small and medium enterprises effectively participate in the mass customization game. IEEE Trans. Eng. Manage. **54**, 86–97 (2007)
4. Kokla, M.: Case study regarding effectiveness and profitability of automating the welding process. Ann. DAAAM Proc. (2011)
5. Pine, B.J.: Mass Customization: The New Frontier in Business Competition. Harvard Business School Press, Boston (1993)
6. Salvador, F., De Holan, P.M., Piller, F.: Cracking the code of mass customization. MIT Sloan Manage. Rev. **50**, 71–78 (2009)
7. Silveira, G.D., Borenstein, D., Fogliatto, F.S.: Mass customization: literature review and research directions. Int. J. Prod. Econ. **72**, 1–13 (2001)
8. Svensson, C., Barfod, A.: Limits and opportunities in mass customization for "build to order" SMEs. Comput. Ind. **49**, 77–89 (2002)
9. Welsh, J.A., White, J.F.: A small business is not a little big business. Harv. Bus. Rev. **59**, 18 (1981)

Chapter 11
Machine-Part Formation Enabling Reconfigurable Manufacturing Systems Configuration Design: Line Balancing Problem for Low Volume and High Variety

Mads Bejlegaard, Thomas Ditlev Brunoe, Kjeld Nielsen, and Jacob Bossen

11.1 Introduction

To cope with the challenges of high variety, Group Technology and the principles of a Reconfigurable Manufacturing Systems (RMSs) can be applied. Group Technology and Reconfigurable Manufacturing Systems deal with the challenges of multivariant, small-lot-sized production and gives the manufacturers the possibility to configure line-oriented layouts if commonalities in both products and machines are utilized.

Group Technology has been broadly applied with the purpose of increasing the production efficiency by grouping products/parts with similar design and/or manufacturing characteristics [11]. Cellular manufacturing is mentioned as a derivative of Group Technology [10], and in the 1970s, cellular manufacturing became a common element of just-in-time production. The manufacturing industry has evolved through several paradigms, but GT has been applied since the 1970s [11] with motivation from lean manufacturing and later mass customization. Actually, cellular manufacturing has been recognized as the second generation of Group Technology, going from part-family formation based on geometric shape without changing the physical layout to producing batches of large variety and physically changing machine layout based on both machine and part families [3]. Thus, Group Technology can be considered to effective formation of part families and rational layout of the manufacturing cell/line. For this reason, Group Technology is mentioned as the key in the successful implementation of Flexible Manufacturing Systems (FMSs) enabling mass customization [12]. Moreover, GT would also be a suitable technique for implementing Reconfigurable Manufacturing Systems, as the system is based on customized flexibility, which means that systems are designed

M. Bejlegaard (✉) • T.D. Brunoe • K. Nielsen • J. Bossen
Department of Mechanical and Manufacturing Engineering, Aalborg University, Aalborg, Denmark
e-mail: bejlegaard@m-tech.aau.dk

J. Bellemare et al. (eds.), *Managing Complexity*, Springer Proceedings in Business and Economics, DOI 10.1007/978-3-319-29058-4_11

for a particular part or product family. Reconfigurable Manufacturing Systems is seen as the new manufacturing system paradigm coping with exactly the challenges of high variety by focusing on families of parts as one important characteristic [7]. At the same time, Reconfigurable Manufacturing Systems is designed for rapid change in capacity and functionality which copes with the line balancing problem of variety.

The outcome of Group Technology can be rather promising, and in this perspective, Kusiak [8] presented the following list of advantages when implementing cellular manufacturing. Later [5] and [9] did familiar conclusions observing the advantages of cellular manufacturing.

- Reduced production lead time (20–88 %)
- Reduced work in process (up to 88 %)
- Reduced labor (15–25 %)
- Reduced tooling (20–30 %)
- Reduced setup time (20–60 %)

The machine-part formation is the first problem faced when implementing Group Technology, which concerns the formation of part families and the identification of machines on which these parts are to be processed [3]. Then decisions need to be taken in relation to allocation of resources on each machine, layout decisions, and balancing of the capacity [3]. In order to manufacture parts or products in significant varieties on the same line, approaches considering variety will be necessary to apply. The most important step for making a Reconfigurable Manufacturing Systems is considered to be the possibility of manufacturing variants in the same system [1]. Thus, if it is possible to get the advantages of producing part families with common characteristics, this will significantly simplify the line balancing problem.

Numerous contributions have been done within the field of Group Technology including machine-part formation [12] testing the efficiency of clustering techniques [3]. However, examples of bringing these results to the next step of configuring a manufacturing system layout to assess line balancing problems based on part-family formations have to our knowledge not been carried out in one step.

The objective of this paper is to make a method that includes both identification of machine-part formations and thus deduce a part-family with the purpose of exemplifying the configuration of an RMS to assess the line balancing problem of high variety.

11.2 Methodology

In order to address the research objective stated above, a case study has been conducted to verify the applicability of machine-part formation in multivariant, small-lot-sized production. Group Technology is first applied in relation to machine-part formation, and second layout choices for variety are outlined for the purpose of layout configuration and assessing the line-balancing problem for high variety.

11.2.1 Machine-Part Formation

The used case study data is partly related to routing information extracted from an ERP system in a large Danish manufacturer of industrial equipment. The extracted data contains all own-produced parts of current products, but in order to avoid unnecessary disturbances from rarely produced products, spare parts for previous product versions are not considered.

The present dataset consists of a binary machine-part incidence matrix representing relationships between the 1989 parts and 36 work centers, describing each part routing. Due to the relatively large number of parts, a manual approach is not applicable for clustering. Furthermore, it is infeasible to have an algorithm testing all possible part families associated with all possible machine cells, and hence the different cell formation procedures are based on heuristics and near-optimal solution approaches [3]. For this reason, *k*-means is applied to cluster groups of machine-part relations and is carried out using the statistical and graphical language R. The "Amap Package," containing standard hierarchical clustering and k-means is applied to explain the relation between the number of clusters and sum of squared error (SSE). This is done to be able to determine the needed number of clusters that sets the starting point of the analysis. Due to the relatively large amount of data, *k*-means is applied because the results then are summarized in compressed overviews although the number of parts is high. To further analyze the results of the computed clusters, it is necessary to acquire knowledge of included processes (i.e., classifying machine functions) such that groups of parts undergoing the same process technology but assigned to different machines undergo the exact same routing. Eventually satisfying formations are formed and outliers made visible, and thus, the design criteria implicitly appear in order to adapt to the established routings.

11.2.2 System Configuration Design

In assembly system configuration, two main aspects exist with regard to decisions on layout choice for variety: (1) decisions at a physical level (i.e., arrangement of stations/machines) and (2) at a logical level (assignment of operations to stations joined to a resource planning activity) [2].

These main aspects are the foundation of the method applied. The physical layout choice for variety is based on the layout configuration code specified in Table 11.1. The physical layout choice depends on the composition of the part families identified during the machine-part formation.

Then the logical layout choice for variety is defined, considering the different kinds of line balancing problems. These balancing problems are listed in Table 11.2.

The next step during configuration design is line balancing which entails to optimally allocate processes to stations which is directly linked to the homogeneity of the products on the production line. In [1] the possibility of manufacturing, different

Table 11.1 Layout configuration characteristics [6]

Layout characteristics	Attributes
Shape	Line, U shaped, loop, network
Segment flow direction	Unidirection, bidirection
Segment flow control	Synchronous, asynchronous
Flow control junction	One to one, one to many or many to one

Table 11.2 Classification of line balancing problems [4]

Kind of product line	Characteristics
Mixed model line	Several models are manufactured on the same production system
Multi-model line	Batches for specified products are launched with setup between
Single product line	Possibly also a multi-product line with products almost identical

variants in the same system are considered as the most important step for making an RMS, which aims to respond to market changes cost-effectively. Additionally, commonality between parts causes substantial simplification of the line balancing problem, why this is a focal point producing multiple parts at the same line.

11.3 Results

As mentioned above, it is infeasible to have an algorithm test all possible part families, which is why the cell formation took outset in a heuristics and near-optimal solution approach defined by [3]. As illustrated in Fig. 11.1, 22 clusters could expectedly explain approximately 80 % of the machine-part relations. This was used as input for the algorithm as an appropriate number of clusters to speed up the process and hereby avoid testing all possible solutions. Executing the algorithm resulted in 22 clusters which could be further combined into 13 clusters due to common functionality across machines.

One of the final 13 clusters is shown in Table 11.3 which and is derived from 3 of the initial 22 clusters. Regarding this cluster, the derivation was possible due to the shared functionality represented in different processes. It increased the volume of the part group but increased though also the complexity. This complexity is caused by the increased number of geometric shapes, which the manufacturing equipment must be able to handle to avoid the costly changeovers. In order to achieve commonality in interfaces, these 95 different parts were divided into four subgroups based on their geometric commonalities.

Table 11.3 shows how strong the relations between the included machines and parts are. This indicates which machines should be the starting point of future routings. Almost every part undergoes the first three processes in the future routing of

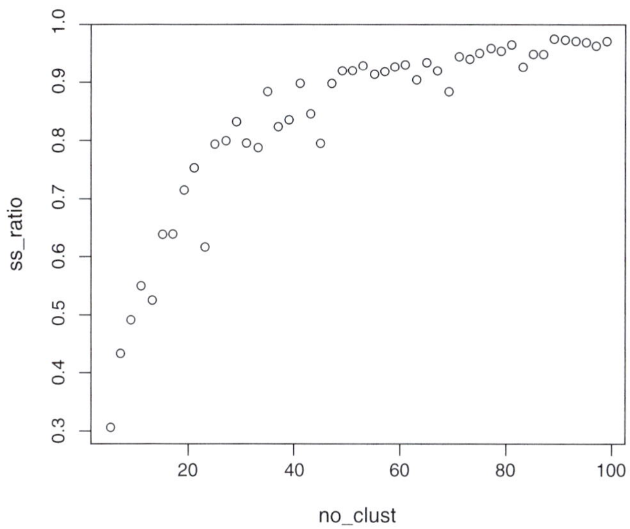

Fig. 11.1 Relations between the number of clusters and the number of machine-part relations explained

Table 11.3 One of 13 clusters derived

Machine	1201	1202	1203	1208	1209	1210	1211	1303	1304	1306	1313
SSE	0.05	0.93	0.96	0.14	0.5	0.31	0.02	0.05	0.04	0.28	0.36
Quantity	5	88	91	14	47	30	2	5	4	27	34
%	5	94	97	15	50	32	2	5	4	29	36

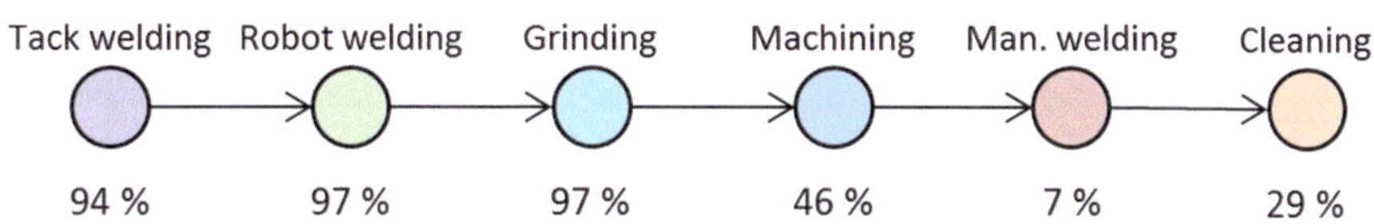

Fig. 11.2 Future routing for all included parts

parts included in the concerned cluster, as shown in Fig. 11.2. This changing of parts routing has been done in accordance with the included machines' different operating cost, but even with higher operating costs, changing routing can be the best solution due to the value of Group Technology.

The number of machines needed in the future line after changing parts routing is then calculated. The reliability of the manufacturing equipment is assumed to be 100 %. This gave a result of 7.9 machines which is rounded to the next larger integer. Calculation of the minimum number of machines needed in the line, N, is calculated according to the following equation (11.1) adopted from [7]:

Table 11.4 Division of machines based on time consumption in each functional stage

Function	Tack weld.	Robot weld.	Grinding	Machining	Man. weld.	Cleaning
% time	25	13	49	12	1	–
Machines	1975~2	1027~1	3.92~4	0.96~1	0.08~0	–

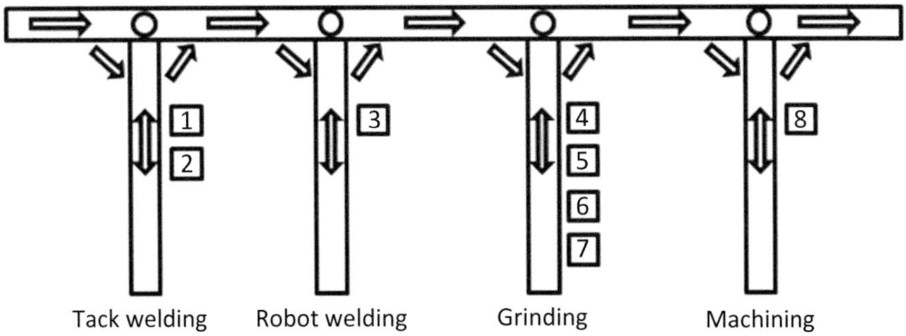

Fig. 11.3 RMS configuration with crossover connections after each stage [7]

$$N = \frac{\text{Daily demand} \times \text{min/part}}{\text{Min/day} \times \text{Machine reliabilty}} \tag{11.1}$$

These eight machines are then divided, respectively, in the functional stages according to the average consumption of process time needed in each stage. This is however only possible because the relative ratio (shown in Table 11.4) between the parts processing time on each stage is almost equal. One of the functional stages must be excluded because it was not possible to relocate.

In order to avoid queuing, the process sequences are divided corresponding to the four subgroups since parts included in each of these subgroups currently almost require the same processing time in each functional stage.

To be able to respond to changing capacity need and need for changing functionality in the single functional stages where parts can be transferred to any machine in any functional stage, the RMS configuration is used. "Mixing different types of machines that perform exactly the same sequence of tasks in the same manufacturing stage is absolutely impractical" [7]. Instead these are divided into functional stages with only identical machines in each stage to support the idea of a modular manufacturing system (Fig. 11.3).

In contrast to RMS configurations, cell configurations impose limitations since all stages of the cell configuration must be equal to be balanced. This is however not the case in order to achieve a balanced RMS configuration where only the relation between the number of machines (N_{si}) and processing time per machine (t_{si}) in each stage (i) needs to be satisfied. Equation (11.2) explains the relation between machines and processing time to get a balanced Reconfigurable Manufacturing System configuration, adopted from [7]:

$$t_{S1}/N_{s1} = t_{S2}/N_{s2} = t_{Si}/N_{si} \tag{11.2}$$

11.4 Discussion

Even though most of the machine-part relation could be explained in 22 clusters, a considerable amount of outliers was identified during the clustering. Thus, it will take a tremendous effort to turn the current manufacturing system into an RMS based on 13 balanced lines. The applied example outlines the importance of the commonalities between parts since this will simplify the line balancing problem. This leads to the issue of defining the interface between the high variety of products and the manufacturing equipment such as fixtures. In order to accomplish commonality in interfaces, the 95 different parts were divided into four subgroups based on their geometric commonalities. As vital as commonalities in products is, defining the right functionality in each functional stage of a line turns out to be important likewise. This allows manufacturers to accomplish homogeneous platforms which set the frame of new product introduction in a line-oriented environment with high variance and thus acquire economy of scale besides the benefits of reuse and reusability in a larger perspective.

Forthcoming actions creating common production platforms based on part commonalities identified have a major influence on the line balancing problem. The more different the equipment, the more changeovers and thus the complexity increase. As manufacturer of high variance in low volumes, many different parts will be manufactured on the same equipment to achieve a proper utilization of the capacity. Of this reason the functionality on lines in some cases should be comprehensive. This requires a continuous discipline of Integrated Product Development. In other words, a technique for defining platforms that includes enough functionality to accommodate requirements both now and in the future is essential to the line balancing problem for high variance.

Another important aspect is the processing time. It becomes clear that dividing parts into subgroups in the light of commonalities does not on its own solve the line balancing problem; it is just smoothing the flow. In order to further simplify the balancing problem and simplify the planning of process sequence, adjusting the processing time by compromising on product features could be an approach, but that might be too great a compromise.

11.5 Conclusion

The objective of this paper is to make a method that includes both identification of machine-part formations and thus deduce a part-family with the purpose of exemplifying the configuration of an RMS to assess the line balancing problem of high variety.

Through clustering techniques, and product and process knowledge, it was possible to form 13 clusters of machine-part formations. Of these, one part-family was selected to be the starting point for configuration design of a Reconfigurable Manufacturing System. The purpose was to identify the line balancing problems of high variety in such environments. The main findings were the vitality of commonality across both products/parts and the manufacturing equipment.

In fact a modularization of the manufacturing system formed in accordance with product and part families appears to be the obstacle of getting the full advantage of grouping parts. If commonalities between parts and machines in time could be integrated platforms, the framework of acquiring economies of scale for high variety is established. It is in this context that the whole idea of Group Technology can support the creation of manufacturing platforms. Considerable work has been done in relation to get full advantages of product and production architectures, e.g., platform-based co-development, Integrated Product and Production Design Development, Collaborative engineering, and Concurring product and Product development. Further contribution within the field of product architectures is an important aspect of creating manufacturing equipment based on customized flexibility with the purpose of acquiring economy of scale.

Acknowledgement The author wishes to thank Associated Professor P. Nielsen for his help in establishing the basis for running an algorithm in the statistical language R.

References

1. Belmokhtar, S., Dolgui, A., Bratcu, A.: Modular machining line design and reconfiguration: some optimization methods. INTECH Open Access Publisher (2006)
2. Benkamoun, N., Huyet, A., Kouiss, K.: Reconfigurable Assembly System Configuration Design Approaches for Product Change. pp. 1–8 (2013)
3. Boutsinas, B.: Machine-part cell formation using biclustering. Eur. J. Oper. Res. **230**, 563–572 (2013)
4. Boysen, N., Fliedner, M., Scholl, A.: A classification of assembly line balancing problems. Eur. J. Oper. Res. **183**, 674–693 (2007)
5. Chu, C.: An improved neural network for manufacturing cell formation. Decis. Support Syst. **20**, 279–295 (1997)
6. ElMaraghy, H.A.: A complexity code for manufacturing systems. pp. 625–634 (2006)
7. Koren, Y., Shpitalni, M.: Design of reconfigurable manufacturing systems. J. Manuf. Syst. **29**, 130–141 (2010)
8. Kusiak, A.: Intelligent manufacturing systems. Prentice Hall Press, 200 Old Tappan Road, Old Tappan, NJ 07675, USA, 1990, 448 (1990)
9. Wemmerlov, U., Johnson, D.J.: Cellular manufacturing at 46 user plants: implementation experiences and performance improvements. Int. J. Prod. Res. **35**, 29–49 (1997)
10. Zhang, Y.: Cellular Manufacturing Systems. Methods in Product Design: New Strategies in Reengineering, 1027 (2013)
11. Zhou, Z.: Manufacturing intelligence for industrial engineering: methods for system self-organization, learning, and adaptation. IGI Global (2010)
12. Zhou, Z., Wang, H., Lou, P.: Group technology

Chapter 12
Engineering Change Management and Transition Towards Mass Customization

Simon Haahr Storbjerg, Thomas Ditlev Brunoe, and Kjeld Nielsen

12.1 Introduction

A common trend of today's markets is the increased demand for customized products and services meeting the individual customer's needs. At the same time, it is common for all industries that globalization has led to increased competition and shortened product life cycles. A recent study by Roland Berger documents that the average product life cycle has been reduced by 24 % over the last 15 years [17]. Consequently, today's manufacturers are in pressure for finding more efficient approaches for bringing customized products to the market.

12.1.1 Mass Customization (MC)

Perfectly suited to the challenge described in above, MC arose as a concept and an operations strategy in the late 1980s, in a response from the US car manufacturers to the new competition from low-cost, high-quality Japanese car manufacturers [3]. MC enabled a new approach to competition, combining the ability to deliver products meeting the individual customer's needs, with an efficiency similar to mass production [15]. By introducing higher variety as a competitive parameter in the market, the US car manufacturers were able to gain a competitive advantage over the Japanese manufactures. Since then, research on MC has focused on clarifying the defining characteristics of the companies that successfully adopt the MC strategy. This has led to the introduction of three fundamental dimensions in enabling

S.H. Storbjerg (✉) • T.D. Brunoe • K. Nielsen
Department of Mechanical and Manufacturing Engineering, Aalborg University, Fibigerstraede 16, 9220 Aalborg East, Aalborg, Denmark
e-mail: shs@m-tech.aau.dk

J. Bellemare et al. (eds.), *Managing Complexity*, Springer Proceedings in Business and Economics, DOI 10.1007/978-3-319-29058-4_12

the MC ability. The three dimensions are by Salvador et al. [18] framed as the three fundamental MC capabilities: solution space development, robust process design and choice navigation.

Companies who decide for an MC strategy can have very different starting points. Traditionally, MC has been described as companies mass-producing standard goods, introducing higher variety, thus customizing each product. However, as MC gained popularity, companies who originally produced high variety, such as engineer-to-order companies, began to recognize the potential of the MC approach. Over the last decade, a number of studies have also been reported on ECO companies adopting an MC approach, to increase profitability [2, 6, 7].

Regardless of the starting point of the transition process towards MC and the approach at MC, the process of designing starts rarely from scratch, but rather by modification of existing design [4, 5, 12]. Eckert et al. [4] also describe two extremes for customizing by modifying existing design: changing existing base to meet individual customer's needs or developing a modular product range which can be configured to the individual customer's need [4]. Another classification of customizing by modifying existing design is proposed by Muntslag [14], who has developed a classification scheme for how much of the design that can be done order-independently in a customization effort [14].

Since designing or product development in a customization scheme inevitably involves modifications to existing design, it is evident that engineering changes and their management (ECM) are an inevitable part of mass customization. This implies furthermore that for companies transitioning towards MC, it is assumed that the effective and efficient management of engineering changes becomes even more important.

12.1.2 Engineering Change Management

As a consequence of the increased demand for customization and that product life cycles are growing shorter, the design of new products based on adaptions from existing products, or by the combination of existing design elements, is today becoming a dominating approach to new product development [8, 23]. On this background ECM, which addresses management of the technical changes, i.e. engineering changes, in products and related value chain processes, is also becoming increasingly important, especially when dealing with complex engineered products. ECM has not only received increased attention due to the shift in approach and focus of product development but also due to the increasing product and value chain complexity [4]. Combined with this, globalization has led to increased complexity in supply chains and to increased variation in market and documentation requirements [17]. All of this increases the need for managing engineering changes in order to ensure efficiency and profitability.

Different definitions for engineering changes can be found in literature. Based on an overview on some of the most prevailing definitions, Hamraz et al. [5] propose the following definition: 'ECs are changes and/or modifications to released structure (fits, forms and dimensions, surfaces, materials etc.), behaviour (stability, strength, corrosion etc.), function (speed, performance, efficiency, etc.), or the relations between functions and behaviour (design principles), or behaviour and structure (physical laws) of a technical artefact'. This definition delimits engineering changes to concern changes to already released structure. Based on experience from practice, this paper takes point of departure in a broader understanding of engineering changes, also concerning the release of new structures. The introduction of a new product family in a product program, or the introduction of new design elements to an existing product family, is also considered as engineering changes. The definition used in this context is therefore:

> ECs are changes and/or modifications to released structure (fits, forms and dimensions, surfaces, materials etc.), or product programs (release of new structure), behaviour (stability, strength, corrosion etc.), function (speed, performance, efficiency, etc.), or the relations between functions and behaviour (design principles), or behaviour and structure (physical laws) of a technical system.

ECM refers to the organization, control and execution of ECs and can, as topic and research field, be regarded as one of the five sub-elements of the broader field configuration management [8]. The goals of ECM are to avoid or reduce the number of engineering change requests (ECRs) before they occur, to select their implementation effectively when they occur, to implement required ECs efficiently and to learn from implemented ECs. Research on ECM has for years focused on specific and isolated practices in relation to change handling, e.g. change impact assessment [13], reasons for initiating EC [10] and effects of change propagation [11]. Only lately research on ECM has focused on giving a more comprehensive overview of the practices and capabilities needed for successful management of engineering changes. Based on an extensive literature review, Jarratt et al. [8] have established an overview on the key practices of ECM. Another extensive and recent literature review has been conducted by Hamraz et al. [5], who focuses on bringing a holistic categorization framework for literature on ECM. In a study by the Aberdeen Group of 135 enterprises from different industries, a number of capabilities, characterizing best in class companies in relation to change handling, are identified [1]. Based on a case study and a review of literature on ECM, a capability framework for ECM has been developed by the authors of this paper [21]. The framework introduces a number of ECM-related capability areas and has in a recent study been further developed into an ECM maturity grid, which provides a comprehensive overview of the capability areas within ECM [20].

12.1.3 Objectives of the Paper

From the overall introduction to ECM and MC, it is clear that the two research domains are related, both from a practical level and in sharing the aim of ensuring an efficient introduction of variants. Despite of the close relation of these fields, none of the contributions within ECM addresses how ECM can support a more efficient customization. The objectives of this paper are on this background to add clarity to the relation between ECM and MC and building on previous research [19–21] to develop new knowledge on what capabilities within ECM a company should mature to support the transition towards MC.

The transition towards MC can generally be described by two extremes, either as a transition from mass production in which the manufacturer pursues a greater adaptability to customer demand by increasing the solution space or as a transition from complete customization, where the manufacturer pursues efficiency by delimiting the solution space. This paper is based on a case study in a global engineering-oriented manufacturer which is managing customization to some degree, but at a cost which is not fully competitive. The case company is based on this striving to achieve an increased and more efficient product customization. The case company is therefore not to be categorized in either of the two extremes, but rather as a mix.

12.1.4 Research Questions

Based on the objective introduced in above, this paper takes point of departure in the following research questions:

1. Which practices of ECM are especially challenged by an increased adaption to customer-specific requirements?
2. What are the relations between the fundamental capabilities of MC and the capability areas of ECM?
3. What capability areas within ECM are central to support the transition towards MC?

Research question 1 is answered in Sect. 12.3 based on findings from a case study. Building of the findings from this and by a literature review and a focus group interview, research questions 2 and 3 are answered in Sect. 12.4.

12.2 Research Methodology

In this section, the research methodology is introduced together with choices in regard to the methods applied. Due to the scarcity of literature addressing the relation between ECM and MC, a combination of case study, literature review and focus group interview was considered necessary to answer the research questions.

12.2.1 Literature Review

In order to clarify the relation between ECM and MC, a thorough review of literature on ECM and MC was conducted. Through title and topic searches on Thomson Reuter's 'Web of Science' and Elsevier Scopus, and by using combinations of the search strings 'mass customization' and 'engineering change management' and 'configuration management', 96 contributions were identified. By title and abstract review against the following criteria, the list of relevant contributions was narrowed down to only three relevant contributions. The criteria used in this were literature that satisfied the following criteria: (a) literature defining or describing MC and ECM capabilities and (b) literature addressing the relation between MC and ECM.

Through detailed review of the relevant papers, the relations between the capabilities of MC and ECM were clarified by the following approach. First the mechanisms of ECM reported in literature as important for MC were identified. From these, ECM-related capability areas were then identified using the ECM capability framework of Storbjerg et al. [20]. Finally the relations between the ECM capabilities reported in literature and the fundamental MC capabilities were identified by deducing for each capability, with logical reasoning, the higher goal or purpose that the ECM capability serves. The findings from the literature review are introduced in Sect. 12.4.

12.2.2 Case Study

In order to identify which capability areas of ECM are especially challenged by the transition towards MC, a longitudinal case study was conducted in a global manufacturer pursuing the MC benefits. Through observations and participation in new product development projects, key challenges and enablers of ECM in achieving the MC benefits were identified. The case study and the case company are further introduced in Sect. 12.3.

12.2.3 Focus Group Interview

It is generally accepted that organizational researchers can improve the accuracy of their judgements by collecting different kinds of data bearing on the same phenomenon. This method is also referred to as triangulation [9]. In order to allow for greater reliability and validity of the results of the paper, it was decided to supplement the literature review and case study with a focus group interview with five configurations responsible from the case company. Focus group interview was selected as it is generally acknowledged that this interview method allow for more efficient data generation, due to the interaction in the group [16]. The participants of the focus group interview were selected based on their experience from practice with configuration and change handling. The questions for the focus group interview are introduced together with the findings from this in Sect. 12.4.

12.3 Case Study: Barriers of ECM for the Transition Towards Mass Customization

This section introduces the capability areas of ECM, which, based on the longitudinal case study, were identified as especially challenged by the transition towards MC. First a brief introduction is given to the case study and the case company, followed by a section on the key findings from the observations and interviews.

12.3.1 Introduction to Case Study and Case Company

The case study was conducted over the period from 2011 to 2015 in a large industrial manufacturer of power plants. The case company is a global organization of 20,000+ employees having activities in more than 70 countries. The internal value chain involves everything from sales, engineering, sourcing, manufacturing, transportation and construction to service and decommissioning. The products of the case company, i.e. the power plants, are complex engineered products having a lifetime of up to 25 years. The development of new products is based on a platform-based approach. Being faced with varying customer requirements and a demand for customization of the product offering, the case company has for years been pursuing the MC benefits. The approach to MC has primarily been relying on building the solution space development capabilities, by a focused effort on modularization, and building the choice navigation capabilities, by a number of initiatives on building the capabilities concerning product and sales configuration. The case company is, in addition to the products, offering service and repair solutions and therefore also required to manage changes throughout the entire product lifetime.

All of this is having a significant impact on the complexity of change handling, e.g. in evaluation of change impact and interchangeability. Operating in a maturing market, the case company is facing an increased competition and is based on this in an emergent need for improving its competitiveness. As a consequence, the design strategy has over the last years increasingly focused on further developing and improving the existing product platforms, in order to introduce more cost-efficient product variants. As a result, the handling of engineering changes, which in the case company is formalized in the ECM process illustrated in Fig. 12.1, has gained increased importance and management attention.

12.3.2 Findings from Case Study

By participating in several new product development projects and ECM improvement initiatives, and from interviews with configuration experts, a number of challenges in ECM have been identified as barriers for the transition towards MC. Using

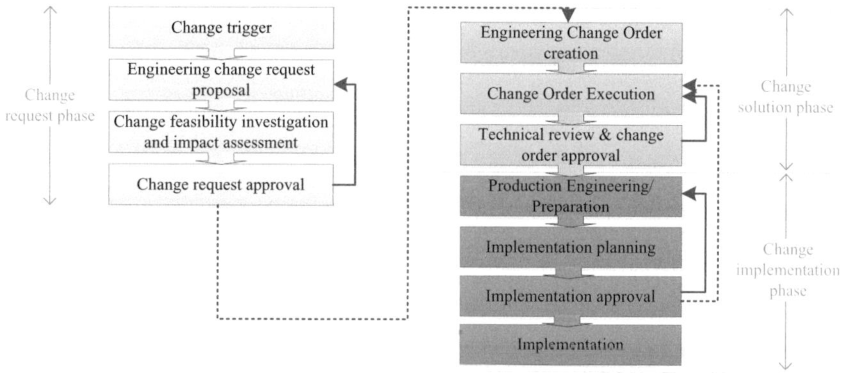

Fig. 12.1 Process for handling engineering changes in case company, i.e. ECM process

the capability framework of Storbjerg, Brunoe and Nielsen [20], the challenges have been characterized. From this, the following four capability areas of ECM have been clarified as important for the transition towards MC:

ECM roles and responsibilities

One of the challenges identified through the case study is how to organize around change execution. This topic has also been noted by other scholars; e.g. Hamraz et al. [5] highlight this as one of the core themes of ECM. From the observations and the interviews, it is clear that the transition towards MC and an increased adaption to customer-specific needs call for changes to the organization around change execution, especially if the MC strategy followed builds on a modular and platform-based approach, as in the case company. In this the product development activities are organized according to a matrix-based organization, in three product platforms and 11 generic technical systems, which carry all functionality of the products. The organization around change execution is based on the product platform driving the need for the change and typically as part of a product development project to create a new platform variant. For each platform project, staffing of the project organization is done by having resources from relevant systems in the line organization of the R&D function allocated to the project. Depending on the scope, one or several changes are initiated for each system, and a case handler, typically the responsible design engineer, is assigned. The case handler is responsible for the change execution until design finalization, where an implementation responsible from the receiving business unit, which often is a manufacturing unit, takes over. This organization has in the case company proven to work well for years. However as the case company during the last years have intensified the adoption of a modular and platform-based approach to product development, the coordination of change execution has become an increasing challenge.Based on the increased reuse of components and modules across products and product families, changes are now not only impacting one configuration but more often several configurations. Planning and coordination are consequently increasingly required across the platforms. On top of this, it is in the case company more the rule than the exception that multiple development projects for each platform are running at the same time with separate timelines. It has based on this become an increasing challenge to handle change execution and change implementation without conflicting priorities and needs. The coordination is up to the responsible case handler which often does not have the full overview of the change and its relations to other ongoing projects and changes. The organization around change execution is on this background identified as an important capability for the success of the transition towards MC.

Change structuring and planning

Closely related to above challenges, structuring and planning change execution are significantly challenged by the transition towards MC. Especially planning and controlling the effectivity of changes have become a challenge. Changes have different priorities and timelines, e.g. due to differences in supply lead time and supply setup across the divisions of the company. As the sequence of the change implementation matters, the structuring of change creation to support an efficient and coordinated change implementation, which is also respecting the diverse timeline requirements, has become an increasing challenge.

Change approval process

Another practice of ECM in the case company, which is becoming increasingly challenged by the transition towards MC, is the change approval process. As a consequence of the increased reuse of components and modules across platforms, changes are increasingly impacting multiple platform variants or configurations. This is a challenge for the change approval process in the case company, which is based on change approval boards organized per product platform. Changes having impact across product platforms are most often only reviewed at one of the platform change boards, with the consequence that key stakeholders are not involved in the change approval process. Consequently it is in several cases experienced that conflicting or non-authorized changes are implemented.

System integration and use of product data management (PDM) systems and product life cycle management (PLM) systems

Customization calls for not only customizing the product but also the related information, e.g. documentation, specifications, etc. The case company uses an own-developed change management system which has been integrated to some degree with the PDM system and the enterprise resource planning (ERP) system. Based on the increased number of product variants in the case company, the handling of changes in documentation and specifications as, e.g., bill of materials and master data have been a significantly growing challenge. Especially the lack of system integration in between, e.g., the CAD, ECM, PDM and ERP system has generated not only a massive administrative task due to the inefficient workflow but also a significant data quality issue. The use of integrated PDM and PLM systems has on this background been identified as a key capability to handle an increasing product variety.

12.4 Enablers of ECM for the Transition Towards MC

In this section, the relation between ECM and MC is described based on a combination of literature review, case study and a focus group interview. By this research question 2 is answered. First, the findings from a review of relevant literature are reported using the method outlined in Sect. 12.2.3. Following this, the findings from a case study and a focus group interview are described. In closure of this section, the results are summarized, and the capability areas of ECM which are central to support the transition towards MC are listed, by which research question 3 is answered.

12.4.1 Literature Review: Mass Customization and ECM

It is, from the introduction, clear that MC and ECM are closely related concepts and research domains. This relation is both on an operational level, e.g. depending on the same activities, and at a strategical level, e.g. sharing the aim of ensuring an efficient introduction of variants. In spite of the close relation of between these fields, the relation between ECM and MC has not been studied. Following a literature search according to the method introduced in Sect. 12.2, the below three contributions were identified and reviewed.

The relation between MC and ECM is noted by Hamraz et al. [5] in a comprehensive review of literature on ECM. In this it is highlighted that ECM has gained increasing popularity benefiting from the rise of attention towards MC among other concepts [5]. No further notions or references are however given the exact relation of ECM and MC.

Eckert et al. [4] have studied relation between MC, change and inspiration with focus on the process of changing design to meet new needs. Based on observations and interviews with engineers in three companies engaged in customization and handling of engineering changes, Eckert et al. [4] propose a classification of change processes in three different types of changes. The classification is used to describe the relation between old and new products. Building on the classification, Eckert et al. [4] describe MC as a change process with the argument that MC can be described as design for change, where change is explicitly aspired to and answered by flexibility. Following this view, the common challenges of MC and ECM are discussed. One of the common challenges highlighted is change propagation, i.e. that changes propagate to other parts of the system. In relation to this Eckert et al. [4] highlight the importance of predicting the scope of change, i.e. knowing the impact of change as early as possible in the process. Another challenge highlighted as common of MC and ECM is planning the change process, e.g. in running parallel challenges with interfaces to each other. The study of Eckert et al. [4] concludes that common methods for predicting impact of changes, assisting in planning of changes and handling parallel changes could be developed with benefit for both ECM and MC. The study does however not bring any further detailed clarification of which capabilities of MC these support or how the methods should be developed.

In a recent study, Veldman and Alblas [22] also argue for that the relation between ECM and MC has only grown more important with what they characterize as the MC wave. MC is in this study viewed as a balancing act of reuse and distinctiveness, from which it is clear that Veldman and Alblas [22] primarily focus on the solution development capabilities of MC. Based on a multiple-case study in two capital-goods manufacturers, the study contributes with knowledge on how to structure and manage the ECM process in order to support the balancing of reuse and distinctiveness [22]. The paper contributes based on this with knowledge on the relation between especially the solution development capability of MC and the capability areas of ECM. The following mechanisms are identified: (a) managing generic design information; (b) isolating large engineering changes, which concern the planning and structuring of changes; (c) managing process variety; and (d) designing and executing change process, which concern the ability to develop a process for controlling changes. In addition to the four mechanisms, the following practices in relation to ECM are identified as important for the management of design variety: minimizing complexity, change approval, change planning, managing generic design types, avoiding engineering changes by standardization and categorizing engineering changes. Veldman and Alblas [22] conclude that the importance of the practices and capabilities differs according to product delivery strategy and, based on this, that different configurations should exist for the ECM setup and each of the four mechanisms [22].

From the above literature review, a number of mechanisms within ECM have been identified as enablers for the transition towards MC. In total five enablers have as illustrated in Fig. 12.2 been identified.

12.4.2 Findings from Focus Group Interview

Seeking the same goal as with the literature review, i.e. to identify capabilities of ECM that are central to support the transitions towards MC, a focus group interview was conducted with configuration responsible from the case company. The interview was conducted by the following approach:

1. Firstly, based on the ECM capability framework by Storbjerg and Brunoe [20], the interviewees were asked to identify which capabilities in relation to ECM they experienced especially challenged by an increasing variety and hence important for the ability to offer new variants efficiently.
2. Secondly, the interviewees were asked to complete and validate the list of case company-specific MC-related capabilities and initiatives in Table 12.1.
3. Finally the interviewees were asked to evaluate the degree of influence of the ECM-related capability areas on each of the case company-specific MC-related capabilities using a scale from 1 to 4. These were done taking point of departure in the capabilities identified as important for MC based on case study, literature and interview (Table 12.1).

Capability area	Sub-capability area	Important for MC: Identified from focus group interview	Important for MC: Identified from literature	Important for MC: Identified from case study	Related mass customization capabilities: Product management process	Modularization process	Standardization	Engineering process	Scoping process	Sales Configuration tools	Sales process	Product configuration setup	New product introduction processes	Flexible manufacturing	Total score
(A) ECM process	(1) Change identification, prediction & proposal	X	X		3	3		3	2				2	1	14
	(2) Change approval	X	X	X		1	3	1	4				1		10
	(3) Change impact analysis	X	X			4		3	3	3		4	3	3	23
	(4) Change prioritization	X			1			3	4			2	2		12
	(5) Change definition and categorization														
	(6) Change cost management	X			3						3				6
	(7) Change solution development, selection and	X						3				2	2	2	9
	(8) Change implementation	X											3	3	6
	(9) Documentation of change														
	(10) Coordination & integration with internal stakeholders	X			2	1	1	4		1	1	1		3	14
	(11) External integration; customers, vendors etc.	X			3	2	1	1	5	3	2	3	1	1	22
	(12) Process for emergency change handling														
	(13) Planning of change handling and implementation	X	X	X	2	3		3	1	2	2	1	4	4	22
	(14) Process of controlling engineering changes		X												
	(15) Scaling and tailoring of ECM process														
(B) Change monitoring	(16) Change status reporting	X						3			1		4		8
	(17) ECM performance management	X						3					4		7
(C) IT tools	(18) ECM system														
	(19) Use of PDM&PLM systems	X		X	3	1	2	2	1	4	1	4	3		21
(D) Management & communication	(20) Communication	X			1	1	1	1	1	1	1	1	1	1	10
	(21) Front-loading of ECM														
	(22) ECM strategy														
	(23) Resource management	X						2	2			1	2		7
(E) People, skill & competencies	(24) ECM roles & responsibilities			X				3	1				4		8
	(25) ECM competence management & training	X			1	2	2	2	2	1	1	1	2	1	15
	(26) Continuous improvement & learning														
1 = 1 slightly influential, 2 = 2 Somewhat influential, 3 = 3 Very influential, 4 = 4 extremely influential					Solution Space Development					Choice Navigation		Robust Process Design			

Fig. 12.2 Mapping of ECM capabilities with the fundamental MC capabilities

The result of the focus group interview is presented in Fig. 12.2. Based on the input from the configuration responsible, 17 ECM capability areas were identified as important for the ability to offer new variants efficiently and thus important for the transition towards MC. From the interviewees' evaluations of the degree of influence, which are summarized in the rightmost column in Fig. 12.2, the following four capability areas are especially attributed as influential:

1. **Change impact analysis**: The capability of early detecting change and predicting the impact of the change before change approval is as indicated in Fig. 12.2 especially impacting the solution space development capabilities. Uncontrolled change propagation is indisputably impacting the ability to efficiently develop a customer-required solution space. Eckert et al. [4] also highlight the importance of predicting the scope of change.

Table 12.1 List of initiatives and capabilities specific for the case company

MC capabilities	Case company-specific MC capabilities and initiatives
Solution space development	• *Modularization*: process and initiative aiming to build an optimal modular product architecture • *Standardization*: process, governance setup and initiative to build a standard product catalogue • *Engineering process*: framework and process to ensure a well-defined and structured engineering design process • *Scoping process*: process, IT and governance setup initiated to control the scoping of new variants
Choice navigation	• *Sales configurator*: IT setup to support the product customization and building of specification and documentation in sales • *Sales process*: process to support customer solution selection
Robust process design	• *Product configuration setup*: IT setup to build the complete set of product specifications based on customer selection • *New product introduction processes*: process to prepare specifications for manufacturing • *Flexible manufacturing*: process, IT setup and initiative aiming to build the needed supply chain flexibility

2. **External integration, customers, vendors, etc.**: ECM is integrating many functions and activities across the value chain; it does based on this not come as a surprise that the ability to integrate to customer, vendors, etc., is important. Dealing with changes to customized products, the configuration responsible reports in the group interview that integration of vendors and customers in the change handling is increasing in importance. Much quality management literature have for decades also highlighted the importance of clearly understanding the voice of the customers. Similarly research on product development has also for years advocated an integrated approach.
3. **Planning of change handling and implementation**: Thorough planning of change handling and implementation is not only evaluated important of the ability to develop a new variant but also for ensuring a robust process. In the group interview, it was reported that challenges in the case company in change implementation originating in poor planning had only increased, as the product variety increased.
4. **Use of PDM and PLM systems**. Another ECM capability that, based on the interviewees' experiences, was significantly impacting the ability to customize efficiently is the use of IT tools to support the change handling. Based on insufficient system integration, it was reported that multiple manual data entry and lack of transparency caused the workload of change execution to explode, with an increasing product variety.

Based on the findings above, it is evident which capability areas of ECM that can be considered as enablers for the transition towards MC and therefore especially important to focus on if pursuing the MC benefits. The results are presented in Fig. 12.2 which also indicate the relation between the fundamental capabilities of MC and the capability areas of ECM.

12.5 Conclusion

With an intensified competition and an increasing demand for customization, manufacturing companies are today more than ever challenged to develop the capabilities, which can ensure an effective and efficient customized product introduction, i.e. the MC capability.

Although engineering changes, which are central for the customization process, has been studied for years, it has until now not been studied how ECM can support MC. On this background, the purpose of this paper is to support practitioners in the academia and industry, with clarity on the relation between ECM and MC.

Based on a combination of review of literature on ECM and MC, a longitudinal case study and a focus group interview in a global manufacturer pursuing the MC benefits, 17 capability areas of ECM have been identified as important for the transition towards MC. From this it is evident that MC and ECM are related concepts and that the transition towards MC also depends on ECM-related capability areas.

The degree of influence of the ECM capability areas to the MC capabilities has been evaluated by a group of domain experts involved in product configuration and change handling. From this it is evident that the following four ECM capability areas are evaluated to have significant importance, for the ability to introduce customized solution effectively and efficiently: change impact analysis, use of PDM and PLM systems, planning of change handling and implementation and external integration with customers, vendors, etc. These ECM capability areas are therefore especially important to focus on, in order to support the transition towards MC.

This paper contributes on this background with knowledge of the relation between the fundamental capabilities of MC and the capability areas of ECM. Furthermore the paper contributes to existing knowledge in ECM and MC with knowledge on which capability areas of ECM are central to support the transition towards MC.

Based on these results, a number of opportunities exist for further research. One opportunity for further research is to extend the group interviews to other companies involved in ECM and pursuing the benefits of MC, in order to strengthen the findings of the research. Another opportunity for further research is to further study and validate the relations between ECM and MC, e.g. by obtaining data from multiple-case studies in companies pursuing MC benefits, in order to investigate the correlation of company performance with presence of ECM capabilities.

References

1. Brown, J., Boucher, M.: Engineering Change Management 2.0: Better Business Decisions from Intelligent Change Management. Aberdeen Group, A Harte-Hanks Company, Technical Report (2007)
2. Brunoe, T.D., Nielsen, P.: A case of cost estimation in an engineer–to–order company moving towards mass customisation. Int. J. Mass Cust. **4**, 239–254 (2012)

3. Davis, S.M.: From "future perfect": mass customizing. Strateg. Leadersh. **17**, 16–21 (1989)
4. Eckert, C., Pulm, U., Jarratt, T.: Mass Customisation, Change and Inspiration-Changing Designs to Meet New Needs. (2003)
5. Hamraz, B., Caldwell, N.H., Clarkson, P.J.: A holistic categorization framework for literature on engineering change management. Syst. Eng. **16**, 473–505 (2013)
6. Haug, A., Ladeby, K., Edwards, K.: From engineer-to-order to mass customization. Manage. Res. News **32**, 633–644 (2009)
7. Hvam, L.: Mass customisation of process plants. Int. J. Mass Custom. **1**, 445–462 (2006)
8. Jarratt, T., Eckert, C., Caldwell, N. et al.: Engineering change: an overview and perspective on the literature. Res. Eng. Des. 22 (2011)
9. Jick, T.D.: Mixing qualitative and quantitative methods: triangulation in action. Adm. Sci. Q. 602–611 (1979)
10. Kanike, Y., Ahmed, S.: Engineering Change During a Product's Lifecycle. (2007)
11. Koh, E.C., Caldwell, N.H., Clarkson, P.J.: A method to assess the effects of engineering change propagation. Res. Eng. Des. **23**, 329–351 (2012)
12. McMahon, A.C.: Observations on modes of incremental change in design. J. Eng. Des. **5**, 195–209 (1994)
13. Morkos, B., Shankar, P., Summers, J.D.: Predicting requirement change propagation, using higher order design structure matrices: an industry case study. J. Eng. Des. **23**, 905–926 (2012)
14. Muntslag, D.R.: Managing customer order driven engineering. An Interdisciplinary and Design Oriented Approach (1993)
15. Pine, B.J.: Mass Customization: The New Frontier in Business Competition. Harvard Business School Press, Boston (1993)
16. Rabiee, F.: Focus-group interview and data analysis. Proc. Nutr. Soc. **63**, 655–660 (2004)
17. Mastering Product Complexity: Mastering Product Complexity. (2012)
18. Salvador, F., De Holan, P.M., Piller, F.: Cracking the code of mass customization. MIT Sloan Manage. Rev. **50**, 71–78 (2009)
19. Storbjerg, S.H., Brunoe, T.D., Nielsen, K.: Maturity assessment & engineering change management: a review of relevant models. IEEE Trans. Eng. Manage (2015, in press)
20. Storbjerg, S.H., Brunoe, T.D., Nielsen, K.: Towards an engineering change management maturity grid. J. Eng. Des. (2015, in press)
21. Storbjerg, S.H., Sommer, A.F., Brunø, T.D., et al.: Development of an Engineering Change Management Capability Framework for Enterprise Transformation. (2013)
22. Veldman, J., Alblas, A.: Managing design variety, process variety and engineering change: a case study of two capital good firms. Res. Eng. Des. **23**, 269–290 (2012)
23. Vonderembse, M.A., Uppal, M., Huang, S.H., et al.: Designing supply chains: towards theory development. Int. J. Prod. Econ. **100**, 223–238 (2006)

Chapter 13
The Evolutionary Process of Product Configurators

Paul Blažek, Monika Kolb, Clarissa Streichsbier, and Simone Honetz

13.1 Introduction

Product configurators are one of the enabling forces in mass customization [1]. Publications like the *Configurator Database Report* (2013 [2] and 2014 [3]) and the *The Customization 500* [4] list various B2C product configurators from many different industries. These studies emphasize the huge amount of product configurators that can be found online. The focus of this paper is to examine whether or not there is a development in terms of newly launched, existing, and removed product configurators over the last 3 years. Furthermore configurators of the automotive industry are analyzed according to changes in their user interface designs and interaction functionalities in the same time period. To get a better idea of the status quo, the changes are outlined by case examples. The data for this research is gathered from the Configurator Database (www.configurator-database.com) [5].

13.2 Development of Product Configurators

In 2007 the Configurator Database was launched, covering initially 600 web-based B2C product configurators. Since then the amount of configurators in the database has increased up to the number of 1034 in 2015. To be more specific, companies located in 28 different countries offer these 1034 configurators. As shown in Fig. 13.1, with 487 almost half of the companies (47%) are settled in Germany and 35% are from the United States (355 companies).

P. Blažek (✉) • M. Kolb • C. Streichsbier • S. Honetz
cyLEDGE Media GmbH, Vienna, Austria
e-mail: p.blazek@cyledge.com; m.kolb@cyledge.com; c.streichsbier@cyledge.com; s.honetz@cyledge.com

J. Bellemare et al. (eds.), *Managing Complexity*, Springer Proceedings in Business and Economics, DOI 10.1007/978-3-319-29058-4_13

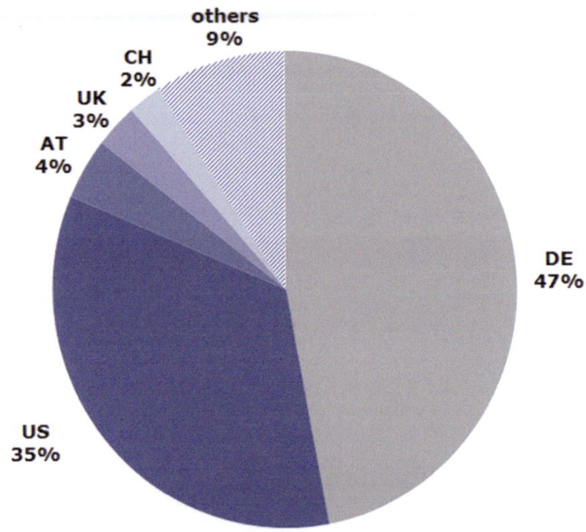

Fig. 13.1 Ranking of entries by country in percent ($n = 1034$)

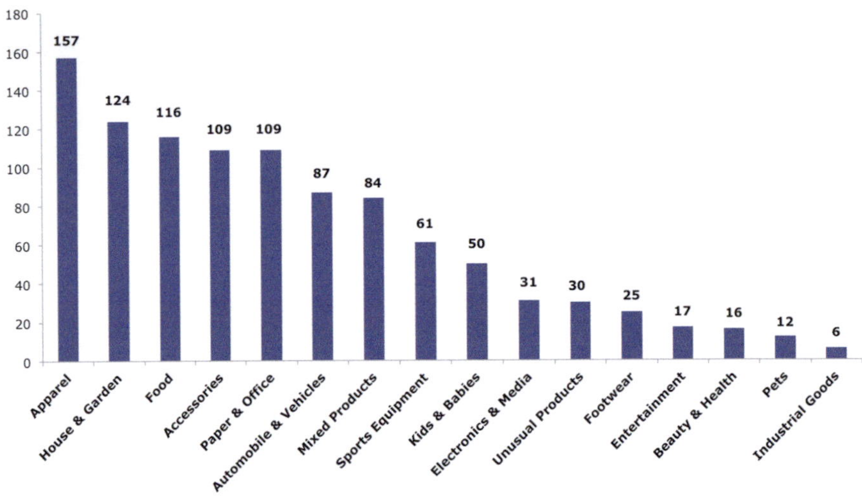

Fig. 13.2 Industry ranking of product configurators ($n = 1034$)

The product configurators in the Configurator Database are classified into 16 different industries. Figure 13.2 illustrates that the industries apparel (157) and house and garden (124) cover most of the configurators in the database, closely followed by food (116) and accessories (109) equal with paper and office (109).

In each industry, different product types can be found, whereby in total 261 different products can be customized. The graphic below depicts that cars (73) and mixed clothing (69) are the most common customizable products (Fig. 13.3).

The first Configurator Database Report was published in 2013 giving an overview about 900 product configurators followed by the Configurator Database Report 2014

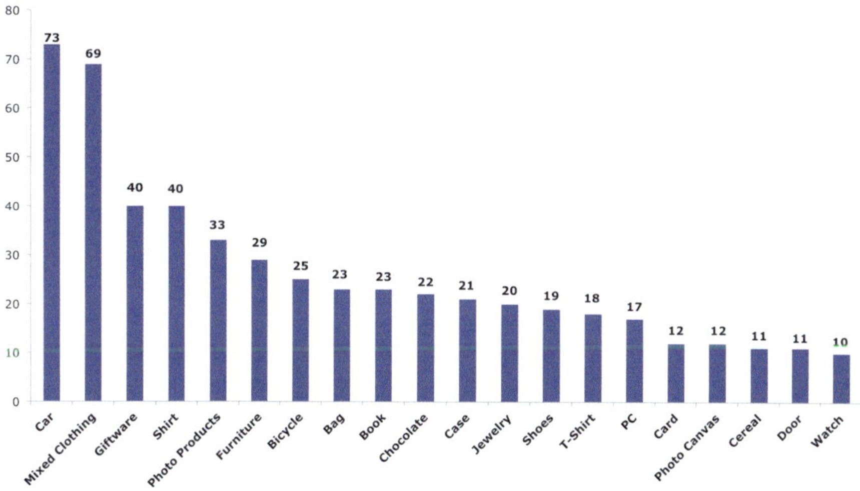

Fig. 13.3 Top 20 of the most popular products

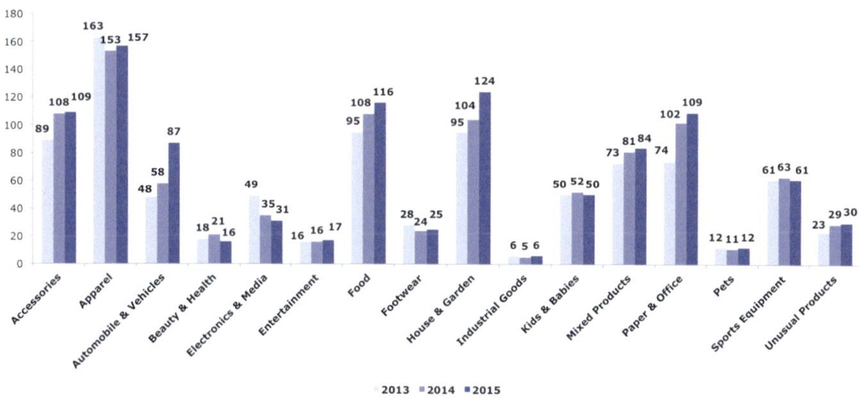

Fig. 13.4 Number of product configurators per industry (2013 $n=900$, 2014 $n=970$, 2015 $n=1034$)

listing 970 product configurators. Both reports already indicated that the field of mass customization is constantly changing.

As seen in Fig. 13.4, the biggest growth in the number of product configurators is in the automobile and vehicle industry, whereas products in the electronic and media as well as beauty and health industry declined.

From 2014 to 2015 in total 92 (9%) of the 970 configurators were removed and 153 (16%) new product configurators were added. The industry with the biggest growth is automobile and vehicles with 52% new and only 2% removed product configurators. The beauty and health (29%) and pets (27%) industry show the highest decline (Figs. 13.5 and 13.6).

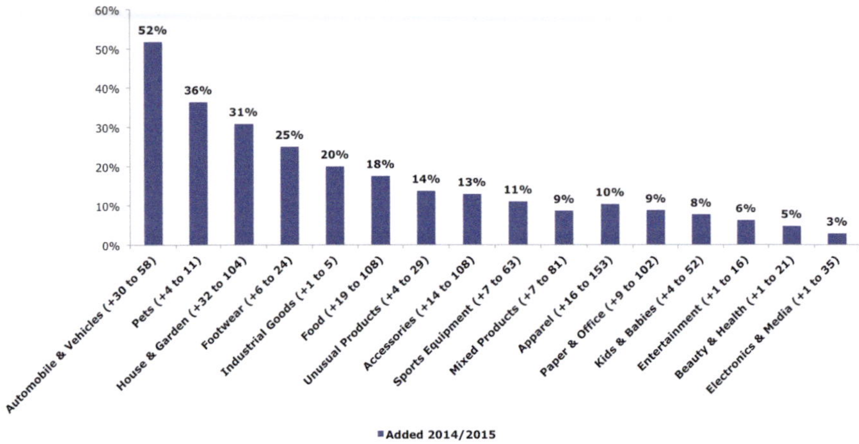

Fig. 13.5 Added configurators 2014–2015

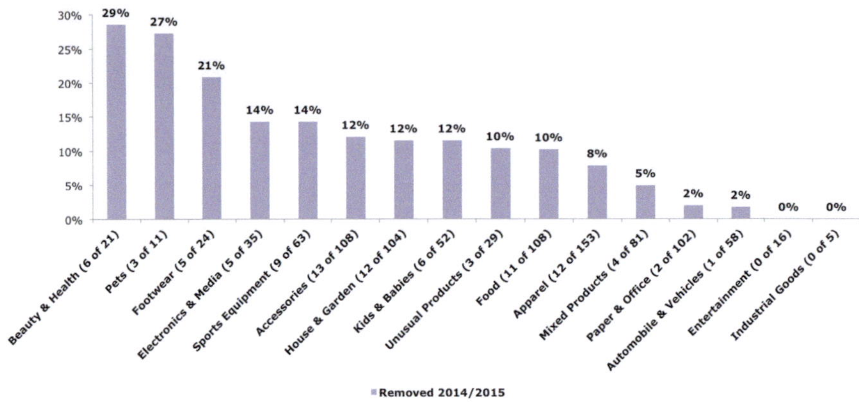

Fig. 13.6 Removed configurators 2014–2015

13.3 Development of the Automotive Industry

As product configurators within the automotive industry show the most significant change in terms of numerical growth, the aim of the following analysis is to investigate the development of configurators in this industry (see Appendix) listed in the Configurator Database. From overall 87 car configurators in 2015, 33 were also listed 2013. Every configurator of these 33 was analyzed in 2013 and 2015 based on the following criteria, which are also illustrated in five case examples:

- Steps in configuration process: How many steps have the consumer to fulfill?
- Positioning of toolboxes: Where and how are the toolboxes placed?

- Visual appearance: How does the configurator look like?
- Product visualization: How is the product depicted?
- Social media: Which social media interaction possibilities are used?

13.3.1 Steps in Configuration Process

Every configuration process consists of different steps which guide the user through all tasks he has to fulfill [6]. The analysis of 33 automobile configurators shows that 22 out of 33 changed their configuration steps. No clear pattern concerning a preferred number of steps could be identified: seven configurators have now more steps than in 2013. On the other hand, four configurators have a reduced number of steps. Seven configurators display the same number of steps in 2013 and 2015 but renamed the label of the steps. Compared to 2013, in 2015 many configurators separate the selection of the car model from the actual configuration process.

The case of Mini illustrates these developments. Figure 13.7 depicts the configurator in 2013 with the 5 steps: *design*, *packages*, *performance*, *instruments*, and *accessories*. In 2015 the steps have increased to 8 separate tabs as shown in Fig. 13.8. Moreover some of the step labels changed, namely, *color*, *packages*, *exterior*, *wheels*, *interior*, *performance*, *audio and tech*, and *accessories*.

13.3.2 Positioning of Toolboxes

In regard to the toolboxes, the evaluation indicates that 23 of 33 configurators changed the layout or positioning between 2013 and 2015. Half of all analyzed configurators modified the position of their toolboxes. Mainly the toolboxes are shifted to the lower part of the UI of the configurator in 2015, whereas in 2013 the positions where mainly right, left, or a combination of below and right/left.

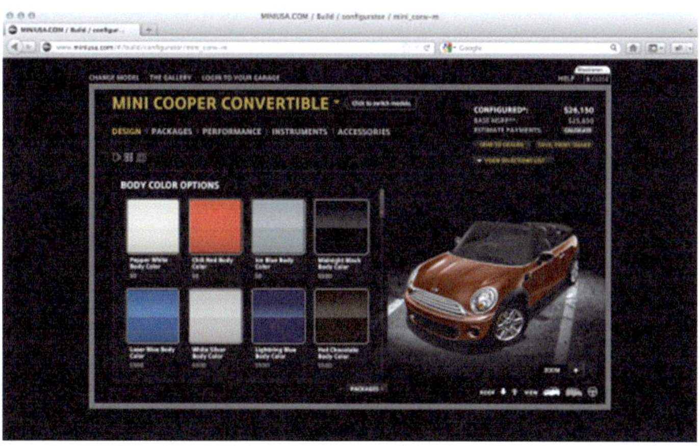

Fig. 13.7 Mini Configurator 2013 (www.miniusa.com)

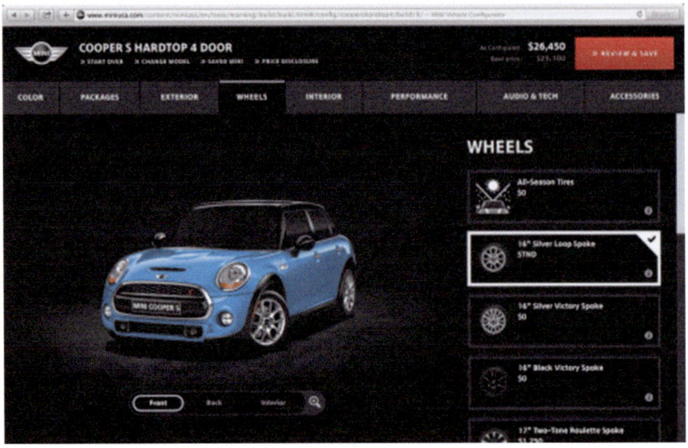

Fig. 13.8 Mini Configurator 2015 (www.miniusa.com)

Fig. 13.9 Toyota Configurator 2013 (www.toyota.com)

Sixteen of the configurators modified the styling of the boxes. Among these style modifications, mainly a more modern or clear look, for example, angular-shaped boxes instead of rounded ones, can be identified. Moreover the toolboxes are often bigger presented in 2015 compared to 2013. Another identified change is the increased usage of icons or symbols to illustrate the toolboxes or steps.

The example of Toyota shows some of the above findings. Figure 13.9 shows a fragmented toolbox environment filled with tiny interface elements.

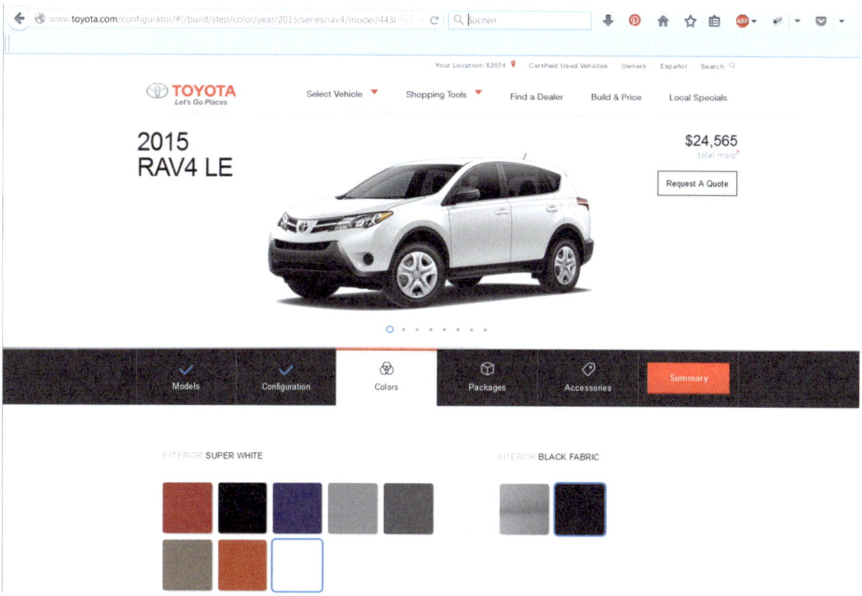

Fig. 13.10 Toyota Configurator 2015 (www.toyota.com)

Figure 13.10 presents the configurator 2 years later with toolboxes placed in the lower part with bigger interaction elements and the additional use of an iconography.

13.3.3 *Visual Appearance*

The investigation reveals that nearly all configurators enhanced the overall visual appearance. Twenty-one car manufacturers provide their users with a more modern, more structured, and more playful interface. In 2013 many configurators had a more technical look, but in 2015, user-friendly, well-structured, and good-designed user interfaces are mainly seen in the automotive industry. Also the color schemes of the configurators and their elements changed during the last 2 years. Thirteen of the 33 configurators use other colors now or made some adaption concerning button or font colors. Furthermore the screen size of the configurators is different than in 2013, as in 2015 most of the configurators have a full screen instead of a boxed layout which allows a better allocation of toolboxes and the product.

The case of Seat illustrates the visual developments of configurators in the automotive industry. The airy design creates a showroom atmosphere and uses the whole available screen. The cleaned up toolbox support orientation and important information for users, like the car price with the particular configuration, are presented in a more dominant way, whereas known information, like the name of the car model, is displayed in a reduced form (Figs. 13.11 and 13.12).

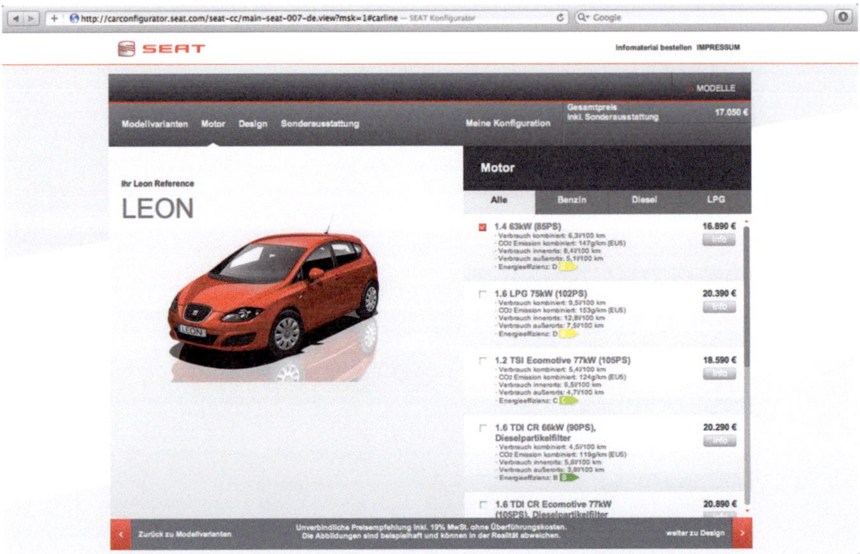

Fig. 13.11 Seat Configurator 2013 (www.seat.de)

Fig.13.12 Seat Configurator 2015 (www.seat.de)

13.3.4 Product Visualization

Especially when configuring a valuable product like a car, a realistic image of the product is very important for the customer and his imagination of the final product [7]. It supports the customer not only in making decisions but also helps to make the configuration process playful.

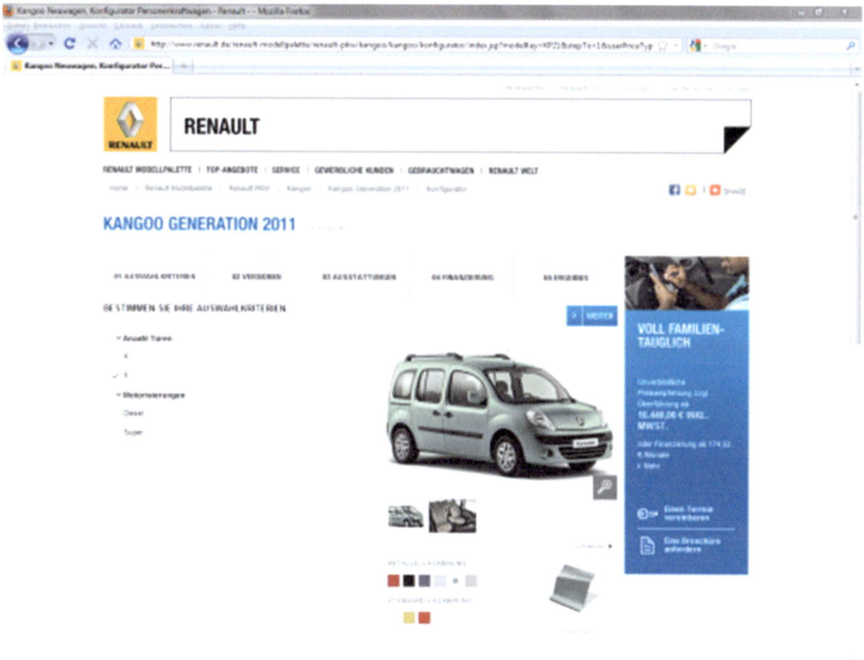

Fig. 13.13 Renault Configurator 2013 (www.renault.com)

In the automotive industry, visualizations in 3D are quite common for many years [8]. Nevertheless 20 configurators have worked on the product visualization compared to 2013. In 14 cases, an enlargement of the depicted car can be identified. Eight configurators improved the visualization to a more realistic illustration of the car and four configurators changed the positioning of the visualization and moved the product to the center of the user interface. Furthermore some user interfaces changed the scene in which the car is presented. Some use a more neutral background and some developed a room or street view.

Figures 13.13 and 13.14 show the difference of the Renault product visualization in 2013 and 2015. In 2015 the car is much bigger, more realistic, and in the center of the user interface. Additionally, the car is presented in a real-life scenario, whereas in 2013, it is shown on a plain white background.

13.3.5 Social Media Sharing

The possibilities of sharing content via social media are quite common on websites and web shops and also often provided in connection with configurators [9]. Frequently used social sharing features include Facebook Share, Tweet, Google Plus Share, Pin It and sharing via mail.

Fig. 13.14 Renault Configurator 2015 (www.renault.com)

The analysis indicates that sharing possibilities are not very common in the automotive industry. In 2013, only five configurators offered the possibility of sharing the created car via social media. In 2015, four more companies added sharing buttons to their configurators. Social media buttons in the footer area of the website linking to the respective channel of the company are not included in this statistic.

The example of the VW configurator shows that the company integrated sharing possibilities to their configurator in 2015, which can be seen on Fig. 13.15 on the right lower corner. Two years before, the configurator looked different and no social media icons were provided (Fig. 13.16).

13.4 Conclusion

The analysis detects that there is an overall growth of web-based B2C product configurators in nearly all industries which are listed in the Configurator Database. Taking a closer look at the automotive industry from 2013 to 2015 reveals that there is an evolutionary process of the configurators' user interfaces. The analyzed 33 car configurators changed their focus to improved and bigger product visualization as well as a better structure of the interaction steps. This may lead to a perceived reduction of complexity in the configuration process and improve the overall usability. To summarize, companies of the automotive industry constantly develop their product configurators to keep up with the state of technology and the changing media affinities of their customers.

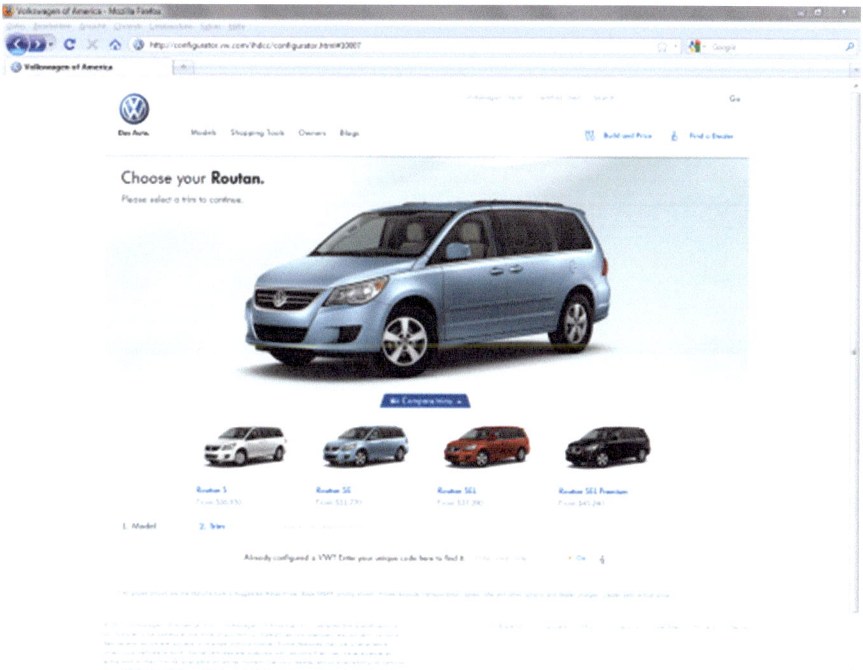

Fig. 13.15 VW Configurator 2013 (www.vw.com)

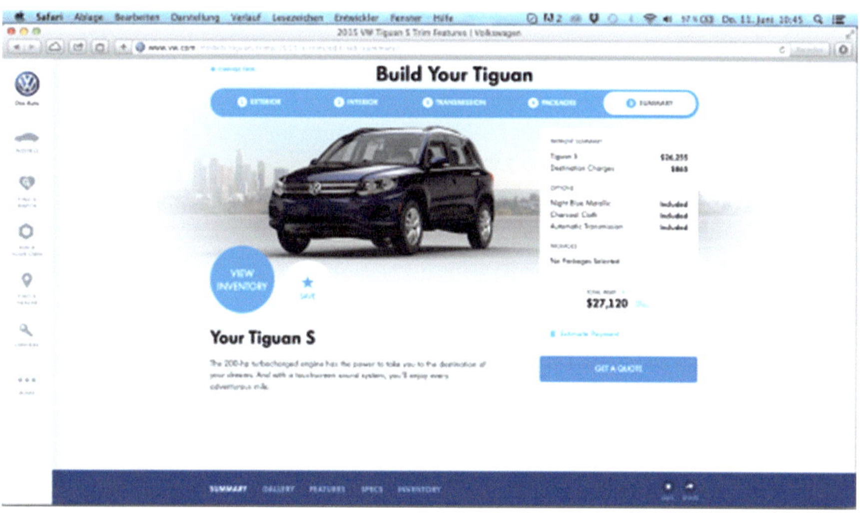

Fig. 13.16 VW Configurator 2015 (www.vw.com)

Appendix

Table 13.1 URLs of the 33 analyzed automobile configurators

www.alfaromeo.de	www.mercedes-benz.de
www.audi.de	www.mini.at
www.bmw.de	www.miniusa.com
www.bugatti.com	www.nissan.de
www.chevrolet.de	www.opel.de
www.daihatsu.de	www.peugeot.at
www.ferrari.com	www.porsche.at
www.fiat.de	www.renault.de
www.ford.de	www.seat.de
www.honda.de	www.skoda-auto.de
www.hyundai.de	www.smartusa.com
www.indicar.de	www.suzuki.at
www.jaguar.com	www.toyota.com
www.lexus.com	www.volkswagen.de
www.lincoln.com	www.volvocars.com
www.maserati.com	www.vw.com
www.mazda.co.uk	

References

1. Piller, F., Blazek, P.: Core capabilities of sustainable mass customization. In: Felfernig, A., Hotz, L., Bagley, C., Tiihonen, J. (eds.) Knowledge-Based Configuration: From Research to Business Cases. Morgan Kaufmann-Elsevier, Waltham (2014)
2. Blazek, P., Partl, M., Streichsbier, C.: Configurator Database Report 2013. Lulu Inc., Raleigh (2013)
3. Blazek, P., Partl, M., Streichsbier, C.: Configurator Database Report 2014. Lulu Inc., Raleigh (2014)
4. Walcher, D., Piller, F.: The Customization 500: An International Benchmark Study on Mass Customization and Personalization in Consumer E-Commerce. Lulu Inc., Raleigh (2012)
5. Configurator Database, cyLEDGE Media GmbH, www.configurator-database.com
6. Blazek, P., Pilsl, K.: The impact of the arrangement of user interface elements on customer satisfaction in the configuration process. In: Brunoe, T., Nielsen, K., Joergensen, K., Taps, S. (eds.) Proceedings of the MCPC 2014, the 7th World Conference on Mass Customization, Personalization, and Co-creation. Lecture Notes in Production Engineering. Springer, Cham (2014)
7. Streichsbier, C., Blazek, P., Partl, M.: The Impact of the product user interface on customer purchase decisions. Proceedings of the 6th International Conference on Mass Customization and Personalization in Central Europe (MCP-CE 2014). Novi. Sad. (2014)
8. Streichsbier, C., Blazek, P., Partl, M.: The influence of de-facto standards on users dealing with B2C configuration systems. Proceedings of the MCP-AP 2010. Taipei (2010)
9. Blazek, P., Kolb, M., Partl, M., Streichsbier, C.: The usage of social media applications in product configurators. Int. J. Ind. Eng. Manage. **3**(4), 179–183 (2012)

Chapter 14
Co-creation and Design Thinking to Envision More Sustainable Business Models: A Foresight Design Approach for Organizational Sustainability of SME Manufacturers

Alexandre Joyce

14.1 Introduction

"If you want to go fast, go alone; if you want to go far, go together" goes the old saying. This might hold true for travelling in the wild; however, this research finds that it doesn't quite fit when it comes to envisioning the future of an organization.

At one point in time, any organization, big or small, reaches a crossroad where strategic planning points towards reinvention [1, 2]. There is a way to avoid ending up in a dilemma by developing and proactively orchestrating a transition [3, 4]. For manufacturers, there is an imperative to always have new and improved products on the market. Behind the scenes, that implies managing an innovation process to filter ideas into desirable products [5]. One of the techniques that innovative companies use is foresight or advanced design concepts [6]. The most common example is concept cars that demonstrate new design ideas, new technologies and potential markets segments. We see the next opportunity in foresight design to apply itself to imagining better business models.

A business model is nothing more than the means by which a business creates, delivers and captures value [7]. Although many business model definitions and ontologies can be found in research literature, the notion of exploiting a business model often remains tacit for organizations [8, 9]. In other words, the organization operates without consciousness of its own operating system. This brings an added level of difficulty when seeking to understand business model transformation. Then again, stories of transforming a business model are nothing new. IBM sold their computer manufacturing operations to focus on their consultancy services in the 1980s [10]. Xerox went from an industrial manufacturer of photocopiers to a document printing service provider in the 1990s [11]. The carpet manufacturer Interface envisioned offering a more convenient flooring service in the late 1990s [12]. Such examples of

A. Joyce (✉)
Concordia University, Montreal, QC, Canada
e-mail: alexandre.joyce@gmail.com

J. Bellemare et al. (eds.), *Managing Complexity*, Springer Proceedings in Business and Economics, DOI 10.1007/978-3-319-29058-4_14

business model transformation have led to a growing field of research on the different phases of business model innovation such as experimentation, acceleration and transition, as described by Johnson [13]. Others see the practice of business model innovation following a similar process to product innovation from idea to launch [11]. In retrospect, one thing all three of these business model transformations—IBM, Xerox, Interface—have in common is that they happened to be improving towards sustainability when transforming their business model. However, we argue that most business model changes aren't specifically designed for sustainability.

There is a growing field of research that is specifically focused on the ends of having integrated a sustainability dimension to business models [14–17]. If we were to look at the timescale of a business model transformation, it can take decades before becoming business as usual. This explains why there is little empirical research on the means or the process of transforming existing business models towards sustainability [18]. This drives our research to focus on the early strategic phase of business model innovation [19]. Our interest lies in the creative process of business model innovation when an organization first conceptualizes future business models. At that moment, we believe there is a need for a conscious design approach.

In this quest for organizations to be more sustainable (i.e. in economic, environmental and social terms), we wonder: How can small- and medium-sized businesses envision their future? We focus our research on existing organizations, mainly manufacturers, because we see their transformation as more beneficial and more challenging than the creation of cleaner start-ups [20]. We posit that existing organizations such as manufacturers have had little time and little experience to reflect upon what could make them significantly more sustainable. We perceive that they don't set aside time or resources to think about such challenges. And lastly, seldom do small and medium businesses have the internal capacity or knowledge to design such visions of the future. It isn't part of their ongoing innovation practices to generate a long vision nor to guide the content of that vision towards being more sustainable.

Our goal is to lay the foundations for the emerging practice of consciously applying a design approach to the creation of business models for sustainability. In this paper, we first ask what the business models of small- and medium-sized manufacturers would look like if they undertake a design approach to imagine business models for sustainability. Second, we ask how applying two design approaches, co-creation and then design thinking, does influence the resulting concepts of more sustainable business models.

14.2 Theoretical Framework

14.2.1 Sustainable Business Models

When retracing the genesis of the concept of a sustainable business model, we look to Stubbs and Cocklin's seminal article [14]. They initiated a description of the characteristics of what makes a business model potentially sustainable. They cite the works of Wicks to describe the effect of sustainability on a firm's business

Table 14.1 Principles of a sustainable business model (SBM)

1	An SBM draws on economic, environmental and social aspects of sustainability in defining an organization's purpose
2	An SBM uses a triple-bottom-line approach in measuring performance
3	An SBM considers the needs of all stakeholders rather than giving priority to shareholders' expectations
4	An SBM treats nature as a stakeholder and promotes environmental stewardship
5	Sustainability leaders drive the cultural and structural changes necessary to implement sustainability
6	An SBM encompasses the systems perspective as well as the firm-level perspective

Reproduced from Stubbs and Cocklin (2008)

model as playing "an integral role in shaping the mission or driving force of the firm and its decision making" [21] (p.104). They research the idea of a sustainable business model, but they remain very broad in a potential application in practice. They refer to a combination of features, conditions, processes and/or narratives. Nevertheless, Stubbs and Cocklin did address the problems with the "neoclassical economic worldview" of organizations by establishing a few principles of a sustainable business model. The authors' main contribution is their definition of a sustainable business model (SBM) in the six principles in the following table (Table 14.1).

Another approach was that of Lüdeke-Freund who took the broad areas of business models and instilled a notion of sustainability [22]. This results in a definition closer to Osterwalder and Pigneur's business model canvas: "A business model for sustainability is the activity system of a firm which allocates resources and coordinates activities in a value creation process which overcomes the public/private benefit discrepancy. That is, a business model for sustainability is the structural template of a business logic which creates the business case for sustainability" (p.43).

Although the initial definition of a business model is centred around the notion of value, it is implied that economic value is the only dimension of value that matters enough to be measured. That was the founding critique that drove Joyce, Paquin and Pigneur to create two additional layers for the original business model canvas by Pigneur and Osterwalder [23]. They added an environmental life cycle layer and a social stakeholder layer in a tool called "the triple-layered business model canvas". Subsequently they defined a sustainable business model as the means of creating, delivering and capturing three types of value: economic, environmental and social.

In order to integrate sustainability into innovation management, Nidumolu, Pralahad and Rangaswami empirically documented how 30 organizations move through five stages in transforming towards sustainability [24]. The authors argue that sustainability has now become the key driver of innovation and not simply the reverse: "Executives behave as though they have to choose between the largely social benefits of developing sustainable products or processes and the financial costs of doing so. But that's simply not true. We've been studying the sustainability initiatives of 30 large corporations for some time. Our research shows that sustainability is a mother lode of organizational and technological innovations that yield both bottom-line and top-line results" (p.3) (Fig. 14.1).

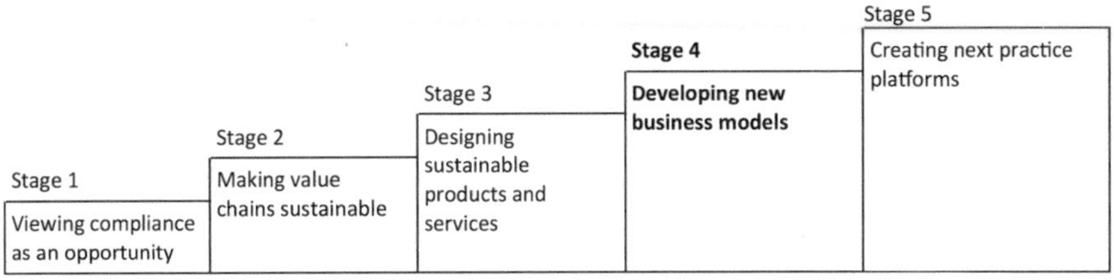

Fig. 14.1 Five stages of sustainable innovation. Reproduced from Nidomulo et al. (2009)

In short, we view the model proposed by Nidumolu et al. favourably and wish to position our present research as a means to move from level three, product and service innovation, to level four, business models. Nidomolu et al. speak of a design as a process to arrive at sustainable products and services in the third level. We contend that a design approach should also lead the way to developing new business models for sustainability. The authors do describe the central challenge of the fourtth stage as "to find new ways of delivering and capturing value, which will change the basis of competition" [24] (p.9).

From our literature review on sustainable business models, we find that the authors describe their goals mostly in terms of reaching an end. We see a gap in research as there yet has been any light shed on the means or the process by which we can arrive at these sustainable business models. Moreover, we believe that there will always be room to make a business model more sustainable. In other words, there is no finish line or end state where one can determine a business model to be sustainable in the absolute. Business model innovation is always improving in relative terms, towards becoming more sustainable. In all, our research begins to study this gap in the means towards envisioning more sustainable business models by undertaking a design approach.

14.2.2 A Design Approach

We seek to study business models for sustainability through a creative process best described as a design approach. Defining design has often been a difficult task because it is comprised of many intangible elements such as intuition, imagination and creativity. Still today Herbert Simon is recognized as having provided a starting point when defining design when he stated that design was "moving from existing situations to preferred situations" [25]. Moreover, the process of design is a difficult undertaking because it deals with more elusive elements such as ambiguity and uncertainty. Although these human intangibles can be seen as barriers to scientific epistemologies, we will later present our action research method that enables a design approach to confront and ultimately shape new business model concepts. Henceforth, when speaking of design, we understand it as a means to produce knowledge in order to move towards preferred situations.

First, we posit that a design approach implies a purposeful process. Many different researchers proposed their interpretation of the design process [26, 27]. Research on the design process has been described most proficiently in the 1980s under the name of "design methodology". In the end, the presence of divergence, emergence and convergence is central in all interpretations. Design methods are essentially a forward spiral of analysis-synthesis-realization steps. Any design method must permit multiple kinds of logical, ethical or creative thoughts to coexist within the iterative progress of the project [28]. For this research, we chose to combine two design processes, co-design and design thinking, to arrive at more creative and more pertinent outcomes. We now dive deeper into each process.

Co-creation. The roots of co-creation begin in the 1970s with participatory design approaches. These approaches comprised of tools and cooperative techniques used in activities such as workshops, prototyping and planning. Participatory design practices such as co-creation were developed to provide users the means to take an active part in the design process [29]. As an expert within the participatory field, Caroll defines participatory design as "the direct inclusion of users within a development team, such that they actively help in setting design goals and planning prototypes" [30]. The difference between participatory design and co-creation lies in that co-creation doesn't assume that any stakeholder has a more important role to play in the participatory design process. Co-creation as a design process was defined by Sanders and Stappers as "any act of collective creativity, i.e., creativity that is shared by two or more people" [31]. By defining co-creation in a broad meaning, the authors point to a collective creativity that can be applied in a design process with applications ranging from the physical to the metaphysical.

The diversity of all participatory practices has not led to a single theory, paradigm of study or common approach to practice [32]. Rather, different perspectives focus on certain aspects of user involvement, and most of participatory design theories and practices require simply the combination of multiple perspectives [33]. We argue that this need for multiple perspectives should also enable multiple design processes. This explains why we choose to enhance a co-creation process with a design thinking process.

Vicente brings to the table some limitations to participatory design: leaving possibilities of new technologies unexplored, the use of incomplete design methods such as scenarios or prototyping and the lack of purpose when analysing the design's progression [34]. To palliate to these deficiencies, Bødker and Iversen offer a frame set to facilitate the development of the project [35]. They present an interesting hybrid approach where the designer facilitates the process. They propose that the designer must envision a strategy for the entire process. This strategy should evolve and develop itself depending on the users, the situation and the progress of the design activity.

Bødker and Iversen state that users need to be implicated in the design process, but they proposed that the participative process requires the planning and intervention of the designer to insure its success. They call this professional participative design (proPD). In developing proPD, the authors respond to the limitations foreseen by Vicente. They propose using scenarios and prototyping and reflecting on the initial problem to have a sense of perspective upon the process. The authors state that the advantage of a professional participative design process is that it remains always in context because the designers implicate problem owners directly in the solution process. Yet, the problem owners might not have the capacity, experience or time to envision future solutions to complex problems such as business model sustainability. In this research, we hypothesize that some shortcomings of a co-creation process in creating more sustainable business models can be levelled out if the designer can also play the role of an external consultant specialized in design thinking.

Design thinking. In 2008, a new avenue for the design process emerged in the management journals oriented towards practitioners. Instead of applying itself to the creation of material products, design methods are used to rethink any type of business problem. Called design thinking, this movement was captured by management to solve their seemingly dialectical problems by using creativity and empathy for the customer [36]. Designer Marty Neumeier calls them "Problems you can't manage your way out of" [37]. By rejecting a closed choice between two options, design thinking creates new solutions starting from consensual forms of reasoning. This was coined as integrative thinking [38]. Creating a third way requires a new approach that can combine the analytical approach taught by management schools and the situational understanding taught in social science schools [39]. Although design thinking was popularized in a business setting, it is now being applied by many types of professionals such as health experts, community organizers, participatory political workers, social workers and teachers of all kinds.

Design thinking remains an intuitive approach taught in industrial design schools. Like a reflex or a part of their DNA, designers use these methods tacitly. Moving from analysis to synthesis in successive iterations, the design practitioner evolves his understanding of the situation. A design thinker can critique his own practice and adjust it in reaction to the dynamics of the context [40]. Complex problems require designers to create ways into them by conducting empirical experiences situated in the context of real-life conditions. This approach is the essence of design thinking.

A debate remains as to whether design thinking remains part of industrial design. Today, some designers are moving past the industrial aspect of product design for broader outcomes. By expanding on the different actions of the design process as well as on the different outcomes, Buchanan describes how design has evolved from creating symbolic forms to harmonious functions, then meaningful interactions and now purposeful organizations [41]. The Industrial Designers Society of America currently (IDSA) defines industrial design as: "the professional service of creating and developing concepts and specifications that optimize the function, value and appearance of products and systems for the mutual benefit of both user and manufacturer" [42]. Beyond the confines of industrial design, the practice of design is expanded by an expertise on the creation of products, services, systems and experiences. To that list of deliverables, we have added the design of businesses [43]. "That designers work for or with organizations is a familiar concept. That design can have an impact upon organizations and that design thinking can shape organizational behavior in productive ways is less well established within the literature devoted to design and design practice" (p.1).

To sum up, we began this section by reviewing existing literature in the field of sustainable business models. We then point to a gap in the process of arriving at more sustainable business model. We propose a design approach as means to research the subject further. In a design approach, there are multiple methods from which we select two and combine them. We hypothesize that this combination will lead to more profound results. In the next section, we seek to demonstrate how we undertook a dual process of co-creation and design thinking within an action research method.

14.3 Method

Action research. Our research method in general can be summed up as action research. Action research finds its roots in Dewey's approach of "learning by doing" [44]. We define action research as a method to extract knowledge from empirical practice. One of the advantages of action research is that it generates knowledge set within a real-world context. According to Reason and Bradbury, the goal of action research is to "produce practical knowledge that is useful to people in the everyday conduct of their lives" [45, p.2]. Thus, there are two simultaneous objectives to action research. The first is to try to solve the problem at hand. The second is to provide a field of research with new knowledge that addresses gaps within that field. O'Brien speaks of a dual commitment to study a system while collaborating with members of the system towards preferable outcomes [46]. Within action research methods, specific approaches vary widely although the criteria of relevance and rigour are upheld [47]. Chisholm and Elden distinguish three types of action research: instrumental, theoretical and emancipatory [48]. Our research fits mostly in the category of instrumental because it first seeks to guide organizations. It is also emancipatory because we hope to guide organizations in a design process to imagine a potential future.

The complicity between the design process and action research was discussed in depth by Swann [49]. Swann elaborates on how a design approach to problem solving and the action research method of generating knowledge are similar in vocation. Many aspects of action research are also present in a design process as they are both iterative and integrative. The choice of action research is coherent with our project to design preferable situations. In this research, the action is the conceptualization of an organizations' business model as more sustainable.

Because the researcher is knowingly participating in creating the data, action research provides a setting to take this bias into account. In the end, there are many types of roles for researchers in action research such as an expert or a collaborator [50]. Here, the researcher plays the role of the expert insuring the quality and effectiveness of the research design, data collection, analysis and induction while maintaining a collaborative relationship by facilitating the activities with organizational participants.

The most important limits of action research lie in the researcher. His participation and facilitation are embedded with tacit and subconscious thinking that influences the generation of knowledge. The only way out of this dilemma is to attempt to reveal presuppositions and cognitive processes. This issue of trustworthiness is well explored by Guba when discussing naturalistic inquiries such as phenomenology or ethnography [51]. Because action research is interpretative by nature, the role of the researcher in this study is effectively under consideration. The values, judgement and biases of the researcher inevitably influence the results of action research. The literature review is one of the main strategies to avoid bias and misinterpretation throughout the study. Moreover, judgements and interpretations that arise in the upcoming analysis of the results will be grounded in past theory.

Research protocol. Our research protocol consisted in undertaking multiple workshops with a consistent design approach. In both cases, we used the same tools and we gathered the same type of outcomes, but we didn't follow the same process. The design tools we used were the triple-layered business model canvas by Joyce, Paquin and Pigneur and playing cards with existing business model patterns [23]. The outcomes of our workshops are more sustainable business model ideas and concepts. When it comes to the process, we devised two different design workshops. First we guided participants in a co-creation process, and second we undertook a design thinking process as external designers. We now further expose the different processes.

For the co-design process, we organized a series of workshops with clients of the Institut de développement de produits (IDP). They are a non-profit organization whose mission is to teach manufacturers how to improve their innovation practices. In all, 13 different manufacturing companies were represented by 17 research and development professionals who participated in our full-day workshops. The participants were all part of the innovation function of their companies, but they all had various backgrounds such as marketing or engineering.

The main researcher played the role of a facilitator during the co-creation workshop which was split in two parts. In the morning, the participants focused on illustrating the current business model of the organization with the triple-layered business model canvas. This provided a baseline to reflect upon the current business model and its shortcomings. In the afternoon, the participants were then put into teams of three or four. They were handed a set of 24 cards that illustrated existing business model patterns to use while imagining a more sustainable business model. In turns, they each had 45 min to first choose 3 cards (one for economic, environmental, social), second generate ideas and third conceptualize a more sustainable business model. We ended the workshop with the participants sharing their broader business model ideas and discussed the virtues of the tools, their outcomes and the design process. The design researcher (i.e. the author) prepares and facilitates the co-creation process in following with the theory on participatory design practices. What brings novelty to this research protocol is that the design researcher, who is a professional industrial designer, also participates in a subsequent design thinking process.

For the design thinking process, we choose to further 5 of the 13 manufacturers' cases. We presented to a professional industrial designer, working as a principal in a well-established design agency, a very short design brief. The professional designer was instructed to undertake a design thinking approach to imagine sustainable business model concepts and sketch out products to support those ideas. The brief contained an image of the first page of the 5 case manufacturers' website. The professional designer used on average 30 min per case to brainstorm original ideas. The design researcher then joined with the professional designer building on the product sketches to fill out the triple-layered business model canvas for another 15 min per case. Together they acted as an external design team by completing the business model concepts.

It is important to note that, in the end, both processes used the same tools and ended up with the same outcomes of triple-layered business model canvas. However, the content of the ideas and concepts varies greatly as presented in the results section.

The results of this research are presented as five cases. Each case presents four distinct elements. First, we expose the basic profile of the organization with quantitative data from external sources such as the number of employees and general qualitative markers from the researcher's perspective such as sustainability maturity. The participating organizations' names have been changed in order to preserve the participants' preference for corporate anonymity. Second, we shortly describe their current business model. Third, we share the main elements of the business model concept they co-created with their fellow participants. Fourth, we share the main elements of the business model concepts resulting from the design thinking process.

14.4 Results

1. **Rainpipe**
   Main industry: Rainwater management
   Product: Rainwater pipes
   Number of employees: 170
   Date of establishment: 1978
   Innovation maturity: Intermediate
   Sustainability maturity: Beginner
   B2B

Rainpipe's current business model is based on selling a variety of products that answer all storm water management needs such as collecting, conveying, treating and storage. They design, manufacture and distribute high-quality products, primarily made in HDPE plastic resin. Considering themselves as leaders in eastern Canada and northeastern USA, they serve 5 major sectors from agricultural, natural resources, infrastructure, residential and commercial.

The co-creation design business model concept for Rainpipe begins with the idea of helping cities finance the purchase of their higher-quality product. They are in competition with low-cost cement solutions, and a financing plan could help extend the purchase on a longer period which goes hand in hand with the product's long-lasting life of more than 100 years. This long life is also part of the environmental strategy as it would encourage efficiency. In the long term, the city would reduce the amount of materials and more importantly the amount of heavy maintenance work required to upkeep the water management system. From a social point of view, this concept would look to serve the needs of smaller cities who don't necessarily have the engineering expertise in-house. Rainpipe could create partnerships with local engineers to be part of the project as entrepreneurs.

The professional design thinking approach dreamed up a vision of a company that not only provides storm water solutions but one day generates electricity with micro-turbines within the storm water pipes. The company could create partnerships or expand its role from a manufacturer to a contractor capable of building street infrastructures. We propose for the municipal client to rent out the space beneath its streets so that utility companies would pay a fee for the access. In return, the utility companies would charge the city for the services it provides such as data, electric, gas or water management. Part of the environmental and social benefits are that citizens could have access and even become owners of a locally distributed form of renewable energy. In all, this concept of a more sustainable business model for Rainpipe revolves around the idea to both sell the energy it generates and the service of storm water management while playing a larger role in the construction and management of the infrastructure.

2. **Offurniture**
   Main industry: Commercial furniture
   Product: Worktable
   Number of employees: 850
   Date of establishment: 1983
   Innovation maturity: Advanced
   Sustainability maturity: Advanced
   B2B

Offurniture's current business model is based on selling a complete furniture solution to large organizations whom wish to personalize or customize their workspaces. Offurniture's large production capacity allows them to supply dealers for a distribution across North America. They roll out new products regularly, pay close attention to design details and aim for a price point under industry giants.

The co-design for Offurniture added to the current manufacturing business model a new location service. This could attract smaller client organizations with 3–4-year leases. The materials could come from refurbished furniture which would reduce the material production volume while leveraging the existing employee resources. The transaction could take place over the Internet and allow for the client to choose preset customization options. This alternative to the current marketplace could allow an outreach into developing countries for clients to get access to higher-quality workstations.

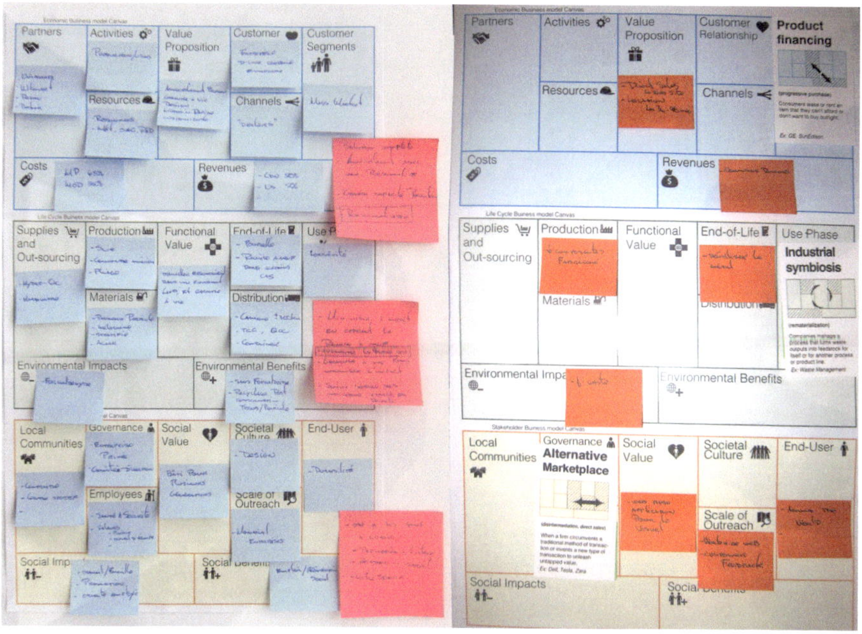

Fig. 14.2 Illustration of Offurniture's present business model and co-creation concept

In the design thinking process, the professional designers also based the business model concept on a leasing service through a transactional website. However, the business model focused on offering a B2C approach for the employees of the client organization. The business model concept makes it possible for the employer to create an individual budget for each of his employees. In turn, they each can select a furniture solution tailored to their individual needs and even the amount of refurbished or new products. This concept can offer similar environmental benefits from remanufacturing, but it encourages sufficiency by building an emotional relationship when selecting a custom solution. The business model pushes for a behaviour change in employers who share the responsibility with their employees and greatly enhance their autonomy (Figs. 14.2 and 14.3).

3. **PaperLam**
   Main industry: Pulp and paper
   Product: Laminated paper packaging
   Number of employees: 75
   Date of establishment: 1992
   Innovation maturity: Advanced
   Sustainability maturity: Leader
   B2B

PaperLam's current business model is mostly based on selling laminated paper to package and protect large paper rolls during transport. Their main client base is the pulp and paper industry where demand for newspapers is declining. An alternative market

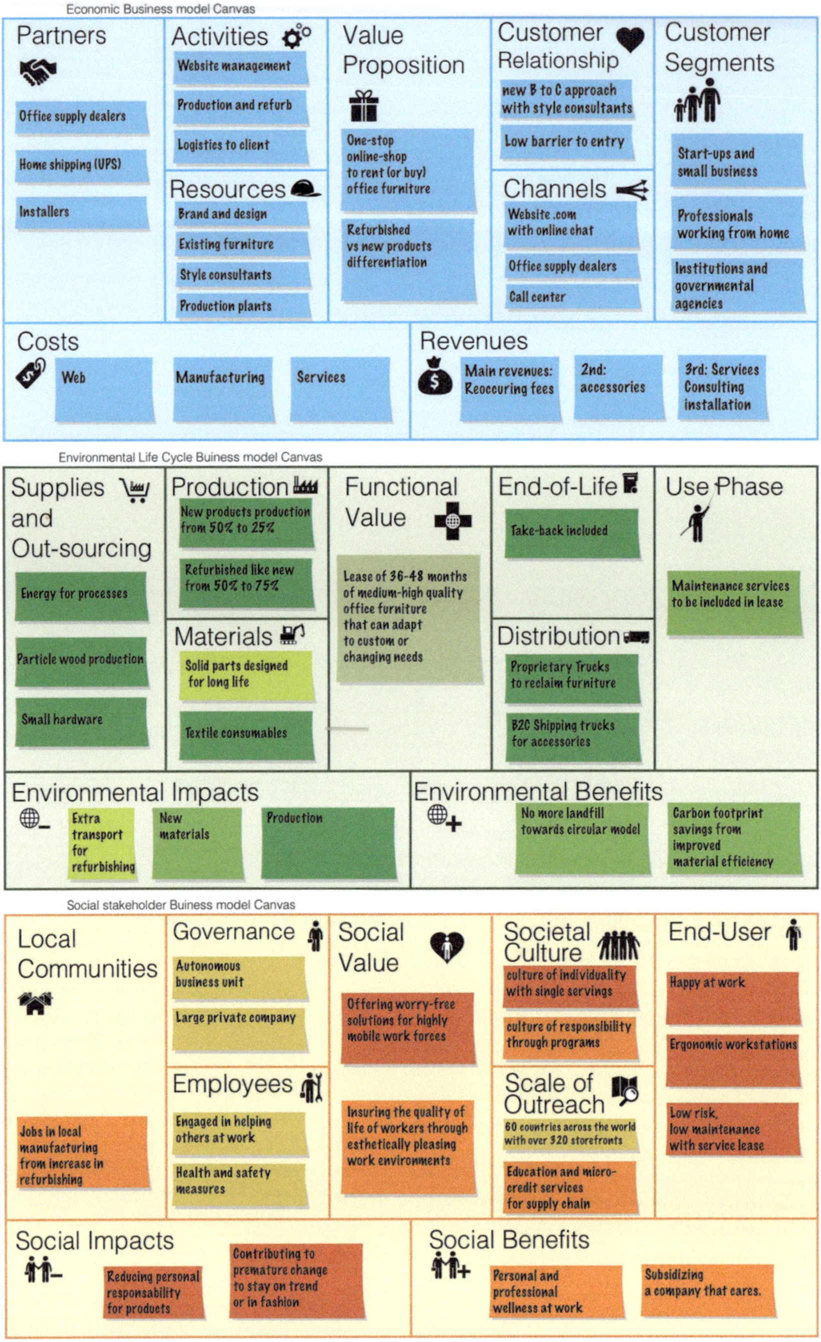

Fig. 14.3 Illustration of Offurniture's more sustainable business model concept resulting from the design thinking process using the triple-layered business model canvas

they are also present in is the food packaging industry. PaperLam generates value by transforming materials from the other business units of PaperLam group such as recycled paper fibres or recycled cardboard into specialized and customized packaging solutions.

The co-design gave birth to a business model concept where PaperLam earns a part of the savings they are able to generate in partnership with the client. This idea begins with PaperLam seeking to understand its client's activities in order to better design and even operate new technologies at the client location. Inversely, this concept moves the business model away from the client buying and owning paper volume towards monetizing the functional service provided by PaperLam. Also from an environmental perspective, the new packaging processes will be built to create financial savings by optimizing time and materials. By including the client in the improvement process, they become part of the solution.

The external design team's business model concept starts with the idea that packaging material could provide more value functioning as a data point more than as a protective device. The fibres would be "smart" in the sense that they can be identified and tracked with a simple app. The business model proposes a hidden revenue model coming from its clients in the food industry paying for access to logistical databases about where and how their products are being shipped, used and recycled. To improve the rate of recycling, each kilo of fibres would be given a financial value depending on the age of the fibre materials. This co-creates a new marketplace that incentivizes people to recycle more and faster. This take-back program could even become an alternative full-time living wage for unemployed or low-income earners.

4. **ProBeauty**
   Main industry: Beauty products
   Product: Make-up and skin cream
   Number of employees: 175
   Date of establishment: 1949
   Innovation maturity: Intermediate
   Sustainability maturity: Beginner
   B2C

Currently, ProBeauty's business model is based on selling beauty products and creams at a lower price than industry giants. They market the added value of all their products being hypoallergenic. Their manufacturing processes involve little automation and function in batches. They are agile in answering local trends and fashions. They have also segmented their client base by creating a sister brand aimed at a younger female public.

The co-design process arrived at a business model that centres on a mobile application that takes a picture or even a video of the customer and then proposes a rendering of 3–5 make-up styles in real time. In the app, there is the possibility to order all the beauty products required for a certain result. The revenues come from monthly subscription and a feeling of paying for success. The packaging products for this line would come from renewable sources and would be biodegradable. A marketing campaign would build a partnership with a cancer patient centre and give away the subscription to women patients who have gone through chemotherapy. This would allow the cancer survivors to go back to feeling good about themselves.

The design thinking process arrived at a concept for ProBeauty that also works in a subscription model. The product formula would be split into three components: a base cream, ingredient "A" and ingredient "B". The customer first receives an explorer's kit to test out different proportions. Once satisfied with the right formula, the customer sends the details to ProBeauty who produces batches every 3 months for the time of the subscription. The packaging is to be sent back to the company to be filled up again thus reducing the need for new materials. So this model thrives from mass customization while co-creating directly with the customer. From a social perspective, the business model allows for disintermediation by creating a stronger relationship between the customer and the provider.

5. **Maverick**
   Main industry: Construction
   Product: Bath and showers
   Number of employees: 265
   Date of establishment: 2004
   Innovation maturity: Advanced
   Sustainability maturity: Beginner
   B2C and B2B

Maverick is currently running a business model that sells a differentiation of exclusive bathtubs to multiple clients at a time. They distribute certain products through outlets such as big box stores and specialty retailers. But they also contract multiple units for commercial and institutional clients. The company is a large employer in their home town playing an important role in that community.

The business model concept of the co-creation workshop came up with the idea of providing a design expertise for the customer. By customizing the product to the specifications of the customer, the company could cater to the experience the user seeks. When installing a new bathtub, the company takes back the old bathtub. This has positive repercussions on the community and employees.

The design thinking team's idea for the bath business model is to create a complete solution by creating partnerships with other manufacturers in the region. By offering a standard solution (razors) to housing complexes, the consumption behaviours (blades) of each owner or tenant can be compared to the building average. As the complete bathroom solution generates smaller revenues to increase adoption, it is the revenues from energy and water consumption that play a central role in the business model thus encouraging sufficiency.

14.5 Discussion

Findings. Our first research question asks what could more sustainable business models look like for small- and medium-sized manufacturers. It was by combining multiple business model patterns that we can address the economic, environmental and social aspects. We used the triple-layered business model canvas and business model pattern cards to generate more sustainable concepts as 5 cases presented in

Table 14.2 Comparison of business model patterns used during co-creation and external design processes

Case	Patterns used in co-creation	Patterns used in design thinking
Rainpipe	Product financing, encourage sufficiency, social entrepreneurship	Pay for success, substitution for renewables, social entrepreneurship
Offurniture	Product financing, industrial symbiosis, alternative marketplace	Produce on demand, encouraging sufficiency, behaviour change
PaperLam	Pay for success functionality not ownership, inclusive sourcing	Hidden revenue, circularity, co-creating a marketplace
ProBeauty	Subscription, substitution for renewables, social entrepreneurship	Subscription, circularity, alternative marketplace
Maverick	Provide on demand, substitute with renewables, inclusive sourcing	Razors and blades, encouraging sufficiency, co-creating a marketplace

the previous results section. Our results showed great variety in the concepts, and we now interpret them to confirm our initial hypothesis and arrive at new findings.

Since the tools and the outcome format for both design processes were very similar, we can directly compare the underlying business model patterns that were used to generate a more sustainable concept. In the following table, we list for each case three business model patterns (Table 14.2). The first pattern of the three refers to the economic, the second to the environmental and third to the social aspect of the business model.

When analysing this table, we see that there is very little overlapping in both outcomes of each design process. This can be explained by the fact that design thinking team was stimulated to propose original business models building upon the co-creation outcomes. Nevertheless, this table serves to demonstrate the heterogeneity in the choice of business model patterns in the pursuit of more sustainable business models. That is to say that it is not the choice of a specific pattern that generates more sustainability. On the contrary, the sustainable ideals, driving the foresight design approach, made use of the business model patterns to arrive at that end. The varied uses of business model patterns illustrated in this table demonstrate our hypothesis that a co-creation process and a design thinking process are mutually reinforcing when it comes to imagining more sustainable business models.

Our second research question is concerned with how co-creation can in addition to design thinking lead to a vision of organizational sustainability. When analysing the resulting concepts in each of our research cases, we find four advantages to combining both types of design processes.

There are two advantages to proceeding with a co-creation process before undergoing a design thinking process. Firstly, as mentioned by Lüscher and Lewis, "participant engagement is critical to ensuring relevance" [50]. We witnessed how co-creation greatly improves the pertinence of the subsequent design thinking process. Schön spoke of the designer's reflective practice [40]. He states that knowledge arises within the design process as the designer gets "talkback" from the design activity. In this case, the co-creation serves as an initial design iteration from which to generate a "talkback".

For example in the Rainpipe case, both concepts started from the struggle for the client municipalities to finance higher quality. The co-creation concept remains within the current economical context, whereas the design thinking creates a new context to eliminate the initial struggle. Consequently, the environmental and social approach changed from reducing impacts to generating benefits. In both concepts, the environmental and social benefits reflected the level of change that the business model concept involves. The initial co-creation situated the project and created a first iteration as a starting point for the following design thinking process. Moreover, we find that the co-creation process allowed for tacit information to be discovered on the nature of the context and the ambitions of the organization as the participants created their visions of the future. This was useful in fine-tuning the additional design thinking approach towards a more pertinent and more complete concept.

Secondly, the co-creation process allows for the participants to become familiarized with the design tools and the outcomes of the design process. Their participation also provides them with a first-hand experience of the size of the challenge that is imagining sustainable business models. For example, in the ProBeauty case, the concept of personalizing the product is present in both the co-creation and the design thinking outcomes. The co-creation team proposed to make a mobile application and later attempted to address the social value by adding a marketing campaign. The design thinking process led to a "do-it-yourself" where the clients directly enjoy the social benefits of adapting the products to their own needs. Having gone through a design process themselves, ProBeauty was able to better appreciate the designer's concept that addressed the social dimension within the subscription business model.

There are two more advantages we have found in adding a design thinking process when seeking to create sustainable business model concepts. First, design thinking is less concerned with the best solution and more with a viable solution to use as a learning tool. Although the learning organization is an ongoing subject of interest in management [52], the difference with the design thinking process is that it purposely thrives on learning quickly to evolve a project. Tom Kelly instilled the following mantra in the IDEO design consultancy: "fail often to succeed sooner" [53] (p.232). This philosophy is where design and management differ because co-creation builds engagement and usually seeks a form of consensus. Johne writes about how managers seek to "avoid mistakes with new products rather than using them as a means for exploiting market potentials" [54] (p.177). When it comes to the PaperLam case, this notion of a learning tool was also foundational in the design thinking process. The designers leapfrogged over the struggle to sell more material in a declining market and demonstrated the learning potential and the new value of offering access to data. We gather from this example and its root in past theory that a design thinking process frames the different outcomes as a means for learning and not as an end in itself.

Secondly, we have found that the design thinking process can move further into building foresight visions by defining ideals as guidelines. In two cases, Offurniture and Maverick, the co-creation team was concerned with protecting manufacturing jobs as well as maintaining the current business model alive. However, the design thinking team was more concerned with better answering the client's needs and even proposed ideals. In the Offurniture case, the business model concept envisioned boosting the client's employee autonomy. In the Maverick case, the goal was to reduce energy and water consumption. This allows us to conclude that a key difference between co-creation and design thinking is that external design thinking better suits building a foresight vision. We can interpret these last two advantages of an additional external design thinking as a demonstration that organizations are focused on solving short-term problems, whereas a foresight design approach uses outcomes as way to ask long-term questions.

Contributions. As mentioned earlier, the gap we are addressing in the field of sustainable business models lies in the process. To arrive at exploring this process, we chose to apply two design processes by following an action research methodology. Our resulting 5 cases can be considered as a first contribution in that they demonstrate how to design and even communicate a more sustainable business model. These cases also exhibit the usefulness of the triple-layered business model canvas and its capacity to support a design process.

A second contribution comes from how we combined co-creation and design thinking processes to build on the strengths of each other. Following in the path laid out by researchers in participatory design, this study should be considered as another example of a means for designers to perceive tacit needs by allowing for their stakeholders to express their ideas within a co-creation process. As the users create freely and intuitively while expressing their needs, solutions and ideas, the designers can have a third-person point of view on the results. Therefore, we see designers assuring the creative process is right while interacting with participants taking on the role of a facilitator of conversation. After that, we see great value in the additional insight that come about when designers can take advantage of the content generated by the client. To that we demonstrated the value of an additional design process where design thinking allows for the designer to propose visions of the future. This reflects the views put forward by Papanek as he acknowledged that designers have not to design for money but to design for many [55].

14.6 Limits and Future Research

The resulting outcome of our dual design process has been foresight concepts of more sustainable business models. There are a few limits when it comes to interpreting our results. First, we are discussing intangible ideas and concepts. This is just the first phase of a design process when it comes to implementing business model innovation. A longitudinal approach would have to follow an organization over many years to observe the design process in all its iterations, from the initial idea to the market launch.

The second limiting factor is the nature of the results we seeked to gather. When speaking of the limits of natural inquiry such as action research, Guba proposes to clearly define the problem boundaries to determine what to include and exclude from consideration [51]. We oriented this research to arrive at specific outcomes of more sustainable business model concepts. However, the same dual process could be used to generate other types of outcomes such as public spaces, product designs or corporate processes. These different outcomes could tell more about the robustness of this dual design process. This would help identify the transferability of our research results.

When it comes to our research protocol, we didn't allow for much iteration. There was little going back and forth to improve the concepts. As Jones mentions, creativity works best when progressing in multiple back-and-forth motions between the problem and solution spaces [56]. This further promotes the idea that the design process is not a systematic step-by-step sequence of predetermined activities and it can be enriched by multiple processes such as co-creation and design thinking. Nonetheless, our design process did allow for an ongoing loop linking thinking and making as well as inspiration and expiration as described by Findeli [29]. This allowed for the design process to move in successive steps from awareness to critique, to reflection and then to creation.

As we have mentioned, the role of the researcher is highly relevant in action research because he is initiating a pattern of events, with engaged participants or through design, that would otherwise not happen naturally. In this study, the researcher played the role of a facilitator and an external designer. By changing the researcher, additional studies could confirm our results and better assess the influence of the researcher in the process. Inversely, the will, openness and creativity of participating companies also influence our capacity to determine the relationships between the design process and the results. Then again, the goal of the dual process was to surmount the potential lack of creativity by the participants through an additional design thinking effort.

14.7 Conclusion

We arrive at the conclusion that the old saying "If you want to go fast, go alone; if you want to go far, go together" doesn't apply to envisioning the future of organizations such as more sustainable business models. By that we mean that our co-creation process resulted with ideas that were quicker to implement. In other respects, the design thinking process built on the co-creation process to generate far-reaching ideas for the future. So, we would have to change the saying to "if you want to change fast, envision together; if you want to imagine far into the future, design from ideals".

There are many existing organizations that could benefit from establishing a practice of design to continuously improve their business models. In our research, we have witnessed the influence a dual design process can have in five cases where we imagined business models to be more sustainable. We seek to learn more about how a design approach can help society and its organizations evolve past the industrial age. In the future, we encourage organizations and designers to use the strengths of co-creation and design thinking to set in motion an innovation process that can lead to the transformation into a sustainable paradigm. The end goal of our research efforts is to help build the tools, develop the methods and provide examples of business models for sustainability.

References

1. Hart, S.L.: Capitalism at the Crossroads: The Unlimited Business Opportunities in Solving the World's Most Difficult Problems. Pearson Prentice Hall, Saddle River (2005)
2. White, M.D.: The Reinvention Imperative. The Magazine, Harvard Business Review (2013)
3. Christensen, C.M.: The Innovator's Dilemma: When New Technologies Cause Great Firms to Fail. Harvard Business Press, Boston (1997)
4. Gilbert, C., Eyring, M., Foster, R.N.: Two routes to resilience. Harv. Bus. Rev. **90**(12), 66 (2012)
5. Cooper, R.G.: Winning at New Products. Basic Books, New York (1988)
6. Manzini, E., Jegou, F.: Scenarios for sustainable households. Paper at the Greening of Industry Network Conference, Rome, November 1998, pp. 15–18
7. Osterwalder, A., Pigneur, Y.: Business Model Generation: A Handbook for Visionaries, Game Changers, and Challengers. Wiley, Hoboken (2010)
8. Zott, C., Amit, R., Massa, L.: The business model: recent developments and future research. J. Manag. **37**(4), 1019–1042 (2011)
9. Teece, D.J.: Business models, business strategy and innovation. Long Range Plann. **43**(2–3), 172–194 (2010)
10. Jetter, M., Satzger, P.D.G., Neus, A.: Technological innovation and its impact on business model, organization and corporate culture—IBM's transformation into a globally integrated, service-oriented enterprise. Bus. Inf. Syst. Eng. **1**(1), 37–45 (2009)
11. Chesbrough, H.: Business model innovation: opportunities and barriers. Long Range Plann. **43**(2–3), 354–363 (2010)
12. Olivia, R., Quinn, J.: Interface's evergreen services agreement. Harvard business school case 9-603-112. Harvard University, Cambridge (2003)
13. Johnson, M.W.: Seizing the White Space: Business Model Innovation for Growth and Renewal. Harvard Business Press, Boston (2010)
14. Stubbs, W., Cocklin, C.: Conceptualizing a 'sustainability business model'. Organ. Environ. **21**(2), 103–127 (2008)
15. Verhulst, E., Boks, C.: Sustainable design strategies in practice and their influence on business models. In: Matsumoto, M., et al. (eds.) EcoDesign 2011 International Symposium. Springer, Dordrecht (2012)
16. Boons, F., Lüdeke-Freund, F.: Business models for sustainable innovation: state-of-the-art and steps towards a research agenda. J. Clean. Prod. **45**, 9–19 (2013)
17. Upward, A.: Towards an Ontology and Canvas for Strongly Sustainable Business Models: A Systemic Design Science Exploration. York University Toronto, Ontario (2013)
18. Sommer, A.: Managing Green Business Model Transformations. Springer, Berlin (2012)
19. Casadesus-Masanell, R., Ricart, J.E.: From strategy to business models and onto tactics. Long Range Plann. **43**(2–3), 195–215 (2010)

20. Hockerts, K., Wüstenhagen, R.: Greening Goliaths versus emerging Davids—theorizing about the role of incumbents and new entrants in sustainable entrepreneurship. J. Bus. Ventur. **25**(5), 481–492 (2010)
21. Wicks, A.: Integrating strategic decision making & business ethics: a cognitive approach. Best papers proceedings… annual meeting of the academy of management (1996), p. 352
22. Lüdeke-Freund, F.: Business Model Concepts in Corporate Sustainability Contexts From Rhetoric to a Generic Template For 'Business Models For Sustainability'. Centre for sustainability management, Lüneburg (2009)
23. Joyce, A., Paquin, R., Pigneur, Y.: The triple layered business model canvas: a tool to design more sustainable business models. Presented at the ARTEM Organizational Creativity International Conference, Nancy, France (2015)
24. Nidumolu, R., Prahalad, C., Rangaswami, M.: Why sustainability is now the key driver of innovation. Harv. Bus. Rev. **87**(9), 56–64 (2009)
25. Simon, H.A.: The Sciences of the Artificial. MIT Press, Cambridge (1969)
26. Jones, J.C.: The state of the art in design methods. In: Design Methods in Architecture. AA papers, London (1969)
27. Cross, N., Naughton, J., Walker, D.: Design method and scientific method. Des. Stud. **2**(4), 195–201 (1981)
28. Findeli, A.: Rethinking design education for the 21st century: theoretical, methodological, and ethical discussion. Des. Issues **17**(1), 5–17 (2001)
29. Bødker, S.: A for alternatives. Scand. J. Inf. Syst. **15**(1), 87–89 (2003)
30. Carroll, J.M.: Dimensions of participation in Simon's design. Des. Issues **22**(2), 3–18 (2006)
31. Sanders, E.B.-N., Stappers, P.J.: Co-creation and the new landscapes of design. Co-design **4**(1), 5–18 (2008)
32. Slater, J.: Professional misinterpretation: what is participatory design. Presented at the proceedings of PDC, vol. 98 (1998)
33. Muller, M.J.: Participatory design: the third space in HCI. Hum. Comp. Interact. Dev. Process **4235** (2003)
34. Vicente, K.J.: Cognitive Work Analysis: Toward Safe, Productive, and Healthy Computer-Based Work. CRC Press, London (1999)
35. Bødker, S., Iversen, O.S.: Staging a professional participatory design practice: moving PD beyond the initial fascination of user involvement. In: Proceedings of the second Nordic conference on human-computer interaction, pp. 11–18. ACM, New York (2002)
36. Brown, T.: Change by Design: How Design Thinking Transforms Organizations and Inspires Innovation. Harper Collins, New York (2009)
37. Neumeier, M.: The Designful Company: How to Build a Culture of Nonstop Innovation, 1st edn. New Riders, Berkeley (2008)
38. Martin, R.L.: The Design of Business: Why Design Thinking Is the Next Competitive Advantage. Harvard Business Press, Boston (2009)
39. Collopy, F.: Managing as Designing. Stanford University Press, Stanford (2004)
40. Schön, D.A.: The Reflective Practitioner: How Professionals Think in Action, vol. 5126. Basic books, New York (1983)
41. Buchanan, R.: Branzi's dilemma: design in contemporary culture. Des. Issues **14**(1), 3–20 (1998)
42. IDSA.: What is industrial design? IDSA. http://www.idsa.org/what-is-industrial-design (2011)
43. Brown, T.: Design thinking. Harv. Bus. Rev. **86**(6), 84 (2008)
44. Dewey, J.: Moral Principles in Education. Houghton Mifflin Company, Boston (1909)
45. Reason, P., Bradbury, H.: Handbook of Action Research: Participative Inquiry and Practice. Sage, London (2001)
46. O'Brien, R.: An overview of the methodological approach of action research. In: Richardson, R. (ed.) Theory and Practice of Action Research. Joao Pesso, Brazil (2001)
47. Melrose, M.J.: Maximizing the rigor of action research: why would you want to? How could you? Field Meth. **13**(2), 160–180 (2001)

48. Elden, M., Chisholm, R.F.: Emerging varieties of action research: introduction to the special issue. Hum. Relations **46**(2), 121–142 (1993)
49. Swann, C.: Action research and the practice of design. Des. Issues **18**(1), 49–61 (2002)
50. Lüscher, L.S., Lewis, M.W.: Organizational change and managerial sensemaking: working through paradox. Acad. Manage. J. **51**(2), 221–240 (2008)
51. Guba, E.G.: Criteria for assessing the trustworthiness of naturalistic inquiries. ECTJ **29**(2), 75–91 (1981)
52. Senge, P.M.: The fifth discipline. Meas. Bus. Excell. **1**(3), 46–51 (1997)
53. Kelley, T., Littman, J.: The Art of Innovation: Lessons in Creativity from IDEO, America's Leading Design Firm, 1st edn. Crown Business, New York (2001)
54. Johne, A.: Succeeding at product development involves more than avoiding failure. Eur. Manage. J. **14**(2), 176–180 (1996)
55. Papanek, V.: The Green Imperative: Natural Design for the Real World. Thames and Hudson, New York (1995)
56. Jones, J.C.: Design Methods: Seeds of Human Futures. Wiley, London (1970)

Chapter 15
Mass Customization in SMEs: Literature Review and Research Directions

Stig B. Taps, Thomas Ditlev, and Kjeld Nielsen

15.1 Introduction

Mass customization was introduced more than two decades ago and was popularized by Pine in his book from 1993 [32]. In the research and literature published since then, it seems there is a focus primarily on large enterprises. The examples that are most often highlighted as successful implementations of mass customization are usually from high-volume consumer electronics or automobiles.

However, as the approaches to implement mass customization have matured, including IT tools, methods, and flexible manufacturing equipment, mass customization has become more accessible to companies. Whereas mass customization was initially applied mostly by companies, which were former mass producers, more companies producing one-of-a-kind products, or engineer-to-order products, have recently recognized the potential in utilizing mass customization tools. Since many small and medium enterprises (SMEs) produce high variety in lower volumes, mass customization is intuitively an attractive business strategy.

One example of where the tools of mass customization have become more accessible is product configuration software. Product configuration software is the software which is used for clarifying customer requirements and translating these into a product specification from which a product can be manufactured. In the early days of mass customization, companies would have to develop product configurators from scratch, using ordinary programming languages, which implied that only large companies with sufficient IT competences and resources would be able to develop product configurators. Today, however, affordable commercial off-the-shelf product configurator software is available which allows companies to easily develop a configurator. There is even an open-source product configurator available for free. This

S.B. Taps • T. Ditlev (✉) • K. Nielsen
Department of Mechanical and Manufacturing Engineering, Aalborg University, Aalborg, Denmark
e-mail: Brunotaps@m-tech.aau.dk

J. Bellemare et al. (eds.), *Managing Complexity*, Springer Proceedings in Business and Economics, DOI 10.1007/978-3-319-29058-4_15

means that product configuration is no longer only for large companies but can now be implemented by SMEs as well. As pointed out by Salvador et al.[34], mass customizers need three fundamental capabilities to be successful: (1) solution space development, identifying the attributes along which customer needs diverge; (2) robust process design, reusing or recombining existing organizational and value chain resources to fulfill a stream of differentiated customer needs; and (3) choice navigation, supporting customers in identifying their own solutions while minimizing complexity and the burden of choice [23, 34]. These are general capabilities which must be present in large companies as well as SMEs. However, we suspect that the approach to obtain and maintain these capabilities will differ from large companies and SMEs. Looking into previous literature reviews within mass customization reveals that hardly any literature identified in those reviews focuses directly on mass customization and SMEs [12, 35]. It is hence relevant to investigate to what extent any other previous literature has this focus. The research question of this paper is thus:

What literature has focused on SME-related issues of mass customization, and within which specific areas has this research had its focus?

15.2 Methodology

To address the research question above, we chose to perform a systematic literature review [4], where an explicit procedure is followed. No restrictions were imposed on the publication year; however, only peer-reviewed articles from journals and conference proceedings were included in the review. The literature review was performed by the steps outlined below:

1. The databases Thomson Reuters Web of Science and Elsevier Scopus were queried using the search string: ("mass customization" or "mass customization") and (SME or SMEs or "small and medium enterprises"). This initial search returned 32 records from Scopus and 25 records from Web of Science.
2. Lists of references from previous literature studies regarding mass customization were analyzed for SME relevant literature.
3. Forward and backward search was applied on the identified literature.
4. The results from the steps above were combined into one data set, and a number of records were excluded based on the following criteria: (1) duplicates; (2) hits in sub-strings, e.g., querying for SME returns results on SMED; (3) retracted articles; and (4) editorials and conference descriptions. In this step, abstracts were also reviewed to assess relevance.
5. The full text of the papers was read to further ensure relevance and to be able to perform a categorization.
6. Classification in categories:

    (a) Solution space development
    (b) Choice navigation and information systems (these two categories are joined since it is difficult to distinguish between these in a mass customization context and there are string relations between choice navigation systems and other information systems in mass customization companies)

(c) Robust process design
(d) General mass customization and strategy research
(e) Mass customization in specific industries
(f) Case studies

The categories were defined through an iterative process, where categories a through c were the initial categories taken from Salvador et al. [34]. Categories d through f were established during the categorization process, when literature appeared which did not fit into any other predefined category. Papers were allowed to be present in more than one category as some papers address multiple issues.

15.3 Results

A total of 39 papers were included in the review after the exclusions performed in step 4. These papers were categorized, and the results of the categorization can be seen in Fig. 15.1. Furthermore, Table 15.1 shows which papers are included in

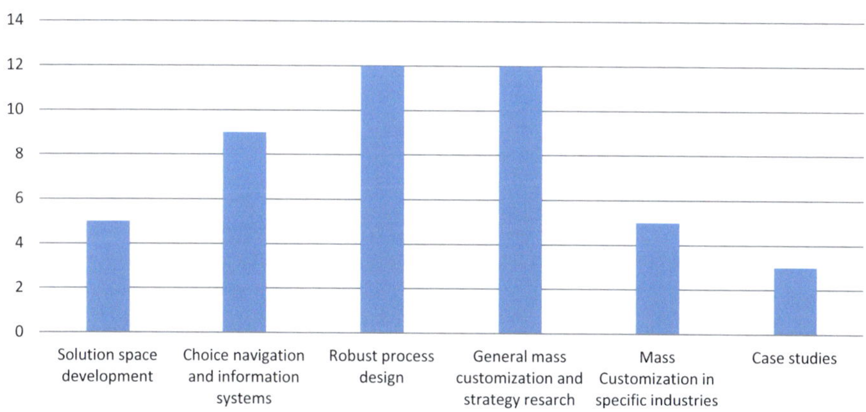

Fig. 15.1 Distribution of contributions on categories

Table 15.1 Overview of categories

Category	References to papers in category	Number of papers
Solution space development	[2, 17, 18, 22, 36]	5
Choice navigation and information systems	[8–11, 21, 22, 25, 27, 37]	9
Robust process design	[1, 3, 5–7, 9, 14, 16, 19, 26, 28, 43]	12
General mass customization and strategy research	[13, 18, 20, 31, 33, 36, 39–42, 44]	12
Mass customization in specific industries	[15, 24, 25, 29, 38]	5
Case studies	[17, 19, 30]	3

which categories. In the subsections below, the characteristics of the individual papers are briefly reviewed.

15.3.1 Solution Space Development

Within solution space development, which is the concept of identifying how customer requirements diverge and developing products which meet these diver requirements, five contributions were identified. The technologies of rapid prototyping and its implications for mass customizers are explored by Ayyaz et al. [2].

The research presented by Stojanova et al. [36] presents a way to measure customer preferences, defining configurable attributes, and finally a way to measure results of implementing mass customization implemented in an information system. A decision support system is proposed by Liu et al. [22], which can support companies in defining how a product should be customized. A feature component matrix is introduced by Ismail et al. [18] as a means to balance diverse customer requirements with manufacturing efficiency. Finally, by Ismail et al. [17], it is described how component commonality can be measured and assessed in order to reduce the internal variety to enhance efficiency.

15.3.2 Choice Navigation and Information Systems

Choice navigation, which is the capability to guide customers to purchase a product which matches their unique requirements, was in this literature review combined with general information systems for mass customization as these two areas are often closely integrated. Within this category, ten contributions were identified.

Some of these contributions are quite technology focused. An ASP-based product configuration system is described by Su et al. [37], and Mourtzis et al. [27] look into how mobile telephone apps can be utilized for product configuration.

Issues related to how configuration system and customization can be utilized in the food industry are investigated in Mertins et al. [25].

General information systems for mass customization in SMEs are discussed by Dean et al. [8], Little et al. [21], and Liu et al. [22]. Similarly, general information system is discussed by [11] where it is also addressed how electronic data interchange can be utilized. Furthermore, development of information systems supporting automation of engineering tasks in mass customization SME settings is addressed [9].

Finally, by Durá et al. [10], it is described how product configurators can be applied in relation to customization of clothing for people with special needs and how this can aid eco-efficient production.

15.3.3 Robust Process Design

The term robust process design covers how to enable production and business processes to handle the increased variety which is introduced with mass customization. Within this category, 12 contributions were identified.

Two papers focus on the automation aspect of robust process design, where Kokla [19] investigates effectiveness and profitability of automating welding and Bi et al. [3] focus on reusing industrial robots in mass customization SMEs.

Two papers focus on supply chain and network issues. Chiu and Okudan [6] address the issue of optimizing supply chain decisions in relation to product variety, by, e.g., altering suppliers, and Antonelli et al. [1] investigate how SMEs may benefit in a mass customization context from collaboration in a production network.

Five papers address various planning issues. Gillies et al. [14] evaluate how SMEs may benefit from implementing agile manufacturing, whereas Villa [43] discusses how configuration of products and processes can be coordinated. Closely related to this, the papers by Mourtzis et al. [28], Hvolby et al. [16], and Mleczko [26] discuss more in detail how scheduling and simulation can be carried out in mass customization SME contexts.

On a production system level, Cedeno-Campos [5] discusses how self-configurable systems can be developed to create systems which can rapidly form a layout for manufacturing customized products.

The contribution from Dai et al. [7] is also quite technology focused and investigates how RFID can be utilized in a mass customization manufacturing execution system. Finally, it is addressed how engineering automation can be applied in SMEs to make the process of preparing manufacturing more efficient [9].

15.3.4 General Mass Customization and Strategy Research

Being one of the categories in which the highest number of papers is identified, the literature within general mass customization and strategic aspects counts 12 papers.

Two papers discuss intercompany collaborations, where Li and Liu [20] discuss industrial mass customization clusters and Wiendahl et al. [44] investigate industrial networks. Implementation issues of mass customization in SMEs are discussed by Stojanova et al. [36], where the process from identifying customer preferences to implementing product configuration is described. Similarly Ismail et al. [18] address how to choose the right product strategy in terms of introducing just the right level of product variety.

An approach to address performance measurement systems is SME mass customizers presented by Gamme et al. [13]. Knowledge management in relation to mass customization and SMEs is discussed by Tsakalerou and Lee [42].

The differences in success factors when implementing mass customization in SMEs compared to large enterprises are discussed by Suzić et al. [39]. Similarly,

Svensson and Barfod [40] discuss how build-to-order-oriented SMEs may benefit or experience challenges in applying elements of mass customization.

By Pizmoht et al. [33], it is discussed how the concept of mass customization may fit with SMEs in relation to their core competences and limitations. Similarly, in Taps et al. [41], it is discussed how SME subcontractors may benefit from mass customization.

Finally, an entirely new manufacturing paradigm, dubbed fit manufacturing, is presented by Pham and Thomas [31], which combines the efficiency of lean manufacturing with the ability to break into new markets, enabling mass customizing SMEs to become more economically sustainable.

15.3.5 Mass Customization in Specific Industries

In the literature review, five papers were identified, which investigate how mass customization can be applied in SMEs in specific industries. Two papers focus on the furniture industry, where Suzić et al. [38] focus on a single case and Ollonqvist et al. [29] focus on networks and on how to stimulate innovation. Other industry-specific papers include one paper on the food industry and how it can benefit from customization and enhance interoperability [25], one paper on clothing and fashion customization [15], and finally one paper on customization within the tourism industry [24].

15.3.6 Case Studies

Only three case studies focusing specifically on mass customization in SMEs were identified during the literature review. One case study focused narrowly on how automation of welding processes could be utilized in a metal furniture manufacturer [19]. By Ismail et al. [17], two cases are presented, involving companies manufacturing children's playground equipment and luxury domestic showers illustrating how tools from mass customization can benefit SMEs. Finally by Orsila and Aho [30], a case study was presented of how an e-commerce system could improve business processes in a manufacturer of custom semiconductor products.

15.4 Research Activity Over Time

Figure 15.2 shows the historical development in research activity related to mass customization in SMEs, measured by the number of published contributions per year. It is obvious that the research activity has been much higher during the last decade, compared to the early years of mass customization (introduced in 1993); however, this can be attributed to a general adoption of mass customization in industry. The number of papers published in 2015 is low, since the literature study was done in early 2015, and this

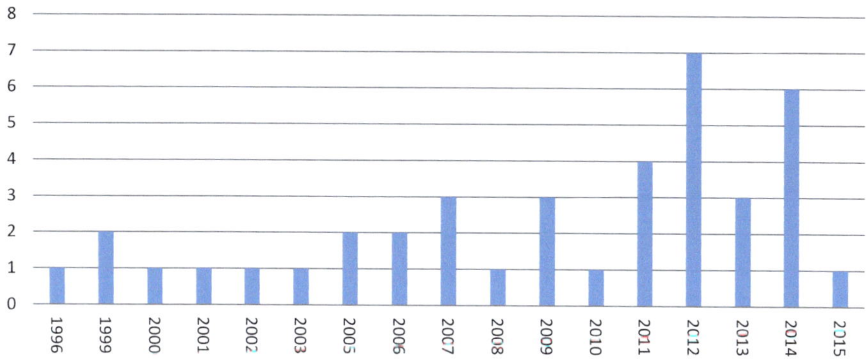

Fig. 15.2 Papers published on mass customization in SMEs per year

number is thus somewhat misleading. Although there is a clear long-term trend in the publication activity, it is not deemed possible to interpret the data to indicate a short-term trend, as the relatively low number of publications per year is somewhat fluctuating.

15.5 Conclusion

In this paper, a literature study was done, searching for contributions related to mass customization in SMEs. A total of 39 papers were identified which had this particular focus. If a search is made, using only the keywords "mass customization," 2077 results are found on Web of Science and 2821 results are found on Scopus, which covers all literature published on any subfield of mass customization indexed in the same databases. It is hence only a very small fraction of the available mass customization literature, which is focused on SMEs. Looking into the time aspect of publications, it was observed that in a long-term perspective, there was an increase in the publication rate of research in SME mass customization; however, on short term, the frequency is too low to conclude anything.

Looking into which categories had received the most attention from scholars, there was a rather large range between the categories with the most and the categories with the least literature. The categories "robust process design," "general mass customization and strategy," and "choice navigation and information systems" were the categories with the most papers, with 12, 12, and 9 papers, respectively. The categories with the least papers were "solution space development," "mass customization in specific industries," and "case studies," where the categories contained five, five, and three papers.

Given that it is generally acknowledged that SMEs are of vital importance for growth in the national societies and given that mass customization in many cases can be a catalyst for growth, the authors believe that mass customization in SMEs is a very important topic to stimulate in society and to do research on. Given the very small amount of research on mass customization in SMEs identified in this study, there is hence a major potential in doing more research within this area. The author recommends that more research is done within each of the identified categories, on short term especially case studies, which can help understand what benefits and challenges SMEs can expect when implementing mass customization.

References

1. Antonelli, D., Pasquino, N., Villa, A.: Mass-customized production in a SME network. In: Anonymous Advances in Production Management Systems, pp. 79–86. Springer, Boston (2007)
2. Ayyaz, A., Maruf, H., Van Voorthuysen, E.J.: RP technology: why limited uptake in SME's? In: Annals of DAAAM for 2008 & Proceedings of the 19th International DAAAM Symposium, pp. 43–44 (2008)
3. Bi, Z., Liu, Y., Baumgartner, B., et al.: Reusing industrial robots to achieve sustainability in small and medium-sized enterprises (SMEs). Ind. Robot **42** (2015)
4. Bryman, A.: Social Research Methods. Oxford University Press, Oxford (2012)
5. Cedeno-Campos, V.M., Trodden, P., Dodd, T.J. et al.: Highly Flexible Self-Reconfigurable Systems for Rapid Layout Formation to Offer Manufacturing Services, pp. 4819–4824 (2013)
6. Chiu, M., Okudan, G.E.: A Small to Medium-Size Enterprise Oriented Methodology for Optimizing Product and Supply Chain Design Decisions, pp. 27–38 (2010)
7. Dai, Q., Zhong, R., Huang, G.Q., et al.: Radio frequency identification-enabled real-time manufacturing execution system: a case study in an automotive part manufacturer. Int. J. Comput. Integr. Manuf. **25**, 51–65 (2012)
8. Dean, P., Tu, Y., Xue, D.: An information system for one-of-a-kind production. Int. J. Prod. Res. **47**, 1071–1087 (2009)
9. Direction, S.: Engineering automation for the SME: a viable tool for "mass" customisation. Strategic Direction, vol. 21 (2005)
10. Durá, J. V., Caprara, G., Ballarino, A., Cavallaro, M., Kaiser, C., & Stellmach, D. (2014, June). New technologies for the flexible and eco-efficient production of customized products for people with special necessities: Results of the FASHION-ABLE project. In Engineering, Technology and Innovation (ICE), 2014 International ICE Conference on (pp. 1–7). IEEE
11. Fellner, K.J., Turowski, K.: Component Framework Supporting Inter-Company Cooperation, pp. 164–171 (1999)
12. Fogliatto, F.S., da Silveira, G.J.C., Borenstein, D.: The mass customization decade: an updated review of the literature. Int. J. Prod. Econ. **138**(1), 14–25 (2012)
13. Gamme, I., Seim, E.A., Lodgaard, E.: Performance measurement systems for craft-oriented small enterprises. In: Anonymous Advances in Production Management Systems. Competitive Manufacturing for Innovative Products and Services, pp. 265–272. Springer (2013)
14. Gillies, J., Nelder, G., Ip-Shing, F.: The Applicability of Agile Manufacturing to Small and Medium-Sized Enterprises, pp. 538–543 (1996)
15. Gonzalez, J.C., Olaso, J., Gil, M. et al.: Fashion-able: Needs and Requirements for Clothing, Footwear and Orthotics of Consumers Groups with Highly Individualised needs, pp. 1–10 (2012)
16. Hvolby, H., Svensson, C., Steger-Jensen, K.: Simulation of production setup changes in an SME. Procedia Technol. **5**, 643–648 (2012)
17. Ismail, H., Reid, I., Mooney, J., et al.: How small and medium enterprises effectively participate in the mass customization game. IEEE Trans. Eng. Manage. **54**, 86–97 (2007)
18. Ismail, H., Reid, I., Poolton, J. et al.: Mass customization: balancing customer desires with operational reality. In: Anonymous Mass Customization: Challenges and Solutions, pp. 85–109. Springer (2006)
19. Kokla, M.: Case study regarding effectiveness and profitability of automating the welding process. Annals of DAAAM & Proceedings (2011)

20. Li, J., Liu, C.: Mass Customization Model in Cluster Supply Chain Based on 2P&2BP, pp. 729–735 (2005)
21. Little, D., Hall, D., Rhymer, S. et al.: The development of a late product configuration tool and procedural ethos within SMEs. 16th National Conference on Manufacturing Research, pp. 203–207 (2000)
22. Liu, H., Ng, W.K., Song, B. et al.: An Intelligent Information System for Enabling Product Mass Customization, pp. 316–321 (2007)
23. Lyons, A.C., Mondragon, A.E.C., Piller, F. et al.: Mass customisation: a strategy for customer-centric enterprises. Customer-Driven Supply Chains, pp. 71–94 (2012)
24. Main, H.C.: Emerging Technologies and Their Role in Developing a Marketing Information System (MKIS) for Tourism and Hospitality Products. Springer (1999)
25. Mertins, K., Jaekel, F., Deng, Q.: Towards information customization and interoperability in food chains. In: Anonymous Enterprise Interoperability, pp. 92–103. Springer (2012)
26. Mleczko, J. (2011). Improving the efficiency of moving bottleneck by dynamic classification of tasks. In 4th international conference on advanced computer theory and engineering: ICACTE
27. Mourtzis, D., Doukas, M., Vandera, C.: Mobile apps for product customisation and design of manufacturing networks. Manuf. Lett. **2**, 30–34 (2014)
28. Mourtzis, D., Doukas, M., Vlachou, A., et al.: Machine availability monitoring for adaptive holistic scheduling: a conceptual framework for mass customization. Procedia CIRP **25**, 406–413 (2014)
29. Ollonqvist, P., Nord, T., Pirc, A. et al.: 14 Networks and local milieus as a furniture industry innovation platform. Innovation in Forestry: Territorial and Value Chain Relationships, p. 233 (2011)
30. Orsila, S., Aho, M.: Re-qualifying delivered devices and inventory for a new product specifications, a case study. Frontiers of e-Business Research-FeBR (2006)
31. Pham, D.T., Thomas, A.J.: Fit manufacturing: a framework for sustainability. J. Manuf. Technol. Manage. **23**, 103–123 (2011)
32. Pine, B.J.: Mass Customization: The New Frontier in Business Competition. Harvard Business School Press, Boston (1993)
33. Pizmoht, P., Polajnar, A., Vujica, N.: Opportunities in mass customization. Annals of DAAAM for 2003 & Proceedings of the 14th International DAAAM Symposium, p. 363 (2003)
34. Salvador, F., De Holan, P.M., Piller, F.: Cracking the code of mass customization. MIT Sloan Manage. Rev. **50**, 71–78 (2009)
35. Silveira, G.D., Borenstein, D., Fogliatto, F.S.: Mass customization: literature review and research directions. Int. J. Prod. Econ. **72**, 1–13 (2001)
36. Stojanova, T., Suzic, N., Orcik, A.: Implementation of mass customization tools in small and medium enterprises. Int. J. Ind. Eng. Manage. **3**, 253–260 (2012)
37. Su, Y., Liao, W., Guo, Y., et al.: Key technologies for ASP-based product customization service system for SMEs: a case study. Int. J. Adv. Manuf. Technol. **42**, 381–397 (2009)
38. Suzić, N., Anišić, Z., Forza, C.: Introducing Mass Customization to SMEs in Furniture Industry: A Case Study, pp. 287–300 (2014)
39. Suzić, N., Anišić, Z., Orčik, A. et al.: Company Size and Successful Mass Customization (2012)
40. Svensson, C., Barfod, A.: Limits and opportunities in mass customization for "build to order" SMEs. Comput. Ind. **49**, 77–89 (2002)
41. Taps, S.B., Brunoe, T.D., Nielsen, K. et al.: Mass Customization as Innovation Driver of International Competitiveness in Peripheral Regional SME Subcontractors, pp. 349–357 (2014)
42. Tsakalerou, M., Lee, R.: Intellectual Capital Practices of SMEs and MNCs: A Knowledge Management Perspective, p. 447 (2013)
43. Villa, A.: Joint Planning of Product/Process Configuration, vol. 13, 58–62 (2009)
44. Wiendahl, E.H., Engelbrecht, A., Hamacher, O.: From single enterprises to complementary networks. In: Anonymous Global Engineering, Manufacturing and Enterprise Networks, pp. 66–73. Springer (2001)

Chapter 16
Reconfigurable Manufacturing Systems in Small and Medium Enterprises

Thomas Ditlev Brunoe, Ann-Louise Andersen, and Kjeld Nielsen

16.1 Introduction

Manufacturing companies today face challenges that have followed globalization, e.g., the fragmentation and change of customer demands, increased need for customized products, fast developing technologies, and focus on environmental sustainability [12–14, 47]. Therefore, in order to remain competitive, they must respond to these challenges and efficiently produce a wide range of products that fits different customer needs and continuously includes new product technologies [19, 25]. Mass customization is a widely adopted strategy for this, where individually configured products are delivered at a cost near mass production [16, 32, 38]. One of the key enablers of mass customization is modular product design, where end variety is achieved through configurations of standardized modules [18, 43, 44]. However, simply introducing modular products is not enough for manufacturing companies to gain competitive advantage, as the products need to be produced and delivered to the market at the right time [26, 36]. This implies that manufacturing companies need to incorporate responsiveness to change and ability to handle high variety at various levels. On an operational level, the assembly system, machines, and stations must be able to switch quickly between the production of different modules, parts, and subassemblies in the product family, in response to differences in product configurations, variety of modules, and unpredictability and variety in demanded quantities. For the entire manufacturing system, this ability is denoted as reconfigurability in terms of capability and capacity [23]. At the same time, manufacturing costs must continuously be reduced, which is a particularly compelling problem for Western manufacturing companies due to high labor cost. Additionally, the manufacturing systems also need to incorporate ability to change on a tactical level and in the longer term, as products change over time and entirely new

T.D. Brunoe (✉) • A.-L. Andersen • K. Nielsen
Department of Mechanical and Manufacturing Engineering, Aalborg University, Aalborg, Denmark
e-mail: tdp@m-tech.aau.dk

J. Bellemare et al. (eds.), *Managing Complexity*, Springer Proceedings in Business and Economics, DOI 10.1007/978-3-319-29058-4_16

generations and variants are introduced, due to rapid advancements of technology and demand for new features [9, 10]. Therefore, in order for the manufacturing systems to be economically viable, these must be developed with the ability to adapt to various product generations in order to exploit market potentials [11, 48]. All of these conditions require that traditional approaches to manufacturing system design must be evaluated and new methods and concepts created for the development of manufacturing systems that are able to handle both capability and capacity changes in a cost-efficient way, through modularity, reconfigurability, and closer integration to the product architecture.

16.1.1 Reconfigurable Manufacturing Systems

Over time, manufacturing concepts have evolved in response to changing conditions [25], and research in manufacturing systems has evolved accordingly. Most recently, research has focused on new manufacturing concepts that incorporate the ability to handle broad product assortments and rapid changes, instead of being dedicated and optimized for one specific product model [13, 23, 30, 31, 37]. The FMS has the goal of providing efficiency through automated transfer lines and flexibility through the CNC machinery, in order to efficiently produce high-variety products in low to medium volumes [7]. The reconfigurable manufacturing system concept was later introduced as an extension of the FMS, with the goal of combining the efficiency of the dedicated manufacturing lines and the high flexibility of the FMS [27, 31]. One of the main differences between the FMS and the RMS is that the RMS is able to be continuously reconfigured in order to contain the exact flexibility, functionality, and capacity needed to produce a given product family, which avoids the issue of FMS in regard to excess flexibility, low production rate, and low return on investments [31, 50].

The heart of the reconfigurable manufacturing concept is the RMS characteristics: customization, convertibility, scalability, modularity, integrability, and diagnosability. Customization refers to the machine and system flexibility, which is limited and customized to part or product families. Convertibility and scalability refer to modifying the functionality and capacity of the existing system and machines, which is achieved through modularity and integrability. Finally, diagnosability refers to the ability to read the state of the system, which is particularly important in the ramp-up phase after the reconfiguration. In essence, these characteristics make the RMS adaptable to the changing market conditions and allow for cost-efficient reuse and prolonged lifetime of existing manufacturing, which is the reason why it is widely labeled the manufacturing paradigm of the future [52].

Reconfigurable manufacturing has mainly been described in literature through the RMS concept, but additional concepts such as holonic manufacturing [45], evolvable production [28], modular manufacturing systems [20], and focused flexible manufacturing systems [42] have been introduced as well. Even though present research lacks a thorough comparison of these reconfigurable concepts, some similar characteristics are evident, in particular the application of manufacturing system modularity as a means for reconfigurability. The concept of changeable manufacturing has been introduced as an umbrella term for manufacturing concepts that allow for rapid and cost-efficient change in accordance with the environment [13]. Additionally, the concept of changeable manufacturing extents focuses to covering both physical and logical aspects of changeability, as well as all structuring levels of the factory, e.g., machines, cells, systems, plants, and networks [48]. Thus, reconfigurability, flexibility, and changeability can be dealt with at different structuring levels, where both practical issues and research focus differ [3].

One of the central areas of reconfigurable manufacturing research is on the lowest structuring level and concerns how to develop reconfigurable machines (RMs) that embed the RMS characteristics and are able to be quickly converted between varieties within product families [4]. RMs cover reconfigurable machine tools (RMTs), reconfigurable fixturing systems, reconfigurable assembly systems, reconfigurable inspection machines, and reconfigurable material handling systems. These RMs are essential to the RMS paradigm, as they provide customized flexibility and ability to reconfigure on equipment level through combinations of basic and auxiliary modules [22]. However, even with current research contributions on RMs, their effective implementation is limited, and the RMs are currently not broadly available as they are still in development [4, 8]. Therefore, it is important to consider that reconfigurability not only can be achieved by introducing RMs on the factory floor but also on system level due to the modularity of the RMS. By rearranging, adding, and removing the modules of the system, new configurations can be developed, which changes the functionality and capacity of the entire system [24]. A critical issue is how to determine the optimal configurations of the system given a part or product family and its volume [1, 49, 50]. This optimal configuration selection problem is a significantly complex problem that involves multiple aspects of the configuration, e.g., arrangement of machines, equipment selection, and assigning of machines, in order to accurately model the feasibility and cost of the system. Moreover, the practicality and feasibility of RMS configurations depend largely on the design phase of the RMS, where critical decisions are made regarding the degree of reconfigurability and scalability of the system [17, 37].

Collectively, the design and planning of the RMS present a major academic and practical challenge that requires consideration of multiple variables and high integration with information on products and product design. Moreover, in terms of designing reconfigurable systems, a major issue is to determine the degree of reconfigurability and flexibility to build into the system, in order to effectively trade off the objectives of flexibility and productivity. In regard to this, research lacks a systematic procedure for determining the needs for reconfigurability and explicitly stated decision parameters that should be addressed in the design process [5, 41]. Moreover, there appears to be a lack of research on the transition of companies from having non-configurable manufacturing systems to developing reconfigurable systems [5], including lack of research on which manufacturing tasks are suited for reconfigurability [41].

Another critical issue in the reconfiguration is the ramp-up time, which defines the time from a given reconfiguration to the point where production reaches its planned output in volume, variety, and quality [27]. First of all, this ramp-up should be as quick as possible, in order to reduce time to market. However, the ramp-up process is connected to considerable investments that are subject to risk and uncertainty, where critical trade-offs must be made. Reconfigurable manufacturing is in research stated as being a means for managing the ramp-up process, but an explicit operationalization of these parameters and a quantification of ramp-up trade-offs are not currently present [29]. Moreover, management of the ramp-up process and continuous reduction of time and cost are essential in order to reap the benefits of reconfigurable manufacturing [26]. However, research on ramp-up is currently primarily focused on classifying generic challenges and their impact [2, 40] and has not been considered explicitly from a RMS perspective. In fact, the lack of systematic approaches to ramp management in a RMS context is one of the main barriers toward its effective implementation [8].

16.1.2 RMS and SMEs

It has long been acknowledged that SMEs are fundamentally different from large enterprises in terms of strategy, operations, etc. [46]. This implies that tools and methods, which are found useful in the context of large enterprises, may not necessarily be as useful in SMEs. Even if they are useful, it is likely that they must be adapted and implemented in a different way than in large enterprises.

A literature search in Thomson Reuters Web of Science revealed six different papers addressing the combination of SMEs and reconfigurable manufacturing. One paper by Rakesh et al. presented a framework for a SME subcontractor to identify part families and process families and subsequently perform production planning for the involved parts [35]. Strasser et al. took a more detailed approach and suggested a design approach for machine tools to enable reconfiguration on machine level, thus supporting the development of RMS in SMEs [39]. Two contributions concerned the development of manufacturing execution systems (MESs) for reconfigurable manufacturing systems in SMEs [6, 15]. Jules et al. described an ontology for holonic manufacturing SME networks, which in concept is quite similar to RMS [21]. Finally, Rahimifard et al. investigated how various IT tools can aid SME metal manufacturers to approach the holonic manufacturing paradigm [33].

After reviewing the sparse literature addressing RMS in SMEs, it can thus be concluded that no contributions focus directly on which benefits can be achieved by SMEs, when applying RMS, as compared to large enterprises. This leads to the research question of this paper: *How can SMEs benefit from RMS compared to large enterprises and what are the major challenges to overcome*?

To address this research question, a case study is performed, where the case is an SME from Danish industry. The observations from the case company are compared to what is described in literature regarding RMS.

16.2 Case Study

The case studied in this paper is a Danish manufacturer of large industrial equipment that employs around 150 people. Products are manufactured using assemble-to-order strategy and consist primarily of large metal components which are cut, welded, and machined. Other components, such as electronics, are produced by sub-suppliers. The production is somewhat influenced by seasonal variations; however, due to the large degree of customization of the products, it is not possible to manufacture to stock in order to level the production.

The case company has long had an ambition to have a one-piece flow, in order to reduce stock and to reduce the lead time of manufacturing components, since manufacturing in large batches results in large stock. The annual production volume in pieces is relatively low (a few hundred, depending on product family), and a large number of components are used for each product. Hence, large batch sizes yield undesirably large stock. However, reducing batch sizes to a one-piece flow has proven impractical, due to the currently long changeover times. Many changeovers are influenced by the fact that welding large steel components requires large and heavy fixtures. Thus, changing from producing one component to another requires a change in the fixtures, as there is typically one fixture per component. In this particular case, this involves removing the previous fixture with a forklift, driving it to a warehouse, locating the new fixture, driving it to the welding station, and setting it up before the actual production can begin. These operations can take a significant amount of time compared to the actual welding time. This may seem as a classic example of balancing productivity and stock sizes, which can be addressed by lean methods, including single-minute exchange of die (SMED). However, traditional methods such as SMED cannot address the fact that the heavy fixtures are difficult and time consuming to handle, and thus a challenge remains in relation to this in the case company.

Due to the challenges outlined above, the case company has begun looking into applying the principles of RMS. The company expects that by applying RMS principles, it will be possible to introduce dedicated flexibility in welding cells, implying that a changeover from producing one component to another can be handled by reconfiguring a fixture rather than replacing it with a new fixture. The benefits of this are both in terms of more efficient handling of variants in production and improved ability to introduce new products.

Nevertheless, there are significant differences between realizing RMS in SMEs compared to large enterprises. One feature often described in literature for RMS is the use of parallel similar manufacturing lines, which produce components belonging to the same part family. These lines can be reconfigured from producing one variant to another variant within a certain period when market demand changes. However, the precondition for doing this is that the company produces a sufficient volume to sustain production of one single component on a manufacturing line over a longer period. This is not the situation in the case company, since the production volume is quite far from being sufficient for continuous production on even one line. This is expected by the authors to be the case in many other SMEs, producing low volume and high variety.

The RMS literature also describes principles for designing reconfigurable machines. This is highly relevant in the case company, especially reconfigurable fixtures. By reconfiguring fixtures rather than replacing fixtures every time a new component is to be produced, changeover time could be significantly reduced, as well as the time and resource usage for introducing new products or parts. As an additional advantage, this could potentially reduce the investment in fixtures, by distributing the fixture cost over multiple components. Furthermore, introduction of new products can be handled more efficiently and faster if the reconfigurable fixtures can be utilized in future part generations. It must be noted, however, that the frequency of reconfigurations must be expected to be much higher in SMEs with low volume compared to larger enterprises with higher volume. This implies that when designing the reconfigurable fixtures, increased focus must be on minimizing the reconfiguration time, as this would be a daily event in the case company, whereas large enterprises with higher volume may experience months or even years between reconfigurations, in which case a reconfiguration duration of several hours or even days may be acceptable. Although fixtures represent the most promising part of the production system to enable reconfigurability, this can highly likely be generalized to any other type of reconfigurable machine, e.g., machine tools, material handling, and inspection that are of similar relevance in other low-volume SMEs.

Since the case company has a limited production volume and high variety, the variety which is necessary to handle in one part of the reconfigurable manufacturing system is likely to be higher compared to a higher-volume large enterprise. This implies that reconfigurable machines, reconfigurable fixtures, etc. designed for SMEs are required to be reconfigurable across a much larger part variety, which must be taken into consideration.

Finally, there are large organizational differences between SMEs and large enterprises, which influence the feasibility of implementing RMS. Large manufacturing enterprises typically have production engineering departments addressing production system design, machine design, and tool design, whereas SMEs obviously have more limited capacity in production engineering and may rely on a handful of people when developing the production system or parts of it. Furthermore, SMEs may also rely on external consultants or machine developers to introduce new equipment in the manufacturing system. This may represent a challenge in terms of implementing reconfigurability in the production systems in the SME.

Changing a company's production system into a reconfigurable manufacturing system is in some way quite similar to introducing a product platform in product development—a large investment is made up-front to reap large benefits on longer term. Similarly, developing an RMS will imply a larger short-term investment, but will ideally give large benefits on the longer term. However, this investment may seem to represent a too high a risk for a SME compared to large enterprises, which also may occupy too big a part of the smaller capacity in the production engineering department of the SME, which would also be a barrier toward implementing RMS.

16.3 Conclusions

By offering dedicated flexibility in manufacturing systems, RMS represents an attractive trade-off between efficiency and flexibility, which is required when product variety and the rate of product introductions increase. Much literature has been published concerning different aspects of RMS; however, the majority of contributions report results from large manufacturing enterprises, which are expected to differ very much from SMEs, in terms of benefits and challenges of implementing RMS. Therefore, this issue was investigated by conducting a case study in a Danish SME that currently considered implementing reconfigurability, due to facing problems with high and resource-intensive changeover times. One of the key findings in the case company is that there might be a significant potential in implementing reconfigurable fixtures, as variety and new product introductions thus could be handled much more efficiently. This implies that SMEs with low volume and high variety in general may benefit from implementing reconfigurability on workstation level, rather than on production line level, because implementing on production line level requires sufficient volumes to sustain production of one variant over a longer period. In addition, reconfigurations are likely to occur much more frequently in the low-volume SME, which requires extensive focus on reducing the time for reconfigurations and production start-up. In addition, implementing RMS in the SME is likely to represent a higher risk than in the large enterprise, due to more limited resources in production engineering and the large investments required for the implementation. These findings suggest that the challenges that RMS addresses in SME and large enterprises are significantly different, partly due to the relation between volume and variety and partly due to the organizational capabilities and development capacity.

References

1. Abdi, M.R., Labib, A.W.: A design strategy for reconfigurable manufacturing systems (RMSs) using analytical hierarchical process (AHP): a case study. Int. J. Prod. Res. **41**, 2273–2299 (2003)
2. Almgren, H.: Pilot production and manufacturing start-up: the case of Volvo S80. Int. J. Prod. Res. **38**, 4577–4588 (2000)
3. Andersen, A., Brunoe, T.D., Nielsen, K.: Reconfigurable manufacturing on multiple levels: literature review & research directions. Proceedings of APMS 2015 (2015)
4. Bi, Z., Lang, S.Y., Verner, M., et al.: Development of reconfigurable machines. Int. J. Adv. Manuf. Technol. **39**, 1227–1251 (2008)
5. Bi, Z., Lang, S., Shen, W., et al.: Reconfigurable manufacturing systems: the state of the art. Int. J. Prod. Res. **46**, 967–992 (2008)
6. Bo, L., Zhenghang, C., Ying, C.: Research on reconfigurable manufacturing execution system (2004), 157–161
7. Browne, J., Dubois, D., Rathmill, K., et al.: Classification of flexible manufacturing systems. FMS Mag. **2**, 114–117 (1984)
8. Chaube, A., Benyoucef, L., Tiwari, M.K.: An adapted NSGA-2 algorithm based dynamic process plan generation for a reconfigurable manufacturing system. J. Intell. Manuf. **23**, 1141–1155 (2012)

9. ElMaraghy, H.A.: Changing and evolving products and systems—models and enablers. In: ElMaraghy, H.A. (ed.) Changeable and reconfigurable manufacturing systems, pp. 25–45. Springer, London (2009)
10. ElMaraghy, H.A.: Reconfigurable process plans for responsive manufacturing systems. In: Cunha, P.F., Maropoulos, P.G. (eds.), Digital enterprise technology—perspectives and future challenges, pp. 35–44. Springer (2007)
11. ElMaraghy, H.A.: Flexible and reconfigurable manufacturing systems paradigms. Int. J. Flex. Manuf. Syst. **17**, 261–276 (2005)
12. ElMaraghy, H., Schuh, G., ElMaraghy, W., et al.: Product variety management. CIRP Ann. Manuf. Technol. **62**, 629–652 (2013)
13. ElMaraghy, H.A., Wiendahl, H.P.: Changeability—an introduction. In: ElMaraghy, H.A. (ed.) Changeable and reconfigurable manufacturing systems, pp. 3–24. Springer, London (2009)
14. ElMaraghy, H., ElMaraghy, W.: Variety, complexity and value Creation. In: Zaeh, M. (ed.), pp. 1–7. Springer International Publishing (2014)
15. Gaxiola, L., Ramírez, M.d.J., Jimenez, G. et al.: Proposal of holonic manufacturing execution systems based on web service technologies for Mexican SMEs. In: Anonymous holonic and multi-agent systems for manufacturing, pp. 156–166. Springer, Heidelberg (2003)
16. Gilmore, J.H., Pine, B.J.: Markets of one: creating customer-unique value through mass customization. Harvard Business School, Boston (2000)
17. Gumasta, K., Kumar Gupta, S., Benyoucef, L., et al.: Developing a reconfigurability index using multi-attribute utility theory. Int. J. Prod. Res. **49**, 1669–1683 (2011)
18. Harlou, U.: Developing product families based on architectures: Contribution to a theory of product families. ORBIT (2006)
19. Hu, S.J., Ko, J., Weyand, L., et al.: Assembly system design and operations for product variety. CIRP Ann. Manuf. Technol. **60**, 715–733 (2011)
20. Joergensen, S.N., Hvilshøj, M., Madsen, O.: Designing modular manufacturing systems using mass customisation theories and methods. Int. J. Mass Cust. **4**, 171–194 (2012)
21. Jules, G.D., Saadat, M., Li, N.: On designing a unified ontology for holonic manufacturing networks. In: Anonymous integration of practice-oriented knowledge technology: trends and prospectives, pp. 207–220. Springer (2013)
22. Katz, R.: Design principles of reconfigurable machines. Int. J. Adv. Manuf. Technol. **34**, 430–439 (2007)
23. Koren, Y.: General RMS characteristics. Comparison with dedicated and flexible systems. In: ElMaraghy, H.A. (ed.) Reconfigurable manufacturing systems and transformable factories. pp. 27–45, Springer (2006)
24. Koren, Y.: The rapid responsiveness of RMS. Int. J. Prod. Res. **51**, 6817–6827 (2013)
25. Koren, Y.: The global manufacturing revolution: product-process-business integration and reconfigurable systems. Wiley, Hoboken (2010)
26. Koren, Y., Shpitalni, M.: Design of reconfigurable manufacturing systems. J. Manuf. Syst. **29**, 130–141 (2010)
27. Koren, Y., Heisel, U., Jovane, F., et al.: Reconfigurable manufacturing systems. CIRP Ann. Manuf. Technol. **48**, 527–540 (1999)
28. Maffei, A., Onori, M., Neves, P. et al.: Evolvable production systems: mechatronic production equipment with evolutionary control. In: Camarinha-Matos, L.M., Pereira, P., Ribeiro, L. (eds.) Emerging trends in technological innovation, pp. 133–142. Springer (2010)
29. Mehrabi, M.G., Ulsoy, A.G., Koren, Y.: Reconfigurable manufacturing systems and their enabling technologies. Int. J. Manuf. Technol. Manage. **1**, 114–131 (2000)
30. Mehrabi, M.G., Ulsoy, A.G., Koren, Y., et al.: Trends and perspectives in flexible and reconfigurable manufacturing systems. J. Intell. Manuf. **13**, 135–146 (2002)
31. Mehrabi, M.G., Ulsoy, A.G., Koren, Y.: Reconfigurable manufacturing systems: key to future manufacturing. J. Intell. Manuf. **11**, 403–419 (2000)
32. Pine, I., Joseph, B., Victor, B.: Making mass customization work. Harv. Bus. Rev. **71**, 108–117 (1993)

33. Rahimifard, S., Bagshaw, R., Newman, S., et al.: IT tools to improve the performance of metalworking SMEs. Int. J. Prod. Res. **40**, 3589–3604 (2002)
34. Rakesh, K., Jain, P.K., Mehta, N.K.: A dynamic RMS framework for a highly dynamic environment of a subcontracting manufacturing enterprise. In: Anonymous mechatronics and intelligent manufacturing, vol. 1, pp. 145. Nova Science Publishers, Inc. (2012)
35. Rösiö, C.: Supporting the design of reconfigurable production systems. (2012)
36. Rösiö, C., Säfsten, K.: Reconfigurable production system design—theoretical and practical challenges. J. Manuf. Technol. Manage. **24**, 998–1018 (2013)
37. Salvador, F., De Holan, P.M., Piller, F.: Cracking the code of mass customization. MIT Sloan Manage. Rev. **50**, 71–78 (2009)
38. Strasser, T., Fessl, K., Hämmerle, A., et al.: Rapid reconfiguration of machine-tools for holonic. Manuf. Syst. **6** (2005)
39. Surbier, L., Alpan, G., Blanco, E.: A comparative study on production ramp up: state-of-the-art and new challenges. Prod. Plan. Control **25**, 1264–1286 (2014)
40. Terkaj, W., Tolio, T., Valente, A.: Focused flexibility in production systems. In: ElMaraghy, H.A. (ed.) Changeable and reconfigurable manufacturing systems, pp. 47–66. Springer (2009)
41. Terkaj, W., Tolio, T., Valente, A.: A review on manufacturing flexibility. In: Tolio, T. (ed.) Design of flexible production systems, pp. 41–61. Springer (2009)
42. Ulrich, K.: The role of product architecture in the manufacturing firm. Res. Policy **24**, 419–440 (1995)
43. Ulrich, K.T., Tung, K., Sloan school of management: fundamentals of product modularity. Sloan school of management, Massachusetts Institute of Technology (1991)
44. Van Brussel, H., Bongaerts, L., Wyns, J., et al.: A conceptual framework for holonic manufacturing: identification of manufacturing holons. J. Manuf. Syst. **18**, 35–52 (1999)
45. Welsh, J.A., White, J.F.: A small business is not a little big business. Harv. Bus. Rev. **59**, 18 (1981)
46. Westkämper, E.: New trends in production. In: Dashcenko, A.I. (ed.) Reconfigurable manufacturing systems and transformable factories, pp. 15–26. Springer (2006)
47. Wiendahl, H., ElMaraghy, H.A., Nyhuis, P., et al.: Changeable manufacturing-classification, design and operation. CIRP Ann. Manuf. Technol. **56**, 783–809 (2007)
48. Xiaobo, Z., Jiancai, W., Zhenbi, L.: A stochastic model of a reconfigurable manufacturing system part 1: a framework. Int. J. Prod. Res. **38**, 2273–2285 (2000)
49. Youssef, A.M., ElMaraghy, H.A.: Optimal configuration selection for reconfigurable manufacturing systems. Int. J. Flex. Manuf. Syst. **19**, 67–106 (2007)
50. Zhang, G., Liu, R., Gong, L. et al.: An analytical comparison on cost and performance among DMS, AMS, FMS and RMS. In: Dashcenko, A.I. (ed.) Reconfigurable manufacturing systems and transformable factories, pp. 659–673. Springer (2006)

Chapter 17
KBE-Modeling Techniques in Standard CAD-Systems: Case Study—Autodesk Inventor Professional

Paul Christoph Gembarski, Haibing Li, and Roland Lachmayer

17.1 Introduction

Enhanced use of carry-over-parts, shorter product life cycles and product customization lead to a frequent modification and the adaptation to new functional or design requirements of digital product models [1]. Basis for this is the ability of parametric modeling in today's CAD-systems which is the use of variable values (parameters) for dimensions and variable constraints between objects and models in a CAx-system [2].

Knowledge-Based Engineering (KBE) extends this approach in order to implement explicit design knowledge into the virtual product model [3]. The overall goal is to transform a design problem into a configuration problem using, e.g., the link to dimensioning or calculation formula, design rules or manufacturing restrictions.

17.1.1 Motivation

The product development process is a structured sequence of creative activities, which aims at translating technical and design requirements into a product specification with its corresponding geometric models. On the one hand, the exploration and limitation of the possible solution space is dependent of the designer's experience and his ability to make design knowledge explicit. The answer to the question why a product looks the way it actually does, has not only to be answered but documented. On the other hand different steps in the design process contain various routine tasks.

P.C. Gembarski (✉) • H. Li • R. Lachmayer
Institut für Produktentwicklung und Gerätebau, Leibniz Universität Hannover, Welfengarten 1a, Hannover 30167, Germany
e-mail: gembarski@ipeg.uni-hannover.de; li@ipeg.uni-hannover.de; lachmayer@ipeg.uni-hannover.de

J. Bellemare et al. (eds.), *Managing Complexity*, Springer Proceedings in Business and Economics, DOI 10.1007/978-3-319-29058-4_17

According to Verhagen, one objective of KBE is to reduce time and cost of product development, which is generally realized through automation of repetitive design tasks as well as capture and re-use of design knowledge [4]. In business models like mass customization, where a product tailored to a customer's needs has to be developed and manufactured with mass production efficiency, the demand for such supporting methodologies and the corresponding modeling techniques is high. One instance would include, e.g., technical product configurators or design configurators. Nevertheless, different authors report that KBE still has not achieved a considerable breakthrough beside different automotive and aerospace applications [5].

Different aspects of KBE have already been discussed for decades, e.g., the product configuration paradigms [6, 7] or process models for creating KBE applications like MOKA and KNOMAD [4]. Other contributions aim at pointing out the delimitations to other research fields such as knowledge engineering and knowledge management. Regarding the impact on product modeling and virtual prototyping, there exist only a little number of contributions, which aims at establishing theoretical foundations for KBE. Most authors do not present concrete design methodologies, modeling principles, or detailed application examples. To date, no scientific books can be found that are dedicated to this topic.

We will bridge a part of this gap by showing how different methods and functionalities, that are already implemented and accessible in the standard package of the CAD-system Autodesk Inventor Professional, can be used for setting up virtual prototypes and their solution spaces as KBE models.

17.1.2 Structure of the Paper

The following Chap. 2 provides the theoretical background for KBE with regard to geometry manipulation and reasoning techniques. In Chap. 3 different modeling techniques in the CAD-system Autodesk Inventor Professional are exemplarily introduced and discussed in context of KBE. Chapter 4 then presents different application examples. Closing the paper, Chap. 5 contains a brief summary and drafts further research questions for future studies.

17.2 Theoretical Background

According to Chapman et al., "KBE represents an evolutionary step in computer-aided-engineering (CAE) and is an engineering method that represents a merging of object-oriented programming (OOP), artificial intelligence (AI) and computer-aided-design (CAD) technologies, giving benefit to customized or variant design automation solutions" [8] (Fig. 17.1).

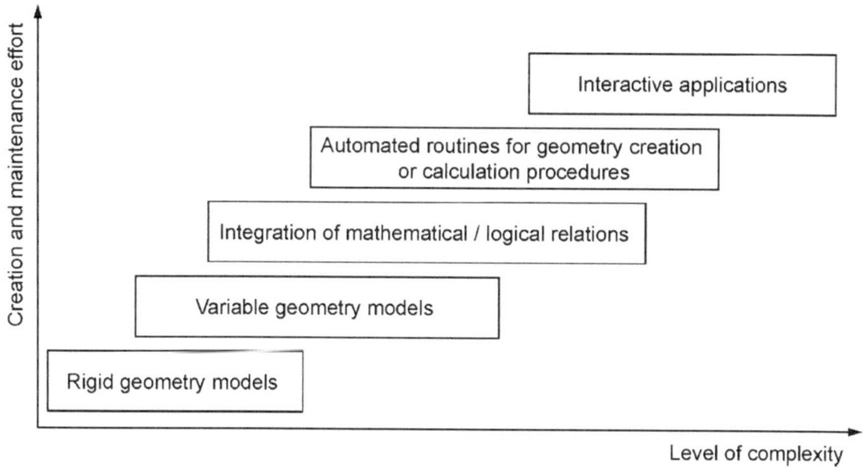

Fig. 17.1 Different types of knowledge-based design applications according to Hirz [3]

Hirz emphasizes that "knowledge-based design supports design processes by reusing predefined methods, algorithms or results, and it is integrated into specific tasks or workflows that are involved in the design processes" [3].

17.2.1 Parametric Design

Basis for knowledge-based design is the application of parametric CAD. There are three major benefits from using parametric design in opposite to rigid geometry [9]:

1. Automatic change propagation
2. Geometry re-use
3. Embedding of design/manufacturing knowledge with geometry

It is commonly accepted that the parameterization of a virtual prototype leads to the individual description of the geometry and its defining parameters and constraints. According to Vajna basically four different parameter types have to be differentiated [10]:

- *Geometric parameters* define the shape of a part or assembly. To these belong all kinds of dimensions and positioning constraints.
- *Topology parameters* have to be understood as structural parameters which can control, e.g., the suppression state of a component in an assembly or that of a feature in a part model
- *Physical parameters* determine the physical properties of the design
- *Process or technological parameters* contain, e.g., manufacturing restrictions like minimum bend radiuses or the angle of mold release slopes for cast designs.

A model's parameters are linked by arithmetical or logical constraints. Another class of constraints is geometric ones like setting two sketch lines parallel to each other or placing a component's connection point coincident on the origin of an assembly.

17.2.2 Parameter Control

Defining mathematical relations between parameters offers the possibility of differentiating leading and driven parameters. Thus, the designer not only models the geometry but also had to plan the configuration concept and the parameterization of the specific component he is drafting.

The use and integration of dimensioning formula and equations in the CAD model supports automated geometry definition and change propagation [3]. Prerequisite is that the CAD system has the ability to create (user) parameters not only for length or angular dimensions, but also supports calculation and processing of all other kind of units, e.g., for stresses, forces, or moments of inertia.

The more complex the component, the more complicated may be the configuration concept, which calls for structuring parameters at different levels. Therefore, assemblies can include a skeleton model, which defines component positioning or superordinate geometrical characteristics, e.g., based on the structural design [11]. The corresponding parameters either can be transferred via design rules within the top-level assembly or exported into the respective part models, which establishes a permanent data link between skeleton and part document.

Another way of structuring parameters is the possibility of externalizing the calculation and input of relevant parameters, which then drives the geometry within the CAD model. This can either be done through the import of text files or the link to commercially available spreadsheet software. The latter commonly offers additional mathematical and statistical operations compared to those implemented in the CAD system itself [3]. Another important fact is that relevant data for the definition and specification of components can be stored on different worksheets and then be linked by use of matrix-operations like VLOOKUP in MS Excel. Such a formulated knowledge base has to be understood as significant element within a CAE environment [12].

17.2.3 Design Rules

The implementation and formulation of design rules strongly depends of the CAD system. Only the minority of systems are able to set up and compute design rules within the functionalities of the standard configuration, most of these systems need extensions like the Knowledge Workbenches for CATIA or Knowledge Fusion for Siemens NX [5].

Basically, the rule concept is grounded upon the IF-THEN-ELSE-notation known from software development. Rules are fired procedurally and can be used to execute subordinate rules or delete them temporarily from the working memory [13].

The rule concept is very well known as reasoning mechanism of the expert systems from the 1980s [6].

17.2.4 Intelligent Templates

An intelligent template has to be understood as a parametric, updatable, and reusable building block within a digital prototype. For specific components, templates include all necessary design rules and features. So, templates can support the collection of expert knowledge and integrate existing knowledge from former development projects into the current design process [3].

According to Cox the creation of an intelligent template involves four steps [14]:

1. Use past experience, define the boundaries for the solution space
2. Map the product development process backwards into the context of the template
3. Develop a generic parametric model of all necessary products and artifacts
4. Map the specific model parameters into a common set of configuration parameters

17.2.5 Automation Routines and Macros

According to Hirz, "the ability to create macros can be very helpful for enabling automatic sequences of features and actions" [3]. Two approaches have to be distinguished; on the one hand the code can be implemented either internally in the CAD system or in single CAD models and drawings. Depending on the application programming interface (API) and its implementation technology the macros are interpreted row by row, class concepts or inheritance like known from object oriented programming may not be fully available. On the other hand the code can be written externally into a compiled software package which then drives the CAD system remotely. There, all functionalities of the used software development environment can be addressed.

Since the automation within a CAD system strongly depends on the system's functionality itself and the corresponding API model, it will not be considered further in this article.

17.2.6 Reasoning Techniques

A completely different approach to classify KBE methods is with regard to their abilities and problem solving methods in order to derive new configurations based on existing designs. This dates back to the expert systems of the 1980s and 1990s, where similarly to a human expert a knowledge-based system makes use of a reasoning mechanism which is also called inference engine [5].

This implies that all necessary engineering knowledge from all participating experts, be it process and simulation specialists, designers or production engineers, has to be formulated in a domain specific knowledge base, which extends the geometric product model [15].

Basically, the following three different paradigms can be distinguished [6]:

- *Rule-based reasoning*: The knowledge representation relies to design rules like mentioned in Sect. 17.2.3. A major disadvantage of this kind of systems is their lack of separation between domain knowledge and control strategy. As reported by McDermott, this results in bad maintainability when the system exceeds a certain amount of rules [13].
- *Model-based reasoning*: The limitation of the possible solution space is done based upon a physical and/or logical model (constraint-based) or by representation of resource consumption and allocation (resource-based) [16].
- *Case-based reasoning*: In this approach, the knowledge representation is not explicitly modeled in form of rules or constraints. The knowledge necessary for reasoning is stored in cases that represent former configurations. Depending on the degree of maturity of the inference engine the system either is limited to search for existing solutions, which match exactly to a given requirements profile, or the system is able to assort a set of existing cases, which represent the best-fit. Highly developed case based systems are able of mixing or altering exiting cases in order to adapt them to new situations.

Differently than traditional knowledge-based systems, a KBE system shows no crisp separation between knowledge base and inference mechanism. The control strategy for accessing and manipulating the knowledge base and the domain knowledge itself are strongly intertwined [5].

17.3 KBE Modeling in Autodesk Inventor

In this chapter different modeling techniques in the CAD-system Autodesk Inventor Professional are exemplarily introduced and discussed in context of KBE. In the following subsection it is evaluated what parameter types as referred in Sect. 17.2.1 can be implemented in Inventor part and assembly models. Afterwards we present possibilities to link these parameters via equations and rules. Using these functionalities dynamic assembly models and intelligent templates can be set up.

The implementation of MS Excel spreadsheets allows the integration of a behavioral model in the background of the geometric one which leads to setting up technical product configurators.

17.3.1 Integration of Different Parameter Types

Within Inventor every parameter regardless of its type (model, reference, or user parameter) has a specific set-up and consists of the four parts: name, value, unit, and comment. Model parameters are introduced by Inventor with every dimension, as

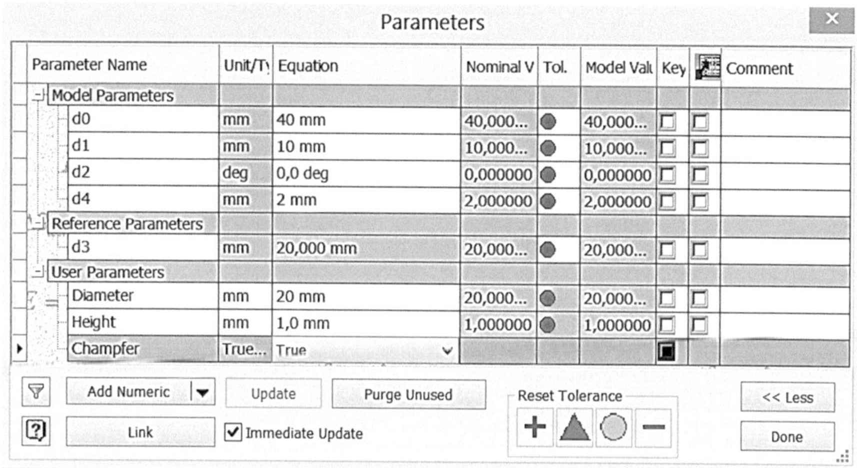

Fig. 17.2 Inventor Professional 2016: parameter table

well as reference parameters. The latter indicate either a driven dimension (element is over-constraint) or an imported parameter.

All parameters are managed in the parameter table as depicted in Fig. 17.2. In this dialog, the user can change all parameter values including adding names and comments to each of them or set new user parameters. Those can either be numerical or Boolean, a third type of user parameters is text (not shown). If a numerical parameter has to contain a value list instead of a single value, e.g., to choose between certain thicknesses for a sheet metal part, the parameter may be defined as multi-value. The input format then changes from text box to dropdown box.

As unit types Inventor can use basically all physical units with all suitable prefixes. This includes units for length (mm, inch, nautical mile, etc.), angularity (radian, degree) but also for mass, forces, power, velocities, electrical, or luminosity to name only a few categories.

17.3.2 Equations and Design Rules

Setting up an equation within Inventor can either be done directly at dimensioning or centralized in the parameter table. As mathematical operators, all basic arithmetic operations, trigonometric functions, roots and powers as well as rounding operations may be used.

In the following example, the diameter of a bold has to be calculated according to the dimensioning formula, as in (17.1), with k as fitting factor, K_A as application factor, *Force* as applied force, and *Bend stress* as maximum bend stress depending on the material of the bolt:

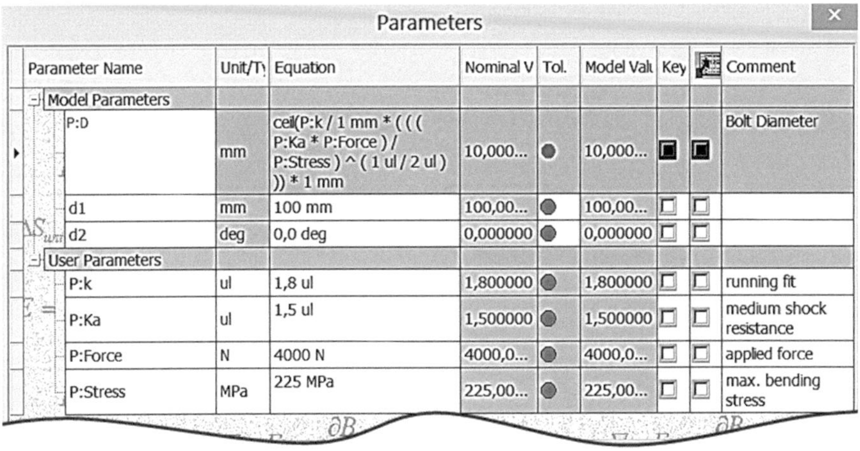

Parameters

Parameter Name	Unit/T	Equation	Nominal V	Tol.	Model Valu	Key		Comment
Model Parameters								
P:D	mm	ceil(P:k / 1 mm * (((P:Ka * P:Force) / P:Stress) ^ (1 ul / 2 ul))) * 1 mm	10,000...	●	10,000...	■	■	Bolt Diameter
d1	mm	100 mm	100,00...	●	100,00...	□	□	
d2	deg	0,0 deg	0,000000	●	0,000000	□	□	
User Parameters								
P:k	ul	1,8 ul	1,800000	●	1,800000	□	□	running fit
P:Ka	ul	1,5 ul	1,500000	●	1,500000	□	□	medium shock resistance
P:Force	N	4000 N	4000,0...	●	4000,0...	□	□	applied force
P:Stress	MPa	225 MPa	225,00...	●	225,00...	□	□	max. bending stress

Fig. 17.3 Dimensioning formula implemented in an Inventor part document

$$d \approx k\sqrt{\frac{K_A \text{ Force}}{\text{Bendstress}}} \tag{17.1}$$

The equation is entered in the parameter *P:D* which defines the bold diameter as depicted in Fig. 17.3.

Note that it is necessary to cancel down the unit within the ceiling operation since Inventor expects a dimensionless factor which then has to be multiplied by 1 mm again.

Such an equation is computed every time when a rebuild to the model geometry is processed.

Another way of linking parameters is the use of design rules which can be done within the iLogic environment as depicted in Fig. 17.4. The iLogic programming language is similar to script languages. Common constructs like if-then-else or select-case decision trees, while loops, the use of sub procedures and a class concept are usable. As command library the snippets include code templates for almost every modeling context within Inventor.

In the example above, the two parameters *d0* and *d1* are linked to prior defined user parameters via equations. In addition, the suppression state of a chamfer feature is linked to a Boolean parameter. All parameters from each component or feature can be addressed through the model tree within the iLogic rule editing dialog. This is the same for assemblies, where parameters of different parts can be linked to each other (this will be shown in Sect. 17.3.3 in context of dynamic assembly models).

In contrast to parameter equations, iLogic rules are not automatically computed at every rebuild of the model. The computation either has to be triggered manually or linked to certain events like geometry update and closing a file.

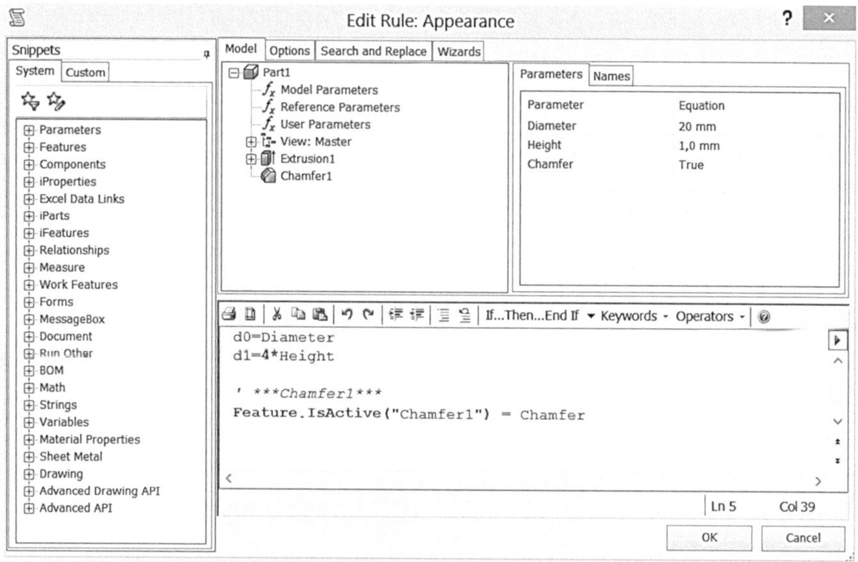

Fig. 17.4 iLogic rule editor

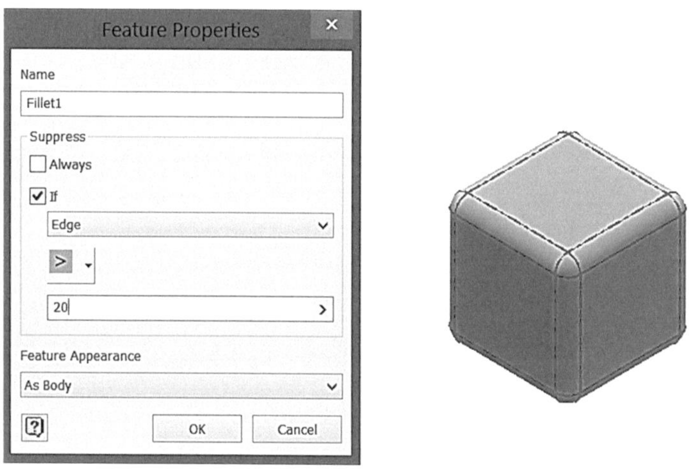

Fig. 17.5 Suppression state definition in feature properties dialog

Within the part document exists another way of using classical design rules for topological parameters. Therefore, a dimension parameter is linked directly to the suppression state in the feature properties of the feature to be controlled.

In the example shown in Fig. 17.5 the cube's fillet is suppressed when the length of the edge (described in a parameter named *edge*) exceeds 20 mm.

17.3.3 Dynamic Assembly Models and Intelligent Templates

In this subsection we present two different methodologies for modeling of a dynamic assembly model for use as intelligent template. The first approach uses a skeleton behind the geometric models, whereas the second one links the necessary parameters via iLogic rules. The running example for this subsection is the base frame of a pipe support as depicted in Fig. 17.6.

For the skeleton model a part document is created which contains the four basic configuration parameters height over ground of both decks, the width of the rack and the thickness of the flange plates. Additionally the cross section for the used beams and the layout of the flange plate are defined as sketches (Fig. 17.7).

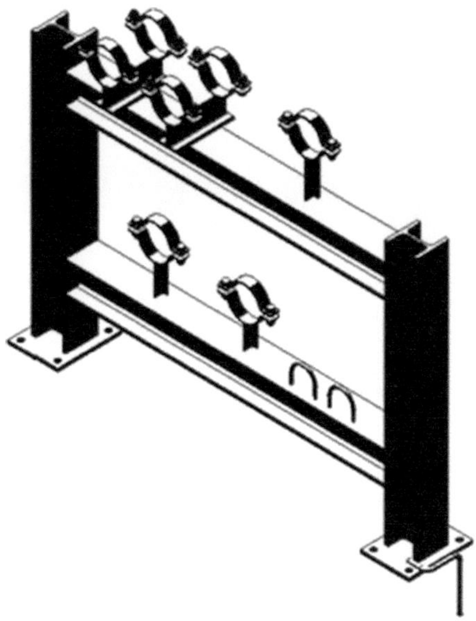

Fig. 17.6 Pipe support

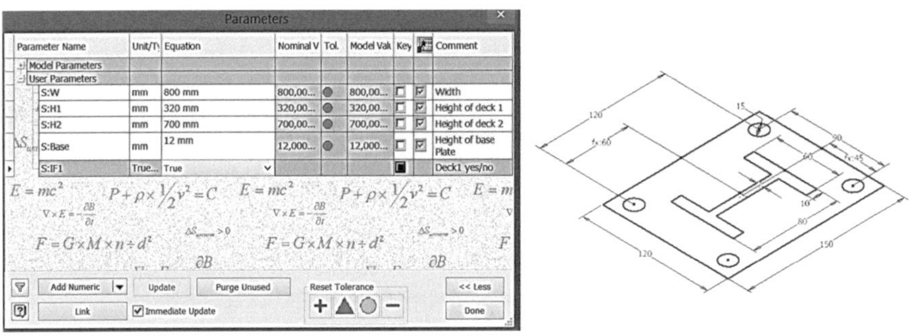

Fig. 17.7 Skeleton model for the base frame

Via the derive component command the skeleton is imported into the corresponding part documents for the beams and the flange plate. Within the derivation dialog only the necessary parameters and sketches are marked for import (Fig. 17.8).

Afterwards the extrusion for the first beam is set up. Therefore, the length dimension is calculated from the imported parameters. The procedure is similar for the second beam and the flange plate.

The next step is building everything together in the assembly document. Here, the parts are traditionally placed and constrained to each other. To adapt the assembly to new geometric boundaries only the configuration parameters within the skeleton have to be modified (Fig. 17.9).

Advantages of this approach are a short set-up time and the change propagation within the derived part. A change not only to the configuration parameters, but also to the sketch dimensions of cross sections and flange are immediately processed to the part documents then. Disadvantage is the fact that each part document has a permanent data link to the skeleton part what might have a negative effect on rebuild and document loading times. The link can be broken, but an artifact of the skeleton remains in the documents. Removing this would result in rebuild errors. Additionally, the parameterization has to be planned beforehand because all necessary parameters and sketches have to be defined in the skeleton.

The second approach to model the base frame is linking parameters by iLogic rules. Here the beams and the flange plate are modeled and assembled traditionally without skeleton or other knowledge implementation. The next step is to add the configuration parameters to the assembly file and then to append the iLogic rules. This is not completely comparable to the skeleton variant above since the skeleton also controls the dimensions for the beams' cross section. If the dimensions would also have to be driven by iLogic rules, then additional parameters have to be linked (Fig. 17.10).

If the geometry alteration should not be made by change of the user parameters in the parameter table, a graphical user interface can be built as iLogic form like depicted in Fig. 17.11. Here, different input controls, e.g., textboxes, sliders, or checkboxes, can be used. Additionally, when one of the input parameters is multi-value, the user may choose between combo box, list box, or radio buttons. Other controls, e.g., command buttons for running macros, can also be implemented.

The advantage of this method is that for a known model set-up the linking of parameters via rules is very comfortable and can be applied to existing components without restructuring parameters or rebuilding features. Since all rules are defined in context of the assembly itself, no data link to other documents like the skeleton exists.

Disadvantages of this approach are the lack of geometry transfer and the transferability of the iLogic rules. If in the example above the beams' cross section also should be modified via iLogic, eight more rules for manipulating the sketch geometry are necessary. Replacing a component within the assembly leads not only to re-constraining but also to adapting the corresponding rule to the new component since name and possibly parameter names have changed.

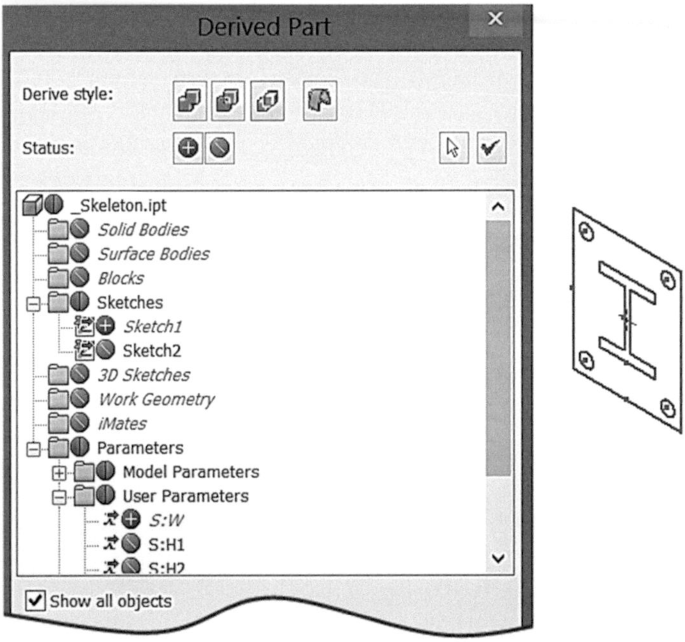

Fig. 17.8 Derived geometry and parameters for the beam model

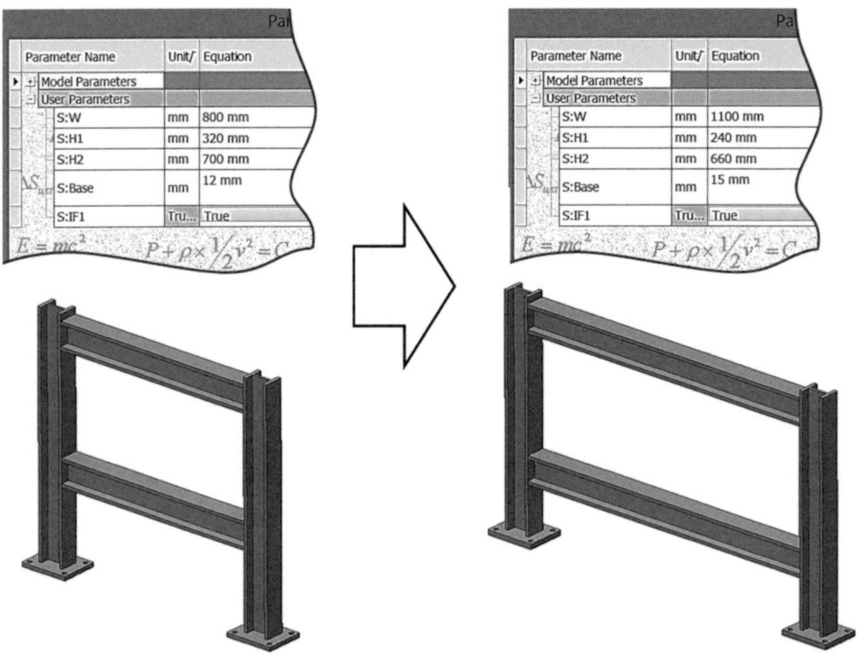

Fig. 17.9 Skeleton controlled assembly document

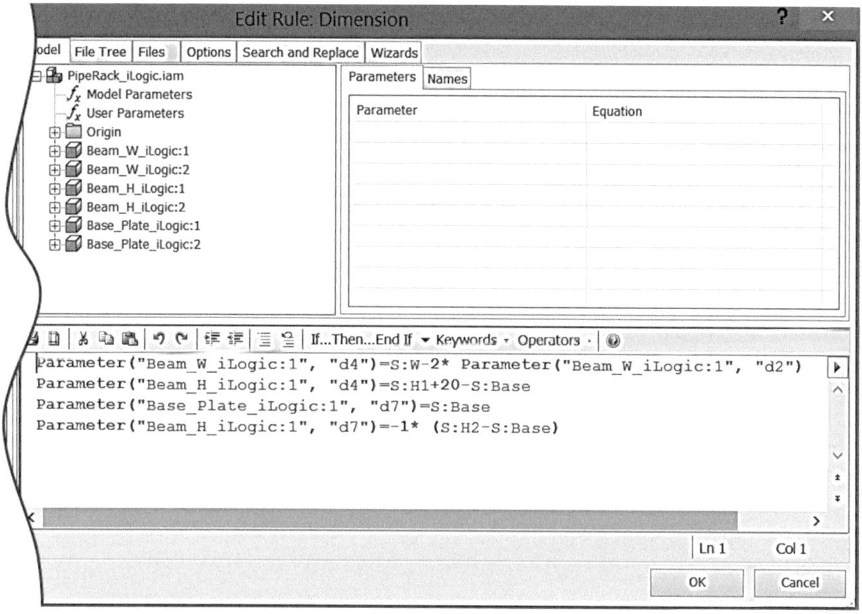

Fig. 17.10 iLogic rules for the configuration of the base frame

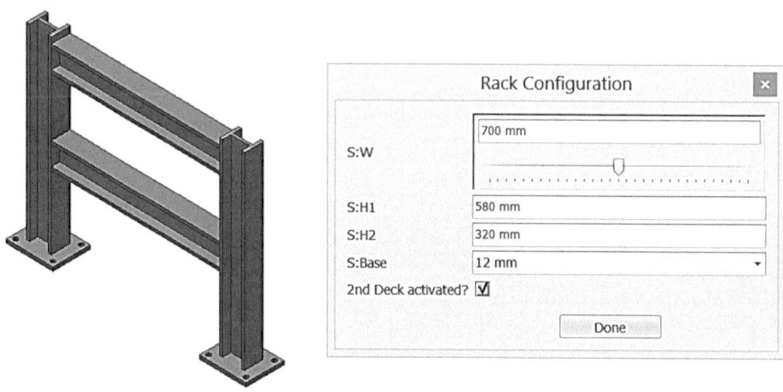

Fig. 17.11 iLogic form

Both of the assemblies mentioned previously can be used as intelligent templates or assembly building blocks. To choose which approach is the best for modeling or if an intermediate approach has to be used depends on the specific modeling task, the degrees of freedom regarding parts as well as topology and the manifold of the solution space. A detailed investigation is beyond of the scope of this paper.

Name	Value	Unit	Comment
S:W	700	mm	Width
S:H1	620	mm	Height Deck 1
S:H2	300	mm	Height Deck 2
S:Base	12 mm		Thickness Base Plate

Fig. 17.12 Inventor embedded Excel table

17.3.4 Spreadsheet Driven Design and Design Configuration

As discussed in Sect. 17.2.2 an externalized parameter management within a spreadsheet application offers the possibilities to use extended mathematical and statistical operations. In order to use MS Excel for parameter processing it is necessary to define the parameters correctly. Since the format of an Inventor parameter is strictly defined the spreadsheet has to be built up as depicted in Fig. 17.12 with name, value, unit, and comment. Additional columns can also be used for user interaction, pictures, etc. but these will be ignored by Inventor.

When the spreadsheet is set-up it can be imported in Inventor in the parameter table dialog. The implementation can be done in two different methods. On the one hand linking the spreadsheet is possible. On the other hand the spreadsheet can be embedded in the Inventor file. The latter is favorable since the data link can easily be broken by moving the Excel-file to another folder, etc. The start cell is another important option. In the example above the first row contains only column titles but no parameters. So the starting cell has to be *A2* since Inventor has to start data processing from here.

Afterwards the parameters are listed as embedded parameters and can be linked via equations or design rules.

The use of MS Excel is a powerful tool for creating engineering or design configurators. Since Inventor reads only the first four columns of the first worksheet all other cells may be used for parameter calculation, linking parameters of standards or user interaction, e.g., plausibility checks or interactive diagrams. Some of the possibilities are shown in the application examples in Sect. 17.4.

17.3.5 Intermediate Result

Regarding the basic parameter types named in Sect. 17.2.1, geometric and physical parameters can be entered directly within the parameter table in Inventor. In addition, process and technological parameters may be used since they usually express boundaries for geometric and physical parameters such as minimum bend radius or a hardening depth. Topological parameters control the suppression state of features and components. Since the state can either be true or false, a user parameter of the type Boolean is suitable.

All methods of parameter control as discussed in Sect. 17.2.2 are available in Inventor. Equations can directly be entered at dimensioning or in the parameter table. The externalization of input and calculation is possible due to the implementation of MS Excel. Additionally, the use of design rules is implemented in the iLogic concept.

This has to be understood as prerequisite for generating dynamic assemblies or intelligent templates as introduced in Sect. 17.2.4. Nevertheless, depending on the modeling task and the solution space, which has to be mapped into the model, an appropriate modeling technique has to be chosen.

Implementing reasoning techniques as presented in Sect. 17.2.6 is partially realizable. While rule-based and model-based approaches can be set up by use of the corresponding parameter control functionalities, the application of a highly developed cased based reasoning is to date not reported.

17.4 Application Examples

In the following subsections three application examples are shown. Some of them represent a combination of the functionalities mentioned above in order to address different control strategies and input formats (Fig. 17.13).

17.4.1 Base Frame

The above assembly is used in multiple manufacturing stations as base frame. It consists of various rectangular tubes, connections plates, mounting feet, and a front cover. On top of the frame, a mounting plate with the assembly system and tools for the particular station is installed.

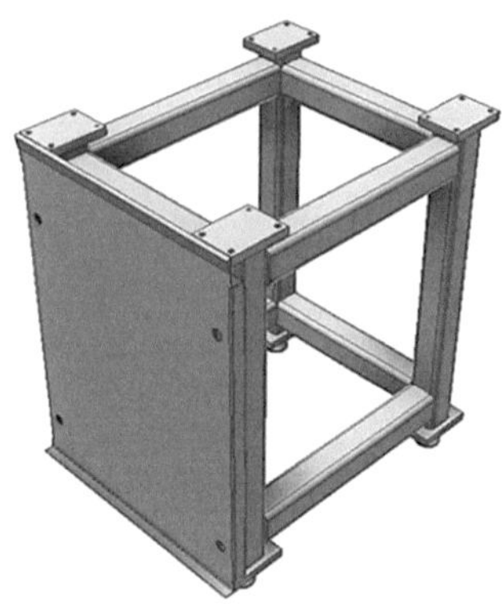

Fig. 17.13 Intelligent template of a base frame

The base frame is modeled as template with iLogic rules for parameter calculation. In order to equip a new assembly station the frame is copied and the configuration parameters for height, width, depth, and clearance height are entered. If a change to the dimensions of the mounting plate occurs the assembly of the base frame can easily be adapted to the new boundaries. All drawings for the frame and its parts are updated automatically as well.

The design and documentation of a single frame was an 8 h task before automation. Every change to the mounting plate has led to the adaption of the single parts of the frame and took in average 1 h.

After automation the generation and maintenance can be done in minutes, the model is stable for all dimensions that usually occur.

17.4.2 Crank

The crank depicted in Fig. 17.14 has to be used as dynamic assembly model since changes in the corresponding transmission design often occur. The assembly consists of the crank itself, a ball bearing and a retaining ring on the one shaft and two fitting keys on the other shaft which can be replaced completely by a spline shaft.

The assembly is modeled with a skeleton model behind the geometric one. In order to facilitate parameter input and definition, a spreadsheet is implemented in the skeleton model with a simple configuration user interface. As input, the geometric boundaries and type as well as size of the standard parts are defined. As output, the spreadsheet generates a report with strength calculation based upon entered loads. Additionally, iLogic rules have been defined to alter the size of the standard parts which have been set up as part families as well as the corresponding interfaces of the crank.

Design and documentation of the depicted assembly was done in 6 h before automation. Afterwards the configuration and reporting takes 10 min.

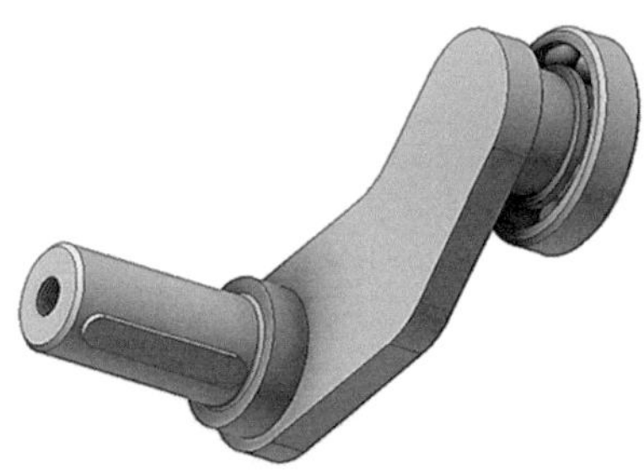

Fig. 17.14 Dynamic assembly model of a crank

17.4.3 *Platform*

Platform design is a common task in plant engineering. For quickly assembling a platform like illustrated in Fig. 17.15 a modular design kit was developed. It consists of building blocks for the base frame, stairs, connection bridges, and handrails.

Here also a combination of all of the above control strategies was implemented. The basic configuration is done via an Excel spreadsheet (Fig. 17.16) which calculates also most of the model parameters. Plausibility checks where defined in order to check whether the configuration is according to existing standards or has to be modified. The parameters are passed to a skeleton model for further computing and definition for the sketch geometries of beams and other parts. Within the main assemblies multiple iLogic rules are implemented for changing parts according to the configuration. Over 300 parameters and 138 part and assembly documents are managed.

The design of such a platform can be done in 40 working hours, after automation it is possible to assemble the above platform in less than 4 h. Some manual tasks remain since the layout plan has to be drafted and dimensioned for manufacturing.

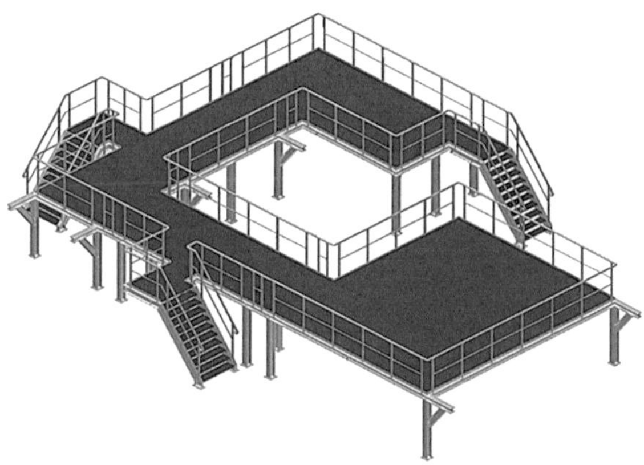

Fig. 17.15 Modular design kit for platforms

Name	Value	Unit	Comment	Part Name	Hint1	Hint2		
M00:01	6000	mm	Length of the platform in x		OK!	if possible choose multiples of 1000	size	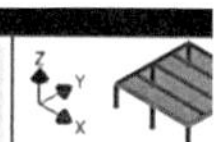
M00:02	9600	mm	Width of the platform in y		OK!	if possible choose multiples of 800		
M00:03	2800	mm	heigth of the platform in z		OK!			
M01:01	180		I-Profiles for Frame		OK!		beams and dimensions	
M01:02	200		I-Profiles for basement		OK!			
M01:03	100x65x10		brackets					
M01:04	3	oE	count of supports in x					
M01:05	4	oE	count of supports in y					
M01:06	15	mm	thickness of base plates					
M01:07	40	mm	distance to support					
M01:08	SP-30 - 50,8 - T 800 x 1000		grating		OK!			

Fig. 17.16 Spreadsheet configurator

17.5 Conclusion

In the present paper different methods and functionalities for knowledge-based engineering have been introduced and discussed in a case study using the CAD system Autodesk Inventor Professional. We showed that building technical product configurators or engineering configurators within a CAD environment is possible.

Nevertheless, it is important to note that KBE is not suitable for all design tasks. Whenever the design is repetitive and can be expressed in explicit rules and so formulated as configuration problem, KBE is one of the best technologies for implementation.

There are still a lot of research questions to be answered. On the one hand we already pointed out at the application examples that choice and combination of different knowledge integration techniques and KBE functionalities is depending of the modeling task and the manifold of the solution space. Here guidelines and modeling principles as well as performance measures still have to be investigated. Another question is the implementation of reasoning techniques into the CAD models. Current research projects at our institute examine whether case-based reasoning might be implemented directly into a CAD configurator.

References

1. Abramovici, M., Stekolschik, A.: Methodische Ansätze für die Weiterverwendung der digitalen Produktmodelle in der Produktentwicklung, Tagungsband 2. Gemeinsames Kolloquium Konstruktionstechnik, Verlag Saxoprint, Dresden (2004)
2. VDI: VDI 2218—Informationsverarbeitung in der Produktentwicklung—Feature-Technologie, Beuth Verlag, Berlin (2003)
3. Integrated Computer-Aided Design in Automotive Development, Springer Science & Business Media (2013)
4. Verhagen, W., et al.: A critical review of knowledge-based engineering: an identification of research challenges. Adv. Eng. Inform. **26**(1), 5–15 (2012)
5. La Rocca, G.: Knowledge based engineering—between AI and CAD. Review of a language based technology to support engineering design. Adv. Eng. Inform. **26**(2), 159–179 (2012)
6. Sabin, D., Weigel, R.: Product configuration frameworks—a survey. IEEE Intell. Syst. **13**(4), 42–49 (1998)
7. Hvam, L., Mortensen, N.H., Riis, J.: Product Customization. Springer Science & Business Media (2008)
8. Chapman, C.B., Pinfold, M.: The application of a knowledge based engineering approach to the rapid design and analysis of an automotive structure. Adv. Eng. Softw. **32**(12), 903–912 (2001)
9. Shah, J.J.: Designing with parametric cad: classification and comparison of construction techniques. Geometric Modelling, pp. 53–68. Springer, New York (2001)
10. Vajna, S.: CAx für Ingenieure: eine praxisbezogene Einführung. Springer Science & Business Media (2009)
11. Saufhoff, B., Lachmayer, R.: Generative design approach for modelling of large design spaces. In: Proceedings of the 7th World Conference on Mass Customization, Personalization, and Co-Creation (MCPC 2014), February 4th–7th, 2014. Lecture Notes in Production Engineering, pp. 241–251, Aalborg, Denmark (2014)

12. Peng, C.C., Rigdway, K.: Integration of CAD/CAM and spreadsheet data processing. Integ. Manuf. Syst. **4**(4), 29–36 (1993)
13. McDermott, J.: R1: a rule-based configurer of computer systems. Artif. Intell. **19**(1), 39–88 (1982)
14. Cox, J.J.: "Product Templates." A parametric approach to mass customization. In: CAD Tools and Algorithms for Product Design, pp. 3–15. Springer, Heidelberg (2000)
15. La Rocca, G., van Tooren, M.: Development of design and engineering engines to support multidisciplinary design and analysis of aircraft. Delft Science in Design–A Congress on Interdisciplinary Design. Delft University of Technology Faculty of Architecture Chair of Product Development Berlageweg 1 2628 CR Delft the Netherlands acjm eekhout@ bk. tudelft. nl/m. eekhout@ octatube. nl (2005)
16. Heinrich, M., Jüngst, E.W.: A resource-based paradigm for the configuring of technical systems from modular components. In: Proceedings of the Seventh IEEE Conference on IEEE, vol. 1 (1991)

Chapter 18
A Business Typological Framework for the Management of Product Complexity

Paul Christoph Gembarski and Roland Lachmayer

18.1 Introduction

Managing product variety in the stage of order acquisition as well as in product development and manufacturing is a key factor to a company's success [1]. Technical products are becoming more complicated and so the demands for product documentation. On the other hand, new requirements for flexibility in product development and manufacturing arise due to deregulated and global competition and the trend of product customization, which Bliss characterizes as new market dynamics [2]. Thereby, it is generally accepted that the customer's desired and perceived diversity as well as the desired individuality of products has to be dealt with a minimum of organizational efforts.

Commonly, the term complexity is used synonymously for product variety in this context. A generally accepted definition for complexity is yet not at hand, but most approaches include organizational effects and take into account that high variety leads to problems and uncertainties in forecasting demands and control of manufacturing and operations. Furthermore, complexity is considered to be strongly company specific.

In the present chapter, we present a business typological framework for management of complexity. Key element is the Hannover House of Complexity, which defines the effects of certain complexity management tools and methods on distinct complexity measures on the one hand. On the other hand, the interdependencies of these tools are documented.

Related to the business model of mass customization product, configuration systems are exemplarily discussed and classified in the Hannover House of Complexity.

P.C. Gembarski (✉) • R. Lachmayer
Institut für Produktentwicklung und Gerätebau, Leibniz Universität Hannover, Welfengarten 1a, Hannover 30167, Germany
e-mail: gembarski@ipeg.uni-hannover.de; lachmayer@ipeg.uni-hannover.de

J. Bellemare et al. (eds.), *Managing Complexity*, Springer Proceedings in Business and Economics, DOI 10.1007/978-3-319-29058-4_18

18.1.1 *Motivation*

According to the process-related management methodologies such as quality management, configuration management, and risk or environmental management we are developing an integrated approach for complexity management. It is not our aim to develop a general theory of complexity, but a scientifically well-grounded process model, which is applicable and implementable in engineering science and mainly focuses on product development. Our work is based on three basic assumptions:

1. A certain amount of complexity is beneficial in today's market environment.
2. There exists an ideal complexity—an increase will not lead to higher revenues.
3. Ideal complexity and complexity management are depending on business type and marketing strategy.

By means of business typologies, models are developed for the assessment of complexity management methods and tools according to a company's specific requirements. A main focus is the formulation of standard sets of methods and tools for different business types and marketing strategies such as business to business (B2B) or business to consumer (B2C). We focus on the product development process since product development is the central transfer site for information in a company [3].

18.1.2 *Structure of the Chapter*

Section 18.2 provides the theoretical background of complexity and complexity management. As framework the Hannover House of Complexity is presented in Sect. 18.3. In Sect. 18.4 the product-process change matrix is characterized as business typology with focus on the business model of mass customization. Section 18.5 then introduces product configuration systems in context of knowledge-based systems and classifies them into the Hannover House of Complexity. Closing the chapter, Sect. 18.6 contains the conclusion and drafts further research questions.

18.2 Complexity and Complexity Management

Cybernetics and system theory can be identified as origin of complexity theory [4]. These approaches have already been adapted to and further developed for various scientific disciplines such as, e.g., natural science and social and labor science [5]. The analysis of these shows that general definitions or modeling principles do not exist. Instead, complexity is mapped and reduced on the particular problem statement.

Generally accepted is the fact that complex systems can be represented at least in the dimensions of diversity, variety, and dynamics. Nevertheless, in the most approaches, dynamics can also be translated into diversity since it is the count of different system states and change possibilities [6].

Approaches in engineering science are also founded on cybernetics and broken down to complexity of products as well as development and production processes. With respect to mechanical engineering, external and internal product complexity is differentiated. External product complexity is understood as diversity of a company's offering, which is perceived and stipulated by the customer. In contrast, internal product complexity is defined as the number of subassemblies and components as well as their design and combination rules in order to assemble them to end products [7].

A lot of authors emphasize that product complexity and process complexity are strongly intertwined. Multivariant products thus lead to an increase of complexity in all operational structures and processes since the high quantity of end products, alternatives, components, and the corresponding documents for each project and each customer has to be managed in operations and the whole supply chain [8].

From our point of view, a system can be considered as complex when it is composed by a large number of components which are arbitrarily linked together. The system can have different states over time, but the system's behavior is diffuse since it cannot be simulated or fully predicted. The system elements are not limited to the use in one single system but can also be used in different systems in the sense of commonality.

18.2.1 Complexity Management

According to Schuh, the management of complexity is "the design, development and control of business activities regarding products, processes and resources. By managing complexity it is aimed to dominate diversity along the whole value chain so that customer satisfaction as well as organizational efficiency gets maximal" [9].

Generally, different aspects of complexity management and single tools can be found in literature. Bliss concludes that the major process management schools of the 1990s can also be regarded as complexity management methods. So lean management is an answer to increased complexity of the production program and the manufacturing techniques, whereas business process reengineering focuses on organizational complexity. Variant management as the third method concentrates efforts of product complexity and customer complexity. Here, e.g., modularization is a valuable building block. Nevertheless, an integrated model for complexity management is still not at hand [2].

From our point of view, this argumentation leads to three basic views of complexity management:

- Management of product complexity: Measures in different areas of the company, which purpose is designing and controlling the complexity of end products as well as their components and individual parts depending on their functional and design requirements
- Management of resource complexity: Methods in order to design and control the complexity of production resources, raw materials as well as knowledge and personnel in the value chain
- Management of process complexity: Approaches which aim at design and control of complexity of operational and organizational structures

As basic strategies for complexity management a literature review shows three basic approaches. On the one hand, different authors name reduction of complexity as the first basic goal. This strategy aims at streamlining the existing product and process portfolio for a short-term complexity reduction. Here, product variants with low demand and overlaps in the overall offering have to be identified and eliminated.

The second step is complexity control which means dealing with strategic planning and development of necessary complexity. Here, approaches for product family design, modular design kits and solution space modeling in general are subsumed. Additionally, an according setup of the manufacturing organization and of order processing has to be implemented.

The third approach is prevention of complexity where new further product and process variants have to be assessed regarding additional benefits for company and customer before realization and implementation.

18.2.2 Measuring Complexity

The lack of a common definition of complexity is continued in measuring it. But as prerequisite for managing complexity, it is necessary to determine an ideal amount of complexity or to differentiate between good and bad complexity. The early attempts of finding describing dimensions failed and resulted in a multitude of measures which could not exactly assess complexity [10].

After a wide-ranging literature review, Bandte condenses different complexity management approaches of various scientific areas and derives properties of complex systems such as variety and diversity, dynamics, feedback from the environment, nonlinearity, self-organization, limited rationality, and emergence to name only a few [5].

For his complexity management approach, Schuh uses the so-called complexity drivers which are diversity on the one hand and dynamics on the other hand. His concept of diversity encompasses both the diversity of system elements and the diversity of relations between these elements as well as the variety of system states over time [9].

Gießmann uses a compact approach from point of view of logistics and describes complexity in the dimensions of variety, heterogeneity, diversity and uncertainty. All these dimensions are dependent since, e.g., an increase of dynamics results in an increase of uncertainty because the prediction of future developments and system states is more difficult. So it is not enough to measure a single aspect of complexity or to consider only a limited count of system elements but to examine the whole system and all possible occurrences [11].

Broken down to manufacturing organizations, Frizelle reduces this to even two dimensions by the consideration that complexity arises out of the presence of variety since increasing variety generates uncertainty so that the system's behavior cannot be completely predicted. According to him "variety can be seen in terms of trajectories—the path a system traces over time; the greater the variety, the more trajectories are open to the system. Uncertainty comes from not knowing which trajectory the system will follow" [10].

Focused on product development, the consideration of commonalities of components between different systems is a relatively new measure. This is important at designing modular design kits since modules should not only be restricted for use in only one system in order to use economies of scale. Hence, a change to a component is more critical when different configurations and end-product variants have to be checked and taken into account. In our first approach, the dimension of nonlocality (later trans-connectivity) was ought to reflect this. Other dimensions in this approach are variety and heterogeneity of components, dynamics in the sense of likelihood of change over time, and uncertainty of system states and system development [6].

Since the measures mentioned above are not fully independent the approach can be simplified from point of view of the possible solution space a product can be developed from (Fig. 18.1).

Hence, one dimension for product complexity is the size of the possible solution space with respect to diversity and the predictability of the boundaries of the solution space regarding uncertainty. The second measure is the predetermination of the

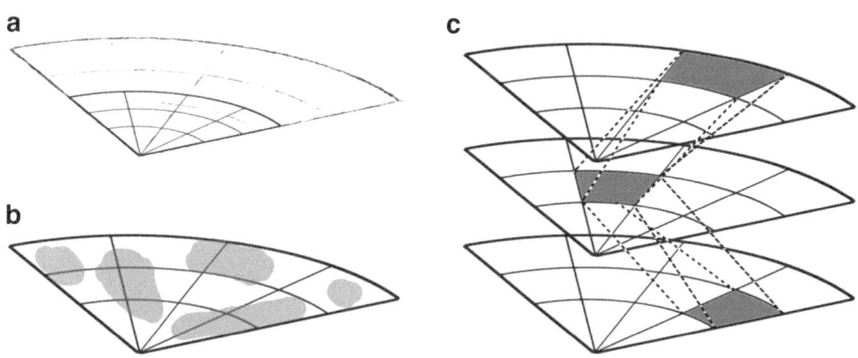

Fig. 18.1 Complexity measures of a solution space for product development: (**a**) size and determination of the solution space, (**b**) degree of exploration of the solution space, and (**c**) interaction between multiple solution spaces

solution space itself. Regarding diversity, we define the degree of exploration as dimension which means that either all possible solutions are calculated and documented beforehand or only part of them. The latter leads to higher uncertainty, since the validity of all end-product variants is not checked. So, possible conflicts are not completely foreseeable. As the third dimension, the commonality of components between different solution spaces is introduced. The more solution spaces are addressed, the more complicated is the prediction of the effects when components change.

Note that at this point we only focus on the possible solution space in product development. The interactions between this type of product complexity and other occurrences of complexity, e.g., requirement complexity or manufacturing complexity, are beyond the scope of this chapter.

18.3 Hannover House of Complexity

The Hannover House of Complexity has to be understood as framework in which different methods, tools, etc. are classified with regard to their effect on distinct complexity dimensions. The basic concept of the House of Complexity is depicted in Fig. 18.2. In principle, the design is similar to the house of quality known from quality function deployment.

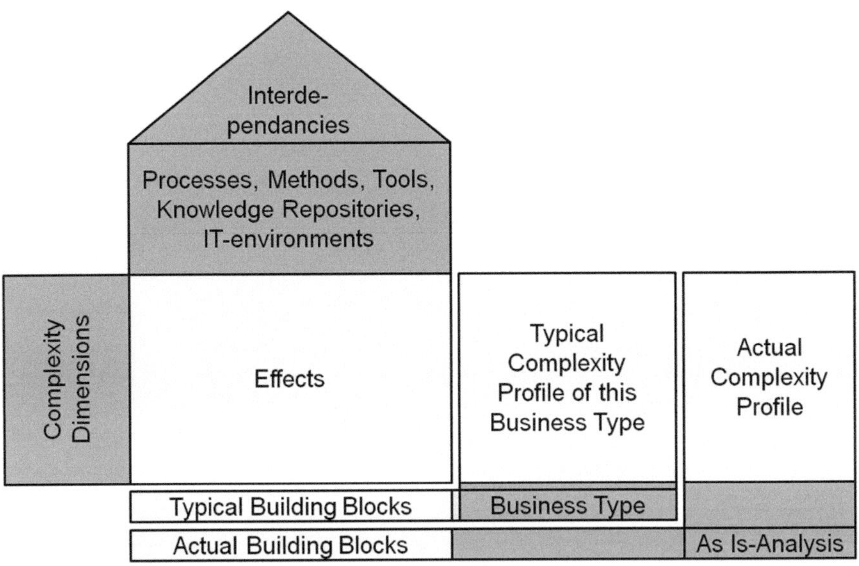

Fig. 18.2 The Hannover House of Complexity—architecture

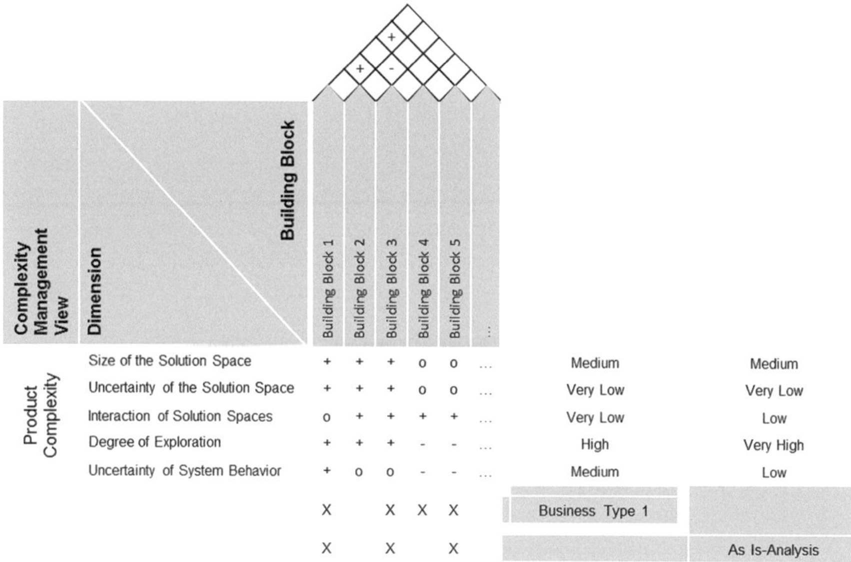

Fig. 18.3 The Hannover House of Complexity—framework

In opposite to QFD, the major areas are not the mapping of customer requirements to functions or properties of the product but the mapping of different building blocks for complexity management and their particular effects on different complexity dimensions. The roof of the House of Complexity documents the interdependencies of these building blocks to estimate whether two of these building blocks intensify the benefit of or, on the contrary, extenuate each other. Since the framework is set up as aid for decision making, a reference to a standard company of an according business type is given for comparison. This includes the choice of typical building blocks on the one hand. On the other hand, it also allows the assessment of the complexity profile which can be seen as usual at this particular business type. The architecture of the House of Complexity is completed by the fields for the as-is analysis. An example of the detailed framework is given in Fig. 18.3.

In the example above, the effect of different building blocks for complexity management on the dimensions of product complexity is shown. Based on a business typology, a company assigns itself to a business type 1. Comparing both complexity profiles, it can be seen that in contrast to the benchmark, the interaction of solution spaces, the degree of exploration of these solution spaces, and the overall uncertainty of the system's behavior differ. This is due to the missing of a complexity management building block which is yet not implemented at the company. Furthermore in the roof the mutual effects of building blocks one to five are depicted.

As can be seen from this example it is not the aim of minimizing every complexity dimension. In the example above, the uncertainty of the systems behavior increases.

18.4 Business Typology: The Product-Process Change Matrix

The product-process change matrix was introduced by Boynton et al. in 1993 and can be understood business typology (Fig. 18.4). The two dimensions for differentiating the single business types are product change and process change. The first focuses on the demand for new products and services; the latter addresses all procedures and technologies to develop, market and manufacture them [12].

Both types of change can either be stable, which means slow and foreseeable, or dynamic in the sense of fast, revolutionary, and generally unpredictable. Within the fields of the matrix, the four basic business models invention, mass production, continuous improvement, and mass customization are differentiated.

Invention refers to organic or job-shop design, where permanently new products and the according processes for development and production are invented which have to compete in market through differentiation and innovation. After the product developed to a certain degree of maturity and market transformed to mass market, the business type changes to mass production. Here the manufacturing processes have to be kept stable in order to achieve economies of scale. Boynton et al. point out that there exists a critical synergy between invention and mass production since the mass production model is incapable of developing completely new products and the invention model has to deliver new products and processes to the mass producer.

The third business model is named continuous improvement and is based upon the improvement of processes and product quality while reducing costs. Known approaches are TQM and kaizen [13].

Mass customization is the fourth business model. The idea behind is that customer-specific products can be tailor-made by the use of flexible but stable processes with mass production efficiency. Taking into account that only the customer himself is able to formulate his specific needs and requirements, Piller suggests that "MC refers to a customer co-design process of products and services, which meets the needs of each individual customer with regard to certain product features. All

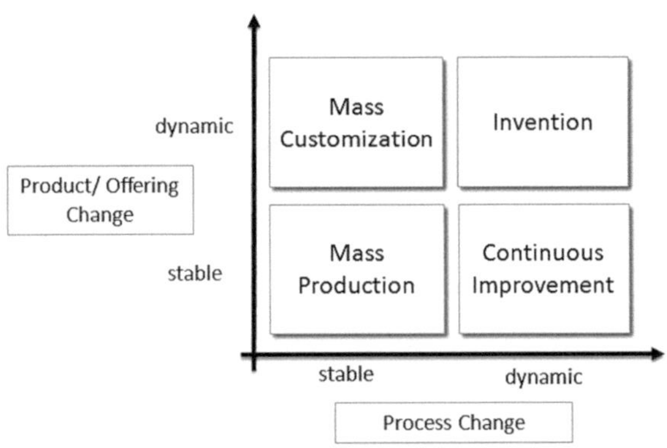

Fig. 18.4 Product-process change matrix acc. to [12]

operations are performed within a fixed solution space, characterized by stable but still flexible and responsive processes" [14].

In order to become a mass customizer, a company has to transform its business model along the so-called right path. This means that all business models have to be traversed without skipping any, especially since the transformation from mass production to MC cannot be done without continuous improvement since the mass production processes cannot stand the high change ratio and flexibility of mass-customized goods.

18.5 Application Example: Classifying Product Configurators

In this section we focus on the business type of mass customization and describe the classification of product configuration systems in the House of Complexity. Therefore, we define configuration as a problem-solving task of knowledge-based technologies. Afterwards we present different approaches for sales configuration systems and engineering configurators. Finally, these are classified in the House of Complexity with some auxiliary building blocks for complexity management.

18.5.1 Configuration as Problem-Solving Task

Sabin states that "configuration is a special case of design activity with two key features: The artifact being configured is assembled from instances of a fixed set of well-defined component types and components interact with each other in predefined ways" [15]. Since configuration systems are more than just filters applied on the portfolio of capabilities, a knowledge base has to be implemented to define possible combinations of components or restrictions.

18.5.2 Sales Configurators

When considered as sales support, the main tasks of a sales configuration system are providing a technically complete and correct product specification, commercial quotation costing, automatic generation of quote documents, and visualization. Another capability of current sales configurators is data collection since the system is able to store all information according to the configuration process, i.e., the time for each configuration step, configuration history, or abort of configurations. Hence, these systems can complement activities of marketing regarding trend scouting and preference analysis [13].

One of the most important characteristics is a sales configurator's ability to translate customer requirements into a valid product specification. On the one hand, these plausibility checks assure a working end product for the customer. On the other hand, decision support is realized. Nevertheless, if the manifold of configuration items is too big, the customer might not be able to choose the right components, which is expressed by the concept of mass confusion [13].

Choice navigation systems add a bidirectional communication component to an online sales process. In contrast to the sales configurator, a choice navigator is able to guide a customer to a certain popular solution [16]. On the basis of detailed customer information, a recommended default configuration is presented which then can be modified by the user. The idea behind is to use statistical data or data from social networks to forecast customer preferences or take influence on the customer in the sense that "other people who define themselves as stylish or sportive have chosen this or that product." The inference mechanisms of such systems rely on case-based reasoning so that the system permanently learns about other configurations [15]. First experiences with those systems are made in automotive or clothing industry. Nevertheless, research in this context is still in the beginning.

18.5.3 Engineering/Design Configurators

While sales configurators aim at managing external product complexity, engineering configurators focus on the internal complexity. Here, engineering configuration has to be considered as knowledge-based engineering (KBE) approach for transforming a design problem into a configuration problem, e.g., by implementation of dimensioning or calculation formula, design rules or manufacturing restrictions.

This implies that all necessary engineering knowledge has to be formulated in a domain-specific knowledge base, which extends the geometric product model. New configurations are calculated and processed by an inference engine where basically the following paradigms can be distinguished [15, 17]:

- *Rule-based reasoning*: The knowledge representation relies to design rules, which are formulated as IF-THEN-ELSE statements. Rules are fired procedurally and can be used to execute subordinate rules or delete them temporarily from the working memory.
- *Model-based reasoning*: The limitation of the possible solution space is done based upon a physical and/or logical model (constraint based) or by representation of resource consumption and allocation (resource based).
- *Case-based reasoning*: In this approach, the knowledge representation is not explicitly modeled in form of rules or constraints. The knowledge necessary for reasoning is stored in cases that represent former configurations. Depending on the degree of maturity of the inference engine, either the system is limited to search for existing solutions, which match exactly to a given requirement profile, or the system is able to assort a set of existing cases, which represent the best fit.

Highly developed case-based systems are able of mixing or altering exiting cases in order to adapt them to new situations.

18.5.4 Complexity Management Using Configurators

In the following, some of the configuration systems mentioned above are exemplarily classified and discussed in the House of Complexity (Fig. 18.5). Therefore, modular product architectures and the specification technique of degrees of freedom of shape attributes [18] are added. A detailed examination of all interdependencies and effects on the complexity dimensions is beyond the scope of this chapter and still part of our actual research. For better readability, the area for the as-is analysis has been left away in the picture below.

As can be seen in Fig. 18.5, we estimate that rule-based configuration systems either for sales or engineering usually do not affect the possible solution space for product development. The rule concept is adequate for documentation of an existing solution space since all configurations have to be predetermined as well as all restrictions, which decreases uncertainty and sets the degree of exploration to 100 %. The uncertainty of the system's behavior with respect to the end product

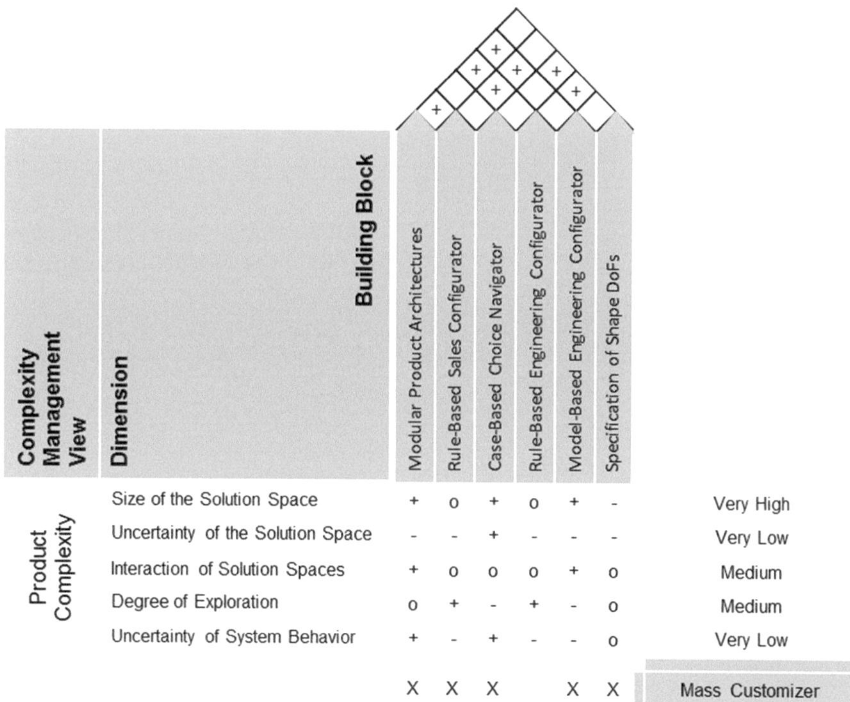

Fig. 18.5 Configuration systems and auxiliary tools in the House of Complexity

vanishes. Since rule-based engineering configurators are too inflexible for efficient use in mass customization, this is not considered as typical building block for complexity management here.

Model-based configurators have different effects. Here, the size of the solution space is usually enlarged but also reducing uncertainty since the end product is represented by a stable model. The degree of exploration of the solution space decreases since not every possible end-product variant has to be planned beforehand. When coupled with modular product architectures, the effects of the configuration systems will be amplified.

On the other hand, certain specification techniques can limit the possible solution space by predefining or limiting change possibilities and thus restricting product variants.

18.6 Conclusion

In the present chapter, we introduced various complexity management dimensions and measures and set up the Hannover House of Complexity Management as a framework for managing complexity. In the application example, we classified different configuration systems in the House of Complexity.

Further research points on the implementation of other complexity occurrences. Already mentioned was requirement complexity, which basically leads to the solution spaces we have discussed for product development. Another extension is manufacturing complexity which also has to be according to the possible solution space.

As second research question, the operationalization of the complexity dimensions is currently investigated. Until now we can identify whether a particular dimension can be either high or low which can be in case subjective. Real and transparent calculating complexity would grade this approach up and simplify decision making.

References

1. Herlyn, W.: PPS im Automobilbau, Produktionsprogrammplanung und Steuerung von Fahrzeugen und Aggregaten. Carl Hanser, Munich (2012)
2. Bliss, C.: Management von Komplexität. Springer Science & Business Media (2000)
3. Collin, H.: Management von Produkt-Informationen in kleinen und mittelständischen Unternehmen. PhD-Thesis, Technical University of Munich, Intitute for Product Development (2001)
4. Ashby, W.R.: An Introduction into Cybernetics. Chapman & Hall, London (1961)
5. Bandte, H.: Komplexität in Organisationen. Springer Science & Business Media (2007)
6. Gembarski, P.C., Lachmayer, R.: Unternehmenstypologisches Komplexitätsmanagement. Design for X—Beiträge zum 24. DfX-Symposium, Hamburg (2013)
7. Ghoffrani, M.: Entwicklung und Einführung eines flexiblen Softwaresystems zur Konfigurierung virtueller Produkte. Shaker, Aachen (2008)

8. Jania, T.: Änderungsmanagement auf Basis eines integrierten Prozess- und Produktdatenmodells mit dem Ziel einer durchgängigen Komplexitätsbewertung. Ph.D. Thesis, University of Paderborn, Institute for Product Development (2005)
9. Schuh, G., Schwenk, U.: Produktkomplexität managen. Hanser, Munich (2001)
10. Frizelle, G.: The Management of Complexity in Manufacturing. Business Intelligence Limited, London (1998)
11. Gießmann, M.: Komplexitätsmanagement in der Logistik. Eul, Cologne (2010)
12. Boynton, A., Victor, B., Pine, J.: New competitive strategies: challenges to organizations and information technology. IBM Syst. J. **32**(1), 40–64 (1993)
13. Reichwald, R., Piller, F.: Interaktive Wertschöpfung. Gabler, Wiesbaden (2009)
14. Piller, F.: Mass customization: reflections on the state of the concept. Int. J. Flex. Manuf. Syst. **16**, 313–344 (2004)
15. Sabin, D., Weigel, R.: Product configuration frameworks—a survey. IEEE Intell. Syst. **13**(4), 42–49 (1998)
16. Gembarski, P.C., Lachmayer, R.: Degrees of Customization and Sales Support Systems—Enablers to Sustainability in Mass Customization. Accepted for ICED2015, Milano (2015)
17. Hvam, L., Mortensen, N.H., Riis, J.: Product Customization. Springer Science & Business Media (2008)
18. Gembarski, P.C., Lachmayer, R.: A method for specification of multi-variant products using degrees of freedom of shape attributes in product structures. In: Proceedings of the 7th World Conference on Mass Customization, Personalization, and Co-Creation (MCPC 2014), Aalborg, Denmark, Springer International Publishing, S. 27–37 (2014)

Chapter 19
Cognitive Computing and Managing Complexity in Open Innovation Model

Robert J. Freund

19.1 Introduction

More and more organizations are confronted with highly dynamic external organizational environments [1]. The drivers of change are globalization, sustainable development, new technologies, and the aging population. The pressure on organizations forces them to continuously adapt to the environmental shifts [2] and to create organizational forms able to provide faster and innovative response to market threats and opportunities [3]. Therefore in today's world, innovation is a subject of great importance because it stimulates sustainable growth in a highly competitive market [4].

Theories and definitions of innovation changed during the last century: "The early 1900s witnessed the birth of the first theories of innovations. Since the second half of the twentieth century the concept of innovation started to spread over the different fields of science. The time span between 1960s and 1990s can rightly be called the golden age in the study of innovation. However in the last 10 years the concept of innovation began to gradually shift from strong scientific definitions to management concepts, slogans and buzzwords" [5]. And this is sometimes confusing for executives, practitioners, and researchers. Table 19.1 shows some assumptions and describes the reality and related core principles of today's successful innovation management.

Based on modern statistical practice, several types of innovation classification can be distinguished (Multiple classifications, multilayer classifications, and dichotomical classifications) and controversial pairs of innovation types can be identified [5]:

- User-driven/supply-side innovation,
- Open/closed innovation,

R.J. Freund (✉)
Finkenweg 6, D-35099, Burgwald, Germany
e-mail: info@robertfreund.de

J. Bellemare et al. (eds.), *Managing Complexity*, Springer Proceedings in Business and Economics, DOI 10.1007/978-3-319-29058-4_19

Table 19.1 Assumption, reality, and core principles [5] and [6]

Assumption	Reality	Core Principles
Innovation as it is currently practiced is good enough	Current innovation practices don't reliably deliver breakthroughs. There is a lack of a set of reliable tools and methods for creating real breakthroughs rather than incremental or random improvements	Principle 1: Build innovations around experiences
Innovation is for executives	Practitioners "on the ground" are most often the source of breakthrough ideas, but they need structure and processes to help them plan and define innovation	Principle 2: Think of innovation as Systems
Innovation is for practitioners	Innovation isn't just for practitioners. Practitioners need work with executives to be able to integrate innovation tactics into a larger strategy	Principle 3: Cultivate an Innovation Culture
"Innovation planning" is an oxymoron	Measured scientific approaches to innovation do exist and can make it a systematic process	Principle 4: Adopt Disciplined Innovation Process

- Product/process innovation,
- Incremental/radical innovation (and other examples of "strong"/"weak" classification of innovation),
- Continuous/discontinuous innovation,
- Instrumental/ultimate innovation,
- True/adoption innovation,
- Original/reformulated innovation,
- Innovation/renovations

When we look closer to "closed innovation—open innovation" dichotomy in the context of organizations, the core aspects of closed innovation and open innovation should be clear. On the other hand, the question is, why should it be a dichotomy? Isn't it possible to bridge the gap between closed innovation and open innovation with a new, hybrid model or framework? The next section highlights the key elements of closed innovation and open innovation.

19.2 Closed Innovation and Open Innovation

19.2.1 Closed Innovation

In the last decades, organizations were primarily concerned with their own ideas, their own manufacturing processes, their own machines, their own scientists and workers. These enterprises couldn't believe in a network of exchanging information and knowledge among other companies, suppliers, universities, customers, etc. There were strategic partners and alliances with severe contracts, protecting the secrets of the company, like ideas, inventions, or innovations. In this context, research teams should cooperate with development teams for accomplishing the company's innovation, but problems were revealed from this communication between the two departments. So, many companies put their ideas coming from the research team on the shelf and after a long time perhaps the development team uses these ideas.

Many dangerous factors came from this inventory, ideas sitting on the shelf, such as many scientists, watching their ideas to wait, could not afford it and resigned from their current position and went to another company with better conditions of working. This means a transfer of ideas and innovation from one company to another. By the same way, there were exchanges between the partners, the suppliers even the customers of the companies [7]. But the whole system cannot protect its own parameters and values in closed smaller systems of a company and its supply chain. In this model of closed innovation, firms relied on the assumption that innovation processes need to be controlled by the company [8].

Because of market pressure it was obvious to improve this closed innovation model. The FORA Report [9] highlighted 9 innovation principles, each based on evidence of new innovation behavior: Co-creating values with customers, users' involvement in innovation processes, accessing and combining globally dispersed knowledge, forming collaborative networks and partnerships, dynamics between large companies and entrepreneurs, environmental concerns drive innovation, needs in developing countries drive innovation, welfare system concerns drive innovation, technology's role as an enabler of innovation.

In "many organizations, especially those with a traditional approach, innovation is often only seen as valid when it is completely 'homemade'. The traditional view of managing innovation (closed innovation) completely disregards the growth market of demand-driven innovation" [10]. This is the main reason for creating a new system of exchanging ideas and information, mostly knowledge, even components from products. An expression of this kind of system is the paradigm named open innovation. Today, there are five erosion factors driving the shift to the open innovation paradigm [11]: (1) Increasingly mobile trained workers, (2) more capable universities, (3) diminished US hegemony, (4) erosion of oligopoly market positions, and (5) enormous increase in venture capital. But there are several (different) definitions of open innovation.

19.2.2 Open Innovation

Co-creation, user involvement, environmental and societal challenges increasingly drive innovation today. Collaborative, global networking and new public private partnerships are becoming crucial elements in companies' innovation process [9]. It is against this background that cooperation's are engaging in forms of open innovation [12–17]. But open innovation today has a much broader application than first proposed by Chesbrough [15], e.g., Reichwald and Piller [18] use the notion "interactive value chain" and the Ministry of Employment and the Economy [19] distinguishes "user innovation" from "user-driven innovation" and "users as collaborators," which is closer to the idea of von Hippel [12] who used the terms "lead user concept" or "user-centered-innovation." His consumer survey in the UK found that "8 percent of UK consumers created or modified one or more of the consumer products they use to better address their needs" [20] and 2 out of 100 said that their products had been taken up by other users or adopted and manufactured by producers [21, 22]. But this kind of open innovation—open collaborative innovation—differ from open innovation for organizations (Chesbrough): Chesbrough and von Hippel use different definitions of open innovation [11].

Open innovation according to Chesbrough [13] is "(…) the use of purposive inflows and outflows of knowledge to accelerate internal innovation, and expand the markets for external use of innovation, respectively." Chesbrough and Bogers updated this definition of open innovation [23]: "Open innovation is a distributed innovation process based on purposively managed knowledge flows across organizational boundaries, using pecuniary and non-pecuniary mechanisms in line with each organization's business model." But for von Hippel open collaborative innovation "(…) is 'open' in our terminology when all information related to the innovation is a public good—non-rivalrous and non-excludable, and"… involves contributors who share the work of generating a design and also reveal the outputs from their individual and collective design efforts openly for anyone to use" [24]. The key differences between the definitions are listed in Table 19.2.

On the one hand, "open innovation" entails purposefully managing knowledge flows across the organizational boundary as well as the associated business model as defining features. On the other hand, "open collaborative innovation" and related notions refer to an innovation model that emphasizes low-cost or free production of public, non-rivalrous, non-excludable goods [23].

Open innovation works from external ideas and knowledge in conjunction with the internal research and development activities. This bidirectional relationship offers new ways to create value. The existence of many smart people outside a company is not a regrettable problem for the prosperity of the company. It indicates also an opportunity for the company. In a better system, the internal research and development occurs awareness, connection, and information from outside research and develop-

Table 19.2 Different definitions for open innovation [23]

Chesbrough et al.:	von Hippel et al.:
• Ideas can come from anywhere	• Users are the source of many innovations
• Ideas must be commercialized through business models	• Users benefit directly from sharing
	• No need for a business model

ment. The innovation process is more profitable, valuable and the effort is multiplied many times through the inspiration of the system. It becomes a value creation engine, value according to the customers, so it is essential for a company to learn from its customers and commercialize their ideas through business models.

Some researchers in Europe published an open innovation 2.0 framework, which integrates different perspectives (definitions) of open innovation. The so-called open innovation 2.0 has some fundamental principles, which lead to needs for new skills among all the actors in the innovation process. Modern innovation spaces span beyond clusters mainly in two dimensions: firstly, the traditional triple helix innovation model with enterprises, research and public sector players (being often top-down) is replaced by the co-creative quadruple helix innovation model where users have an active role too, in all phases of the innovation, from the early ideation to the co-creation of solutions. Secondly, the ecosystem drives for multi-disciplinarity rather than clusters, which tend to be quite monolithic [1]. Open innovation 2.0 was published as a new innovation paradigm in a white paper by Curley and Salmelin, at the open innovation 2.0 2013 conference in Dublin. The original paper was elaborated further in the open innovation 2.0 yearbook 2014 [25]. The twenty characteristics of open innovation 2.0 are the foundation of the proposed approach to increase creativity in innovation processes" [1] (Table 19.3).

For further discussions in this chapter, only the new updated definition of open innovation according to Chesbrough is relevant [23]. In the next section, complexity and uncertainty in innovation process and management is analyzed.

Table 19.3 20 characteristics of open innovation 2.0 [1]

Shared Value and Vision	User-Driven Innovation	Sustainable Intelligent Living	Full Spectrum Innovation	Innovation Capability Management
Quadruple Helix Innovation	Openness to Innovation and culture	Simultaneous Innovation	Mixed Model Technologies	High Expectation Entrepreneurship
Ecosystem Orchestration and Management	Adoption Focus	Business Model Innovation	Network Effects	Social Innovation
Co-Creation and Innovation Platforms	Twenty-first Century Industrial Research	Instructional Innovation	Servicitation	Structural Capital Innovation

19.3 Complexity, Uncertainty, and (Open) Business Models

19.3.1 Complexity and Uncertainty

Market is no longer a target, it is more a forum [26] to "tap into the knowledge of participants in the social ecosystem to create a freer flow of information, engage people more wholeheartedly, and enable richer, fuller stakeholder interactions" [27]. Further, in such a complex system knowledge is unevenly distributed [28] and the direction of flows of knowledge and information cannot be predetermined [29] (For further information on socio-technical systems, see Chatzimichailidou et al. [30]).

The social world, like most of the biological world and a good part of even the physical world, is populated by highly contingent, context-sensitive, emergent complex systems [31]. Complexity and historicity mean above all that human action inevitably takes place in the face of an uncertain future (Reflexive Modernization) [32]. Haken [33] characterizes complex systems like this: "In a naive way, we may describe them as systems which are composed of many parts, or elements, or components which may be the same or of different kinds. The components or parts may be connected in a more or less complicated fashion. Systems may not only be complex as a result of being composed of so many parts but we may also speak of complex behavior. The various manifestations of human behavior may be very complex as is studied, e.g., in psychology (…). An important step in treating complex systems consists in establishing relations between various macroscopic quantities. These relations are a consequence of microscopic events which, however, are often unknown or only partially known."

In business organization, it's about complex/uncertain problem solving for customer. "Complex problem solving (CPS) occurs to overcome barriers between a given state and a desired goal state by means of behavioral and/or cognitive, multistep activities. The given state, goal state, and barriers between given state and goal state are complex, change dynamically during problem solving, and are intransparent. The exact properties of the given state, goal state, and barriers are unknown to the solver at the outset. CPS implies the efficient interaction between a solver and the situational requirements of the task, and involves a solver's cognitive, emotional, personal, and social abilities and knowledge" [34]. Therefore, knowledge must be applicable to different, new, and complex situations and contexts [8, 35–39]. When we look at activities in innovation processes as highly dynamic, complex, nonlinear and with many positive and negative feedback-loops not only innovation policy is limited in what it can change [40], but analogously organizations are limited too in managing this complex system.

In literature, we can find the statement that "managing uncertainty can be regarded as a core practice of successful innovation management" [41]. If so, we need a deeper understanding of the term "uncertainty." Ninety years ago, Knight [42] described in detail: "*It will appear that a measurable uncertainty, or 'risk' proper, as we shall use the term, is so far different from an unmeasurable one that it is not in effect an uncertainty at all. We shall accordingly restrict the term 'uncertainty' to cases of the non-quantitative type.*" It is this dimension of risk—uncertainty—which is an "under-investigated feature of organizations in late modernity" [43].

Creative work models—like open innovation—are likely to become more prevalent. Such models use a distributed problem-solving approach to tap into large pools of people with unique skills, each of whom can contribute to a final solution [44, 45]. In this new world of work, the barriers between work and life have been eliminated [46]. In this context, it is particularly important that the traditional technology and product-oriented perspective on innovation evolves into a more holistic one in which the key role of people and their working conditions is acknowledged [4].

19.3.2 (Open) Business Models

Chesbrough [16]) states that the first book [15] treated the business model as static and utilized open innovation to find more ways to create and capture value within the given business model. In his second book, the business model itself could be innovated, enabling new ways to obtain more value from the company's innovation activities. A business model is a framework to link ideas and technologies to valuable economic outcomes. At its heart, a business model performs two key functions: (1) it creates value and (2) it captures a portion of that value [47] (Table 19.4).

"Innovation is a paradoxical process, which requires a leap into the unknown and at the same time complex management processes and efforts for rigorous planning. In an innovation ecosystem it is not possible to manage many aspects of the innovation process. Orchestration is needed; this relates to both: The capacity to create conditions where the diverse parties can work together with the right balance

Table 19.4 A classification of combinations of open Innovation and open business models [47]

	Closed/Stand Alone Business Model	Open/Linked Business Model
Outside-in Open Innovation	Use others' knowledge to develop a new offering	Use others' knowledge to develop a new Business Model
	Early iPod—Apple	iPod/iTunes Store—Apple
	Swiffer—P&G	SkyNRG-KLM
		Better Place
Inside-out Open Innovation	Unused knowledge used by others	Internal knowledge accessible to others to develop a new Business Model
	Food ingredients—P&G/ ConAgra Foods	Amazon WS—Facebook
	Nodax—P&G	Salesforce.com
	Glad—P&G	IBM-Linux
Closed Innovation	Closed Innovation Model	Search for assets owned by others to develop new Business Model

of inner and outer focus, and thus reinforcing both their own work and benefiting the ecosystem as a whole; and the provision of supporting service infrastructure to help sustain effective operation within the system" [1].

If self-organization is the answer to complexity, we need competent knowledge worker who are able to handle uncertainty better than technology [36]. If we look at competencies as self-organization dispositions [48] on the individual level, we need a competence model that fits to the need of open business models. The question is: Which competence model can fulfill the requirements named? Grollmann [49] proposes in this case: "The attribution of human capabilities in a universal competence model is a question that research is dealing with since many years also in competition against the traditional intelligence concept. More honest seem to be contributions that have been developed for example within the debate of multiple competence/intelligence. Here various areas can be considered, in which expertise can be developed and in which talents exist. In the model of Gardner for example specific 'intelligences' will be differentiated [50]. If somebody would transfer different individual competence profiles on these eight dimensions, it would result in a much differentiated images." Open innovation can only be successful if the involved partners have sufficient and symmetric degrees of both motivation and competency: Customer competencies (product, technical, leadership) and firm competencies (disclosure, appropriation, integration) [41]. But these competencies do not really fit to manage uncertainty because traditional theories are based on logical–mathematical dimensions and did not take into account individual feelings, impressions, etc. [51]. The concept of multiple competencies on individual/group/organizational/network level integrates these aspects and can be applied for open innovation business model [8, 36].

19.4 Cognitive Computing and Managing Complexity in Open Innovation Model

19.4.1 Cognitive Computing

Computing can bring open innovation to new levels—but it is not traditional computing, its cognitive computing. The idea of artificial intelligence as "the science and engineering of making intelligent machines" [52] started 60 years ago. During the last decades, many improvements were made and the public heard about it the first time, when the computer system Deep Blue played chess against world class champions. But finally Watson—a jeopardy-winning computer system—changed the game [53]. Today, artificial intelligence—or better: cognitive computing—is able to solve complex problems (CPS: Complex Problem Solving).

The Cognitive Computing Consortium defined several characteristics of cognitive systems [54]:

- Adaptive. They must learn as information changes and as goals and requirements evolve. They must resolve ambiguity and tolerate unpredictability. They must be engineered to feed on dynamic data in real time, or near real time.
- Interactive. They must interact easily with users so that those users can define their needs comfortably. They may also interact with other processors, devices, and cloud services, as well as with people.
- Iterative and stateful. They must aid in defining a problem by asking questions or finding additional source input if a problem statement is ambiguous or incomplete. They must "remember" previous interactions in a process and return information that is suitable for the specific application at that point in time.
- Contextual. They must understand, identify, and extract contextual elements such as meaning, syntax, time, location, appropriate domain, regulations, user's profile, process, task, and goal. They may draw on multiple sources of information, including both structured and unstructured digital information, as well as sensory inputs (visual, gestural, auditory, or sensor-provided).

It's obvious, that cognitive computing systems will be able to substitute jobs of today's knowledge worker in several traditional industries: Finance, retail, health, education, etc. [44–46, 55–58]. But the question is which jobs/competencies will be substituted by cognitive computer systems and how can these systems contribute to innovation process and innovation management?

19.4.2 Cognitive Computing and Innovation

Cognitive computing makes a new class of problems computable. It addresses complex situations that are characterized by ambiguity and uncertainty; in other words, it handles human kinds of problems. In these dynamic, information-rich and shifting situations, data tends to change frequently, and it is often conflicting [54]. For organizations that want to improve their ability to sense and respond, cognitive analytics offers a powerful way to bridge the gap between the promises of big data [59] (Table 19.5).

Table 19.5 Data Science today and tomorrow [60]

Data science today	Data science tomorrow
Developer—knowing data, math/stats, and application development by heart	Professional User—act as a "white layer" to the black box, know and define data input and machine functionalities
Explorer—experience and skill to come up with a profound way to go from data to business value	Interpreter—make sense of machine-generated hypothesis and confidence-weighted results, detect "errors," tune and (re-) run end-to-end
Business-Enabler—connect business with big data and create data products	Business-Supporter—support business data operations and be master of machines

A good example is open evaluation: To handle the huge amount of ideas created by online communities isn't that easy. Google's project 10 to the 100 got 150,000 ideas from more than 170 countries, from general investment suggestions to specific implementation proposals. These ideas were evaluated by 3000 Google employees [61], not by the crowd (community), and not by cognitive computer systems [35].

According to Bitkom [57], cognitive computing can also contribute to user-centered design, user experience design, service design, design thinking, and lean innovation. Design thinking for example "(…) is a discipline that uses the designer's sensibility and methods to match people's needs with what is technologically feasible and what a viable business strategy can convert into customer value and market opportunity" [62]. Table 19.6 shows elements of design think.

When comparing these elements with the above-mentioned characteristics for cognitive computing systems, it is clear that cognitive computing will bring design thinking on a new level. But not only design thinking, it's the whole innovation process that can benefit from cognitive computing [57]:

- Market: Monitoring, Screening, Business-Modelling
- Trends: Trend scouting, (n) Ethnography, User-Insights, Community-Research, Usability-Testing, …
- Creativity and Pattern: Ideation, Co-Creation, Crowd-Sourcing, Brand-Naming
- Technology: Patent, Material screening, Research Projects

But there are some limitations: "While robots are highly efficient at applying math to do routine tasks, humans are able to complement their robot 'colleagues' with non-programmable capabilities, such as the ability to be flexible and adaptable, interact effectively with humans, and use judgment and common sense to solve unexpected problems" [56]. As we know from knowledge management systems, they will "organize all the knowledge in a corporation, but they cannot produce imaginative breakthroughs" [56].

Table 19.6 Design Thinking [63, 64, 65]

Ambiguity	Being comfortable when things are unclear or when you don't know the answer
Collaborative	Working together across disciplines
Constructive	Creating new ideas based on old ideas, which can also be the most successful ideas
Curiosity	Being interested in things you don't understand or perceiving things with fresh eyes
Empathy	Seeing and understanding things from your customers' point of views
Holistic	Looking at the bigger context for the customer
Iterative	A cyclical process where improvements are made to a solution or ideas regardless of the phase
Nonjudgmental	Creating ideas with no judgements toward the idea creator or the idea
Open mindset	Embracing design thinking as an approach for any problem regardless of industry or scope

Table 19.7 Levels of competencies: People-Machine [57]

Basis	Level	People	Machine
Facts and norms	Novice	X	X
	Advanced Beginner	X	X
	Competency	X	X
Holistic cognition	Proficiency	X	–
	Expertise	X	–

When we compare levels of competencies [66] for people and for machines, we have to realize that machines (cognitive computer systems) can work with facts and norms but are limited on the levels proficiency and expertise (Table 19.7). Experts, people with expertise, cannot easily substituted by machines (cognitive computer systems). So workers with advanced degrees—expertise (E)—are together with cognitive computing (CC) an essential starting point for [4] for a new ECC-Open Innovation Model.

19.5 Conclusion

This chapter explains the key elements of closed and open innovation and pointed to different definitions of open innovation. For further discussions, the new updated definition of open innovation according to Chesbrough and Bogers was relevant. In the next section, complexity and uncertainty in innovation process and management was analyzed because managing uncertainty in innovation process can be regarded as a core practice of successful innovation management. It is argued that the concept of multiple competencies on individual/group/organizational/network level can be applied for open innovation business model and that cognitive computing can bring open innovation to a new level. It is shown that Cognitive Computing (CC) can bring innovation management on a new level (ECC-Open Innovation Model). At the end of this chapter, limitations of cognitive computing are outlined. Further research should analyze the whole open innovation model from the cognitive computing and multiple competencies point of view.

References

1. European Commission: Open Innovation 2.0 Yearbook 2015 (2015)
2. Greenwood, R., Hinings, C.R.: Understanding Radical Organizational Change: Bringing Together the Old and the New Institutionalism. Academy of Management Review, 21 October, pp. 1022–1054 (1996)
3. Coleman Jr., H.J.: What enables self-organizing behavior in businesses? Emergence **1**, 33–47 (1999)
4. University, C.: INSEAD, and WIPO: The Global Innovation Index 2014: The Human Factor In innovation. Fontainebleau, Ithaca (2014)

5. Kotsemir, M., Abroskin, A., Meissner, D. (2013): Innovation Concepts and typology—an evolution discussion. Basic Research Program Working Papers Series Science, Technology and Innovation (2013) http://ssrn.com/abstract=2221299. Accessed 03.07.2015
6. Kumar, V.: 101 Design Methods. A structured approach for driving innovation in your organization. Wiley, Hoboken (2013)
7. Weßels, D.: Zukunft der Wissens- und Projektarbeit. Neue Organisationsformen in vernetzten Welten. Symposion Publishing GmbH, Düsseldorf (2014)
8. Freund, R.: Multiple Competencies in Open Innovation Business Model. MCPC2009, 5th World Conference on Mass Customization and Personalization, 04.-08.10.2009, Helsinki, Finland (2009)
9. OECD: New Nature of Innovation. Report (2009)
10. Fredberg, T., Elmquist, M., Ollila, S.: Managing Open Innovation. Present findings and future directions. VINNOVA Report VR 2008:02 (2008)
11. Chesbrough, H.W.: What's Next for Open Innovation Research and Practice? Presentation to R&D Management Conference Pisa, June 2015. (2015)
12. Von Hippel, E.: Democratizing Innovation. MIT Press, Cambridge (2005)
13. Chesbrough, H.W., Vanhaverbeke, W., West, J.: Open Innovation: Researching a New Paradigm. Oxford University Press, Oxford (2006)
14. Drossou, O., Krempl, S., Poltermann, A. (eds.): Die wunderbare Wissensvermehrung. Wie Open Innovation unsere Welt revolutioniert. Heise, Hannover (2002)
15. Chesbrough, H.W.: Open Innovation. The New Imperative for Creating Profiting from Technology. Harvard Business Press, Boston (2003)
16. Chesbrough, H.W.: Open Innovation. The New Imperative for Creating Profiting from Technology. McGraw Hill, London (2006)
17. Kleemann, F., Voß, G.G.: Un(der)paid Innovators: The Commercial Utilization of Consumer Work through Crowdsourcing. Science, Technology & Innovation Studies, Vol. 4, No. 1 (2008)
18. Reichwald, R., Piller, F.: Interaktive Wertschöpfung. Open Innovation, Individualisierung und neue Formen der Arbeitsteilung. 2. Auflage. Gabler, Wiesbaden (2009)
19. Ministry of Employment and the Economy: Demand and User-Driven Innovation Policy. Ministry of Employment and the Economy, Helsinki (2010)
20. Flowers, S., von Hippel, E., de Jong, J., Siozic, T.: Measuring user innovation in the UK: The Importance of Product Creation by Users. NESTA, London (2010)
21. Von Hippel, E., de Jong, J., Flowers, S. (2010): Comparing Business and Household Sector Innovation in Consumer Products: Findings from a Representative Study in the UK (2010). http://ssrn.com/abstract=1683503. Accessed 3 July 2015
22. Gault, F.: User innovation and the market. UNU-MERIT Working Paper 2011-009 (2011). http://www.merit.unu.edu/publications/wppdf/2011/wp2011-009.pdf. Accessed 3 July 2015
23. Chesbrough, H., Marcel Bogers, M.: Explicating Open Innovation. Clarifying an Emerging Paradigm for Understanding Innovation (2014)
24. Baldwin, C., von Hippel, E. (2010). Modeling a Paradigm Shift: From Producer Innovation to User and Open Collaborative Innovation, Harvard Business School Finance Working Paper 10-038, MIT Sloan School of Management Working Paper No. 4764-09 (2010). http://ssrn.com/abstract=1502864. Accessed 3 July 2015
25. European Commission: Open Innovation 2.0 Yearbook 2014 (2014)
26. Prahalad, C.K., Ramaswamy, V.: The Future of Competition. Co-Creating Unique Value with Customers (2004)
27. Ramaswamy, V., Gouillart, F.: The Power of Co-Creation. (2010)
28. Hayek, F.: The Use of Knowledge. American Economic Review, no. 4. pp. 519–530 (1945)
29. Ramaswamy, C., Ozcan, K.: The Co-Creation Paradigm. Stanford University Press, Stanford (2014)
30. Chatzimichailidou, M.M., Freund, R., Dokas, I.: Distributed Situation Awareness as a 'middleware' between the New Economic Sociology and Embedded Open Innovation. Proceedings of the 6th International Conference for Mass Customization, Personalization and Co-Creation, MCP-CE 2014, 23.-26.06.2014, Novi Sad, Serbia (2014)

31. Mitchell, S., Streeck, W.: Complex, Historical, Self-Reflexive: Expect the Unexpected! MPIfG (Max Planck Institute for the Study of Society) Working Paper 09/15 (2009)
32. Lukac, D., Freund, R.: Open Innovation, Social Embeddedness of Economic Action and its Cultural Determinants. Proceedings of the 5th International Conference on Mass Customization and Personalization in Central Europe (MCP-CE 2012), 19.-21.09.2012, Novi Sad, Serbia (2012)
33. Haken, H.: Information and Self-Organization: A Macroscopic Approach to Complex Systems, 3rd edn. Springer, Berlin (2006)
34. Frensch, P.A., Funke, J.: Definitions, traditions, and a general framework for understanding complex problem solving. In: Frensch, P.A., Funke, J. (eds.) Complex Problem Solving. The European Perspective, pp. 3–23 (1995)
35. Freund, R.: How to overcome the barriers between economy and sociology with open innovation, open evaluation and crowdfunding? In: Int. J. Ind. Eng. Manage. (IJIEM), Vol. 1 No 3, pp. 105–109 (2010)
36. Freund, R.: Das Konzept der Multiplen Kompetenz auf den Analyseebenen Individuum. Organisation und Netzwerk. Dissertation. Das Buch ist in der wissenschaftlichen Reihe Wandlungsprozesse in Industrie- und Dienstleistungsberufen und moderne Lernwelten, Schriftenreihe des Graduiertenkollegs der Martin-Luther-Universität Halle-Wittenberg von Prof. Dr. Bernd Dewe erschienen. Verlag Dr. Kovac, Hamburg, Gruppe (2011)
37. Freund, R., Anisic, Z.: How to use complexity and uncertainty for new business? Proceedings of the 3rd International Conference on Entrepreneurship, Innovation and Regional Development ICEIRD 2010, May 27-29 2010. Novi Sad, Serbia (2010)
38. Freund, R., Anisic, Z.: Management of Uncertainty: The Key for Open Innovation. MCPC 2011, Worldcongress on Mass Customization, Personalization and Co-Creation, 15.-19.11.2011, San Francisco, USA (2011)
39. Freund, R., Chatzopoulos, C., Lalic, D.: Reflexive Open Innovation in Central Europe. 4th International Conference for Entrepreneurship, Innovation, and Regional Development (ICEIRD2011), 05.-07. May, Ohrid, Macedonia (2011)
40. Gault, F.: Social impacts of the development of science, technology and innovation indicators. UNU-MERIT Working Paper 2011-008 (2011) http://www.merit.unu.edu/publications/wppdf/2011/wp2011-008.pdf. Accessed 3 July 2015
41. Piller, F., Ihl, C.: Open Innovation with customers. Foundations, Competences and International Trends (2009)
42. Knight, F.H.: Risk, Uncertainty, and Profit. Boston, MA: Hart, Schaffner & Marx; Houghton Mifflin Co. (1921)
43. Gephart, R.P., Van Maanen, J., Oberlechner, T.: Organizations and Risk in Late Modernity. Org. Studies **30**(02&03), 141–155 (2009)
44. Intel: Future of Knowledge Workplace Transformation. An Outlook on the Changing Nature of the Work Environment. White Paper (2014)
45. IBM: A New Way to Work: Futurist Insights to 2025 and Beyond (2015)
46. Deloitte University Press: Global Human Capital Trends 2015. Leading in the new world of work (2015)
47. Vanhaverbeke, W., Chesbrough, H.W.: A Classification of Open Innovation and Open Business Models (2014)
48. Erpenbeck, J.: How New Are the New Approaches of Competency Development Really? 2. International SIL Congress, Workshop 6: New types of competency development, 11./12.10.2005, University of St. Gall, Switzerland (2005)
49. Grollmann, P.: BIBB-Workshop Zur Konstruktion von Qualifikationsrahmen. 16.11.2005, Bonn, Germany (2005)
50. Gardner, H.: Frames of Mind: The Theory of Multiple Intelligences. Basic Books, New York (1983)
51. Akerlof, G.A., Shiller, R.J.: Animal Spirits: How Human Psychology Drives the Economy, and Why It Matters for Global Capitalism. (2009)
52. McCarthy, J.: What is Artificial Intelligence? CA, Stanford (2007)

53. Baker, S.: Final Jeopardy. The story of Watson, the Computer that Will Transform Our World, 1st Mariner Books ed. Boston, MA (2012)
54. Feldman, S, Reynolds, H.: Cognitive Computing: A Definition and Some Thoughts. KMWorld November/December 2014, vol. 23, Issue 10 (2014)
55. Bundesministerium für Arbeit und Soziales: Arbeit weiter denken. Grünbuch Arbeiten 4.0 (2015)
56. Gold, S.: The Future of Work. Industry Week, May 12 (2015) http://www.industryweek.com/automation/future-work. Accessed 3 July 2015
57. Bitkom: Kognitive-Maschinen. Leitfaden (2015)
58. Brynjolfsson, E., McAfee, A.: The Second Machine Age. Work Progress and Prosperity in a Time of Brilliant Technologies. W.W. Norton & Company, New York (2014)
59. Ronanki, R., Steier, D.: Cognitive analytics. Tech Trends 2014. Deloitte University Press (2014)
60. Bollhofer: Data Science in the Age of Cognitive Computing (2015). First published: https://speakerdeck.com/klabol/data-science-in-the-age-of-cognitive-computing. Accessed 3 July 2015
61. Google Project 10 to 100, http://www.google.com/campaigns/project10tothe100/
62. Brown, T.: Design Thinking. In: Harvard Business Review, June 2008, 86 (6), pp. 84–96 (2008)
63. Baeck, A., Gremett, P.: Design Thinking: Expanding UX methods beyond designers. In: Degen, H., Yuan, X.: UX Best Practices. How to Achieve More Impact with User Experience. McGrawHill, pp. 230–250 (2012)
64. Raimer, S.: Design Thinking, Lean, Agile: Schlüssel zur Innovation. In: Weßels, D. (ed.) Zukunft der Wissens- und Projektarbeit, pp. 169–194 (2014)
65. Martin, R.: The Design of Business. Why Design Thinking is the Next Competitive Advantage. Harvard Business Press, Boston (2009)
66. Dreyfus, H., Dreyfus, S.: Mind over Machine—The Power of Human Intuition and Expertise in the Era of the Computer. TFP (1986)

Chapter 20
Combining Configurator 2.0 Software with Designcrowdfunding

Matthias Kulcke

20.1 Introduction

Designcrowdfunding or short designfunding, belonging to the category of reward-based crowdfunding [1], is and has been used by campaigners to gather information on the solution space desired by customers. Since those who order a product via a designfunding, by buying it as an incentive, are viable sources considering what variations of the product might be wished for by future customers, designfundings constitute a potential cost minimizing pre-phase for launching a product configurator. They may aid in reducing solution space complexity at an early stage. Using designfundings as a tool for information-gathering on the subject of desired product-variations has already been discussed by the author [2].

20.2 State-of-the-Art Designfunding: First Steps of Improvement

Crowdfundings for design-products are generally still run on crowdfunding platforms that are not specifically conceived for designfundings, but are also open for all kinds of projects, i.e. music-CDs, book-projects, movies, theater-projects, and technological innovation, the usual suspects being www.kickstarter.com, www.indigogo.com and national European platforms like the German www.startnext.de. Because of this, most designfundings are launched on a platform that is not specifically suited to sell design-products. For example, on www.startnext.de the user has

M. Kulcke (✉)
Am Batardeau 1, 25348 Glueckstadt, Germany
e-mail: matthias.kulcke@web.de

J. Bellemare et al. (eds.), *Managing Complexity*, Springer Proceedings in Business and Economics, DOI 10.1007/978-3-319-29058-4_20

to click at least two times to reach a picture of one of the incentives, although looks play a vitally important part in strategies for selling design.

Launched solely for dutch design-products www.dutchdesignstarter.com was one of the first to offer a platform specifically for design-product campaigns. For a while, probably due to limited success, it has reduced its service to just posting articles on designfundings running on the bigger platforms. Still, even www.dutchdesignstarter.com does not feature the visual aspects of the design on offer as much as is possible on an online platform for selling design.

Comparing crowdfunding platforms featuring designcrowdfundings among other categories to plain design-selling platforms like www.monoqi.com, www.architonic.de, etc., it becomes apparent that a crowdfunding platform suited for designfundings should be a lot more product oriented. This means, e.g., that images displaying incentives, which are often and advisably versions of the design-product or smaller products related to the main design, have to be reachable for the user with the first mouse-click leading to a designfunding. There should be multiple touchpoints leading the user to a specific design-product, design-product package, or an incentive being a smaller design related to the main design.

20.3 Including 2.0 Configurators in MC-Inquiry Designfundings

Already the comment-sections integrated in crowdfundings have been used for information-gathering through customer feedback that led to a customer-oriented succeeding configurator design of a permanent sales platform for a design-product (e.g., [3]). Configurator-like arrangement of incentives (e.g., for each available color or material one incentive has been listed in some instances) and customer comments further support the thesis that there may be a need for configuration solutions within designcrowdfundings (e.g., [4], see also Table 20.1).

Meeting these needs, the next step would be to use simple configurators already in the designfunding phase (see Fig. 20.1). In 2013, combeenation presented a do-it-yourself configurator that can be individualized for offering a certain product in as short a time as 5 min. This easy-to-handle approach made possible by a configurator 2.0 software (which off course comes with some limitations in smaller versions) is ideal to be integrated in future designfunding platforms that also take an MC-strategy into account (see Fig. 20.2). If incentives offered through designfundings can already be configured by supporters, even if only in simple categories like color-selection, this will probably trigger more propositions, considering what design-categories of the product should be configurable to what extent. These desires can be uttered and discussed in the comment-section of that particular designfunding.

Until such a platform has been realized, workarounds could be used. For example, incentives can be presented parallel on the designers' website and linked to the crowdfunding platform. Similarly, a simple configurator for each incentive could be set up on a different platform parallel to the running designfunding and linked

Table 20.1 MC-relevant product aspects mentioned by supporters in different designfundings

MC-relevant product aspects mentioned by supporters in designcrowdfunding-commentsections (www.startnext.de)	1 = Yes 0 = No					
Campaign-title	Desire for design-variations	Desire for additional functionality	Desire for functional improvement	Desire for product-variations in general	Successful funding	Campaign-duration(yy_mm_dd-yy_mm_dd)
KANCHA—Design Accessories	1	0	0	1	1	13_08_01-13_09_22
Holzbaukasten-Set WOODKIT	1	1	0	1	1	13_08_09-13_09_18
YUNIKUE—Fitted Bag	1	1	1	1	1	13_08_23-13_09_30
COMAKE Shoes—MIT DIR WIRD EIN SCHUH DRAUS!	1	1	1	1	1	13_11_25-14_01_13
Origami Geschirr von moji design	1	0	0	1	1	13_12_10-14_01_31
ROOM IN A BOX	1	1	1	1	1	14_05_26-14_06_30
NUI-Case	0	1	0	1	1	14_07_24-14_09_10
Black Roses Playing Cards	1	0	0	1	1	14_08_12-14_09_21
Tonart-Würfel	0	1	0	1	1	15_01_27-15_02_20
hamaka—die hyperleichte Haengematte	0	0	0	0	1	15_02_05-15_04_08
Webcam Cover—EyePatch	0	0	0	0	1	15_02_09-15_03_23
Total	7	6	3	9	11	

Link overview: http://www.startnext.com/Projekte.html#!design/beliebtheit-d/20

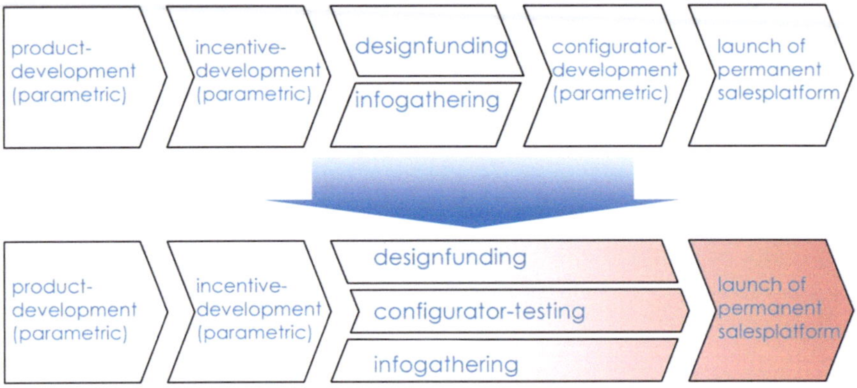

Fig. 20.1 Change of process from product development to configurator launch

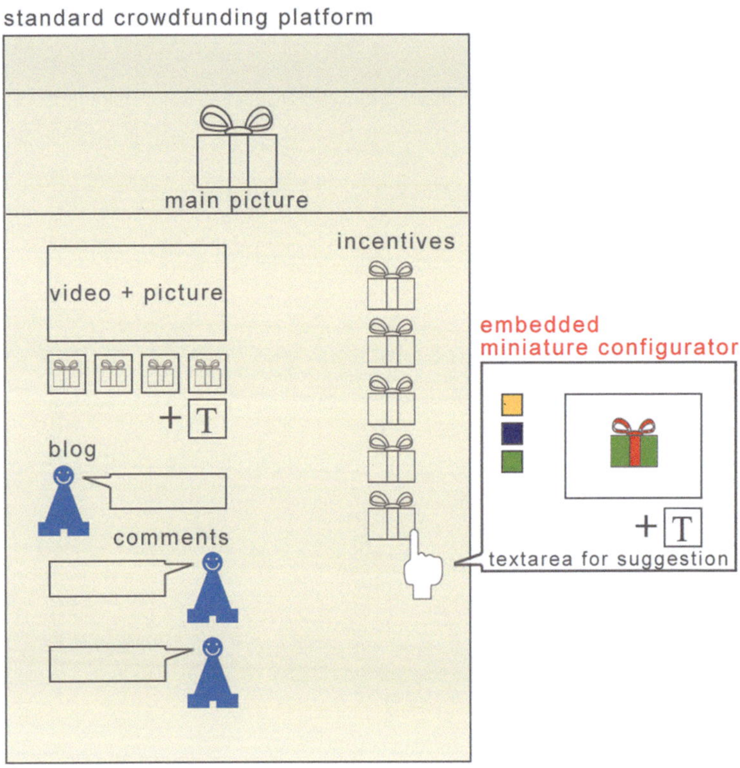

Fig. 20.2 Overview: configurator embedded in designcrowdfunding

accordingly. But all this is rather a crutch than a solution because the flow of configurating, ordering, and commenting will be disrupted by the platform change and the accompanying potential confusion of the customers.

20.4 Conclusion and Outlook

As crowdfundings generally continue to grow more successful [5], for designfundings to pick up momentum using the crowdfunding-technique, there is a need for specialized designfunding platforms which have to mature and develop their specific form of explaining the projects and presenting the incentives. An integration of easy-to-use configurator 2.0 software would also suit the needs for such platforms and may even be vital for their overall success.

The design of configurators as part of designing a product will be part of the task in this year's Hamburger Möbel contest, serving as a working laboratory (see www.hamburgermoebel.com).

The question of the paradox of choice [6] has to be scrutinized in the special context of designfundings considering their function out of the campaigner's perspective as well as regarding the customer's role and desires in the choice process of such designfundings.

References

1. Calveri, Claudio, Esposito, Roberto.: Crowdfunding World 2013—Report, Analysis and Trends. http://www.derev.com [05.06.2015] (2013)
2. Kulcke, Matthias.: Designfunding: an inquiry-tool for mass customisation. In: Proceedings MCPC 2014, pp. 139–144. Aalborg, S. (2014)
3. YUNIKUE ™.: Yunikue Fitted Bag. http://www.nordstarter.org/yunikue-fitted-bag/blog/ [05.11.2013] + http://www.yunikue.com/shop/ [05.11.2013] (2013)
4. Fredbox ™.: Fredbox. http://www.startnext.de/fredbox [05.06.2015] (2015)
5. Crowdfunding industry report—Market Trends, Composition and Crowdfunding Platforms—Maggio. http://www.crowdsourcing.org/research [05.06.2015] (2012)
6. Schwartz, B.: The Paradox of Choice: Why More is Less. Harper Collins, New York (2004)

Chapter 21
Current Challenges for Mass Customization on B2B Markets

Leontin K. Grafmüller and Hagen Habicht

21.1 Introduction

At the beginning of the twenty-first century, more and more customers are demanding individualized products and therefore challenging traditional mass production. Mass customization (MC) offers a solution in order to provide products which are both inexpensive and individualized [39]. The increase of output diversity without losing cost advantages of mass production was considered to be the main challenge of MC. MC relies on a number of principles for its realization. Among the frequently stressed principles are the modularization of production, often combined with technologies which enable a direct conversion between digital and real world (e.g., 3D scanning, 3D printing), and high customer integration, in particular in the process of outcome design which is usually supported by toolkits [12].

Since Pine's seminal work [39], much has been written about MC drivers, success factors, enablers, the customer–manufacturer interaction or the solution space of individualization on B2C markets [12, 25]. However, less attention has been paid to B2B markets. This is striking for two reasons: Early examples of MC are also found in B2B value creation networks [1, 54];hence, it represents a relevant empirical phenomenon. Second, the case of business customers differs fundamentally from the situation of end consumers. For instance, a single business customer is usually much more important than a single consumer. As a consequence, business customers have always been attended to with individualized offers. This means that in made-to-order B2B markets, the solution is created within an individual and personal customer–manufacturer interaction, a time-consuming process with a high degree of product customization [5]. Hence, the promise of MC with respect to business customers is rather the opposite. It revolves around questions such as: How

L.K. Grafmüller (✉) • H. Habicht
HHL Leipzig Graduate School of Management, Jahnallee 59, Leipzig 04109, Germany
e-mail: Leontin.grafmueller@hhl.de; Hagen.habicht@hhl.de

J. Bellemare et al. (eds.), *Managing Complexity*, Springer Proceedings in Business and Economics, DOI 10.1007/978-3-319-29058-4_21

can the established degree of individualization be managed while moving from made-to-order production to a more standardized (supposedly largely IT-based) MC offer? What are the central customer value components of MC for business customers? How can these value components be addressed?

Business customers are very dissimilar to end consumers. They exhibit different characteristics and goals that impose other requirements on the design of the customer interaction process. For instance, whereas toolkits for end consumers are designed for nonexperts, i.e., they represent relatively small solution spaces and focus on intuitive use, business customers are typically experts. Hence, toolkits are much more complex [15]. Furthermore, it has been shown that end consumers value elements that fulfill hedonic desires in multiple ways (cf. [32]). The existence of such value components is at least questionable for business customers.

We take stock of the current challenges of MC on B2B markets. More specifically, we consider differing customer values with respect to the individualization process and the individualized outcome. Based on the customer value MC provides, we consult service-dominant logic (SDL) as a framework of reference that helps us to structure origins of customer value. For our purposes, SDL seems to be a very fitting approach mainly for three reasons. First, SDL emphasizes a process view on customer interaction. This is important because establishing close customer relationships is essential for achieving the high levels of customer integration MC requires. For instance, customers are the main actors in the configuration process [15]. Second, SDL considers value creation to take place in the customer's domain, which is particularly true for business clients. Third, the service perspective is generally helpful for structuring origins of customer value that unfold throughout the periods of customer–provider interaction. The service perspective offers the three dimensions of potential, process, and outcome as origins of customer value. In this respect, Ihl et al. [23] point to the relevance of MC potential and MC process.

21.2 The Focus on a Service Perspective

A decade ago, marketing literature saw a paradigm shift moving from a goods-centric toward a service-dominant view. Developed by Vargo and Lusch [49], the service-dominant logic comprises ten foundational premises which were introduced in Vargo and Lusch [49] and expanded in Vargo and Lusch [50]. It is assumed that service is the essential basis for exchange. Consequently, markets, societies, and economies are focused on exchanging services. It follows that marketing research and practice should be based on a service-centered view. This perspective, since services always imply customer integration to a certain degree, creates various opportunities to redefine the customer's role. Because MC is also seen as a strategy which involves the customer to a large extent, SDL is likely to open the frame for the considerations in this chapter. Therefore, Vargo and Lusch [50] take special account of the role of the customer as a co-creator on their premises. The co-creation process refers to customer interaction, which in turn draws attention to the relationship between the actors. In relationship marketing, this gained particular importance in B2B contexts where long-term loyalty and the relatively intense communication play a major role [10, 44]. In these early studies, it is argued that B2B firms do not have a process view on the interaction, and in more recent studies the necessity of process control within co-creation is still stressed [8, 9]. This can be considered highly relevant with regard to implementing MC on B2B markets. Furthermore, Sheth and Parvatiyar [44] found that many B2B firms lack the competencies to structure the customer–manufacturer interaction in an efficient manner. In SDL research, analyses include also the customers' competencies in the co-creation process [2, 47]. Consumer integration itself can increase satisfaction and create value for the customer [24, 26].

The term of value is inherently connected with SDL because it shifted the former focus on value creation away from output and price to a service-centered view [51]. Value co-creation as its central part is always reciprocal, involving both customer and manufacturer [29].However, they only create a part of that value collaboratively, implying that SDL takes account of value co-creation taking place especially on the part of customers. More precisely, customer value unfolds over time, without the involvement of the manufacturer [20]. This is particularly the case for the B2B sector [10] and is arguably a crucial factor with regard to MC in these markets. In this context, Clauß, Laudien, and Daxböck [3] suggest establishing systems to foster customer interaction and dialogue in general, which is an opportunity to understand value creation taking place exclusively in the customer's domain after the actual sales process. They recommend looking behind the visible customer–manufacturer interaction, moving into the mental life of the customer.

Companies benefit generally from co-creation capabilities, which also have positive effects on customers' perceived value [27]. Due to the changing roles between supplier and business customer, Gummesson and Polese [18] suggest paying systematic attention to the customer's active role in value creation. Accordingly, several studies have been released addressing the customer's perspective [17, 20].

In general, the service perspective is helpful in order to structure the origins of customer value starting with the customer–manufacturer interaction. In this chapter, we focus primarily on the ideas within SDL that concern the co-creation processes and customer value because these points are likely to provide insights into how to expedite MC in B2B markets.

21.3 The Customer Value Proposition of MC

21.3.1 Providing Customer Value in the B2C Domain

In order to develop a differentiated view on its current conception in the MC literature, we will contrast customer value on B2B markets with B2C customer value. In general, MC studies on B2B markets are rare compared to studies in the B2C domain. This seems to be due in part to the relatively high access barriers for data collection. In contrast, consumer behavior can be studied in many more contexts, including samples of potential customers. In line with this, current literature reviews and meta studies in the field consistently show a high percentage of studies based on student samples [12, 32, 33, 48]. As a consequence, the state of the art in the B2C domain can be considered substantially more developed. This makes the current body of B2C-based knowledge interesting as an orientation for the less developed B2B domain. In the following section, we will therefore provide a structured overview of the customer value concept in the B2C domain.

After three decades of MC research in the B2C context, the current conception of customer value is differentiated and multifaceted. Early works have stressed outcome-related components such as increased functionality and individual esthetics (cf. [43]). It is only in the last decade that MC was increasingly conceptualized as a service offer and more particularly as a technology-based self-service (Benedict G. C. [6]). Accordingly, its pivotal customer interaction has been labeled "self-design" [15, 42] or "customer co-design" [14, 31, 36]. As a consequence, process-related value components such as perceived fun and creativity (cf. [31, 42]) and potential-related value components such as the quality of sales personnel and buying environment [23] have been recognized. Furthermore, companies' capabilities to integrate customer knowledge are considered crucially important in regard to the success of the co-creation process [34, 37]. Following this multidimensional view, Table 21.1 provides a systematization of customer value components based on the three service dimensions of potential, process, and outcome.

Altogether, the impact of the value components listed above is usually assessed according to the increase in customers' willingness to pay for custom versus "off the shelf" products and services (cf. [31]). Alternative measures include willingness to buy, willingness to recommend the MC offer to others, customer satisfaction, and customer loyalty [23].

21.3.2 Providing Customer Value in the B2B Domain

Very much like on consumer markets, the benefits of MC on B2B markets depend on the complexity of the design problem. In areas with low to medium complexity in the co-design process, e.g., personal computers, the process of co-creation represents a relatively simple configuration task [55]. We focus on industries with comparably complex outcomes and a complex co-design task. In practice, such markets are typically addressed with one-to-one marketing and personalization, but not necessarily with customization [54]. In such markets, e.g., industrial architecture or technical textiles, MC represents a means of standardizing an otherwise individual made-to-order process.

Furthermore, investigations of B2B markets typically focus on improving joint value creation by integrating flows of information, parts, or components by means of IT-based logistics. For instance, Hong, Dobrzykowski, and Vonderembse [21] provide a structured overview of the benefits of integrating suppliers into an e-procurement

Table 21.1 Customer value of mass-customized products

Offer dimension	Value	Characteristic	Sources
Potential-related customer value	Quality of service personnel	Quality of the personnel in terms of reliability, responsiveness, assurance, and empathy	[4]
	Quality of buying environment (offline shop)	Refers to the SERVPERF dimension of "tangibles," i.e., up-to-date equipment, physically appealing, and appropriate facilities	[4]
	Quality of buying environment (online shop)	Evaluation of the toolkit, e.g., based on the attributes of trial-and-error element, appropriateness of the solution space (degrees of freedom), user-friendliness of the toolkit, number of module libraries, and the ability to "translate" users' designs for production (extent of transferred user insights)	[16]
Process-related customer value	Hedonic value	Value acquired from the experience's capacity to meet needs related to enjoyment, fun, or pleasure	[32, 42]
	Creative achievement value	Value acquired from the feeling of accomplishment related to the creative task of co-designing	[32, 42]
	Perceived complexity of the design process	High perceived complexity represents a burden for customers in the co-design process	[22]
	Perceived delivery value	Time and reliability of delivery	[23]
	Quality of the co-design process (offline shop)	Comprises the perceptions about relevant activities of the offline co-design process. In particular, it ascertains whether they are perceived as necessary effort or exceptional experience.	[23]
	Integration of customer's competences	Relates to companies' capabilities to integrate customer knowledge and to foster the success of his contribution	[34, 38]
Outcome-related customer value	Utility value, perceived preference fit	closeness of fit between outcome characteristics and personal preferences	(Benedict G C [7, 13, 32, 52])
	Uniqueness value	Value acquired from the opportunity to assert personal uniqueness using or consuming the customized outcome, other-oriented value	[32, 42, 46, 53]
	Self-expression value, pride of authorship value	Value derived from the opportunity to possess or consume something that is a reflection of personality, self-oriented value	[32, 33, 42, 45]

system that shares end consumer information in real time with the business customer's suppliers. Moreover, Fauska, Kryvinska, and Strauss [11] find that especially B2B sales of highly complex products cannot be fully switched to e-commerce and highlight the importance of direct customer–manufacturer interaction. The co-design process which precedes joint production is, however, largely neglected. Wind and Rangaswamy [54] provide one of the few exceptions. They formulate a number of new challenges for MC providers based on widely acknowledged market trends—although without providing particular insights about customer value of business customers. Given this void of knowledge and concepts, we draw on the wider area of B2B value co-creation and map this knowledge to the field of MC.[1]

The B2B co-creation literature provides a comparably rich picture about potential-related antecedents of customer value. It addresses the potentials of providers and business customers alike. For instance, the use of IT systems dedicated to the management of customer interaction is found to have a positive effect on the creation of customer value. In particular the use of e-commerce, e-procurement, and ERP systems positively impact MC performance, that is, "competitive manufacturing lead times, delivery speed, unit manufacturing costs, and mix flexibility" ([21], p. 569), which contributes to improved value for customers. On the side of the business customer, the use of IT requires sufficient IT knowledge in order to be able to effectively co-design. In this respect, Komulainen [28] calls for an appropriate level of (technical) knowledge. They identify that the capacity to use technological components of an offer is a prerequisite for customers to generate value from the offer at all. Several authors emphasize the capabilities of both, provider and customer, to successfully interact and learn. In this respect, Hawkins, Gravier, Berkowitz, and Muir [19] show a positive impact of the customer's commitment to collaborate with the provider. With respect to the provider, Salomonson, Åberg, and Allwood [40] determine that communication skills of sales personnel positively impact customer value. According to their exploratory study, sales professionals need to show attentiveness, perceptiveness, and responsiveness. At the same time, business customers need corresponding skills to be able to effectively co-create. In addition, Komulainen [28] emphasizes the importance of the customer's attitude toward learning as a prerequisite of learning how to use a new offer. In this context, they assign a central role to the customer's absorptive capacity. Going one step further, Mele [30] describes joint "value innovation" of customer and provider as a source of value for both. Value innovation encompasses ways in which a provider raises new value potential for customers. Customers employ their own capabilities and resources to realize as much as possible of this potential in a process of value co-creation. In this sense, designing an MC offer always implies designing new value potentials.

With respect to the process dimension, the literature on B2B co-creation focuses on communication and learning. In this respect, the definition and communication requirement is crucial to customer value (cf. [19]). As this can be a moving target,

[1] We are aware that our chapter effectively addresses the co-design process as it is defined in the literature [42]. However, as we largely consult the literature on value co-creation as follows, we use the term co-creation in this chapter.

monitoring becomes an additional and equally important requirement. Monitoring by the customer improves transparency and provides the opportunity to readjust desired outcomes, which is particularly important in the context of internal and external dynamics of the customer domain [19]. Likewise, learning is seen as an issue for both parties (cf. [28]). Customers are required to invest in learning during the co-creation process in order to realize as much value from the provided offer as possible. In order to learn effectively, providers need to support the learning process of customers appropriately.

21.4 Synopsis

This chapter proposes that the main challenges for implementing MC on B2B markets are in the areas of customer value, co-creation, and complexity of individualization. SDL was chosen to structure the origins of customer value because of its emphasis on the customer–manufacturer interaction. Regarding potential-related customer value, we do not find any major dissimilarities. Studies on B2C as well as on B2B markets refer to the quality of service personnel and the quality of the encounter—be it online or offline. The notion of quality can be interpreted as an indicator of appropriateness to the MC offer, which opens the field for further studies on its particular expressions in different product and service markets. In contrast, the process- and outcome-related value dimension showed major differences. B2C studies found hedonic value and creative achievement to play a major role in the co-design process for products and services [32, 42]. In a similar way, perceived uniqueness and self-expression represent outcome-related hedonic customer values. The analyzed literature on B2B markets does not refer to analogue customer value components. Hence, it raises the question of whether business customers show these components too, or whether their behavior is exclusively utility-driven. Accordingly, further studies on the motivations of sourcing behavior of business customers that can be addressed by MC seem to be a fruitful path. This is arguably a decisive factor with regard to redesigning the co-creation process as well as for the customized outcome. Finally, we assume a difference with respect to the overall value emerging at the consumer and business customer side. Unlike end customers, business customers source strategically. Hence, the perception of whether the customizer represents a potential value creation partner in the future should have additional impact on the perceived overall value of the MC offer.

In addition to rethinking the structure and importance of value components, we argue that the co-creation process needs to be redesigned. We see three reasons as follows. First, unlike end consumers, business customers are typically experts themselves and possess a relatively high degree of knowledge [15]. Consequently, the identification of the customers' requirements can draw on a more elaborate picture of actual customer needs. This is in line with Franke and Piller [15] who state that toolkits (as means of eliciting needs) for B2B markets are expert toolkits that allow real innovation [15]. From a broader perspective, the higher level of expertise of business customers calls for conceptualizing the co-design process as interaction among experts. Secondly, end customers clearly value hedonic and creative elements within the configuration process. This is in line with the outcome-related value of uniqueness or self-expression, which we propose to consider "fun" components, since customers make use of MC voluntarily in their free time. In B2B contexts, it is typically the task of procurement personnel to source inputs professionally, which represents a different context for decision-making, hence, a different design of the co-creation process seems necessary, i.e., with customer efficiency as its main goal. Thirdly, in many B2B markets personal sales forces are used to perform one-to-one marketing and personalization. Hence, the co-design process is highly dependent on the personal interaction between sales personnel of the provider organization and procurement personnel of the customer organization. We assume that B2B toolkits can positively affect the transparency and reliability of this process as well as the quality of outcomes as they reduce the dependency on individual expertise and motivation.

The complexity in B2B contexts is a result of the high degree of product customization, which implies a time-consuming customer–manufacturer interaction [5]. While the idea of deploying an online configurator to replace (a part of) the customization process has been successfully implemented in B2C markets for many years [35], it can be expected that these findings cannot be simply copied to B2B markets. In this context, Fauska, Kryvinska, and Strauss [11] found that highly complex products present specific features which cannot be easily described on an online sales platform. Based on this finding, they argue that the selling process can never be fully switched to an online configurator due to the lack of interaction. We suggest that the answer to this question be seen as one of the key components to successfully offering mass customized products and services in B2B markets.

Due to the discussed differences, an "actor-to-actor" approach as presented by [56], which proposes a generic orientation for both B2B and B2C contexts, is not recommended in the present case.

In conclusion, we showed that business customers are highly dissimilar to end consumers and pointed out three hotspots with regard to implementing MC in B2B markets, namely customer value, co-creation, and complexity. As a consequence, we call for further studies in this field—not least because MC as a strategy for B2B companies is considered very promising.

References

1. Berman, B.: Should your firm adopt a mass customization strategy? Bus. Horiz. **45**(4), 51–60 (2002). http://doi.org/10.1016/S0007-6813(02)00227-6
2. Chen, C., Watanabe, C.: Competitiveness through co-evolution between innovation and institutional systems—new dimensions of competitiveness in a service-oriented economy. J. Serv. Res. **7**(2), 27–55 (2007). Retrieved from http://search.ebscohost.com/login.aspx?direct=true&db=bah&AN=27617087&site=ehost-live
3. Clauß, T., Laudien, S.M., Daxböck, B.: Service-dominant logic and the business model concept: toward a conceptual integration. Int. J. Entrep. Innov. Manage. **18**(4), 266–288 (2014). Retrieved from 10.1504/IJEIM.2014.064209

4. Cronin, J.J., Taylor, S.A.: Measuring service quality: a reexamination and extension. J. Mark. **56**(3), 55 (1992). http://doi.org/10.2307/1252296
5. Davies, A., Brady, T., Hobday, M.: Organizing for solutions: systems seller vs. systems integrator. Ind. Mark. Manage. **36**(2), 183–193 (2007). http://doi.org/10.1016/j.indmarman.2006.04.009
6. Dellaert, B.G.C., Dabholkar, P.A.: Increasing the attractiveness of mass customization: the role of complementary on-line services and range of options. Int. J. Electron. Commer. **13**(3), 43–70 (2009). http://doi.org/10.2753/JEC1086-4415130302
7. Dellaert, B.G.C., Stremersch, S.: Marketing mass-customized products: striking a balance between utility and complexity. J. Mark. Res. **17**, 219–227 (2005)
8. Edvardsson, B., Kristensson, P., Magnusson, P., Sundström, E.: Customer integration within service development—a review of methods and an analysis of insitu and exsitu contributions. In: Technovation, vol. 32, pp. 419–429. http://doi.org/10.1016/j.technovation.2011.04.006 (2012)
9. Eichentopf, T., Kleinaltenkamp, M., Stiphout, J. van.: Modelling customer process activities in interactive value creation. J. Serv. Manage. http://doi.org/10.1108/09564231111174997 (2011)
10. Evans, J.R., Laskin, R.L.: The relationship marketing process: a conceptualization and application. Ind. Mark. Manage. http://doi.org/10.1016/0019-8501(94)90007-8 (1994)
11. Fauska, P., Kryvinska, N., Strauss, C.: Agile management of complex goods & services bundles for B2B E-commerce by global narrow-specialized companies. Glob. J. Flex. Syst. Manage. **15**(1), 5–23 (2014)
12. Fogliatto, F.S., da Silveira, G.J.C., Borenstein, D.: The mass customization decade: an updated review of the literature. Int. J. Prod. Econ. **138**(1), 14–25 (2012). http://doi.org/10.1016/j.ijpe.2012.03.002
13. Franke, N., Keinz, P., Schreier, M.: Complementing mass customization toolkits with user communities: how peer input improves customer self-design. J. Prod. Innov. Manage. **25**, 546–559 (2008)
14. Franke, N., Piller, F.T.: Key research issues in user interaction with configuration toolkits in a mass customization system: the foundation of the Idtown user design project key research issues in user interaction with configuration toolkits in a mass customization system: Int. J. Entrep. Innov. Manage. (2003)
15. Franke, N., Piller, F.T.: Value creation by toolkits for user innovation and design: the case of the watch market. J. Prod. Innov. Manage. **21**(6), 401–415 (2004)
16. Goduscheit, R.C., Jørgensen, J.H.: User toolkits for innovation—a literature review. Int. J. Technol. Manage. **61**(3/4), 274–292 (2013)
17. Grönroos, C., Ravald, A.: Service as business logic: implications for value creation and marketing. J. Serv. Manage. **22**(1), 5–22 (2011). http://doi.org/10.1108/09564231111106893
18. Gummesson, E., Polese, F.: B2B is not an island! J. Bus. Ind. Mark. **24**(5/6), 337–350 (2009). http://doi.org/10.1108/08858620910966228
19. Hawkins, T.G., Gravier, M.J., Berkowitz, D., Muir, W.A.: Improving services supply management in the defense sector: how the procurement process affects B2B service quality. J. Purch. Supply Manage. **21**(2), 81–94 (2015). http://doi.org/10.1016/j.pursup.2014.12.007
20. Heinonen, K., Strandvik, T., Voima, P.: Customer dominant value formation in service. Eur. Bus. Rev. **25**(2), 104–123 (2013). http://doi.org/10.1108/09555341311302639
21. Hong, P.C., Dobrzykowski, D.D., Vonderembse, M.A.: Integration of supply chain IT and lean practices for mass customization: benchmarking of product and service focused manufacturers. Benchmarking **17**(4), 561–592 (2010). http://doi.org/10.1108/14635771011060594
22. Huffman, C., Kahn, B.E.: Variety for sale: mass customization or mass confusion? J. Retail. **74**(4), 491–513 (1998). http://doi.org/10.1016/S0022-4359(99)80105-5
23. Ihl, C., Müller, M., Piller, F.: Kundenzufriedenheit bei Mass Customization: Eine empirische Untersuchung zur Bedeutung des Co-Design-Prozesses aus Kundensicht. Die Unternehmung **59**(3), 165–184 (2006)
24. Jaworski, B., Kohli, A.: Co-creating the voice of the customer. In: The Service-Dominant Logic of Marketing: Dialog, Debate, and Directions, p. 453 (2006)
25. Jose, G., Borenstein, D., Silveira, G.D., Fogliatto, H.S.: Mass customization: literature review and research directions. Int. J. Prod. Econ. **72**(1), 1–13 (2001)

26. Kahn, B.E.: Dynamic relationships with customers: high-variety strategies. J. Acad. Mark. Sci. http://doi.org/10.1177/0092070398261005 (1998)
27. Karpen, I.O., Bove, L.L., Lukas, B.A., Zyphur, M.J.: Service-dominant orientation: measurement and impact on performance outcomes. J. Retail. **91**(1), 89–108 (2014). http://doi.org/10.1016/j.jretai.2014.10.002
28. Komulainen, H.: The role of learning in value co-creation in new technological B2B services. J. Bus. Ind. Mark. **3**, 238–252 (2014). http://doi.org/10.1108/JBIM-04-2011-0042
29. Lusch, R.F., Vargo, S.L., O'Brien, M.: Competing through service: insights from service-dominant logic. J. Retail. **83**(1), 5–18 (2007). http://doi.org/10.1016/j.jretai.2006.10.002
30. Mele, C.: Value innovation in B2B: learning, creativity, and the provision of solutions within service-dominant logic. J. Cust. Behav. **8**(3), 199–220 (2009). http://doi.org/10.1362/147539209X469308
31. Merle, A., Chandon, J.-L., Roux, E.: Understanding the perceived value of mass customization: the distinction between product value and experiential value of co-design. Recherche et Applications En Marketing **23**(3), 27–50 (2008). http://doi.org/10.1177/205157070802300303
32. Merle, A., Chandon, J.-L., Roux, E., Alizon, F.: Perceived value of the mass-customized product and mass customization experience for individual consumers. Prod. Oper. Manage. **19**(5), 503–514 (2010). http://doi.org/10.1111/j.1937-5956.2010.01131.x
33. Merle, A., Senecal, S., St-Onge, A.: Whether and how virtual try-on influences consumer responses to an apparel web site. Int. J. Electron. Commer. **16**(3), 41–64 (2012). http://doi.org/10.2753/JEC1086-4415160302
34. Müller, M.: Integrationskompetenz von Kunden bei individuellen Leistungen: Konzeptualisierung, Operationalisierung und Erfolgswirkung. Deutscher Universitätsverlag, Wiesbaden (2007)
35. Piller, F., Harzer, T., Ihl, C., Salvador, F.: Strategic capabilities of mass customization based e-commerce: construct development & empirical test. In: Proceedings of the Annual Hawaii International Conference on System Sciences, pp. 3255–3264. http://doi.org/10.1109/HICSS.2014.403 (2014)
36. Piller, F., Koch, M., Möslein, K.M., Schubert, P.: From mass customization to collaborative customer co-design. Inf. Syst. J. 1497–1509. Retrieved from http://is2.lse.ac.uk/asp/aspecis/20040133.pdf (2004)
37. Piller, F., Müller, M.: A new marketing approach to mass customisation. Int. J. Comp. Integr. Manuf. **17**(7), 583–593 (2004). http://doi.org/10.1080/0951192042000273140
38. Piller, F., Stotko, C.M.: Mass Customization und Kundenintegration. Integrationskompetenz von Kunden bei individuellen Leistungen. Retrieved from http://www.amazon.com/Mass-Customization-Kundenintegration-Christine-Kohl/dp/3936608059 (2003)
39. Pine, B.J.: Making mass customization happen: strategies for the new competitive realities. Strateg. Leadersh. http://doi.org/10.1108/eb054435 (1993)
40. Salomonson, N., Åberg, A., Allwood, J.: Communicative skills that support value creation: a study of B2B interactions between customers and customer service representatives. Ind. Mark. Manage. **41**(1), 145–155 (2012). http://doi.org/10.1016/j.indmarman.2011.11.021
41. Sanders, E.B.-N., Stappers, P.J.: Co-creation and the new landscapes of design. CoDesign. http://doi.org/10.1080/15710880701875068 (2008)
42. Schreier, M.: The value increment of mass-customized products: an empirical assessment. J. Consumer Behav. 327, 317–327. http://doi.org/10.1002/cb (2006)
43. Schreier, M., Mair, A., Franke, N.: Warum "Toolkits for User Innovation and Design" für ihre Nutzer Wert schaffen: eine qualitative Analyse. Die Unternehmung **59**(3), 185–201 (2006)
44. Sheth, J.N., Parvatiyar, A.: The evolution of relationship marketing. Int. Bus. Rev. http://doi.org/10.1016/0969-5931(95)00018-6 (1995)
45. Sirgy, M.J.: Self-concept in consumer behavior: a critical review. J. Consum. Res. **9**(3), 287 (1982). http://doi.org/10.1086/208924
46. Snyder, C.R.: Product scarcity by need for uniqueness interaction: a consumer catch-22 carousel? Basic Appl. Soc. Psychol. **13**(1), 9–24 (1992). http://doi.org/10.1207/s15324834basp1301_3
47. Tokman, M., Beitelspacher, L.S.: Supply chain networks and service-dominant logic: suggestions for future research. Int. J. Phys. Distrib. Logist. Manage. http://doi.org/10.1108/09600031111154152 (2011)

48. Turner, F., Merle, A., Diochon, P.F.: How to assess and increase the value of a co-design experience: a synthesis of the extant literature. In: Mass Customization, Personalization, and Co-Creation: Bridging Mass Customization and Open Innovation, pp. 1–9. Retrieved from http://halshs.archives-ouvertes.fr/hal-00649498/ (2012)
49. Vargo, S.L., Lusch, R.F.: The four service marketing myths: remnants of a goods-based, manufacturing model. J. Serv. Res. http://doi.org/10.1177/1094670503262946 (2004)
50. Vargo, S.L., Lusch, R.F.: Service-dominant logic: continuing the evolution. J. Acad. Mark. Sci. **36**(1), 1–10 (2008). http://doi.org/10.1007/s11747-007-0069-6
51. Vargo, S.L., Maglio, P.P., Akaka, M.A.: On value and value co-creation: a service systems and service logic perspective. Eur. Manage. J. **26**(3), 145–152 (2008). http://doi.org/10.1016/j.emj.2008.04.003
52. Von Hippel, E.: Perspective: user toolkits for innovation. J. Prod. Innov. Manage. **18**(4), 247–257 (2001). http://doi.org/10.1016/S0737-6782(01)00090-X
53. Ward, M.K.: Developing a deeper understanding of scarcity: contextual and individual influences on demand scarcity how consumers are affected by the framing of numerical information contribution of the concept of identity to the understanding of responsible consumer. Adv. Consum. Res. **34**, 384–388 (2007)
54. Wind, J., Rangaswamy, A.: Customerization: the next revolution in mass customization. J. Interact. Mark. **15**(1), 13–32 (2001). http://doi.org/10.1002/1520-6653(200124)15:1<13::AID-DIR1001>3.0.CO;2-#
55. Helander, M.G., Jiao, J.: Research on E-product development (ePD) for mass customization. Technovation, **22**(11), 717–724 (2002)
56. Vargo, S.L., Lusch, R.F.: It's all B2B… and beyond: Toward a systems perspective of the market. Industrial Marketing Management, **40**(2), 181–187 (2011)

Chapter 22
Does the Size of a Fashion Model on a Retailer's Website Impact the Customer Perceived Attractiveness of the Model and Purchase Intention? The Role of Gender, Body Satisfaction and Congruence

Anik St-Onge, Aurelie Merles, Florian Pichonneau, and Sylvain Sénécal

22.1 Introduction

In the USA, retail e-Commerce sales in apparel and accessories totalled $54.2 billion in 2013 [4]. Although approximately 24 % of Internet users report having done online research for products in this category, only 17 % of these sales were completed online [5]. Lack of direct experiential information, or more precisely the difficulty of finding out how well an item fits without trying it on, is one factor that is limiting the proportion of apparel sales conducted online [13, 14, 18]. To address this problem, most of the retailer websites feature a model wearing the clothes. However, although viewing the apparel on models reduces the risk related to buying clothing online, most models are often more attractive and thinner than average.

Numerous studies have explored the impact of models in advertisements and in the field of online retailing [2, 3, 8, 11, 12, 15]. Baker and Churchill [3] have brought to life an important relationship between the perception and the reaction of individuals to others considered physically attractive. It is important to note that a higher level of physical attraction of a model in an advert has a negative impact on the

---

The original version of this chapter was revised. The correction to this chapter is available at https://doi.org/10.1007/978-3-319-29058-4_38

A. St-Onge (✉)
Department of Marketing, University of Quebec in Montreal (UQAM), Montreal, QC, Canada
e-mail: st-onge.anik@uqam.ca

A. Merles
École de Management, Université de Grenoble, Grenoble, France

F. Pichonneau
Sidlee, Montreal, QC, Canada

S. Sénécal
Department of Marketing, HEC Montreal, Montreal, QC, Canada

J. Bellemare et al. (eds.), *Managing Complexity*, Springer Proceedings in Business and Economics, DOI 10.1007/978-3-319-29058-4_22

evaluation, the attention and the perception of value of the exposed individual [3]. Moreover, the literature informs us that the corpulence of models used in advertisements and on the Internet has an impact on self-esteem and body satisfaction [11, 15]. In fact, the corpulence of thin models presented to consumers in real-life purchase situations has a negative impact on self-esteem and body satisfaction [9, 11]. Moreover, Shin and Baytar [15] have demonstrated a negative relationship between the level of bodily insatisfaction and the intention of using tools allowing the visualization of clothes on models.

In the field of clothes retailing, the satisfaction with one's body plays an important role. Indeed, research highlights the sensitivity of people relative to the evaluation of one's body satisfaction in link with apparel products [1, 17]. Moreover, the level of body dissatisfaction contributes to negative behaviour linked to the avoidance of apparel products [17]. There is therefore an important relationship between body satisfaction and the purchase of clothes [7]. It is important to note that men have generally a higher level of personal body satisfaction than women. Nevertheless, it is known that men's body esteem is also negatively impacted by the visualization of commercial models than women's [3]. We also note that women have a tendency to compare themselves more to models they are exposed to if those models and young and of average corpulence [11]. The differences of body satisfaction evaluation between men and women being sustained by research, it is interesting to explore the differences induced by the gender. Furthermore, the more similar the avatar is to the user, the more this one will have a positive attitude towards the avatar [16].

A retailer's website is not only a very interesting distribution channel for sales of apparel and accessories, but it is also a vehicle of the company brand that lets consumers experience the brand. Antecedents of consumers' online attitude have also been identified in the literature. Notably, it has been demonstrated that attitude toward a merchant's site positively influences attitude toward the retailer [6]. Heanlein and Kaplan [6] show that online attitude influences attitude in a traditional buying situation and hence purchase intention. Even though most fashion retailers let consumers buy their products online, to date no study has examined the impact of the presentation of 2D fashion models on attitude toward the retailer's brand. In addition, many studies of the use of tools to view apparel online had exclusively female samples [11, 12, 15].

In a clothes buying situation, research has also shown the influence of the consumer's body of his attitudes. In fact, consumers having a large girth perceive more utility to shopping with a model. However, they experience more negative attitudes relative to enjoyment and leisure [10]. Furthermore, persons having a thin body are more motivated by these aspects of online shopping, based on pleasure and a positive attitude towards the brand than by their commodity. The research concludes that persons having a thin body have a tendency to have a more positive attitude towards online shopping then overweight persons [10]

The present study answers this question: is it important to personalize the dimensions of a fashion 2D model on a retailer's website? Fig. 22.1.

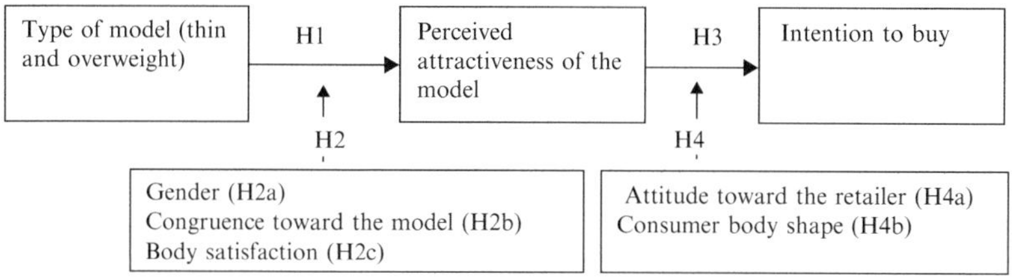

Fig. 22.1 Conceptual framework

22.2 Methodology/Approach

To test the research hypotheses, we did an experimental design (2×2). Four different versions of website models have been created using the website modelmydiet.com. Manipulation was based solely on an increase in size of the male and female models used on the site hm.com. Body mass index is the factor retained to numerically translate an individual's build, namely thin and overweight [15].

A questionnaire was drawn up using measurement scales validated by the literature. The questionnaire comprises three parts. The first part of the questionnaire contained a set of questions that let respondents evaluate their perceived risk, level of buying clothing online and their attitude toward the H&M brand. In the second part of the questionnaire, the following scenario was presented: "You are now shopping for clothes on the H&M retail website. You are looking for pants and a jacket for a special occasion. Finally, you choose the pants and jacket presented on the model below. You really like these products and the whole set of apparel fit you selected. Before reviewing your final cart and proceeding to checkout, you want to check the overall apparel you have chosen". After having been exposed to one of the two replicas of the website with a model corresponding to their gender, participants were redirected to an online version of the survey, where they were asked to complete a second questionnaire evaluating their level of perceived attractiveness of the model shown, level of ideal perceived size of the model, attitude toward the H&M brand and purchase intention. The questionnaire was administered via the online survey platform Amazon Mechanical Turk, we got approximately 45 respondents per cell, for a total of 184 completed questionnaires.

22.3 Results

The T-test results (H1) indicate a significant positive difference in type of model (thin and overweight) and the attractiveness toward the model ($t(182) = -4.115$, $p < 0.05$, Moverweight 4.34 < Mthin 5.13). So the more attractive model is the thin model. Moreover, the results show that this relation is significantly moderated by the gender (H2a: R2. = 0.185, $p < 0.05$), the congruence toward the model (H2b: R2. = 0.180, $p < 0.05$) and slightly moderated by the customer body satisfaction (H2c:R2. = 0.113, $p < 0.10$).

The relation between perceived attractiveness toward the model and intention to purchase intention is also significantly positive (H3: R2. = 0.140, $p < 0.05$). In addition, this relation is positively moderated by the attitude toward the brand (H4a: R2. = 0.3802, $p < 0.05$) and negatively moderated by the customer body shape (H4b: R2. = 0.2302, $p < 0.05$).

22.4 Limitations

One of the limitations of this study is that respondents were exposed to replicas of static sites that do not allow interaction. This lack of interaction is an important limitation.

References

1. Aghekyan-Simonian, M., Forsythe, S., Suk Kwon, X., Chattaraman, V.: The role of product brand image and online storage image on perceived risks and online purchase intentions for apparel. J. Retail. Consum. Serv. **19**(3), 325–331 (2012)
2. Baird, A.L., Grieve, F.G.: Exposure to male models in advertisements leads to a decrease in men's body satisfaction. N. Am. J. Psychol. **8**(1), 115–121 (2006)
3. Baker, M.J., Churchill, J.R.G.A.: The impact of physically attractive model on advertising evaluation. J. Mark. Res. **14**(4), 538–555 (1977)
4. eMarketer.: US Apparel and accessories retail ecommerce sales 2013–2018, consulted October 1, www.emarketer.com (2014a)
5. eMarketer.: US Retail Ecommerce 2013–2018, consulted October 1, www.emarketer.com
6. Heanlien, M., Kaplan, A.M.: Les magasins de marques phares dans les mondes virtuels: l'impact de l'exposition au magasin virtuel sur l'attitude envers la marque et l'intention d'achat dans la vie réelle. Recherche et application en marketing **24**(3), 57–80 (2009)
7. Manuel, M., Connell, Lenda, A. Presley.: Body shape and fit preference in body cathexis and clothing benefits sought for professional African-American women. Int. J. Fash. Des. Technol. Edu. **3**(1), 25–32 (2010)
8. Merle, A., Senecal, S., St-Onge, A.: Whether and how virtual try-on influences consumer responses to an apparel web site. Int. J. Electron. Commer. **16**(3), 41–64 (2012)
9. Micu, C.C., Coulter, R.A., Price, L.L.: How product trial alters the effect of model attractiveness. J. Advert. **38**(2), 69–81 (2009)
10. Park, J., Nam, Y., Choi, K.M., Lee, Y., Lee, K.H.: Apparel consumers' body type and their shopping characteristics. J. Fash. Mark. Manage. **13**(3), 372–393 (2009)
11. Peck, J., Loken, B.: When will larder-sized female models in advertisements be viewed positively? The moderating effect of instructional frame, gender and need for cognition. Psychol. Mark. **21**(6), 425–442 (2004)
12. Reaves, S., Hitchon, J.B., Park, S.Y., Gi, W.: If looks could kill: digital manipulation of fashion models. J. Mass Media Ethics **19**(1), 56–71 (2004)

13. Rosa, J.A., Garbarino, E.C., Malter, A.J.: Keeping the body in mind: the influence of body esteem and body boundary aberration on consumer beliefs and purchase intentions. J. Consum. Psychol. **16**, 179–191 (2006)
14. Shim, S.I., Lee, Y.: Consumer' perceived risk reduction by 3D virtual model. Int. J. Retail. Distrib. Manage. **39**(12), 945–959 (2011)
15. Shin, E., Baytar, F.: Apparel fit and size concerns and intentions to use virtual try-on: impacts of body satisfaction and images of model's bodies. Cloth. Text. Res. J. **32**(1), 20–33 (2014)
16. Suh, K., Kim, H., Suh, E.K.: What if your avatar looks like you? Dual-congruity perspectives for avatar use. MIS Q. **35**(3), 711–729 (2011)
17. Trautmann, J., Lokken, S., Lokken, K.: Body dissatisfaction, bulimic symptoms, and clothing practices among college women. J. Psychol. Interdiscipl. Appl. **141**(5), 485–498 (2007)
18. Zheng, L., Favier, M., Huang, P., Coat, F.: Chinese consumer perceived risk relievers in e-shopping for clothing. J. Electron. Comm. Res. **13**(3), 255–274 (2012)

Chapter 23
"La Chispa de la Ciudad de México": Co-creation of Organizational Innovations and Its Implications for Managing Innovation

Hans Lundberg, Ian Sutherland, Birgit Penzenstadler, Paul Blazek, and Hagen Habicht

23.1 Introduction

In our pilot study [18], the InnoTracing methodology and InnoTrace tool were put in the hands of participants of the Leadership for Innovation Conference of the Peter Pribilla Foundation. Subsequently, the research team established a second study at the Mexico City office of Great Place to Work (GPTW Mxc), a globally operating HR consulting firm. Within the innovation department at GPTW Mxc, the InnoTracing methodology and the InnoTrace tool were used by three employees (Angie, Caroline, Jennifer) on an everyday basis for a period of 3, 5 months (mid-August–end of November, 2013). Participants from the innovation department and their management were given an overview of our project, along with a training session for the participating three employees. Due to the length of the Mexico City

H. Lundberg (✉)
School of Business and Economics, Linnaeus University, Växjö, Sweden
e-mail: hans.lundberg@lnu.se

I. Sutherland
IEDC-Bled School of Management, Bled, Slovenia
e-mail: Ian.Sutherland@iedc.si

B. Penzenstadler
Department of Computer Engineering and Computer Science, California State University, Long Beach, CA, USA
e-mail: birgit@penzenstadler.de

P. Blazek
cyLEDGE Media GmbH, Vienna, Austria
e-mail: p.blazek@cyledge.com

H. Habicht
Center for Leading Innovation and Cooperation, HHL Leipzig Graduate School of Management, Leipzig, Germany
e-mail: hagen.habicht@hhl.de

J. Bellemare et al. (eds.), *Managing Complexity*, Springer Proceedings in Business and Economics, DOI 10.1007/978-3-319-29058-4_23

project, research team members also provided technical support and further training within the first couple of weeks of the study.

"*La chispa*" ("sparkle," "ignition," "lively" in Mexican–Spanish) in the headline refers to two specifics of the context of this case:

The specifics of Mexico City (*Ciudad de Mexico*) *as the scene for the case*: Contrary to most popular beliefs, Mexico is the hardest working country in the world (2237 h per year) according to a recent OECD study [22]. The valley of Mexico and its modern invention, *the megalopolis of Mexico City*, with an estimated population of just above 22 million people, is a concentrated embodiment of such OECD statistics. One of the authors, having lived and worked 11 years in Mexico, has over the years and in many different ways become struck by the frantic entrepreneurial energy one encounters in a myriad of ways. This entrepreneurial energy is less based on "collaboration," "cooperation," or "mobilization of the good forces" to develop something, and more based upon survival instinct, competition-oriented self-organization, strong self-confidence and self-esteem, calculated mutual benefit, and pure raw creative power. In summary, a focus on opportunity creation and exploitation enacted in subtle relational systems and social codes within and between formal/informal dimensions of society and white/dark/gray sectors of economy. All these seemingly "rough characteristics" are simultaneously embedded in gentle, smooth, and highly stylish tone and manners. Ways of acting, talking, and going about differs highly but as a *principle* one must be equally gentle in tone, smoothness, and style towards high and low in various socioeconomic strata. Lida [17] discusses this frantic energy in terms of that Mexico City has everything required to be for the world in the twenty-first century what New York was in the twentieth century and Paris in the nineteenth century. Such a hyper-creative city unfolds constantly is reinventing itself over and over again. This is why "La chispa de la Ciudad de México" should be understood as a *fluid texture* rather than a static context in this project.

The specifics of the case as such: The "everydayness" character of this study is founded in a new idea about the GPTW Mxc office that Angie just had launched before we were entering the scene. Angie is a highly energetic and proactive innovator, change agent, and a bringer of new things within GPTW Mxc. Our project is based on one of her innovations, the reorganizing of their office space into the so-called cubiculos, aiming at facilitating immediate dialogical interaction with each other when in need of that while still safeguarding individual concentration and silence when necessary. At large, it is a version of an open space office that is not very "new" in a Western context but in a more hierarchically oriented Mexico City work life texture, it represents a radical change. Their office and their working methods are furthermore not representative for the GPTW Mxc office as a whole. They are rather seen as the "oddballs" internally that try new things. This sometimes leads to others imitating their behavior and new practices, sometimes to negative reactions. Angie's perspective on this is the classical entrepreneur's: "Hey, we have the ambition to be HR leaders in Mexico and globally; then we must start 'at home' by being at the forefront of HR in our own work place, right?"

With this texture established, we next provide a summary of the methodology InnoTracing and the software InnoTrace and the call for methodological innovation it is a response to (Sect. 23.2) (more in detail in [18]). Next, we shortly link InnoTracing, InnoTrace, and our GPTW Mxc study to the fields of organizational innovation and management of innovation (Sect. 23.3). Thereafter, we provide data results and key insights drawn from the research and analysis made on GPTW Mxc (Sect. 23.4). Last, we draw conclusions from the organizational innovations of GPTW Mxc and its implications for the management of innovation (Sect. 23.5).

23.2 Background: The Need for Methodological Innovation and Our Response to It

Much has been written about the complex, nonlinear, emergent, and to a large extent tacit nature of group processes [27]. Social phenomena such as innovation and leadership as they unfold in groups over time reflect these properties in particular [1, 2, 7, 13, 19, 28]. Their moment-to-moment unfolding on the micro-level of interaction has, to date, largely remained unexplored. Partly this is because conventional study designs face a common methodological problem of distance, that is, data gathering is largely researcher dependent. Hence, it is subject to their biases with respect to identifying significant moments, important events, and central actors as they present themselves to the participants [16]. This is equally true for any "researcher-designed" data gathering technique, e.g., survey methods, interviews, focus groups, participant observation, or any other familiar data gathering processes.

We advocate an alternative approach that empowers participants to generate data on social interactions at a micro-level themselves: InnoTrace is a tool for researchers to put data gathering directly into the hands of participants. It enables them to capture "moments of significance" (MOS) they experience in unfolding processes. Moreover, it helps to aggregate that data for each participant as well as the whole group of participants in the form of cognitive maps [5, 26]. The created cognitive maps serve as a rich visual tapestry of the micro-level interactions within the group.

With this approach we respond to increasing attention placed on the micro-level of innovation and leadership, that is, on their emergence in real time from moment-to-moment [12, 14]. The called-for ethnomethodological approach "pays attention to, and seeks to make visible, the 'ethno-methods' [9] through which the social order of [a] setting is inter-subjectively constructed..." [12] (p. 124). This is in line with Wood and Ladkin [31], who argue for a "lens of process philosophy [which] frames leadership as an unfolding, emerging process; a continuous coming into being." [31] (p. 15). Yet, here we find a methodological quagmire—the ability to investigate the continuously "coming into being" of human interactions is trying to make the invisible of highly complex interactions visible. There are no established methodological tools readymade for this task. For instance, any ethnographic approach will face problems of observer influence and interpretation. In addition, non-ethnographic approaches, which rely on post hoc data, are bound to the recall and confirmative ex post sense making of participants. Hence, we often find ourselves in the realm of ex post facto research when what we really desire is to capture the involved actors' perception of the moment in the moment [31].

We believe that, in order to respond to these calls, there is no alternative to empowering participating actors to document and comment upon their perceptions as they experience them. InnoTracing refers to that as "moments of significance" (MOS). The InnoTrace tool allows participants to capture—via picture, video, text, or voice notes—moments which feel significant in the unfolding, emergent processes of leadership and innovation, as they happen. InnoTracing[1] is thereby a methodological development that combines a unique data gathering and aggregating software tool with social science methods to help researchers and participants open, visualize, and investigate the moments of significance (MOS) of leadership and innovation.

The InnoTrace tool (screenshots, see Fig. 23.1) is designed as user-friendly and user-configurable software, affording participants the ability to capture and trace MOS of innovation and leadership.

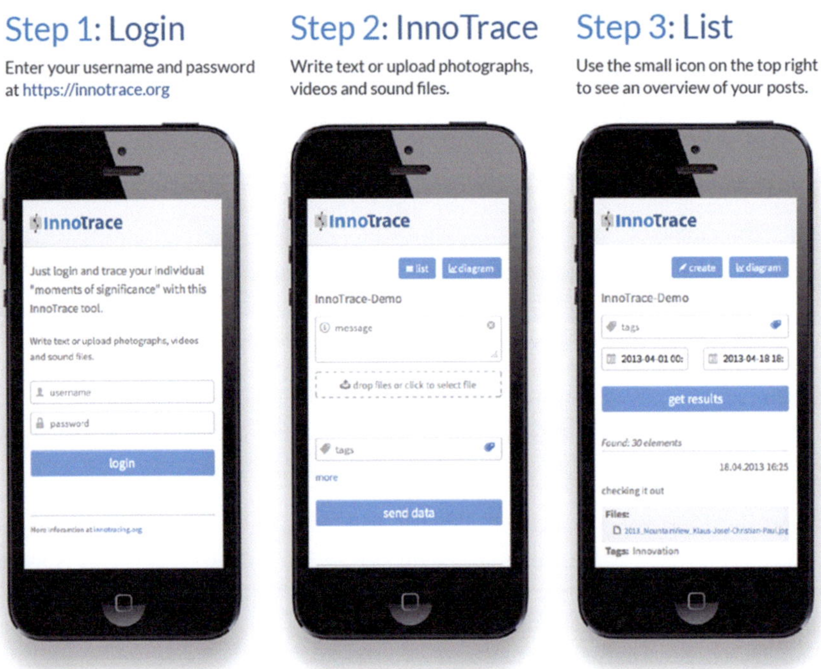

Fig. 23.1 Screenshot of steps to get started with InnoTrace

[1] http://www.innotracing.org

Consequently, the characteristic of significance is based on the participants' perceptions. Whether the significance is of something positive, negative, or even mundane lies in their eyes. It allows documenting these moments as participants feel them occur in real time (through photos, videos, text files, or sound messages). The web-based software collects and organizes this data in a variety of ways. The standard four-step process is as follows (more in detail in [18]):

1. *Phenomena of Interest*. With each project, the researcher(s) indicate to the participants the phenomena of interest around which they would like to gather MOS.
2. *MOS Tagging*. Within the tool, researchers can include a variety of classification options (tags) or leave tagging open to the discretion of research participants.
3. *Participant-Generated MOS*: The tool is made available to research participants who engage in gathering data on the MOS of processes in which they are involved.
4. *MOS Aggregation and Visualization*. As research participants gather data, the InnoTrace software collects and organizes this data by user, time, format, and tag. Using InnoTracing and InnoTrace in studies on Organizational Innovation and Management of Innovation.

23.3 Using InnoTracing and InnoTrace in Studies on Organizational Innovation and Management of Innovation

In this section, we shortly link InnoTracing, InnoTrace, and our GPTW Mxc study to the fields of organizational innovation and management of innovation.

Innovation research is moving from studying coordination issues of research and development activities within particular departments towards an increased interest in collaborative research efforts which cross organizational boundaries [3]. This change has taken on speed due to the development of social software-enabled innovation methods such as communities and contests [16, 21]. Scholars nowadays recognize various forms of collaborative (open) innovation; be it in the form of interorganizational innovation networks, or be it based on crowd sourcing mechanisms.

Although this research has produced a variety of studies on individual characteristics [15, 28] and motivators [11, 20, 29] of participants, on success-relevant management capabilities and organizational characteristics [6, 8, 16, 32], the micro-foundations of collaboration among innovators have to date remained a black box. In particular, studying the in situ unfolding of creative "momentum" [25] on the group level, such as by tracing the actual process of identifying and spanning of boundaries, or self-reporting about direct group-level effects of self-rewarding activities (e.g., group flow [4, 23]), would lead to new insights on the actual foundations of collaborative innovation. Yet, ultimately these concepts and theories encounter a methodological brick wall as researchers face the difficulty of getting to the in situ, socially constructed dynamics of leadership and innovation unfolding in real time. What is missing is the ability to visualize the seemingly invisible, moment-to-moment emergence of such collaborative processes at the situated level of individual and group action as people interact in space and time. We designed the methodology InnoTracing and the software tool InnoTrace as a response to this gap and have in the previous chapter [18] elaborated upon the details.

23.4 Analysis and Findings

The *Phenomena of interest* for GPTW was (in dialogue with them) defined as follows: "As InnoTracer at Great Place to Work, Mexico City, please document in any mode you prefer experiences you have that you see as significant for your everyday creativity." With "any mode you prefer," we emphasized that the InnoTracer chooses the way of capturing MOS (i.e., short text with tags and with/without photo, video, audio message). The *Tags* used at GPTW was a mix between three agreed upon basic tags that the participants should use for each MOS (Person Tags, Place Tags, Project Tags) combined with a fourth free category (see Fig. 23.2).

Below, a snapshot from an MOS diary (Fig. 23.3) to indicate the more *dialogical character* of this study:

This more dialogical feature of InnoTracing shows that the methodology can serve as a looped, feedback-enabling foundation for ongoing reflexivity or "act thinkingly" as Karl Weick [30] put it; ongoing reflections-over-work-as-you-work

Agreed upon "Basic Tags" (=should be used by everyone for every MOS) NOTE: Please reg
1) Person Tag; 2) Place Tag; 3) Project Tag; 4-xx) As many individual tags as you want

Person Tags	Place Tags	Project Tags	Your Choice of Any Tag Beyond Basic Tags
My Self	Office	RKG CA	
Core Team	Home	RKG Comments	
Other Group	Home Office	RKG Audit	
	Coffe Place	RKG Process	
	What's App	BPP	
	Anywhere Else	Indicators	
		CAAF	
		Industry Analysis	
		Publication	
		White Paper	
		Research	
		Success Case	
		Great Lunch	
		Great Team	
		Social Media	
		Events	
		Quick Talk	
		PCH	
		No Project	

Fig. 23.2 Tags used in the Great Place To Work study

Date 2013	MOS	Translation of MOS to English	File/s uploaded	Tags	Hans comments to Angie	Angie's comments to Hans comments	Comments for analysis	Content in file uploaded in 4th column)
Thu Aug 29	18:31 In a meeting with the newspaper "el economista" reviewing the financial sector indexes. Why is she going to ask manpower about the industry? I think we should be experts of the industry in order to explain the phenomena in the financial sector!			Othergroup, Indicators, Meeting, financialsector	Interesting power related reflection. Can you please elaborate upon the following: Why do think it is like this, that Manpower and not you are given the space to "define reality"? How will you change that?	I think we don't have the resources (software and people) to dedicate too much time to this kind of stuff. This year in particularly I had been working on indexes, but I think that we need to establish a more aggressive strategy in order to position GPTW as an expert in HR field, getting advantage of the indexes we have already worked.		
Thu Aug 29	19:09 Too much emails to answer			stress, myself, Office			The price of creativity; mundane work tasks lags behind…	
Fri Aug 30	01:41 Thinking about how to deal with many activities at the same time.			Home, myself, management	Difficult but important to be able to do for creative people; any suggestions for how to solve that?	I had already put down my activities and started to delegate some of them. I think that it helps me to deal with everyday issues and problem-solving.		

Fig. 23.3 Snapshot from an MOS diary used by each participant individually to support and stimulate ongoing reflexivity (act thinkingly)

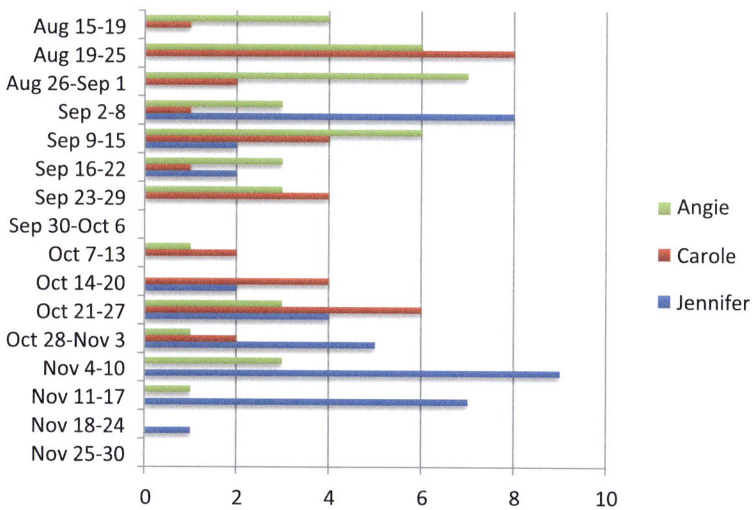

Fig. 23.4 MOS/week per participant during the study period

rather than waiting for the divine moment in tranquility (that never comes) where you hoped to get time for retrospective reflections-over-work-after-work.

The three participants made 116 MOSs, fairly equally distributed over the whole study period (Angie, 41 MOS; Carole, 35 MOS; Jennifer, 40 MOS) but distinctly uneven distributed among the participants during various periods (see Fig. 23.4).

The following three patterns stand out from this overall frequency analysis:

As so often with IT-related processes, there was also here a high level of enthusiasm the first weeks (the honeymoon effect), as clearly indicated by the frequency of Angie and Carole August 15–September 1. For the newest employee at the innovation department, Jennifer, this effect is delayed until September 2 (and ends on September 22), simply because she did not start working at the department until then.

That activity is fading after the initial period is true for all participants, although *when* and *why* it does so, differs individually. *In the case of Angie*, the dip comes halfway (September 30–October 6) and although she recovers, it is on a steadily lower level (0–3 MOSs the second half of the study). The reason for the dip and the following lower level of activity is problems with multitasking as well as problems with defining what actually qualifies as an MOS: "I think I was working on the Gender Equity Ranking. Sometimes for me it is difficult to determine what is a MOS, while working in team and dealing too many things, and given them solution in the moment" (Comment Angie, October 7). The specific Mexico City *texture* here manifest in the data; the fluidity by which everything happens simultaneously, while at the same time having one major task to focus on (Gender Equality Ranking). The ongoingness of "everything else" simply does not allow such a focus, but instead "nags" constantly and demands action and attention. These characteristics are as such nothing extraordinary but a normal feature for modern work life; it is the *intensity* of the fluidity, multitasking, colliding attention spans, etc. that is specific to the Mexico City texture, and it manifest here as a clear dip in MOS activity that Angie never recovers fully (while still testifying having clear intentions to, as she was the participant most committed to this project). *In the case of Carole*, the dips are several (September 30–October 6 and all November) and strongly connected to her travels. At the time, she was the head of the innovation department and, as such, she travelled more than the others, which had strong negative impact on her MOS activity. This is an important finding as using InnoTracing via your smartphone or iPad is supposed to neutralize shifts in physical place and space—but the data testifies on the opposite; change of context/texture = strong dip in activity, as illustrated by this comment by Carole: "My stay in France in August–September 2013 was very messy and busy. […] So i haven't taken the time to register MOS. Also I think I kind of disconnected from Mexico, to prepare my transition and future stay in France. As for creativity, physical distance does impact in collaboration. Even if i think not so much on individual creativity" (Comment Carole, September 6). *In the case of Jennifer*, we analytically have an outlier. It was a bit tough for her already as it was, coming in new at the department for innovation and thereby coming a bit late into this project. Once in, she engaged with strong enthusiasm and "MOS'd" on, until the next blow came; vast technical problems with logging in. All in all, it took us 3 weeks (September 23–October 13) of troubleshooting and problem solving until it worked again for Jennifer. This caused not only a major dip in Jennifer's motivation (of course) but also threw the whole project into a kind of minor legitimacy crisis. Urgent meetings on-site, assurances of various kinds, and compensation promises (holding free workshops, etc.) were required to put things fairly back on track. In a way, Carole and Angie never fully recovered motivation after this (they more stumbled on) while Jennifer met "a new spring," as manifested in her vast activity October 21–November 17.

These first two points shows the roller-coaster character of everyday creativity; moments of glimmering creativity is mixed with the harsh realties of multitasking, technical problems, travelling interruptions, new employees coming (Jennifer), old ones going (Carole) as well as many individual events and situations. To manage everyday creativity and the results and innovations it leads to is not only about the glimmering moments when something productively actually is created ("02:00 I succeeded in a pretty difficult task.... Somehow I did it!!! I'm proud of myself! Special presentation: Emotional Salary," MOS Jennifer, November 5) but also about having management mechanisms that creates reasonable stability and endurance over time, overcoming dips and periods of problems. When everyday creativity passes reality tests, can survive in individual mode when team spirit fades (and vice verse), and manifest itself over longer periods of time, then an organization may claim to have a management of innovation that is suitable and sustainable.

Content-wise, our data points at the following five key findings and accompanying management challenges:

Key finding 1: Team and individual creativity constitute each other; GPTW have succeeded in forming a productive team spirit, which is essential in order to capitalize on individual everyday creativity.

"I'm very proud for the group achievements and all the accomplishments we have made this year [...] I'm proud that the Innovation area finally consolidated this year and all the rest of the areas on GPTW recognizes Innovation as an independent area. Also, Jorge (our director) trusts the area on special tasks and presentations, which are strategic for GPTW. This positions us as an area valuable to GPTW. I think that these accomplishments were possible thanks to the team spirit" (Comment, Jennifer).

"Generally speaking, the area has been much more productive than last year. We are well positioned within the Institute; Jorge, the President, said in some occasion that we were the only area that he trusted to do the industry reports for instance (Comment, Carole).

Main management challenge relative to key finding 1: Forming a good team spirit is one thing, keeping and developing it is another.

Two empirical illustrations of this challenge:

- Pressure from internal changes: "Mon Aug 19 23:53: Carole is leaving today. Even though she will be connected, things won't be the same" (MOS, Angie).
- Pressure from external expectations: "Another challenge is that we sometimes are seen as a non-productive area, that we have fun and so on. Two reasons: the aim of the area and what we do on a day to day basis, the other reason is that we have a really great climate inside of the area which is not so common in the institute, we get on well, we are friends so from the outside it may be confusing" (Reflections, Carole).

Key finding 2: Everyday creativity leads to tangible and important results, but it takes time and it is difficult to communicate the cause-effect to others.

"Thu Aug 29 18:31 In a meeting with the newspaper "el economista" reviewing the financial sector indexes. Why is she going to ask manpower about the industry? I think we should be experts of the industry in order to explain the phenomena in the financial sector!" (MOS, Angie).

"Interesting power related reflection. Can you please elaborate upon the following: Why do you think it is like this, that Manpower and not you are given the space to "define reality"? How will you change that?" (Comment, Hans).

"I think we don't have the resources (software and people) to dedicate too much time to this kind of stuff. This year in particularly I had been working on indexes, but I think that we need to establish a more aggressive strategy in order to position GPTW as an expert in HR field, getting advantage of the indexes we have already worked" (Comment, Angie).

Main challenge relative to key finding 2: To make management understand that awareness and attention is "the main currency" in work life as of today.

Many companies still tend to manage innovation as they manage other more tangible processes (production, sales, distribution, etc.) within the company. The prioritization described by Angie above, where concrete *work with indexes* as a distinct GPTW revenue stream is prioritized over the more abstract task of positioning GPTW as the preferred option in the public *discourse about indexes*, distinctly illustrates this key management challenge. A further breakdown of this challenge can be done via the three concepts of awareness-resources-attention:

- How to raise awareness within senior management of the time required working GPTW into a position where GPTW is given the space to "define reality"?
- How to translate this awareness into concrete resources for the innovation department?
- How to best use these resources to form "a more aggressive strategy in order to position GPTW as an expert in HR field" (Comment, Angie)?

Key finding 3: "Me-time" is essential in order for everyday creativity to be productive for the individual, the team, and for the organization as a whole.

"Myself" was the most common tag used by all three participants (Angie 22, Carole 18, Jennifer 27) which indicates that everyday creativity is less project related (fewer MOSs are tagged to particular projects) and less place related (places differ highly) but more individual and relational, where "me-time" is the reflexive mechanism needed to link the everyday creativity of the individual to benefits for the team and the organization as a whole.

Main challenge relative to key finding 3: How to defend, communicate, and make visible the fundamental need for "Me-time"?

One empirical illustration of this challenge:

- "Fri Oct 25 07:40 At the staff meeting, the organization is asking for engagement, but definitely it has to be reciprocal" (MOS, Angie).

This is once again a tension between the concrete and the abstract. Employees doing tangible things at the office on office hours are concrete, measurable and easily manageable. Employees working with everyday creativity, partly out of office, are not.

Key finding 4: Everyday creativity is not necessarily very glamorous, it emerges mainly out of pushing ones limits, sometimes to the max.

"Thu Nov 14 20:28 AT LAST!!!!! I finally made progress on my article about fun @ work! I think I'm about to finish this… probably tomorrow" (4:33 pm) (MOS, Jennifer).

"Tue Nov 5 02:00 I succeeded in a pretty difficult task.... Somehow I did it!!! I'm proud of myself! Special presentation: Emotional Salary" (2 pm) (MOS, Jennifer).

"This task was very difficult because we didn't have a clue of how to do it. But somehow I managed it and felt very proud of myself" (Comment, Jennifer).

Main challenge relative to key finding 4: How to manage the balance between pushing oneself/being pushed and who owns the agenda?

As shown in our literature review above, there is at large an absence of critically oriented studies within the innovation management literature on the potential dark sides of everyday creativity within corporate structures. Such studies are at large located outside innovation management studies and not activated therein to any larger extent. We think it would be beneficial to change that, but does not have the space here to do so ourselves. If to do it later on, some basic questions out of our data that can serve as a starting point are given as follows:

- How to manage the balance between pushing oneself/being pushed and office time/"me-time"?
- How to create awareness and attention among colleagues and senior management about these delicate balance issues?
- Who is "the manager" here—the individual, managers, or corporate culture (the enterprising self internalizes a particular management discourse)?

Key finding 5: Everyday creativity is constant change, not the change between two constants.

"Finally, a big challenge for the transition that started before I left is Mariana. I think there has not been so much preparation before she started in the area, which has been frustrating for me but I did not want to involve too much so Angie and Jennifer would start to deal with it. Now they are in a process of reinventing the area, new dynamic, new competencies, new history for the area, etc." (Comment, Carole).

Main challenge relative to key finding 5: For new GPTW employees, how to embody the cognitive schemata that correspond to the texture of working at the innovation department of GPTW?

When we left the young innovation department of GPTW, they faced their first major structural change. They now have to face the challenge on how to institutionalize their mentality and "the way we do things here" and transfer this to incoming employees but *without* curbing the individuality of the everyday creativity that the newcomers might bring. We think that innovation management to a larger extent needs to distant itself from traditional management (From-column) and focus more on what we see as more distinct innovation management features (To-column):

From	*To*
Change = between constants	Change as constant
Longing for linearity and order	Embracing fluidity and connectivity
Employee time as main asset	Employee creativity as main asset
"Project finalized"	"-ing is the thing"

23.5 Conclusions

This is the first time we try out InnoTrace and InnoTracing within a Mexican work context and we have not found other studies on micro-level interactions of everyday creativity processes (our fluency in Spanish is limited though, why we might have missed studies written in Spanish). Being that, GPTW showed up to be a beneficial choice for a first study.

The department of innovation at GPTW Mxc is *as such* an organizational innovation that *works with* organizational innovations within GPTW at large. Such an internal role, perceived self-identity and relative self-organizing freedom is rather radical within a Mexican work context in general but less so within the more fluid Mexico City texture described in Sect. 23.1 above. This ambiguity "haunts" the three employees at GPTW Mxc and has concrete implications for the management of innovation. We formulated these in Sect. 23.4 as "five main management challenges" in relation to each key content finding.

Overall, our conclusion is that the combined use of InnoTrace software and InnoTracing methodology in order to aggregate data on individual and group level, creates detailed cognitive maps [5, 26] which serve as a rich visual tapestry of the micro-level interactions of processes underlying leadership and innovation. By visualizing these interactions, empirical support for innovation managers that consider taking a stand in the name of co-creation, everyday creativity, and empowered employees is produced and thereby of use in innovation management and everyday innovation practices.

Acknowledgment We thank the Peter Pribilla Foundation who made this project possible by bringing us together as a team and by funding part of this research.

References

1. Alvesson, M., Sveningsson, S.: Managers doing leadership: the extra-ordinarization of the mundane. Hum. Relat. **56**(12), 1435–1459 (2003)
2. Alvesson, M., Sveningsson, S.: The great disappearing act: difficulties in doing “leadership”. Leadership Quart. **14**, 359–381 (2003)
3. Baldwin, C., Von Hippel, E.: Modeling a paradigm shift: from producer innovation to user and open collaborative innovation. Organ. Sci. **22**(6), 1399–1417 (2011)
4. Bryan-Kinns, N., Hamilton, F.: Identifying mutual engagement. Behav. Inform. Technol. **31**(2), 101–125 (2012)
5. Bougon, M.G.: Congregate cognitive maps: a unified dynamic theory of organization and strategy. J. Manag. Stud. **29**(3), 369–389 (1992)
6. Cohen, W.M., Levinthal, D.: Fortune favors the prepared firm. Manag. Sci. **40**(2), 227–251 (1994)
7. Corbin, J., Strauss, A.: Basics of Qualitative Research: Techniques and Approaches for Developing Grounded Theory, 3rd edn. Sage, Los Angeles (2008)
8. Frey, K., Lüthje, C.: Antecedents and consequences of interaction quality in virtual end-user communities. Creativity Innovation Manag. **20**(1), 22–35 (2011)
9. Garfinkel, H.: Studies in Ethnomethodology. Prentice Hall, Englewood Cliffs (1967)
10. Grint, K.: Problems, problems, problems: the social construction of ‘leadership’. Hum. Relat. **58**(11), 1467–1494 (2005)
11. Harhoff, D., Henkel, J., Von Hippel, E.: Profiting from voluntary information spillovers: how users benefit by freely revealing their innovations. Res. Policy **32**(10), 1753–1769 (2003)
12. Iszatt-White, M.: Methodological crises and contextual solutions: an ethnomethodologically informed approach to understanding leadership. Leadership **7**(2), 119–135 (2011)
13. Kempster, S., Parry, K.W.: Grounded theory and leadership research: a critical realist perspective. Leadership Quart. **22**, 106–120 (2011)
14. Ladkin, D.: Leading beautifully: how mastery, congruence and purpose create the aesthetic of embodied leadership practice. Leadership Quart. **19**, 31–41 (2008)
15. Lakhani, K.R., Von Hippel, E.: How open source software works: “free” user-to-user assistance. Res. Policy **32**(6), 923–943 (2003)
16. Laursen, K., Salter, A.: Open for innovation: the role of openness in explaining innovation performance among U.K. manufacturing firms. Strateg. Manag. J. **27**(2), 131–150 (2006)
17. Lida, D.: First Stop in the New World: Mexico City, The Capital of the 21st Century. Riverhead Books, New York (2009)
18. Lundberg, H., Sutherland, I., Blazek, P., Penzenstadler, B., Habicht, H.: The emergence of creativity, innovation and leadership in micro-level social interactions and how to research it. Int. J. Ind. Eng. Manag. **5**(4), 221–232 (2014)
19. Meindl, J.: The romance of leadership as a follower-centric theory: a social constructionist perspective. Leadership Quart. **6**, 329–341 (1995)
20. Morrison, P.D., Roberts, J.H., Von Hippel, E.: Determinants of user innovation and innovation sharing in a local market. Manag. Sci. **46**(12), 1513 (2000)
21. Neyer, A.-K., Bullinger, A.C., Moeslein, K.M.: Integrating inside and outside innovators: a sociotechnical systems perspective. R&D Manag. **39**(4), 410–419 (2009)
22. OECD: Factbook 2013. Economic, Environmental and Social Statistics. OECD, Paris (2013)
23. Ormont, L.R.: Drawing the Isolate into the group flow. Group Anal. **37**(1), 65–76 (2004)
24. Pearce, C.L., Conger, J.A., Locke, E.A.: Shared leadership theory. Leadership Quart. **19**, 622–628 (2008)
25. Schubert, C., Sydow, J., Windeler, A.: The means of managing momentum: bridging technological paths and organisational fields. Res. Policy **42**(8), 1389–1405 (2013)
26. Tolman, E.C.: Cognitive maps in rats and men. Psychol. Rev. **55**(4), 189–208 (1948)
27. Uhl-Bien, M.: Relational leadership theory: exploring the social processes of leadership and organizing. Leadership Quart. **17**, 654–676 (2006)

28. Von Hippel, E.: "Sticky Information" and the locus of problem solving: implications for innovation. Manag. Sci. **40**(4), 429–439 (1994)
29. Von Hippel, E., Von Krogh, G.: Free revealing and the private-collective model for innovation incentives. R&D Manag. **36**(3), 295–306 (2006)
30. Wieck, K.E.: The Social Psychology of Organizing. Addison-Wesley, Boston (1979)
31. Wood, M., Ladkin, D.: The event's the thing: brief encounters with the leaderful moment. In: James, K.T., Collins, J. (eds.) Leadership Perspectives: Knowledge into Action, pp. 15–28. Palgrave-Macmillan, London (2008)
32. Zahra, S.A., George, G.: Absorptive capacity: a review, reconceptualization, and extension. Acad. Manag. Rev. **27**(2), 185–203 (2002)

Chapter 24
Equity Crowdfunding and the Online Investors' Risk Perception: A Co-created List of Web Design Guidelines for Optimizing the User Experience

Sandrine Prom Tep, Sylvain Sénécal, François Courtemanche, and Valeric Gohier

24.1 Introduction

Equity crowdfunding is a new venue for businesses in search of investment money at low cost in order to secure their foundation. It consists of funding through dedicated online marketplaces, connecting small investors and startups with business projects or recently launched existing products. It implies investment for partial ownership which can translate in a certain percentage of shares in a commercial enterprise, which value depends on their business success.

The original version of this chapter was revised. The correction to this chapter is available at https://doi.org/10.1007/978-3-319-29058-4_39

S. Prom Tep (✉)
Department of Marketing, ESG, University of Quebec in Montréal, Montreal, QC, Canada
e-mail: promtep.sandrine@uqam.ca

S. Sénécal
RBC Financial Group Chair of E-commerce, HEC Montréal, Montreal, QC, Canada
e-mail: sylvain.senecal@hec.ca

F. Courtemanche
Tech3lab, HEC Montreal, Montréal, QC, Canada
e-mail: francois.courtemanche@hec.ca

V. Gohier
Imarklab, Montréal, QC, Canada
e-mail: valerie.gohier@imarklab.com

J. Bellemare et al. (eds.), *Managing Complexity*, Springer Proceedings in Business and Economics, DOI 10.1007/978-3-319-29058-4_24

24.1.1 Context

Equity crowdfunding is related to the rise of social media and Web platforms. Hundreds of these online marketplaces exist in the United States only, and the Quebec business community is following the trend [1]. As a new type of financial investment for business equity, it is subject to securities and financial regulations [2], with the need for the *AMF* Securities Regulators to legislate the practice in the Province of Quebec.

In that context, we investigate how risk is presented to investors on such transactional Web platforms, and to what extent the risk involved in investment crowdfunding is perceived by the investors throughout the process. As the screen content and interface elements communicate the investment procedure, the users' perception of the information related to risk was assessed.

In order to prevent fraud and inefficiency with such form of crowd investment, investors' remarks were gathered and reformulated as usability guidelines to apply to the design of equity crowdfunding platforms, and more specifically regarding the informative elements communicating risk and the sequence of steps to proceed with the investment.

24.1.2 Objectives

In this study, the general goal is to test equity crowdfunding online practices with potential investors in order to optimize the user experience, knowing that content and interface elements convey information for the users to process [3].

In order to assess user perception of the risk involved with this particular type of investment, the study focused on the risk labels and descriptions presented throughout the online investment process. We observed participants' eye movements and how long they fixated areas of interest to assess if they paid attention to the risk warnings displayed. We also used facial emotion coding to assess the emotional valence while gazing at them. Finally, we triangulated the eye movement and emotion data with self-reported evaluations of the risk perception.

To assess the overall online experience, we also asked questions about the ease of use and satisfaction at each step of the investment process, such as the:

- Business/Project presentation
- Investment information
- Investment risk presentation (screen capture featured in Appendix, Fig. 24.2)
- Investment risk acknowledgement (screen capture featured in Appendix, Fig. 24.3)
- Check out
- *AMF* Securities Regulation role

Investigating the objective and subjective usability of screens content and interface elements which communicate the specific risk-related information processed by the users is useful in order to determine how the risk involved is well communicated via design [4]. Within the restricted scope of this short chapter, we mainly focused on reporting the data related to the *Investment risk presentation* and *Investment risk acknowledgement* pages.

24.1.3 Co-creation and Participatory Design

Today, when consumers customize online services while using them [5], they participate in creating their own online experiences, co-creating it along with the brand [6]. Involving users in co-creating the design of their online experience can also be done by incorporating their evaluative feedback into design recommendations. This is called participatory design: "The dogma of participatory design is the direct involvement of people in the shaping of future artifacts. Thus central for designers within this field are the staging of a design process involving participation of people." [7].

In this study, we conducted usability testing of an equity crowdfunding platform. Through the process, we collected objective (direct observation) and subjective (reported) measures of how usable the platform is from a user perspective. Drawing from the evaluative feedback from the users and the measuring of effective user experience, we formulated design recommendations based on both. As a result, we obtained a co-created list of design guidelines for equity crowdfunding website managers wishing to optimize the user experience offered through their online services.

24.2 Methodology

24.2.1 Procedure

Usability testing was conducted during the fall of 2014 at HEC Montreal's Tech3lab. Each interview session lasted approximately one hour and participants received $85 CAD as a compensation. They were asked to make an investment on a prototype website mirroring the equity crowdfunding platform Seedrs.com, launched in Great-Britain in 2012 and with a total of twelve (12) already financed projects and twenty (20) more awaiting for crowd investment (see Fig. 24.1).

After the completion of the goal-oriented task, participants were asked questions in order to assess the online experience and risk perception with the process altogether.

24.2.2 Recruitment and Sample Description

Thirty (30) bilingual adult participants using Internet at least 15+ hours per week were recruited from the Montreal area in Quebec and indicating that they would be "interested" (70 %) and "very interested" (30 %) to invest—no more than 2500$—online to help starting businesses. Eyetracking-related restrictions were applied to

Fig. 24.1 Seedrs.com homepage

the recruiting process as well. A balanced sample across gender (16 women/14 men), age group (31 % [18–34], 36 % [35–49], 33 % [50+]), education (37 % [University degree]), and revenue criteria (40 % [<$50 K], 30 % [$50–100 K], 30 % [$100 K+]) was constituted for the study.

24.2.3 Assessment Method

Eyetracking measures were used to track attention paid to risk labels and descriptions during the navigation task which they had to complete without verbalization. Face reading was used to assess emotional states associated with reading the risk labels and descriptions. Questionnaires were used to evaluate the online experience (subjective usability and satisfaction) and risk perception. The triangulation of the data collected through the combined methods greatly enriched the analytical process [8].

24.2.4 Testing Environment

The Tobii X60 eye tracker was used for measuring gaze fixations with cumulative total time of the fixations on predetermined Areas of Interest (AOIs) in each screen of the online investment process (i.e., risk labels and descriptions in the risk warnings presentation and the risk acknowledgement pages). It is presented in Fig. 24.2 for risk warnings (gaze opacity shows what is mostly seen) and in Fig. 24.3 for risk acknowledgment (heatmaps reveal what is mostly looked at by the participants) (aggregated data for all participants in both cases). Noldus Facereader assessed the emotional valence while fixating AOIs. Audio and video direct observations of the website use along with think-aloud protocols were monitored and recorded via the Techsmith Morae solution. It allowed us to collect objective and subjective usability data while users were completing their investment task on the site.

24.3 Results

The Investment Risk Presentation page. On the *Investment risk presentation* page, which presented a list of seven (7) warnings, participants paid decreasing attention to risk warnings (i.e., more attention to the top ones). The average reading time starts at 3.62 s with the first warning up to 6.22 s for the fourth. On average, risk warning #5 is read 3.18 s, and 1.68 s is all that is left for reading warning #7.

Almost half of the participants (13/30) judged that the warnings were important and helpful in understanding the investment level of risk. The warning descriptions scored 5.93/7 for adequate risk warning and 6.4/7 for risk warning pertinence. A few participants (6/30) got scared by the warnings, judging the risks presented were too high and that important basic technical terms (i.e., dilution and diversifications) should be explained to help the investor's decision-making process via a pop-up definition or access to a glossary.

On average, a negative emotional valence was observed when participants were reading risk warnings. Facial emotion observations also showed that on average, risk warning #4 consisted of the highest negative emotional state while users were reading the warnings. Warning #4 was related to company shares and dilution, a specific legal terminology related to business ownership equity which participants were not familiar with and would have required precisions for them to be able to fully understand and evaluate the risk. These negative emotional valence results stress the fact that the participants were actually reading the warnings and were emotionally reacting to the risks while perceiving them, and knowing the human natural propensity to flee danger and avoid what is unknown or unfamiliar.

The Investment Risk Acknowledgement page. Before consenting, the participants read risk acknowledgements during 11.18 s on average (much longer than on the risk warnings presentations page), and almost half of them (13/30) clicked to open the pop-ups providing more detailed information regarding the risks presented, indicating that they needed to know more to move on with the investment process.

On average, risk perception was higher after reading the risk-related information on the *Investment Risk Presentation* and the *Investment Risk Acknowledgement* pages, rose up to 5.23/7 vs. 4.33/7 before reading. Thus, risks appeared clearer after reading them twice (i.e., Presentation and Acknowledgement), as their scores rose from 5.8/7 to 6.07/7 for evaluating how complete the warning descriptions were, and rose from 5.8 to 6.3 for readability. Again, a negative emotional valence was observed while reading the risk acknowledgements and the supplemental information pop-ups, which confirmed that risk warning is perceived. It is important though, to note that at this stage of the investment process, a few participants (5/30) spontaneously checked "No" to risk consenting questions such as "Can you afford to lose your total investment?", "Do you understand that your investment has not been approved in any way by the securities regulators or by anyone else?", and "Do you understand that you may not have the same legal protections as you would if your investment was made under a different regime under securities laws? If you want to know more, you need to seek professional legal advice?". By checking "No" to any of the acknowledgement questions, the users would automatically provoke a system abortion of their investment process task without any reversibility option (i.e., "undo"). As far as usability is concerned, it is also important to note that half the participants (14/30) reported that it was not clear to them at this stage that answering "No" to any of these questions would prevent them from moving on in the investment. Such observation results indicate poor objective usability. In order for a system to abide by the ergonomic criteria [9], it must offer user control, which calls for clear anticipated consequences of user input as well as reversible user actions.

24.4 Conclusion and Design Recommendations

The goal of this study was manifold. First, it aimed at understanding more thoroughly investors' perception of the risk level associated with equity crowdfunding. Second, it aimed at producing a co-created list of web usability guidelines for equity crowdfunding site managers in order to optimize the user experience offered through their online investment platforms. This set of design recommendations stems from a participatory design approach leveraging new (eyetracking and emotional state reading) and traditional (think-aloud protocol and questionnaires) website usability testing methods [10, 11]. We integrated users' objective and subjective usability evaluations in design recommendations based on both our research and practical expertise in online system evaluation and conception for consumers.

24.4.1 Assessment of Risk Perception as the First Objective of the Study

First, our study evaluated how design and interface elements influence risk perceptions. While the risk warning labels and descriptions were judged adequate and relevant by the participants, the study mainly identified two prohibitive usability issues linked to risk perception:

- A lack of information content detail regarding some specific business equity ownership terminology not sufficiently explained for the participants to understand the risks involved (e.g., dilution).
- As well as a redundant sequence of presenting both the risks warnings and the risks acknowledgement pages successively.

Hence, in order to better support the user experience of the consumer's investment process and the decision-making evaluation based on risk perception, technical term definitions should be clearly provided as information content, as well as made easily accessible and readable. Also, as risk presentation and risk acknowledgement are two immediate sequential steps in the investment process, the study showed that the level of risk perception is increased for the investors—along with their negatively valenced emotional state. In some cases, it eventually led to prevent potential investors from investing in business projects as they answered "No" to some of the risk acknowledgement questions not knowing this would abort the investment process without any option to make the action reversible. Hence, as far as usability is concerned, the successive risk warning screens suggest an effective fraud prevention sequence, but without a clear indication of the consequences associated with the deal-breaker questions and no reversibility options for the users, it could become overly conservative on the safe side of investors' protection, and counter-productive to the crowd investment marketplaces main purposes.

24.4.2 Usability Guidelines as a Second Objective of the Study

As a result of this study, we provide here a list of co-created design recommendations [12] for equity crowdfunding website managers. Aimed at fraud prevention and platform efficiency, these design recommendations are based on participants' direct observation, users' evaluations and suggestions, and our own applied research expertise in online system evaluation and design. Formulated as usability guidelines to wise future crowd investment Web platform development, these recommendations are presented along with the precise study result which allowed us to draw the conclusion:

- Present major risk warnings first and foremost (from the eyetracking fixations duration on AOIs in the risk warnings presentation page).
- Provide access to a glossary of terms or mouse-over definition for complex technical financial terms such as "equity" (from the eyetracking fixations duration on AOIs in the risk warnings presentation page and the think-aloud protocol).
- Avoid the immediate successive sequence of the risk warnings presentation page and the risk acknowledgement page as it increases investors' level of risk perception (from the emotional state reading based on measures of emotional valence on both pages and direct observation of investors' behavior on the risk acknowledgement page).

- Warn users that risk acknowledgement questions are process preventive depending on the answers and provide some explanation as to why the investment process is aborted (from direct observation of task completion while navigating the risk acknowledgement page).
- Reversibility options should be allowed at the *Investment Risk Acknowledgement* page for investors who wish to change their mind after realizing that the investment is prohibited without acknowledging some risk (from direct observation of task completion while navigating the risk acknowledgement page).

Acknowledgments The authors wish to thank the CEFRIO, Center for the use and adoption of digital technology, which initiated the project with the support of the AMF Securities Regulators, The Montreal industrial cluster Grappe Finance and the Quebec retirement savings fund Fondaction.

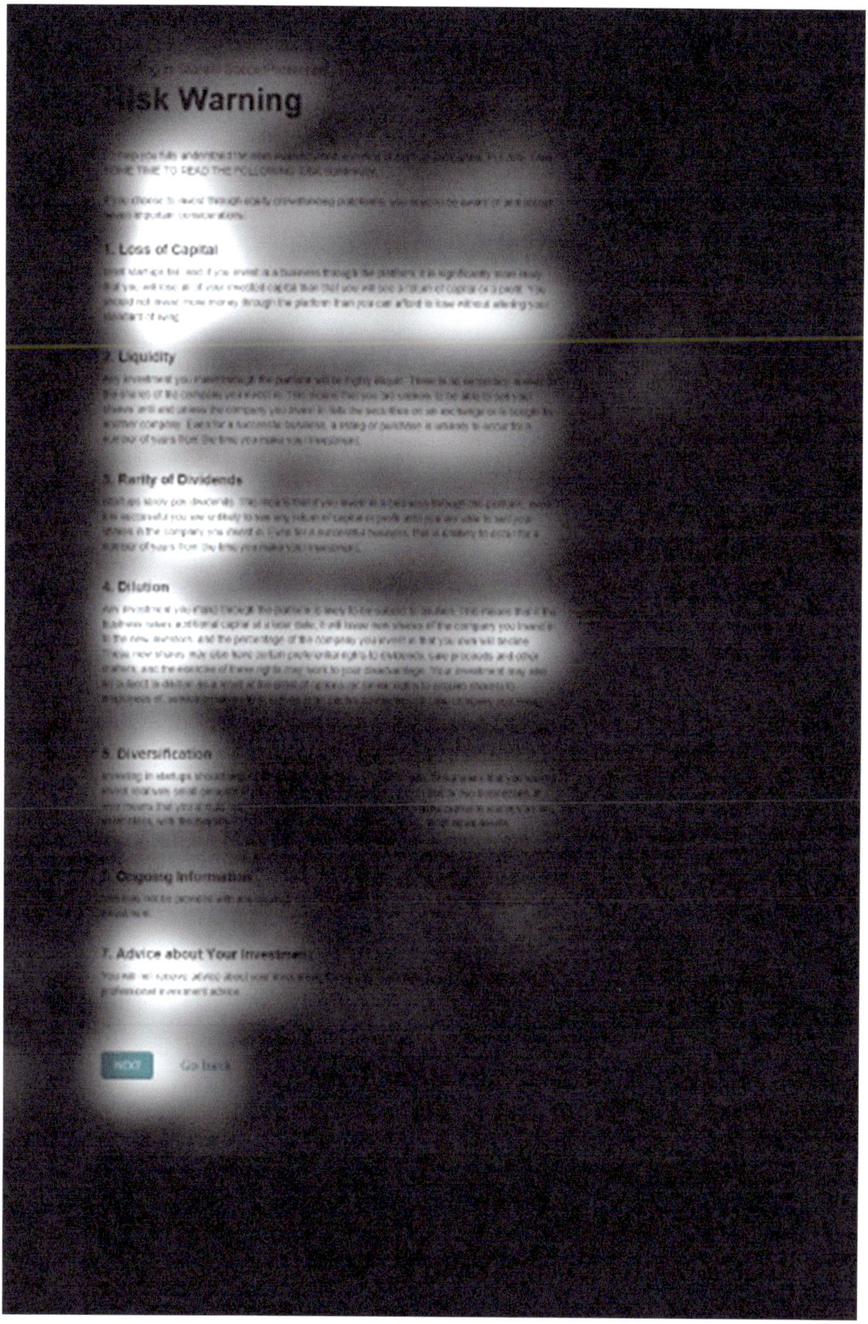

Fig. 24.2 Gaze opacity for the Investment Risk Warning page

Fig. 24.3 Heatmap for the Investment Risk Acknowledgement page

References

1. CEFRIO. Financement participatif en capital: l'émergence d'une nouvelle pratique Web au Québec. Avril 2015. 30p. (2015)
2. Securities and Exchange Commission. Proposed rules for Regulation Crowdfunding. http://www.sec.gov/rules/proposed/2013/33-9470.pdf. (2014)
3. Nielsen, J.: Designing Web Usability: The Practice of Simplicity, p. 419. New Riders, Thousand Oaks (1999)
4. Norman, D.A., Draper, S.W. (eds.): User-Centered System Design: New Perspectives on Human-Computer Interaction. Lawrence Erlbaum, Hillsdale (1986)
5. Wan, H.: Personalized Ubiquitous commerce: an application perspective. dans John Karat Clare-Marie Karat, Jan O. Blom (dir.), Designing Personalized User Experiences in Ecommerce, 360 p. Kluwer, Norwell (2004)
6. Prahalad, C.K., Ramaswamy, V.: Co-creation experiences: the next practice in value creation. J. Interact. Mark. **18**(3), 5–14 (2004)
7. Brandt, E.: Designing exploratory design games: a framework for participation in participatory design? PDC'06 Proceedings of the Nineth Conference on Participatory Design: Expanding Boundaries in Design, vol. 1, pp. 57–66. ACM, New York (2006)
8. Dimoka, A., Pavlou, P.A., Davis, F.D.: NeuroIS: the potential of cognitive neuroscience for information systems research. Inform. Syst. Res. **22**(4), 687–702 (2011)
9. Bastien, C., Scapin, D.: Critères ergonomiques pour l'évaluation d'interfaces utilisateurs. INRIA, Le Chesnay (1993)
10. Nielsen, J., Pernice, K.: Eyetracking Web Usability, p. 456. New Riders, Thousand Oaks (2009)
11. Nielsen, J.: Usability Engineering, p. 362. Morgan Kaufmann, San Francisco (1993)
12. Bertini, P., Plumley, E.: Co-creation: designing with the user, for the user. http://www.uxbooth.com/articles/co-creation-designing-with-the-user-for-the-user. (2014)

Chapter 25
Lean Customisation and Co-creation: Supplying Value in Everyday Life

Alexander Tsigkas and Antonia Natsika

25.1 Introduction

Mass production is dead; long live Mass Customisation (MC). MC as a term has been used in a multitude of different ways in the last 50 years since its creation [1]. The dominant meaning, which has prevailed and survived throughout the years, is that of the production of individualised and personalised products at close to mass production efficiency. When the term was born, mass production was the leading paradigm and the economic benefits from economies of scale achieved through mass production were not only understood, but also heavily pursued. The idea of producing individual products at close to mass production efficiency belonged to avant-garde thinking. Efficiency, once connected to pure productivity, in recent years is a term that has been extended to encompass sustainability. Producing more individualised products with less resources and waste can only be thought of in connection with some sort of interaction with a user or customer [2–5]. Mass Customisation is the new social system of production in the post-industrial era [6] in which authenticity and personal freedom in all aspects of life are becoming more significant than in the industrial era. With economies of scale losing its importance in society (though this is not yet widely accepted), *mass* is getting into a *red* square. With an anti-waste movement growing everywhere, and environmental pressures (as a consequence of mass production thinking) gaining momentum, *mass* becomes synonymous with *waste*. Mass is connected symbolically, to abundance, to more than is needed, to an attitude that calls for anonymity, to commodification, to something not of high value and much more. Today, *mass* is out of context with the different, the other, of the special that may arrive. Mass is a deficient mode of the

A. Tsigkas (✉) • A. Natsika
Democritus University of Thrace, School of Engineering, Xanthi, Greece
e-mail: atsigkas@pme.duth.gr; antonats@pme.duth.gr

J. Bellemare et al. (eds.), *Managing Complexity*, Springer Proceedings in Business and Economics, DOI 10.1007/978-3-319-29058-4_25

tolerance and openness of the human attitude in multicultural societies. Authenticity is the prevailing attitude that is expressed in *just doing* and in so doing one increases awareness of one's own individuality. *Just doing* should be understood in the sense of making or acting, however, not for the sake of *just acting. Just doing* is more in terms of being authentic. It is in opposition to following the masses, or flowing with the stream, and it concerns mainly the individual human yearning for freedom by directing his/her life as it appears to oneself. When one *just does*, then he/she is more aware of himself/herself and if that happens as-is then an individual does not fall prey to the fragmentation forced by modern technology, especially information technology. As the individual gains space in the *topos* of authentic life, mass is displaced. Although the term lean was devised to describe the way Toyota produced cars, unlike mass production [7], Lean fits very well the purpose of what is *just* described. The essence of Lean is being *undividable*; it has value on its own. Individual thinking means lean thinking; it is not mass thinking. Customisation, on the other hand, is a difficult and confusing term, in many ways. Even though it addresses the technical process of a product or service to be adapted and changed according to customer or user demand, its essence is not technical. Customisation is a human *attitude* towards the way it encounters the world and its association with this world. Lean customisation, therefore, means the attitude towards the undividable, the authentic that is not a result of *mimesis*. In production, it leads to individual products designed and produced for specific needs based on user needs and desires. However, research so far has mainly concentrated on the technical level of customisation which deals with the individuation of the mere thing. A lot of creative thinking has been invested in answering how to design individual products for efficient manufacturing. The product is seen as an object, isolated from its environment of use, isolated from the way its user is connected to it in everyday life (Being-in-the-world). Heidegger calls this environment a totality-of-involvements (or references), which is how the human being is related and connected to the stuff that surrounds him/her. In this everyday setting, the human attitude towards things is what Heidegger calls *ready-to-hand*. The human uses the stuff in-order-to do what s/he intends to do without being aware of the individual characteristics and features of the stuff. This recognition is of great importance in the customisation discourse among scholars and practitioners. What is, however, worth thinking about is how to increase human awareness when encountering the World, allowing for more *authenticity* (*individual*) and less *mimesis* (*mass*). On that level, co-creation is of vital importance. Co-creation moves away from the individual product or service as a mere thing and is drawn into the totality-of-involvements of human activity. This type of co-creation has to be distinguished from the usual term *the customer journey* that conventional Design Thinking pushes with great zeal. This type of approach looks at designing customer lived experience that s/he may enjoy through a specific service. Design Thinking hides (in an implicit way) a manipulative desire to direct customer experience towards enjoyment as a kind of *pleasure*. Human experience though, especially the one in which Heidegger is concerned, is the initial experience, not some sort of superficial lived experience of visually recognisable nature that somebody else may design. Exactly the opposite is the case. One needs to understand the essence of how humans encounter the world in order to be able to advise users or individuals of how to become aware of their individuality and how this can be expressed in any kind of activity in which they may get involved. For that reason, a method called De-sign Thinking [8], based on Heidegger's notion of ready-to-hand attitude and hermeneutic phenomenology, has been developed. The hyphen in *De-sign* signals the need to deconstruct the ontological meaning of Design. De-sign thinking means *deconstructing signification* of the human experience. Along the totality-of-involvements, a number of products and services are thought to be produced from a number of suppliers operating in a supply chain. Such supply chains are necessary for serving totalities of products and services, and not delivering mere products as in times of mass production. Lean customised supply chains consist of many different and diverse companies offering specialised products as members of a supply *team* which needs to be designed, organised and managed as a single system. The term Supply Chain System (SCS) is used for this purpose. Another appropriate term to use instead of SCS is the term Extended Enterprise (EE). SCS needs to be flexible, agile and resilient (FAR), terms that need to be clarified and explained in the context of lean customisation.

Taking into account the scholarly and practitioner research on MC as a totality, three different levels of lean customisation have become distinguishable. The first level addresses the mere thing as an object in isolation from its environment. It is argued that the majority of work concerning MC basically addresses this level of customisation. This fact is reflected in the availability of a huge number of research papers. Historically, research starts with the configuration of the design problem [9, 10] and later, the research is directed towards the design and construction of extremely flexible machinery for building individual products. The objective at this level is product design and production for MC. The second level of customisation includes co-creation at the level of the use of a product along a totality-of-involvements. At this level, there is no recognisable research work identifiable as such so far, at least not as a systematic discipline. It is suggested that research is needed on this level which requires more than just engineering skills and technology. An interdisciplinary attempt is needed to combine phenomenology, especially Heideggerian hermeneutics, sociology, design expertise, as well as digital technology and other engineering disciplines. The third level of customisation concerns SCS that will be designed, organised and managed around meaningful structures for authenticity and not mimesis. These SCS must be simultaneously flexible, agile and resilient. Surrounding sustainability can also be added next to resilience of SCS to surrounding disruptions. As lean customisation becomes recognisable within society as a movement of human authenticity, issues of ethics are raised that need to be taken into account. Openness and the readiness for the arrival of the different from whatever is considered established, call for cultural identities in the sense of personalisation. Quality, etymologically, means what characterises a person in its idiosyncrasy (*qua*, who). Personalisation, therefore, refers to ethics and not to some sort of, for example, kitsch representations of oneself, made recently possible through 3D-printers generally treating a person as some sort of object. Lévinas and Derrida philosophies will help to address these issues on a first level.

This chapter sets the context for promoting scholarly as well as practitioner research to address all three levels of lean customisation as an integral whole and not each level individually. It is argued that the three levels should not be seen separately since all three are interdependent. Interdependent means that a mere integration of separate parts is not possible. It is not a matter of defining and resolving interfaces. In the next sections, the second and third levels of customisation are discussed. The first level of customisation has been more than adequately addressed by scholars and therefore is not addressed in this chapter. In Sect. 25.2, important terms such as flexibility, agility, resilience and ethics are clarified using as an example the game of soccer seen from a different perspective. In Sect. 25.3, the second level of customisation is addressed. The discourse is continued at the level of how value is disclosed along the totality-of-involvements. It is presented how De-sign thinking can be used to investigate how a useful thing (Zeug) is encountered by humans along the use chain and what this use means for both the owner and the thing itself as a unity. In Sect. 25.4, a third level of customisation is introduced to address lean customisation in SCS as meaningful structures. At this level, SC are organised as SCS in an ad hoc fashion for delivering products or services for a certain period addressing diverse challenges in the Market. Such SCS must be flexible, agile, resilient as well as ethical. Tsigkas describes such SC as Value Adding Communities (VAC) [11]. In Sect. 25.5, the three levels of customisation are presented as different aspects within a unified, integrally connected whole; that of lean customisation. One cannot exist without the other, in the sense that none can reflect and represent lean customisation individually. The conclusion summarises the findings of the analysis elaborated in this chapter.

25.2 What Supply Chains Can Learn from the Game of Soccer

The essence of the game of soccer is *arriving-home*. It is to dwell after an *Odyssey*. This happens when the ball reaches and finally rests in the net of a goal. The way to home is blocked by dynamically moving obstacles, incorporated and identified as members of the co-team called the opponent team. However, in reality obstacles exist in order to signal that any kind of a human activity and endeavour is an Odyssey. In order to be able to dwell, each team has to *collectively* think and construct a plan of how they ought to do it. Thinking, building and dwelling [12] as a unity discloses the attitude of the team towards the intrinsic desire of reaching home in their own way, thus developing a *collective individuality*. What the outcome will be is unpredictable, as *collective individualities of the involved teams try their own authentic way*. What, how and when something will happen is not a definite actual event that has some sort of measurable probability or frequency of appearance.

Soccer is like a text, written at a particular moment, as collective individualities strive to express their authenticity. It is open at any time to the unknown, to the new, to the other. This *other* is not something known a priori, nor can it be foreseen, planned or calculated, or even simulated. As such, it cannot be considered as part of a risk management issue, since risk concerns a measurable or estimated probability that something known will happen again under certain conditions, thus limiting the scope of risk. A posteriori is always possible to look and analyse the post game video, but even then the *other*, although accepted and welcomed (depending on the situation), cannot be discovered. The disclosure of the *other* would mean that soccer is something that can be a priori planned and performed as planned or, at least, a posteriori recognisable activity as a number of cause and effect relationships that led to the happening of this one and not another one. If this were to be the case, then the so-called better team would only lose because their players made so many mistakes on that day, that due to that fact (cause) they lost the game (effect), while the other team did everything right. If all these do not work, then luck is called upon to fill in the gaps in a logical type of reasoning. Luck is always a complement to reason. However, the history of soccer proves that this is not true. Soccer is not a process and cannot be analysed as such. It is not uncertainty that is involved here that can be calculated in some kind of probability factors. There are plans certainly, there are also interventions from the manager (the coach) during the game in order to influence the state and the development of the game, but all these activities are connected to the attitude of calculative thought that wishes to calculate before something happens in order to control. Soccer is not a process. It does have rules, but these rules address what is not allowed during the game; hence, the need for a referee. But, even the referee is a kind of *obstacle* that needs to be considered as such by both teams. Soccer is not a process. It is similar to a text that is written as it happens and before it happened it did not exist. After the (temporal) completion of the game, the reality ends there in its mystery. Videos that replay the game are only media for sustaining a sort of historic memory of the game and nothing else. How the *other* happens and happened stays closed forever. This fact is even more fascinating because soccer, as well as supplying, is an ongoing activity that steadily and consistently looks and strives for letting the *other* happen. It happens through the capacity of staying open to its arrival and steadily *investing* in the ethical aspect of the activity as staying open as an attitude. Improving and pursuing ethics will improve their capability to become continuously more aware of themselves, becoming more authentic and constantly rediscovering their collective individuality. The *other* comes when consilience prevails in the system, not resilience. Consilience means that the *other* is bound to come when *everything* is ready for it. When the *other* comes, the text disappears and then it starts all over again. There is no written evidence after that of how they got there. There is only speculation and wishful thinking. This is the only kind of text that is constructed while being written and stays there until the *other* comes. For some, it would be enough to say that the objective is the appearance of the *other*. In the language of soccer, we say *only the result counts*. This expression may sound arrogant to many people, but the use of the expression is *because we do not know how it happened exactly*. So, based on the above analysis, *collective individuality* belongs to the second level of lean customisation. This is the reason why soccer is an excellent example for opposing discrimination and racism. Soccer is *de profundis* ethical, as it is open to the event

[13] of the arrival of the new, the different, the other [14]. At the end of the game, the *winner* should not be viewed as a winner in an antagonistic manner, but a winner in a con-petition by showing a higher degree of ethical capacity and capability. The ontological essence of a soccer game is ethics.

Let us look now at the three technical conditions that support the arrival of the *other*. These three conditions are resilience, flexibility and agility. Resilience of the team as a system is how *efficiently*, in terms of limiting impact, the team copes with possible adversity, e.g. a sudden change in the other team's tactics, a red card, a serious injury to a player that has to be substituted, etc. or even the entrance of a new player. The capability of the system to *bounce back* to a different level of operation is called *resilience*. Resilience is similar to the behaviour of self-preservation of a living organism. Human systems exhibit resilience as a mode of care. They enter a mode of operation that could be named *safe mode* in order to continue functioning, while limiting the impact. Resilience may also be positive, not only negative. For example, a goal (dwell) achieved while playing with less players than the co-team has a psychologically positive impact that the team needs to deal with. Obviously, it is not possible for a system to be resilient to everything. Therefore, when dealing with resilience, one must begin by clearly defining resilience in terms of what is to be resilient to what. Carpenter claims that these aspects change, depending on the temporal, social and spatial scale at which it deals with resilience [15]. Thus, it is important to define upfront what should be resilient to what. To cope with the conditions of the game, a team also needs two further abilities; *flexibility* and *agility*. Flexibility refers to the ability of the team to be responsive to varying degrees to their effort demanded by the situation. Agility is the ability of the team to change in a rapid fashion their way of playing in order to *go around* and avoid or neutralise obstacles. Flexibility, agility and resilience (FAR) are the three modes of behaviour that a team adopts in order to take care of itself. It is exactly the FAR abilities that make soccer such a difficult, but fascinating game to play. However, FAR abilities are always connected to an overall organisational scheme or structure, which in the language of soccer is called a system (e.g. 4-4-2, 4-3-3). In this different perspective of soccer, there is no attack or defence. There is no competition; there is only con-petition in ethical terms.

25.3 Second Level of Lean Customisation: Value in Everyday Life

25.3.1 Disclosing Value as a Totality of Use

Marketing denotes agorevin *(αγορεύειν)*, which in Greek means *to announce in the market*. Marketing announces something in the market through disclosure. The essence of marketing is value. What marketing therefore discloses is value. Without marketing, there is no value and inversely without value there is no marketing. Through coming to see how value in the market is *possible*, one arrives at the so-called *value added* in terms of value for money awareness for the customer. However, this very user awareness is not based on optical characteristics of the thing which makes up value, nor is it awareness through cognition, nor a function and neither is it a lived-experience of the single thing. Users interpret experience in *seeking* significance in their *contact* with value. Contacting value means the way users encounter usable things in managing their everyday lives, be it at work, privately or socially. Customers interpret value when encountering useful things in their lifetime, which marketing sets on its way in the market (*εμπορεύειν*). This setting in the market signifies a positioning of value, and in so doing it challenges the market. Challenging is also a way of disclosure. Through challenging, something emerges which has a form within the market as a totality of relationships. When marketing challenges something towards disclosure until form is shown, we talk about *topos*; in this case a *marketing topos*. Topos is the essence of modern marketing, but in itself has nothing to do with marketing [8].

What is under investigation in this discourse is the way customers interpret value when encountering useful things in their daily lives. The investigation leads to the conclusion that added value of a single useful thing is a myth. In reality, useful things and, therefore, their related value are meaningful only within the context of a totality of involvements. When a useful thing escapes from this context, when it becomes a mere thing, value is meaningless. Value cannot be created or added or even destroyed; it can only be interpretatively understood within the context of a totality of involvements. Only in this context has value its significance.

The role of marketing is to create marketing topos as contexts of totalities of involvements of useful things encountered as such by customers. In this way value is disclosed. Going in and out of this topos is as if going in and out of the context of a totality of involvements. This cycle for interpreting value is called the value hermeneutic cycle (vhc). In order to investigate the issue, *Heideggerian hermeneutic Phenomenology* (HhP) is used, as outlined by Heidegger in his seminal book: Being and Time [16, 17].

The following discussion is devoted to the way humans encounter the world, and defends the argument that value is meaningful only within the context of a totality of involvement of a useful thing. Next, the value hermeneutic cycle is presented. Finally, application of the value hermeneutic cycle in interpreting value, and how marketing could benefit from it, is briefly discussed.

25.3.2 Encountering the World and the Meaning of Value

It is important to refer to three terms that make up the central features of HhP. One pertains to the meaning of *Dasein*, the next concerns the term *understanding* and the third refers to the term *totality of involvements*.

HhP involves an analysis of human beings not as epistemic agents, but as *Dasein*. Dasein is Being-in-the-world (*Sein-in-der-Welt*) and the name for *the human being or beings*. By understanding Dasein as Being-in-the-world, Heidegger explicated the question of being in terms of the practical orientation one exhibits towards the world and others. Heidegger's main interest was to elevate the issue of Being, that is, to grasp the human capacity to make sense of things. At the same time, Being-in-the-world is a collapse between Dasein and world. Everyone comes to understand oneself only in light of the everyday life context one finds itself already in. For example, a hammer is not comprehended from a detached perspective as just another epistemic object. Rather, the hammer is known from the contextual significance it possesses in a nexus of instrumental relationships in which it is used. Hence, phenomenology attempts to bring to light that which is concealed or taken for granted. Phenomenological description brings into explicit relief the hidden contexts and purposes that underscore practical interaction with the world. This point can only further be clarified under the term *understanding*.

Dasein is centrally characterised as *understanding* [16, 17, p. 147, 139], but this concept of understanding does not mean understanding as knowledge in the cognitive theoretical sense that epistemologists analyse and consider primordial to human experience. Instead, understanding is the implicit intelligibility that characterises human activities as meaningful and already familiar in practice. Understanding is meant as the capability of the human being to paint different pictures of oneself and the world, including different possible management plans and different possible opinions in relation to the one and the same fact. Understanding relates to views, concepts and plans. Through understanding, the human being gains access to oneself, the surrounding world and the various encounters in this world. Only on that basis does one have such an understanding, an ability to react at all to what is encountered—and in a typically human way and manner, namely, in different ways and manners. A stone, for instance, cannot understand and therefore cannot interpret an event in a different way. Through the activity of understanding, the world is disclosed as a totality of meanings. Understanding objects is understanding them not as objects with external properties.

By interpretation (Auslegung), Heidegger [16, 17, p. 148, 139] means the practically oriented capacity of understanding to bring into view the parts and wholes of an entire possibility and context. In other words, interpretation is the development of the understanding's projection upon what is inherently possible [18]. According to Heidegger, an interpretation is *the working out of possibilities projected in understanding* [16, 17, p. 148, 139]. Thus, one must already have a worked out understanding of possibilities prior to interpretation, since interpretation is grounded in understanding. Understanding is never generated out of interpretation. Instead, understanding is the pre-reflective, pre-linguistic and pre-cognitive practical orientation that makes it possible to interpret the world at all [18]. Humans understand aspects of the world already; humans understand *something as something* [16, 17, p. 149, 139]. When one is engaged in reading a book, one understands the book as something to be read. The book occurs in the *in-order-to* relationship, which constitute the whole world, and possible interpretations of it.

That which is disclosed in understanding—that which is understood—is already accessible in such a way that its *as which* can be made to stand out explicitly. The *as* comprises the structure of explicitness of something that is understood. It constitutes the interpretation. In other words, there is an implicit background to the world, a nexus of practical relationships behind understanding and interpreting the world, which Heidegger termed the *totality of involvements* (*Bewandtnisganzheit*) [17, p. 84]. The human being possesses already an intimate familiarity with many of these practical relationships. For Heidegger, humans are born into a world already under-way within its own historicity and, likewise, all interpretations are an accomplishment of projective understanding in that historicity and totality of involvements.

The totality of involvements is always understood not as a grasping of facts independently of that historicity and already understood contexts of significance. Instead, it is the totality of involvements what Heidegger called *ready-to-hand* (*Zuhanden*) [17, p. 69]. Humans do not apprehend properties about objects inside the interpretively loaded contexts they inhabit. Such an apprehension would exemplify what Heidegger called *present-at-hand* (*Vorhanden*).

According to Heidegger [17, p. 69], an object can be encountered in a twofold manner: in the mode as present-at-hand thing (Ding) and in the mode as ready-to-hand *useful thing* or *equipment* (*Zeug*). *Zeug* is translated in the literature either as *equipment* or *useful thing*. Both translations will be used in this chapter depending on the context. *Useful thing* or *equipment* is used alternatively for the word Zeug. Some translators of Heidegger use the term *useful thing* instead of *equipment* and certainly equipment is a useful thing. However, not all useful things are equipment, and a useful thing or equipment is never alone, it belongs, argues Heidegger [16, 17, p. 68, 64], to an *equipmental totality* or *totality of usable things* (Zeugganzheit). The various structures of a *useful thing* are encapsulated in the term ready-to-hand. Ready-to-hand is a state a useful thing possesses, in contrast to present-at-hand which is the state an entity possesses when regarded as being a *mere thing* [16, 17, p. 69,64]. However, *presence-at-hand* and *readiness-to-hand* are not intrinsic states possessed by the Being of an entity. For example, if it is decided to pick up a rock and start hammering with it, the rock is transformed from something *present-at-hand* into something *ready-to-hand*. The rock becomes a hammer simply by giving it the assignment of a hammer. At this moment and in that context, the stone is of value to the user and, most importantly, a very individual and intimate one. A broken hammer is certainly of no use to the manual worker, but this is not as conspicuous as the work that remains unfinished. Furthermore, a hammer placed in a workshop toolbox is of no value if there is no assignment for it. It stands there as *standing reserve* (*Bestand*) [19, p. 17], as an object, as a mere thing, as a *present-at-hand* entity waiting for an assignment. To insist on mass customising things as mere things is certainly not meaningful.

The above analysis of how humans encounter the world provides an important insight for marketing. Marketing should be looking at the entire value chain, not as a chain of adding mere value to a mere product as an object or entity, but as a *value totality* that extends to the totality of involvements of the useful thing (Zeug) in everyday life. In the next section, the meaning of value within the value hermeneutic cycle as opposed to *value added* is described.

25.3.3 The Value Hermeneutic Cycle: A Meaningful Cycle for Co-creation

The value hermeneutic cycle (Fig. 25.1) consists of recursive turns between ready-to-hand and unready-to-hand mode of a useful thing within the context of a totality of involvements. Every useful thing, every product, sooner or later, gets released by the user into a totality of involvements. The totality of involvements is, in fact, a hermeneutic understanding and, as Heidegger puts it, "in interpreting we do not throw a signification over some naked thing which is present-at-hand, we do not stick a value on it…" [16, 17, p. 190, 150]. In other words, value is presupposed before the useful thing enters the totality of involvements since value is interpreted as ready-to-hand and not as something added to something present-at-hand, lying there. Value is not a separate feature of the product that becomes experienced in an awareness mode of its use in a form of addition or as sticking a value on it as Heidegger argues. Value added is therefore *hermeneutically* a myth.

Interpretation cannot escape from the contextual significance of the totality of involvements. Instead, this hermeneutic threshold holds for value. As Heidegger puts it: an interpretation is never a presupposition-less apprehending [16, 17, p. 191, 150], a presupposition that is being decided by the user. Crucially, it does not follow from this analysis that user behaviour in such contexts is automatic, and that there is no awareness present at all, but rather that the awareness present is not of a subject–object nature. This type of awareness Heidegger calls *circumspection* (Umsicht) [16, 17, p. 69, 65]. Heidegger's notion on *circumspection* as a type of awareness is extremely important for the process of co-creation. In most co-creation settings for customised products, it is implicit the attitude that everything depends on the relation between the subject (user) and the object (product) as a subject–object nature. This attitude allows the creation process to bring to the foreground an almost dogmatic insistence on product details, treating the product as a mere thing, that later when the product makes its way into everyday life as part of a totality of involvements, these details disappear in practice. One of the arguments, for example,

Fig. 25.1 The value hermeneutic cycle

against open innovation in the form of a co-creation process, is the long temporal durations that the process has before arriving at the end of the co-creation process with a final proposition. The problem lies mainly in the (false) assumption that awareness is of a subject–object nature and not of the open innovation [20] nature of co-creation process per se. Often, co-creation processes run into intrinsically antagonistic (instead of co-agonistic) settings where time is expended without even being noticed. Although openness and co-operation is wished as a place ready for the arrival of the new, of *other*, collective individuality is not an issue, because the concentration on the subject–object relation directs the creation process towards technical customisation, i.e. the first level of customisation.

This brief analysis suggests that co-creation settings can essentially be structured using HhP for disclosing value. Specifically, what is needed is for the co-creation settings and organisation to embrace the phenomenological modes of encountering the totality of uses, and build a value hermeneutic circle fitting its targets for innovation. This can be open innovation or targeted open innovation, or any kind of strategy for lean customisation.

25.4 Third Level of Lean Customisation: Supply Chains in Everyday Life

Below, we shall deal with the overall organisational structure for SCS following the hermeneutic analysis of the game of soccer. SCS are similar to soccer teams as systems (scs). One supplier is considered as corresponding to one player in a soccer team in the discourse that follows. Therefore, an SCS is a conjoint of companies forming a SC based on common interests and capabilities or attitudes. Supplying (ἐφοδιάζειν) is a very difficult, but also very fascinating *game* to play. Supplying means to bring something *home*. The essence of supplying is the same as that of the game of soccer. It is all about ethics. Here, the *game* is played in parallel, worldwide with many teams at the same time. The con-petition is not directly visible as in soccer because it spans almost the universe in its totality. Supplying could be described in terms of at least four basic activities as defined by the Supply Chain Operational Reference (SCOR): Plan, Source, Make and Deliver. Supplying as a whole expresses the happening of all these activities in parallel temporally as well as spatially. These four activities are equally basic and mutually interdependent. They pick out different aspects within a unified, integrally connected whole; that of Supplying, and one cannot exist without the other. This is a non-hierarchical relationship. Neither term is more basic than the other. Therefore, individuality for a specific supplier company is the capability of supplying, and acting as a unified, integrally connected whole. Furthermore, *collective individuality* is the capacity and capability of a number of suppliers to form and operate a unified, integrally connected whole and exhibit this wholeness at the level of an SCS. There is certainly nothing new in this statement. The desire of Management is to achieve the best collaboration among the SCS members. However, especially in the area of Supply Chain Management across multiple partners in the chain, daily practice proves that the best collaboration is wishful thinking. Power games, as well as rival business interests, make the attainment of collective individuality unrealistic. Unless collective individuality is achieved, Supply Chain Management is empty as well as blind. Emptiness and blindness are deficiency modes of care. Both do not serve the purpose of a supply chain for serving the customer. Empty means in that respect an SCS without common goals and blindness means without an *eye* for overlooking and managing the wholeness of the SCS. Just as in soccer teams there are rivalries between the team players, it is the duty of the coach to smooth them out and finally eliminate them. On the other hand, the team manager has a complex role of not only choosing the right mix of FAR, but also to support the team in any way he/she can. However, at the moment, SCS do not have such an authority, and it is not even a matter of discussion. Although this is claimed as necessary, it will be rather difficult to achieve since SCS member companies in reality do not work as systems, and the closest to what maybe achievable is a loose chain. An SCS often consists of companies which are in themselves independent profit-oriented companies, and they do not think of themselves as members of a larger *system* when they pursue their individual business interests. While in SCS there is no *opponent* (theoretically) and all the members of the supply chain have a common interest (to serve the final customer), in fact there is, and, in many cases, strong competition because of the supplier (customer)—customer (supplier) duality. Under the condition that this duality is broken throughout the Supply Chain, then the chain has a chance to become a unified integral whole SCS [21]. Once the duality is broken and company relationships in terms of conflict improve, cultural differences and strategies become a major issue in need of special attention from the point of view of managing SCS. Managing SCS means a twofold strategy (similar to scs). The first one addresses the need for an appropriate FAR mix and strategy, depending on the individual situation of a supplying activity (supplying never stops and in that sense it never ends) and the second is strengthening the collective individuality and therefore the ethics of the SCS regarding openness for the arrival of the *other*.

Many may object to the argument that SCS have something in common with the game of soccer for two reasons. The first concerns the non-process character of soccer. They will counter that an SCS is based on activities structured around processes. The second argument concerns the notion of the impossibility of supplying the unknown [13], the different, the *other* [14]. They will counter that what is going to be supplied is an agreed product which is concrete, discrete and has a time tag, as well as a value attached to it. These arguments are both, in reality, superficial or pseudo arguments. In the game of soccer, what should be delivered is also concrete and discrete, and there is also a time limit within which this is allowed to happen. Because of the non-process nature of the game, this cannot be planned and executed as planned. Coming to SCS, the mere and beyond any doubt existence of a huge amount of scholarly and practitioner research papers is a strong indication, if not proof, that the *other* is not easy to touch and, therefore, to make it happen as planned, scheduled or even decided or imagined. The work involved in this research is in many cases epitomic, exhibiting a formidable complexity of mathematical modelling. This modelling includes artificial intelligence, fuzzy logic and, in fact, any logic. On the contrary, the presumption that supplying is a process drives scholars as well as practitioners to assume supplying as a process. Assuming supplying as a process opens the way to algorithmic thinking as to how to plan and manage this process. Here is where the problem lies. However, if supplying is not a process, then what is it? How can one go about designing, organising and managing meaningful SCS which supply value in everyday life? Let us go back to the game of soccer.

Soccer, as well as supplying, is an ongoing activity that steadily looks and strives for *letting the other happen*. It happens through the capacity of staying open to its arrival and steadily *investing* in this ethical aspect of the activity. Improving and pursuing ethics will improve their capacity and capability to become more aware of themselves, becoming more authentic and continuously rediscovering their collective individuality [22] (Ereignis). Continuous improvements gain at that level new insight. The *other* comes when consilience and not only resilience prevails in the system. Consilience is the attitude of many different human and other factors involved for an SCS to work independently from each other (although connected) without being aware of it, with no guidance or control to reach a common end that is reached when the other arrives. Ethics strengthen consilience. There is no command and control, no central place that guides and obliges what to do. There is the manager for the strategy and for corrective interventions, but at the execution level are the team players. Supplying is like texting. When the *other* comes, then the text disappears and starts all over again. There is no written evidence after that of how we got there. There is only speculation and wishful thinking. This is the only kind of text that is constructed while being written and it stays there until the other comes. For some, it would be enough to say that the objective is the appearance of the other. In soccer language, we say *only the result counts*. This expression may sound arrogant to many people, but the use of the expression is *because we do not know how it exactly happened*. Similar holds for supplying value in everyday life.

Following the argument presented so far, the introduction of a coaching team for managing and coaching SCS is proposed. The task of this coaching team is, on the one hand, to train and coach the SC individual members so that they develop into a team with their own *collective individuality* as well their own FAR capabilities. There is a need for these companies that will offer the service of coaching SCS to be born, or simply individual teams put together ad hoc for serving this purpose. We expect that, in this way, several *Schools of Thought* (SoT) will be established at how to best combine ethics with the more technical aspect represented by FAR for SCS. Of course, collective individuality and FAR management can also be applied within an individual company. The objective is to think about the options, build the appropriate capability of collective individuality, and finally guide SCS to dwell, i.e. how to reach the user with products and services.

With SCM, risk is intrinsically connected. Traditional risk management addresses calculable failures of any kind expressed in the probability that a known disruption may occur that will endanger the SC. However, the real meaning of risk is the one that is connected to the non-arrival of the *other*. This cannot be expressed in any kind of statistical probability, since it is not known a priori, and every activity on the SCS is an individual one. For example, cultural differences in SCS cannot be mathematically or otherwise modelled. Therefore, the key in the design, organisation and managing SCS is the improvement of consilience, by strengthening the ethics throughout the SCS for collective individuality further and beyond the material aspect of supplying. Only in this way can meaningful SCS exist. This is the third level of lean customisation.

25.5 The Totality of Lean Customisation and Co-creation

The three levels of customisation should be seen as a unity in their totality as well as equally basic and mutually interdependent. They pick out different aspects within a unified, integrally connected whole; that of lean customisation, and one cannot exist without the other. The one is not the condition for the other. This is a non-hierarchical relation that demands the simultaneous consideration of all three levels since the three levels do not constitute distinct independent parts ready for integration. Quite the opposite is the case. They belong to each other as in a deadlock situation. One of the high impact benefits of this approach (e.g. in time and investment involved especially in Open Innovation or Co-creations settings) is that technical extremity of customisation can be tamed and placed into context if seen within the perspective of the three levels of customisation with many other benefits as by-products. Table 25.1 shows a summary of the three levels of lean customisation.

Table 25.1 The three levels of lean customisation

Level of lean customisation	Collective individuality	Supplying
Level 3 Supplying of the *other* Originary time	Ethics and the arrival of the *other* — existential (for-the-sake-of) — facticity (having been in order to be) — falling (acting with what is occurring) Coaching and managing SCS	Meaningful supply chains in everyday life Risk for the non-arrival
Level 2 Use chain Datable time, that is events located in relation to others	Co-creation Totality of involvements Ready-to-hand and unready-to-hand Value seen in the use chain in everyday life	Initial thoughts on supply chain
Level 1 Technical flexibility Ordinary time	No collective individuality identifiable Co-creation as present-at-hand object User seen detached from the World	Not usually an issue

25.6 Conclusion

Collective individuality and consilience are the preconditions for supplying lean customised value in everyday life. This is the conclusion of the analysis presented in this chapter. The significance of lean customisation was disclosed under the premises of three levels of customisation: the technical, the customer value through co-creation and the supply of the value. Consequently, a new context of thinking is proposed in order to accommodate very diverse management challenges in which ethics are intrinsically incorporated by nature and not artificially in the design, organisation and management of SCS. Risk is meaningful in ethical terms, that is, the non-arrival of the other, and not within a technical calculable context.

References

1. Davis, S.: Future Perfect, 10th edn. Addison-Wesley, Harlow (1996)
2. Piller, F., Christoph, I., Vossen, A.: A typology of customer co-creation in the innovation process. In: Hanekop, H., Wittke, V. (eds.) New Forms of Collaborative Production and Innovation. Universitätsverlag Göttingen, Goettingen (2011)
3. Reichwald, R., Piller, F.: Interaktive Wertschöpfung. Wiesbaden: Betriebswirtschaftlicher Verlag Dr. Th. Gabler GWV Fachverlage Gmbh (2009)
4. Pine, J., Gilmore, J.: The Experience Economy. Harvard Business School Press, Boston (1999)
5. Tseng, M.M., Jiao, J.: Mass customization. In: Gavriel, S. (ed.) Handbook of Industrial Engineering, Technology and Operation Management. Wiley, New York (2001)
6. Tsigkas, A.C.: The factory in the post-industrial era variety instead of flexibility, mass customization: the production system of the future. In: Second Conference CE Conference on Mass Customization and Personalization, Rzeszow, Poland (2006)
7. Womack, J.P., Daniel, J.T., Daniel, R.: The Machine that Changed the World. Rawson, New York (1990)
8. Tsigkas, A.: A method for design thinking. Conference Proceedings of the 13th International Science-to-Business Marketing Conference on Cross Organizational Value Creation, Winterthur, Switzerland, pp. 376–386 (2014)
9. Mittal, S., Frayman, F.: Towards a generic model of configuration tasks. Proceedings of the 11th IJCAI, pp. 1395–1401. Morgan Kaufmann, San Mateo (1989)
10. Wielinga, B., Schreiber, G.: Configuration design problem solving. IEEE Intell. Syst. **12**, 49–56 (1997)
11. Tsigkas, A.C.: Mass customization through value adding communities. In: Third World Wide Conference on Mass Customization and Personalization, Hong Kong (2005)
12. Heidegger, M.: Poetry, Language, Thought, translated by Albert Hofstadter. Harper Colophon Books, New York (1971)
13. Derrida, J.: Une Certaine possibilite impossible de dire l' évènement (2003) by Editions de l' Harmattan. English translation by The University of Chicago (2007)
14. Levinas, E.: Time and the Other. Trans. Richard A. Duquesne University Press, Pittsburgh (1987)
15. Carpenter, S.R., Walker, B.H., Anderies, M.A., Abel, N.A.: From metaphor to measurement: resilience of what to what? Ecosystem **4**, 765–781 (2001)
16. Heidegger, M.: Sein und Zeit. Max Niemayer Verlag, Tübingen (1967)
17. Heidegger, M.: Being and Time. State University of New York, New York (1996)
18. Hackett, E.J.: Scheler, Heidegger, and the Hermeneutics of value. J. Appl. Hermeneutics (2013) ISSN: 1927-4416, (http://jah.journalhosting.ucalgary.ca/jah/index.php/jah/issue/view/14)

19. Heidegger, M.: Die Technik und die Kehre. Klett Cotta, Stuttgart (2007)
20. Chesbrough, H.W.: Open Innovation: The New Imperative for Creating and Profiting from Technology. Harvard Business School Press, Boston (2003)
21. Tsigkas, A.: A research method for the future: Heidegger's Phenomenology in the management of surrounding sustainable supply chains. Philos. Manag. J. under review (2015)
22. Heidegger, M.: (Beiträge zur Philosophie). English: Contributions to philosophy: from enowning I translated by Parvis Emad and Kenneth Maly. Indiana University Press, Bloomington. Indiana (1999)

Chapter 26
Modular Standard in Independent Automotive Aftermarket

Thomas Kampschulte

Worldwide business in the independent Automotive Aftermarket requires a product portfolio that covers the full range for all relevant cars available in the local markets ("from Alfa to Zastawa"—Target 95 %). Variants of cars increase developed in shortening lifecycles by Original Equipment Manufacturers (OEMs) to cover niche markets with the latest technology. All these different car applications split the quantity of identical products to be sold.

The end customer requires 24-h availability of spare parts or a tuning solution for individualization. His or her purchasing decision is answered by the simple question, if his or her individual requirements are fulfilled short term. The thyssenkrupp BILSTEIN brand gets attractive by availability of solutions designed to fit the specific car as functional replacement part in OE quality or to satisfy the individual customer need for a performance product to experience the "BILSTEIN Fahrgefühl."

Speed of Time to Market is essential. Target costs are limited on the one hand by the original spare part offered by the OEM produced in mass production or on the other hand by specialized competitors for tuning parts in niche markets. The challenge is: What can be done to accelerate the availability of solutions for a high variety of individual car applications on cost level of mass production?

26.1 Engineering Concept "Modular Standard"

To answer the Aftermarket challenge, the concept at thyssenkrupp BILSTEIN named Modular Standard is expressed by: "Shock absorbers will be assembled faster and cheaper in all required variations based on a limited set of modules."

T. Kampschulte (✉)
Aftermarket, thyssenkrupp Bilstein GmbH, Milsper Strasse 214,
Ennepetal 58256, Germany
e-mail: thomas.kampschulte@thyssenkrupp.com; kampschulte.bilstein@gmail.com

J. Bellemare et al. (eds.), *Managing Complexity*, Springer Proceedings
in Business and Economics, DOI 10.1007/978-3-319-29058-4_26

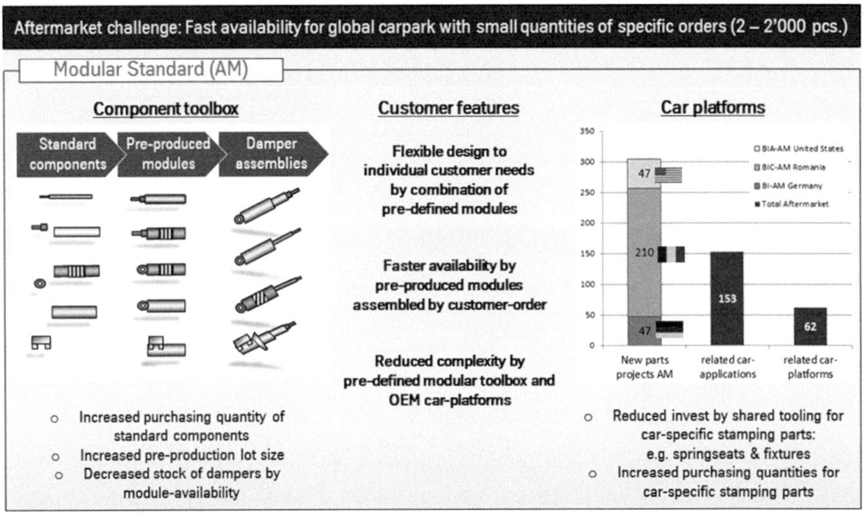

Fig. 26.1 Grouping of existing parts by component classes, identify classic forms of component types with specific variants specified by standard design parameters

The Modular Standard concept requires detailed preparation in the design process. Two approaches support this concept (Fig. 26.1):

- *Component toolbox*: Standardize shock-absorber internal parts to increase purchasing volume including their combinations to standard modules for pre-production.
- *Car platforms*: Utilization of modularity of car platforms to share car-specific stamping parts between different application projects to reduce invest effort and increase quantities.

26.2 Implementation

First step starting in 2007 was to create examples of a Modular Standard component toolbox. It was a decision within the Engineering department, not to start a general discussion about a theoretical concept but to simply prepare a solid foundation of standard parts for future potentials in purchasing and production. To reduce complexity, focus was set on effectiveness by implementing an isolated task and not directly reorganize business processes for overall efficiency improvement [1].

26.2.1 Structuring a Component Toolbox as "Easy to Use" Building Blocks

The pilot project started with an internally used parts catalogue representing the main components for rework- and tuning service offered by local thyssenkrupp BILSTEIN workshops worldwide. Listed part numbers were grouped in component classes specified by parameters and measures relevant for the design process like length, diameter, etc. [2]. This simple logic got time consuming based on a parts range from more than 40 years of design work by generations of engineers in different locations with changing supply chains consisting of handmade drawings, 2D-pdfs up to 3D-CAD models. Many variations of dimensioning methods made it hard to even recognize physically identical parts. Final result was a component catalogue specified by predefined standard parameters as a HTML document available for all thyssenkrupp BILSTEIN Technical Centres for customer service worldwide. On the other hand, the workshop catalogue provided the general structure for the component classes of the Modular Standard toolbox.

A relocation project for all variants of the component class "Piston rod" to centralize production to one thyssenkrupp BILSTEIN plant provided the next step, as designers would work on all of these parts anyhow. The predefined design parameters for piston rods were used to build up sample drawings, where the technical drawing describes the dimensioning by variables and an attached table shows the specific measures of all variants. This also was the chance to elaborate a common understanding of the toolbox approach between the designer teams of two plants. The first example of a component toolbox in our "BILSTEIN language" was prepared.

The transfer of the toolbox structure to other component classes needed an IT tool to handle the complexity as an "easy to use" database for the design process. The requirement was formulated by: "It should be easier to find an existing component than creating a new one." The keyword is "class list of characteristics" (CLC) as a very strong design tool that requires detailed preparation and discipline to unfold its potential. It took about 2 years to prepare and fill an Oracle-based PLM system with all the component classes and types including the transfer of the measures from technical drawings into the CLC list. The structure of the component toolbox was agreed on with all design teams and described in a toolbox manual defining the classes, types, and parameters including their abbreviations now mandatory at thyssenkrupp BILSTEIN Aftermarket. Formal release and transfer from project status into daily design work was done with two-thirds of all relevant components available in the PLM system. Further continuous improvements are focused on eliminating similar parts from all locations and definition of standard parts to be preferred in damper design. First measures are started with purchasing department to establish toolbox suppliers based on a by component type up to ten times increased total annual volume for these standard parts (Fig. 26.2).

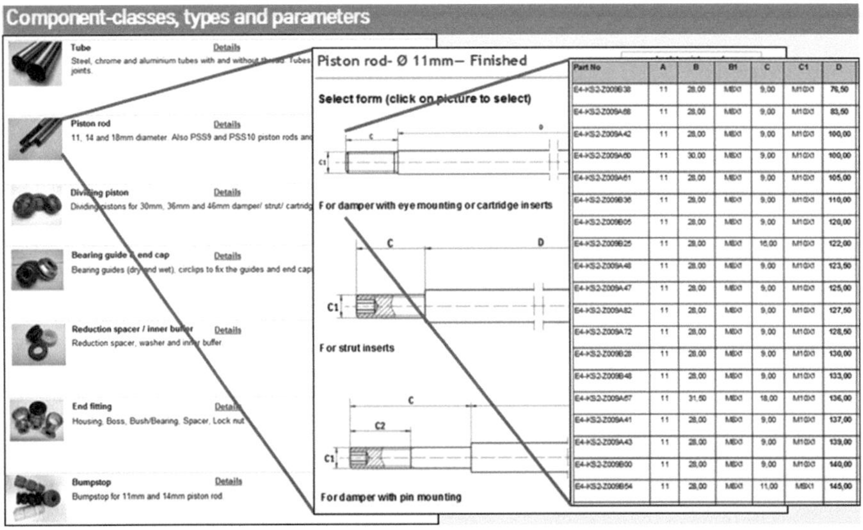

Fig. 26.2 Design complexity in the lifecycle of an OEM car platform

26.2.2 *Usage of Car Platforms to Reduce Design Complexity*

Vehicle manufacturers define car platforms for interchangeability of modules and flexibility in configuration of customer features to reduce complexity and costs. Platforms define a common structure [3] for different cars like the Volkswagen MQB Platform for VW Golf VII, Passat, Touran, Audi A3, TT, Skoda Octavia, Superb, and Seat Leon [4]. More cars on that platform are to come as limousine or station wagon, convertible, four wheel drive, etc. The new VW Passat will be the first car of a revised MQB-B platform optimized for weight reduction [5]. In addition, there are temporary co-operations on project base like Daimler and Renault, e.g., for development of the Smart ForTwo/Renault Twingo, before Renault moved closer to Nissan creating the car platforms BO (Dacia), B (including Infinity LE concept), and C (including Mercedes-Benz Citan). So there are many cars out there even without the same name, brand, or manufacturer, where similar components should fit.

From product management point of view, the need regarding an Aftermarket product for a new car application is expressed by name of a specific car and manufacturer. By adding up the planned sales volumes from relevant countries some well-known car platforms with their already existing car applications might be taken into account. Based on the required technology as spare part and/or tuning kit, the project is split and placed at the design teams competent for the required technologies located at the plant producing this technology. Because OEMs develop several cars of one platform spread over many years, many designers in different design teams work on similar products based on different car names over time or even in parallel.

From OEM point of view, car platforms provide standardized interfaces to exchange modules, components and even suppliers. Starting to work on the VW

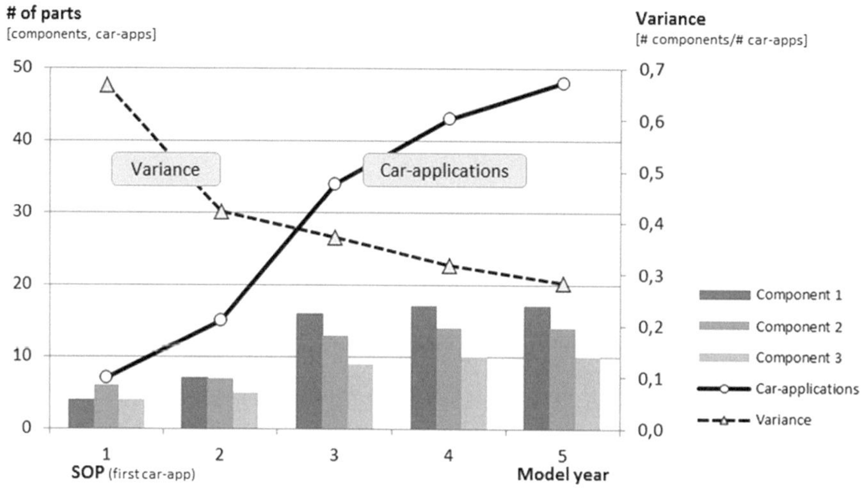

Fig. 26.3 Aftermarket challenge and engineering concept "Modular Standard"

MQB platform at thyssenkrupp BILSTEIN Aftermarket, more than 40 different part numbers for shock absorbers offered by the OEM were identified and there are more to come. The design complexity in the lifecycle of an OEM platform is determined by the quantity of the components used as building blocks for variants of shock absorbers. Figure 26.3 shows an example, how the design complexity is reduced over lifetime even with an increasing number of components and car applications.

But an OEM car platform does not always mean, that one specific design applies to all related cars. So the other way round from supplier point of view a platform is given, if the same module or component does fit into different cars. A specialized engineering team was established at thyssenkrupp BILSTEIN to identify these platform synergies of current application projects compared to already existing products and to other cars that might be technically similar regarding suspension.

Over the years for the former "Golf V platform" (VW PQ35), the relevant 13 car applications at BILSTEIN were realized by 22 suspension kits by combinations of 20 part numbers (Variance = 1.5). By redesign including Ride and Handling tests all 13 car applications could be covered by 7 suspension kits by combinations of 10 part numbers (Variance = 0.77). The design complexity could be reduced by half. For the VW MQB platform, its relevant 16 car applications directly could be covered by combinations of 8 part numbers (Variance = 0.5). Invest effort for a car-specific stamping part could be decreased by half in combining the quantities by using the same design in two project teams at two locations working in parallel for spare part and tuning applications.

Currently running at thyssenkrupp BILSTEIN Aftermarket are 304 projects at three locations for 153 car applications based on 62 car platforms (see Fig. 26.1). Experience proves the cost advantages of the platform approach for the group in many cases based on painful compromises at least at one location even cost wise. So continuous communication and leadership is essential to optimize total costs and not just reduce local effort.

26.3 Experience and Perspective

Implementing Modular Standard in Engineering at thyssenkrupp BILSTEIN Aftermarket is just a first step to enable the organizations flexibility to fulfill individual customer needs short term. Based on a predefined component toolbox and extended usage of platform design, an increased quantity of parts can be purchased with cost effects. This also allows process standardization and production segmentation by preproduction for increased stock availability of standard modules instead of final products already assembled in all variations waiting in the warehouse for the customer order. Quality management is an essential precondition for component toolboxes, platform design, preproduction and module availability because any failure will have an increased impact compared to a car-specific design with special parts in lower quantities.

All of these are just internal steps to build the fundament to be able for flexible (re)action by customer order. A fast final assembly process has to be linked to the specific customer requirement just-in-time when the need is expressed. This includes the location of final assembly and the logistics process. For special parts, the developments of additive manufacturing regarding the current limits in material strength and durability ought to be taken into account. Finally, the co-creative configuration process by the customer for complex systems like car suspension supported by axiomatic design principles, e.g., [6] assuring functionality and safety should be elaborated.

References

1. Maier, H.: Steigerung der Wertschöpfung in der automobile Entwicklung. In: Jochem, R., Landgraf, K. (eds.) Anforderungsmanagement in der Produktentwicklung, pp. 121–140. Symposion, Germany (2011)
2. Hyer, N.L., Wemmerlöv, U.: Group technology and productivity. Harvard Business Review **62**(4), 140–149 (1984)
3. Meyer, M., Lehnerd, A.: The Power of Product Platforms, p. 39. The Free Press, New York (1997)
4. Wikipedia. https://en.wikipedia.org/wiki/Automobile_platform
5. Vehicle Dynamics International, Annual showcase 2015. Volkswagen Passat, pp. 8–9. UKIP Media & Events Ltd. (2015)
6. Suh, N.P.: Axiomatic Design. Advances and Applications. Oxford University Press, Oxford (2001)

Chapter 27
Identification of Profitable Areas to Apply Product Configuration Systems in Engineer-To-Order Companies

Katrin Kristjansdottir, Lars Hvam, Sara Shafiee, and Martin Bonev

27.1 Introduction

Engineer-To-Order (ETO) companies are increasingly showing more interest towards applying Product Configuration Systems (PCSs) in order to support their various specification processes. Specification processes can be defined as a business process where the customer's requirement is analysed and the product is designed to fulfil the customer's needs [1]. PCSs can be defined as an IT system used in a design activity, where a set of components along with their connections are pre-defined and additional constrains are used to prevent illegal combinations and to reduce the solution space [2].

ETO companies that have implemented a PCS have achieved substantial benefits in terms of shorter lead time, improved quality of products and specifications, more on time deliveries, reduced resource consumption, optimization of product and increased customer satisfaction [1, 3]. Furthermore, utilizing PCSs provides ETO companies with the opportunity to increase sales of more standardized products and become more in control of their product range. This can result in higher efficiency and improved quality [4]. In ETO companies, PCSs are usually gradually implemented where they are normally used to only support a specific part of the specification process or a subset of the product families. That is since it requires significant work to acquire and structure the product information that are needed to be modelled into the PCS due to the complexity of products and the specification processes. Therefore, it may not be profitable to formalize the complete product knowledge, especially if the sales volumes are low [4].

K. Kristjansdottir (✉) • L. Hvam • S. Shafiee • M. Bonev
Department of Engineering Management, Technical University of Denmark, Lyngby, Denmark
e-mail: katkr@dtu.dk; lahv@dtu.dk; sashaf@dtu.dk; mbon@dtu.dk

J. Bellemare et al. (eds.), *Managing Complexity*, Springer Proceedings in Business and Economics, DOI 10.1007/978-3-319-29058-4_27

When starting PCSs projects in ETO companies, there are currently no existing guidelines supporting the decision-making processes regarding how to identify both the products and the specification processes. In order to improve the decision-making process, this chapter proposes a three-step systematic framework to identify the most profitable areas for applying PCSs in ETO companies. The first step is to analyse projects in terms of profitability and accuracy of the cost estimations in order to identify the factors causing the deviations from estimated to real cost. The second step is concerned with identifying different areas for applying a PCS and the scope of the system, and finally in the third step cost-benefit analyses are conducted in order to find the most promising scenario and areas for applying PCS. The chapter's aim is to provide answers to following questions based on the framework mentioned above:

1. How to analyse profitability and accuracy of cost estimations in ETO companies?
2. How to identify possible areas where PCS could provide cost savings for businesses?
3. How to assess cost-benefits for potential applications of PCS?

27.2 Research Method

The research methodology in this chapter is structured in two phases. The *first phase* is dedicated to the development of the framework, which is based on both literature and experience from working with PCSs in ETO companies. The *second phase* is concerned with the testing the framework. For that purpose, a project team was formed in an industrial ETO company operating in the oil and gas industry, including two researchers from the Technical University of Denmark and experts from the company. During the period of the case study, weekly meetings were held to validate the processes, access to internal databases was provided and workshop with key employees were held. Aligned with the data collection part, direct method of interviews combined with the researchers' observations were used in this research.

Finally, in order to identify the potential cost of developing and implementing a PCS, seven ETO companies that had implemented a similar system were contacted and asked to provide information regarding the development and maintenance of their PCSs.

27.3 Literature Review

In order to examine the theoretical background of this research, a literature review was conducted in the area of cost analysis in ETO companies and ETO companies that have implemented a PCS. The main purpose with the literature review was to gain insight into the different approaches used to analyse cost in ETO companies and to identify how ETO companies have implemented PCS with regard to scope of products, processes and cost-benefits.

27.3.1 Cost Analysis in ETO Companies

ETO companies providing customized products face the challenge of reaching an acceptable earning before interests and taxes (EBIT) and to achieve the same gross margins from different projects [5]. In the sales phase, the most important decisions regarding profitability of projects are taken. Inaccuracy in the cost estimations in terms of over- and underestimating the project cost can have significant consequences on the project's profitability. By overestimating the cost the risk of losing the customer increases and by underestimating the cost project's profitability is reduced. In the pretender phase, inaccuracy of the cost estimation is often the result of being made within a limited time and when the project scope has not been fully determined [6]. Other factors that can influence the cost estimations are project complexity, technological requirements, project information, project team requirement, contractual arrangement, project duration and market requirements [7].

Several approaches have been developed to estimate the cost of products. Cooper and Kaplan [8] proposed Activity Based Costing (ABC), which has been proven to be a powerful tool to distribute the overhead cost by first distributing the indirect cost evenly to the various activities performed by the company's resources. Thereafter, the cost is assigned to individual orders, customer or products. Walker et al. [9] suggest a Volume-Based Costing and Feature Costing method in order to allocate cost on product's attribute level. Kaplan and Anderson [10] propose a Time-Driven ABC, approach where resources are connected directly to cost objects and where time estimations are used to predict the cost for certain activity. Zhang and Tseng [11] then define a method for assessing products' profitability and cost behaviour from four aspects in order to provide a method for measuring product costs in terms of unit level, batch-level, product-sustaining and facility-sustaining.

27.3.2 Analysis of ETO Companies that Have Implemented a PCS

Several examples can be found in the literature of companies providing customized products that utilize a PCS. This section provides description of processes and products that are included in PCSs followed by cost/benefits from the implementation.

Barker et al. [12] present the case of Digital Equipment Corporation. The PCSs were developed to check the technical correctness, to guide the assembly of customer's order, select part that can be purchased, illustrate the computer room under design and finally to configure clusters. The PCSs have been gradually implemented to support the complete product range, which consist of 42 product families. Main benefits are described in terms of improved quality, optimized performance of the products, increased manufacturing flexibility and increased product development. The development took place over nearly 10 years and the estimated yearly net return is expected to be around $40 million.

Fleischanderl et al. [13] present a PCS for complex telephone switching systems. The general configuration task involves selecting the right components, connecting them together and setting the different parameters. The system supports various functions of the company and the product life cycle, such as sales, engineering, manufacturing, assembly and maintenance. The benefits from the implementation of the PCS are improved quality, identification of errors and increased knowledge sharing. The development time or the cost is not indicated. However, a positive return of investment was achieved in the first year of operation.

Forza and Salvador [4] present a case of company making voltage transformers where a PCS is used to support the information exchange in the sales phase, the data gathering and to ensure validity of the configuration. The technical features are only included in the system for the simplest product family. For the more complex product families, the system supports the design activity by collecting the technical characteristics. The main benefits are listed in terms of reduction in errors, lead time and resources. Furthermore, the correctness of the bill of material generated by the PCS has positively impacted the production. For the development of the system it is mentioned that building up the product model was a very time-consuming activity.

Hvam [14] describes how PCS is used to support complex engineering processes in the sales phase by automating the quotation generation for a cement plant. In the first prototype of the system, the focus was set on 20 % of the parts, which generate 80 % of the cost. The main benefits are described in terms of reduction in lead time for generating tenders and engineering hours for the conceptual design, increased quality of the quotation and optimization of the plant. Furthermore, the company might gain extra sale of cement plant as a result of shorter lead time, which would outsource all other benefits [1]. The development of the quotation processes has lasted for 3–4 years and 1 year was spent on generating the PCS for proof of concept. The development cost of the system was estimated to be 800 million Euro.

Petersen [15] explains how PCS was used to support the sales and engineering process at Aalborg Industries A/S, which produces marine boiler for ships. The PCS was gradually implemented where one to two product families was added at each time. The system is used to support the sales processes. The realized benefits are listed in terms of reduced lead time and resource consumption for making the quotations. The development of the system included evaluation of different systems, standardization of the product programme and implementation of the product knowledge into the selected system.

27.4 The Suggested Framework for Identification of Profitable Areas for Applying PCSs

As revealed in the literature study, ETO companies usually gradually implement PCS to support specific parts of their products range and specification processes. In order to improve the decision-making processes regarding how to select the products and to what extent the system should support the specification processes, a three-step framework is proposed where profitable areas can be identified in ETO companies for applying a PCS. The individual steps and substeps of the framework are shown in Fig. 27.1.

27.4.1 Phase 1: Analysis of Profitability and Accuracy of Cost Estimation

The first phase includes analyses of the projects' profitability and of the accuracy of the cost estimations. Based on those analyses, the main factors influencing projects' profitability and causing the deviations can be identified. To calculate the profitability of the projects, it is suggested to use contribution margins (CM) and contribution ratios (CR). The CM (27.1) and CR (27.2) are calculated as following [16]:

$$CM = Sales\ price - Cost\ price \tag{27.1}$$

$$CR = \frac{CM}{Sales\ price} \tag{27.2}$$

When calculating the cost price, it has to be ensured that the right approach is used. The most common approaches include material and production cost to determine the cost prices [17]. Other factors that might be added to the cost estimations

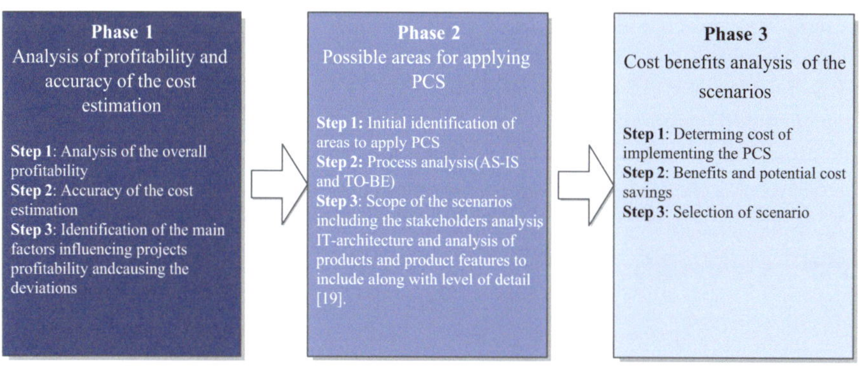

Fig. 27.1 Framework for identifying profitable areas for applying PCS in ETO companies

are labour, machinery and inventory cost [11]. As the fixed cost is not included in the CM, the margins have to be high enough to cover that cost.

In order to analyse the accuracy of the cost estimation, it is suggested to compare the CM and CR from what is expected based on the budgetary offers to the actual CM and CR calculated after the project has been closed. Based on those analyses, the main factors that cause deviation from the estimated cost can be identified in more details.

27.4.2 Phase 2: Possible Areas for Applying PCS and Scope of the System

The second phase includes an initial identification of potential areas to apply PCS based on the findings in the first phase. Thereafter, process analysis is conducted where the current processes (AS-IS) are analysed and the future processes are developed (TO-BE), which includes where to apply a PCS to increase the efficiency of the process. In order to map the processes, it is recommended to use Business Processes Modelling Notation (BPMN) to demonstrate the communication between different actors and the tasks performed by the individual actors [18]. Finally, the scope of the different scenarios is analysed in terms of stakeholders, IT-architecture and products and products' features to include along with level of details [19].

27.4.3 Phase 3: Cost-Benefit Analysis of the Scenarios

In order to estimate whether a company should proceed and invest in the PCS, it is recommended to do cost-benefit analysis. Cost-benefit analyses are carried out to compare different scenarios and are an effective method to compare different results from variety of actions [20]. When estimating the cost of developing a PCS, several factors have to be taken into consideration such as expected time needed from internal and external resources in order to increase standardization of the product range as well as to gather and structure the product information and to model them into the system. Based on this analysis, the company should be able to make informed decision regarding whether it provides value for the business to implement a PCS. Furthermore, in order to keep the level of commitment from the top-level management, economic benefits have to be made very clear from the beginning of the project and emphasized in order to keep the project alive [21].

27.5 Case study

The framework was tested in a global engineering company that provides equipment as well as complete systems and services for the oil and gas industry. Over the last years the company has gone through significant growth that has resulted in greater product variety and higher processes complexity, which has negatively affected the profitability of the company.

The data for the analysis was gathered from the company's internal systems and verified with the company's employees. It was decided to analyse both projects where a complete system solution for rigs is provided and smaller projects where single equipment is sold, as it represents the main activities at the company. The complete system projects require highly complex solution that has to be adjusted to the customer's demands and includes engineering work, manufacturing and commissioning at the customer's site. For the single equipment sale commissioning is not required. The lead time for complete system projects is approximately 4 years. Therefore, to be able to include both pre- and postcalculations for the projects the time scope for the analysis was set to 4 years. This resulted in 116 single equipment projects and 12 complete system projects. The complete system projects were divided into three categories with regard to rigs types, which will be referred as types A, B and C. For the smaller projects, it was decided to not make any categorizations.

27.5.1 Phase 1: Analysis of Profitability and Accuracy of the Cost Estimation

Step 1: Analysis of overall profitability. The first step includes analysis of the projects' profitability both for the complete systems and the single equipment projects. The cost used for the calculations of the CM consists of engineering hours, production cost and material consumption and commissioning.

The calculations of the company's overall projects' CM indicated great deviation between the budgetary offers and the actual margins, calculated after the project had been closed. This deviation caused the company 101,304 million Euro reduction in CM from what was expected, resulting in CR of only 0.65 % instead of 21.74 % for the complete system projects. For single equipment projects, the analysis indicated much less deviation and more profitable business, even though it only accounts for 14 % of the total revenues (Table 27.1).

Step 2: Accuracy of the cost estimations. In order to calculate the accuracy of the cost estimations, the actual CM and the CR are compared to the expected ones in the budgetary offers.

The analysis for the complete system projects revealed that in 11 out of 12 projects, the actual profit is less than the expected in the budgetary offers (Fig. 27.2). However, as previously stated there is a great similarity between projects that are carried out in the same categorize (A, B or C). Therefore as the pattern in Fig. 27.3 shows, due to economies of scope the CM is expected to increase after completing the first project. Nevertheless, the cost reduction is not in proportion to the amount of information that can be reused, which can partly be explained by the fact that incomplete projects were copied as they were overlapping in time.

Table 27.1 Overview of the company's profitability

	Complete system projects			Single equipment sales projects		
	Actual	Budget	Deviation	Actual	Budget	Deviation
Revenue	542,976 million euro	482.16 million euro	60.816 million euro	74,424 million euro	66,192 million euro	**8.232 million euro**
Cost	539,448 million euro	377,328 million euro	162.12 million euro	48,384 million euro	44,352 million euro	**7.224 million euro**
CM	3,528 million euro	104,832 million euro	−101,304 million euro	26.04 million euro	21.84 million euro	**1.008 million euro**
CR (%)	0.65%	21.74%	**−21.09%**	34.99%	32.99%	**2.24%**

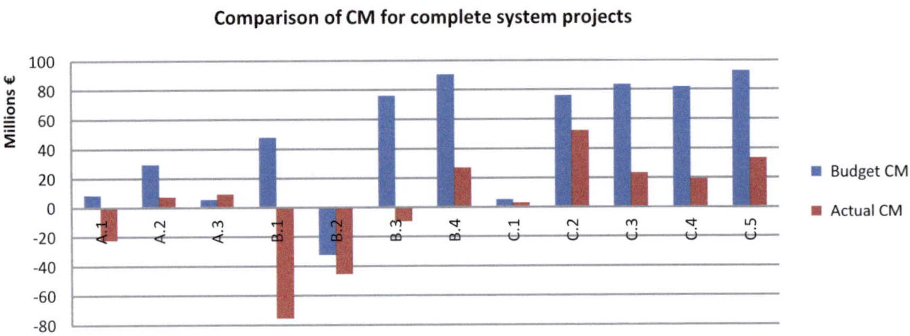

Fig. 27.2 Deviations in CR for complete system projects

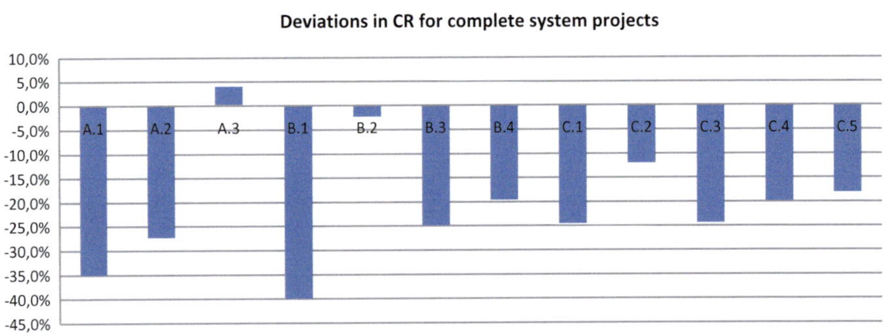

Fig. 27.3 Comparison of CM in the budgetary offers and the actual for complete system projects

For single equipment projects, the analysis revealed also deviation between the expected CM in the budgetary offered and the actual CM (Fig. 27.4) as well for the CR (Fig. 27.5). However, the deviation fluctuates both on the positive and the negative side and the results do therefore not affect the overall profitability. This indicates that the cost estimation is not accurate as the cost is both over- and underestimated. Even though the overestimated cost results in higher CM, the risk of losing the sale increases as the customer might go elsewhere to purchase the equipment. In today's market condition, this is not a problem due to strong market

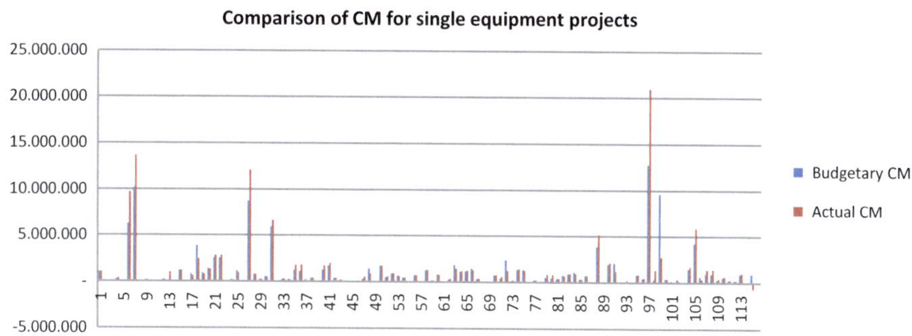

Fig. 27.4 Comparison of CM in the budgetary offers and the actual for single equipment projects

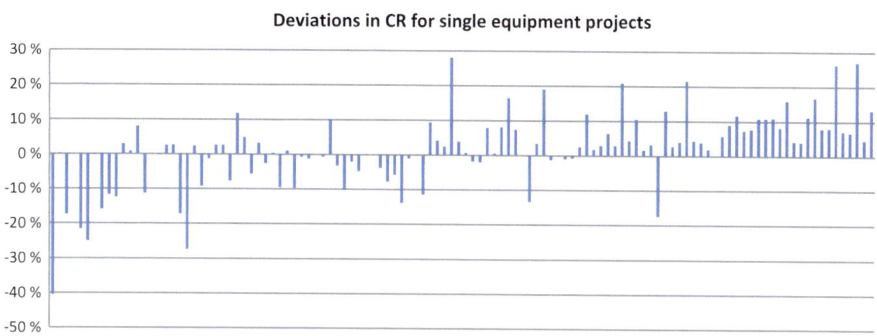

Fig. 27.5 Deviations in CR for single equipment projects

position of the company. However, it is anticipated that the competition on the market will increase and therefore this could become a threat in the future.

Step 3: Identification of the main factors that influence the projects' profitability and causing the deviations. In order to identify the main factors, a brainstorming session was carried out at the company with representatives from the management, project leaders and other key employees. The most important factors identified are listed below.

- *External factors:* The time scope of the project has great impact on their profitability. It was calculated that the external factors were accountable for 30% of the deviations in the CM. From these 30%, 15% could be traced to increased steel price, 12% to increased cost of industrial products and 13% to increased cost of labour (engineering, manufacturing).
- *Cost of carry over work:* The cost of carry over work was extracted from the company's internal system and accounted for 3% of the reduction in the CM for the projects. The carry over work occurs when additional work has to be made at the customer's side as result of defects or other unforeseen factors, which fall under the company's warranty.

- *Workflow and responsibility:* There is unclear workflow and lacking overall responsibility when it comes to the purchase, engineering, production and commissioning. Furthermore, the knowledge transfer between different departments is lacking.
- *Incompleteness or errors in the product's specifications:* Lack of information in the sales phase has resulted in delays and costly changes late in the processes. Furthermore, this also has impact on the production, commissioning and the carry over work when defects in the products have to be fixed.
- *Tendency to sell products not within the company's standard product architecture*: Number of sales persons has grown significantly over the last years and only the most experienced sales persons have the overview of the standard solution and how they can be combined. This has resulted in costly development of new products that the customers cannot be charged for as it was thought to be within the company's standard product range.
- *Product designs:* The focus of the company has been on designing products for specific projects instead of designing the products and adjusting them to different projects. Therefore, product growth is not in control. Furthermore, the master data often contains errors and standard interfaces are not defined.

27.5.2 Phase 2: Possible Areas for Applying PCS

Step 1: Initial identification of areas to apply PCS. The analysis from the first phase indicates that many of company's current challenges are concerned with the early phases of the sales and engineering processes for both projects and single equipment sale. In the early stages of the sales and engineering processes, the most important decisions regarding products capability and 80–90 % of the products cost are determined [1, 22] Therefore, two scenarios are generated in order to determine the scope of the PCS, for both projects and single equipment sales where PCS is used in the early phases of the specification processes.

Step 2: Process analysis. The most important specification processes in the early sales and engineering phase are regarding the encapsulation of the customers' demands, the transformation of the customer's demand into valid solution that can be provided by the company and finally the ability to make cost estimation. In Fig. 27.6, the current (AS-IS) process flow is visualized where the most critical aspects are marked with blue colour and as well as the future (TO-BE) processes where the process is supported with PCS. In the TO-BE processes, the PCS is used by the sales persons where the system should ensure that all relevant information is gathered. For standard products, the system should be able to suggest feasible solution while for none standard products an input from engineer is required before making the budgetary offer.

Step 3: Scope of the Scenarios. The individual step in this phase builds on framework for scoping product configuration project for ETO companies as suggested by Shafiee et al. [19].

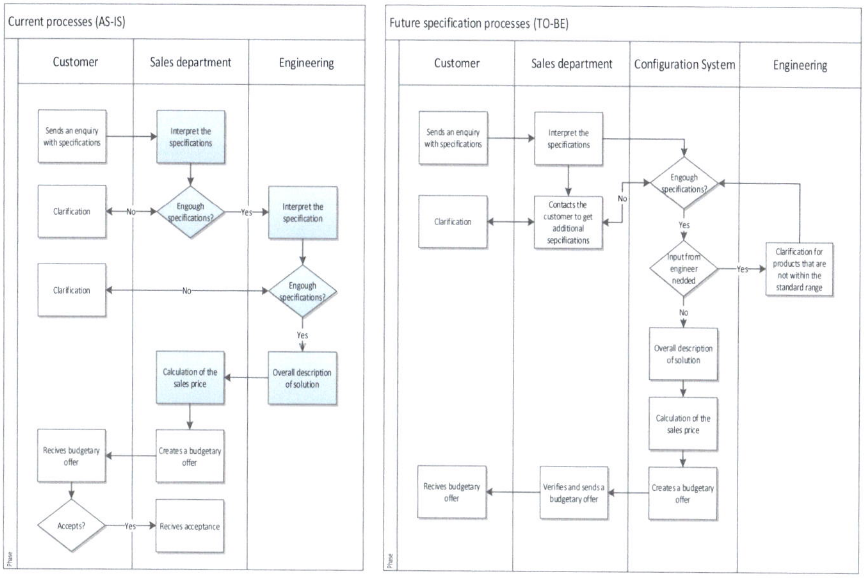

Fig. 27.6 AS-IS and TO-BE process flow for producing a budgetary proposal both for complete system projects and single equipment projects

Stakeholder analysis. The main stakeholders that have been identified are employees from sales and engineering. The sales employees are the main user of the system and the engineers have to provide the information and work on standardization of the product range. The management is also important as they have to support the project and finally employees involved in the manufacturing and commissioning will be affected as they will be working with the product specifications and more standardized product range. Those stakeholders' requirements have to be taken into consideration. Tools such as use case diagrams are used to communicate and determine the stakeholders' requirements [19].

The overall content of the PCS. The overall content of the PCS is described by the main IT-architecture, input/outputs, main functionalities and integrations. In this case it was decided to only include the main functionalities, the input/output and integrations to describe the overall content of the system. The requirements for each of those categories are shown in Table 27.2 for projects and single equipment sale. The overall content of PCS is determined based on the stakeholders' requirements.

Products and product features to include in the PCS and the level of details. In this phase, a description of the products to be considered for implementation into the system and their level of details should be provided. Identification of the right level of details when scoping the system is crucial in order to reduce time and resources when developing the PCS [19]. The product features can be divided into property models, product structure models and other life cycles models [1]. Furthermore, Hvam [14] describes how basic modules consisting of machines and equipment can

Table 27.2 Scope of the scenarios

Complete system projects	Single equipment projects
Inputs/outputs	
The main inputs to the system are concerned with regulations for different regions, space available for installation, required performance, environmental loads and temperature ranges. The main output from the system is a complete budgetary proposal including the calculations of cost, weight and power consumption, list of machines and the overall process flow	The main inputs to the system are based on the customers' requirements regarding performance parameters. The main output from the system is a complete budgetary proposal
Main functionalities	
• Secure that all relevant information are gathered from the customer	• Secure that all relevant information are gathered from the customer
• Capacity calculations with respect to oil power consumption and drilling speed	• Give an overview of different combination within the standard product architecture and suggest a solution that fulfils the customer's requirements
• Suggest an overall solution that fulfils the customer's requirements	• Ability to overrule some of the default options with additional functionalities or additional performance
• Visualize the drilling process and how a different selection affects the overall processes	• Generate bill of material
• Ability to handle complex calculations as well as integration with other calculation systems	• Estimation of cost of engineering hours, material cost and fabrication
• Estimation of cost of engineering hours, material cost, fabrication and commissioning	•Generate budgetary offer
• Generate budgetary offer	
• Integrations with CAD systems to make the engineering diagram generation from PCS	

be used to cover 80 % of the overall specifications for PCSs of complex ETO products, such as a cement factory. The products and product features to include in the PCS and the level of details for the scenarios are shown in Table 27.3.

27.5.3 Phase 3: Cost-Benefit Analysis for the Scenarios

In this phase, a cost-benefit analysis is used in order to identify the feasibility for implementing a PCS for the developed scenarios. Based on this analysis, the most promising scenario can be chosen.

Cost of implementing PCS for the scenarios. The costs of implementing the PCS will require both internal and external resources to build up the required knowledge, to improve the product architecture and to model and gather information for the

Table 27.3 The products and product features to include in the PCS and the level of details for the scenarios

Complete system projects	Single equipment projects
Products	
A complete system solution that is provided in a project consists of 7 main processes units, where each unit consists of several machines that again consist of number of equipments. In general, a complete project consists of 40–80 machines that have to be combined and complex constrains regarding interfaces have to be taken into consideration	The single equipment can vary from equipment provided for machine or it can be a complete machine
Product features and level of details	
Here the basic modules correspond to the seven processes units. An example of a processes unit or basic module is trip out. Here the main focus is on the product properties and the product structure models on a rig level	Product features for the single equipment sales are modelled on machine level. An example of a machine for the trip out processes (that was described as one of the basic modules for project) is a crane. As for the projects the main focus is on product properties and products structure models in this case it is provided in more detail or on a machine level instead of a rig level as for the projects

Table 27.4 Development cost for PCS for ETO companies

Company size (no. of employees)	Complexity of the PCS		No. of PCS	Development cost	Man-months used for development	Man-months used for maintenance	
	Attributes	Constrains				Internal	External
>1000	2000	450	4	923,470 euro	30	24	0
>1000	2000	2001	10	469,137 euro	15	14	0,1
<500	999	499	1	840,828 euro	36	7.5	0
>1000	2001	499	4	1,206,353 euro	12	36	0
>1000	300	350	2	133,145 euro	16	9.6	0
<500	999	2000	1	446,350 euro	30	8	0
<500	2000	999	2	1,072,314 euro	12	6	1

PCS. Furthermore, configuration software licenses have to be purchased for the users. In order to estimate the development and the maintenance cost, seven ETO companies that have implemented a PCS were asked to provide information, which are summarized in Table 27.4.

Based on this, it can be assumed that the development of a PCS in ETO companies is on the scale 133,145–1,260,353 euro (Table 27.4). Those numbers should only provide some rough indications of potential cost range as the projects scope and products complexity vary greatly between these cases.

Table 27.5 Estimated cost of the PCS for the developed scenarios

	Complete system projects				Single equipment projects			
	Development		Maintenance/year		Development		Maintenance year	
	Man-months	Cost (euro)	Man-months	Cost (euro)	Man-months	Cost (euro)	Man-months	Cost (euro)
Internal	50	600,000	16	120,000	15	180,000	5	60,000
External	12	216,000	2	36,000	6	108,000	0.5	9000
Software	–	25,000	–	5000	–	25,000	–	5000
Total		841,000		161,000		313,000		74,000

Table 27.6 Potential annual cost savings from implementing a PCS

	Projects			Single equipment sales		
	Conservative	Realistic	Optimistic	Conservative	Realistic	Optimistic
Material consumption	1 %	3 %	5 %	1 %	3 %	5 %
	556,550 euro	1,669,651 euro	2,782,752 euro	59,270 euro	177,811 euro	296,352 euro
Production hours	3 %	5 %	8 %	3 %	5 %	8 %
	936,633 euro	1,561,056 euro	2,497,690 euro	29,030 euro	48,384 euro	77,414 euro
Engineering hours	3 %	10 %	15 %	3 %	10 %	15 %
	882,336 euro	1,764,672 euro	2,647,008 euro	72,578 euro	21,772 euro	36,288 euro
Commissioning	10 %	15 %	20 %	N/A		
	1,764,672 euro	2,647,008 euro	3,529,344 euro			
Carry over work	20 %	40 %	60 %	N/A		
	814,464 euro	1,628,928 euro	2,443,392 euro			
Total	**4,954,656 euro**	**9,271,315 euro**	**13,900,186 euro**	160,878 **euro**	**247,968 euro**	**410,054 euro**

The cost for the developed scenarios is estimated based on experiences from other projects, interviews with experts at the company and the suggested scope of the PCS. In Table 27.5, the cost estimation is listed for both the scenarios.

Benefits and potential cost savings. By implementing the PCS for the scenarios it is expected that more standardized products will be sold and the quality of the specification will improve. That should have a positive impact on the material consumption, production hours, engineering hours, commissioning and carry over work. It should be noted that the commissioning and the carry over work only apply to the complete system projects. The impact on these factors was estimated in terms of conservative, realistic and optimistic for both the scenarios. In Table 27.6, the potential cost savings on yearly base are indicated for the complete system project and single equipment project.

Selection of scenarios. Based on these analyses, it was decided to select the scenario for the complete system projects as much greater cost savings can be achieved. The conservative case indicates potential savings of 4,954,656 euro while the development cost accounts for 841,000 euro and the yearly maintenance cost for 161,000 euro. Therefore, the potential benefits are much greater than the anticipated cost.

27.6 Conclusion and Discussion

The aim of this chapter is to offer a more comprehensive framework for ETO companies to identify profitable areas for applying PCS to support the specification processes. The framework consists of three phases where the first phase is where analysis of profitability and accuracy of cost estimation is performed and identification of factors causing deviations in the cost estimations and influencing the profitability. In the second phase different scenarios are generated along with the scope of the PCS for the different scenarios. Finally, cost-benefit analyses are made to identify the most promising areas for applying a PCS. The framework was applied in an ETO company where it gave a structured approach. The analysis of projects' profitability revealed a reduction in the CM of 101,304 million euro from what was expected in the budgetary offers for the complete system projects as a result of inaccuracy in the cost estimations. As the analysis in the first phase indicated that the company could benefit from implementing PCS both for complete rig projects as well as single equipment sale, which represent the two scenarios generated. Based on the cost-benefit analysis, significant savings were identified for the complete system projects. Furthermore, the PCS is thought to also influence positively on other factors that could not be quantified such as potential extra sales as the company is able to respond quicker to customer's enquiry, more professional dialogue with the customers and to enable market-driven standardization of the product range. However there are some limitations to this study as the framework has only been applied in one case company and therefore further testing to improve the framework and achieve generalizability is required.

References

1. Hvam, L., Mortensen, N.H., Riis, J.: Product Customization. Springer, Berlin (2008)
2. Felfernig, A., Friedrich, G.E., Jannach, D.: UML as domain specific language for the construction of knowledge base configuration system. Int. J. Software Eng. Knowl. Eng. **10**, 449–469 (2000)
3. Zhang, L.L.: Product configuration: a review of the state-of-the-art and future research. Int. J. Prod. Res. **52**, 6381–6398 (2014)
4. Forza, C., Fabrizio, S.: Managing for variety in the order acquisition and fulfilment process: the contribution of product configuration systems. Int. J. Prod. Econ. **76**, 87–98 (2002)
5. Mortensen, N.H., Hvam, L., Boelskifte, A., Hansen, C.L.: Making product customization profitable. Int. J. Ind. Eng. Theory Appl. Pract. **17**, 25–35 (2010)

6. Aibinu, A.A., Pasco, T.: The accuracy of pre-tender building cost estimates in Australia. Constr. Manage. Econ. **26**, 1257–1269 (2008)
7. Akintoye, A.: Analysis of factors influencing project cost estimating practice. Constr. Manag. Econ. **18**, 77–89 (2000)
8. Cooper, R., Kaplan, R.S.: Measure costs right: make the right decisions. Harvard Bus. Rev. **66**, 96–103 (1988)
9. Walker, M.: ABC using product attributes. Financ. Manag. **69**, 34 (1991)
10. Kaplan, R., Anderson, S.R.: Time-Driven Activity-Based Costing: A Simpler and More Powerful Path to Higher Profits. Harvard Business Press, Boston (2013)
11. Zhang, M., Tseng, M.M.: A product and process modeling based approach to study cost implications of product variety in mass customization. IEEE Trans. Eng. Manag. **54**, 130–144 (2007)
12. Barker, V., O'Connor, E., Bachant, D.E.: Expert systems for configuration at Digital: XCON and beyond. Commun. ACM **32**, 298–318 (1989)
13. Fleischanderl, G., Friedrich, G.E., Haselböck, A., Schreiner, H., Stumptnerm, M.: Configuring large systems using generative constraint satisfaction. IEEE Intell. Syst. **4**, 59–68 (1998)
14. Hvam, L.: Mass customisation of process plants. Int. J. Mass Customisation **1**, 445–462 (2006)
15. Petersen, T.D.: Product configuration in ETO companies. In: Blecker, T. (ed.) Mass Customization Information Systems in Business, pp. 59–76. Igi Global, London (2007)
16. Farris, P.W., Bendle, N.T., Pfeifer, P.E., Reibstein, D.J.: Marketing Metrics: The Definitive Guide to Measuring Marketing Performance. Pearson Education, Upper Saddle River (2010)
17. Myrodia, A., Hvam, L.: Managing variety in configure-to-order products—an operational method. Int. J. Eng. Industr. Manag. **5**, 195–206 (2014)
18. White, S.A.: Process modeling notations and workflow patterns. In: Fischer, L. (ed.) Workflow Handbook 2004, pp. 265–294. Future Strategies, Lighthouse Point (2004)
19. Shafiee, S., Hvam, L., Bonev, M.: Scoping a product configuration project for engineer-to-order companies. Int. J. Industr. Eng. Manag. **5**, 207–220 (2014)
20. Messonnier, M., Meltzer, M.: Prevention Effectiveness: A Guide to Decision Analysis and Economic Evaluation. Oxford University Press, Oxford (2003)
21. Ariano, M., Dagnino, A.: An intelligent order entry and dynamic bill of materials system for manufacturing customized furniture. Comput. Electr. Eng. **22**, 45–60 (1996)
22. Andreasen, M.M.: Integrated Product Development. Springer, Berlin (1987)

Chapter 28
Goal-Oriented Data Collection Framework in Configuration Projects

Sara Shafiee, Lars Hvam, and Katrin Kristjansdottir

28.1 Introduction

A product configurator is a subtype of software-based expert systems or Knowledge-Based Systems (KBS) with a focus on the creation of product specifications [1]. Data collection in configuration projects is one of the most time-consuming tasks due to the different expertise between domain experts and configuration engineers.[1] Therefore it is important to scope the data collection process and use the right tools in order to reduce time and resources. Fleischanderl et al. [2] argue that up to 20 % of the PCS development cost is usually spent for the configuration software system. Early in the implementation of a configuration model, a knowledge acquisition and data cleansing stage is required to centralize the product knowledge, which includes the corresponding product data [3]. Currently there is no systematic methodological framework for the knowledge acquisition processes to guide organizations to select from the appropriate application that can be used for knowledge acquisition [4]. The level of detail of the information gathered and modeled in the PCS determines the

[1] Configuration Engineer models, implement, maintain PCS, and also support and train users. Configuration Engineers have to create accurate product plans and manage design projects. They also are responsible for leading other staff members and keeping their knowledge of the industry as up-to-date as possible. A configuration engineer uses manual drawing tools or computer-aided design programs to create drafts of how his or her company's goods should look and operate. These drawings need to meet product specification standards and be detailed, and the products must be designed well enough to meet customer needs. Engineers in this career area also should have skills with computer programming particularly when dealing with software development, in which case their focus is on tracking and controlling changes in software products [45].

S. Shafiee (✉) • L. Hvam • K. Kristjansdottir
Department of Engineering Management, Technical University of Denmark, Lyngby, Denmark
e-mail: sashaf@dtu.dk; shafieesara7@gmail.com; lahv@dtu.dk; katkr@dtu.dk

J. Bellemare et al. (eds.), *Managing Complexity*, Springer Proceedings in Business and Economics, DOI 10.1007/978-3-319-29058-4_28

complexity of the PCS. The process of configurator's development is built around a very important element, which is to ensure that the required information is available for the project team [5].

28.1.1 Problem Statement

Ohno Taiichi, the father of Toyota production systems and the founder of "Just In Time (JIT)" methodology, states "Making only what is needed, when it is needed, and in the amount needed!" [6]. Based on this choosing the most efficient way of collecting the right data just in the sufficient amount is necessary. Configuration engineers spend both time and energy on gathering information, and sometimes without knowing if the gathered data are the necessary knowledge for PCS or if there is some missing datum. The main difficulties in acquiring knowledge from domain experts are due to variety of the knowledge, the various representations of the knowledge, and due to difficulties in making the knowledge explicit and accessible. Furthermore, the knowledge has to be up to date in order to secure that the PCS will provide only valid configurations. In the light of those problems for the knowledge acquisition and maintenance processes, the performance of the reasoning engine and the availability of tools for knowledge acquisition have to be taken into the account when deciding on technological basis for the configuration application [3].

This paper aims to help in the processes of controlling the exact level of detail to be included in the system, before doing the actual modeling of the PCS that result in less complexity of the PCS. In order to do so this paper defines tools to handle and manage the large amount of complex data in the early phases of the PCS project. Managing the level details in the early phases is of great importance from different aspects as it will increase the understanding, learning and make the modeling task less complex. Data collection in PCS projects is one of the most time-consuming tasks due to the different expertise between domain experts and configuration engineers. There are several researches on knowledge acquisition for PCS projects but few researches have focused on the data gathering before entering and explicating them in the PCS. Therefore this paper aims to pursue that research opportunity by presenting a framework for gathering data in a more efficient way for PCS projects before they are modeled into the system.

This article's aim is to provide answers to following questions based on the framework mentioned above.

1. What are the goals of the project and the stakeholders' requirements?
2. How to categorize data before starting the data collection?
3. How to prioritize the products and functionalities of the PCS?
4. How to validate data?
5. How to maintain and document data?

28.2 Research Method

In accordance with the overall objective, the research has been structured into two phases. The first phase is focused on the development of the framework and the second phase is concerned with testing the framework.

28.2.1 Development of the Framework

The first phase of the research was devoted to selecting a data collection framework for IT projects from pervious literature. In this research the framework is customized based on an available framework developed for IT projects [7], methodologies and requirements for documenting PCSs [8, 9], and stakeholders' analysis based on RUP principles [10].

The framework was developed by researchers with an applied research background in modeling products, product architecture, knowledge engineering and product configuration, software development, combining traditional domains of mechanical engineering with product configuration and software development.

28.2.2 Testing the Framework

The purpose of testing the framework in a company was to explore if the proposed framework would perform as expected. In particular, the test aimed at establishing whether the data gathered based on this framework were sufficient and efficient enough for the configuration team at the case company.

A project team was formed in an industrial company that included two researchers from the Technical University of Denmark, a software developer and configuration engineer from the company working 50 % of their time on the project for 3 months.

28.3 Literature Review

The literature is identified from searching online libraries (such as Science direct, Scopus, etc.) by the use of keywords, such as "modeling techniques," "mass customization," "product configuration," "IT systems," "UML," "data gathering," "knowledge acquisition," "configuration systems structure," "knowledge management," "expert systems," and "product life cycle and data management systems." Additionally, the list of references of each article is used to identify the related bibliography, as well as the names of the researchers in the recognized research groups within this field. The first section in the literature describes the previous research works for data acquisition for IT systems in general and PCS in particular. The literature review revealed lack of structural frameworks for data gathering in PCS. The framework proposed in this article is based on the previous frameworks for data collection in IT projects, which is explained in Sect. 28.3.1. In Sect. 28.3.2 previous studies and tools for stakeholder analysis are introduced, as it is fundamental in determining the goals and deliverables of the project in the early stages. There are lots of research works on efficient maintenance and documentation of the data in PCSs due to the high importance of this step which is provided in Sect. 28.3.3.

28.3.1 Existing Frameworks for Data Acquisition for IT Projects in General and for Configuration Projects in Particular

One of the first steps in most IT projects, including PCS projects, is to collect and organize the required domain experts' knowledge. It should be noticed that a valuable source of information regarding the different aspects of products can also be available in internal software systems such as ERP system, calculation system, and spreadsheet documents [11]. Felfernig et al. [12] describe how to support both goals by demonstrating the applicability of the Unified Modeling Language (UML) for configuration knowledge acquisition. Barker et al. [13] explain the volatility problems and scope expansion in the PCSs, which differentiates PCS from other IT projects. Scope expansion means that the system becomes more integrated with the other systems and as the system is used by different business groups that lead to additional requirements and data. In order to acquire knowledge in a more efficient way for the PCS with minimum time and resources consumption in the early phases of the project, more systematic methods are required. PCSs often use large and complex knowledge bases that have to be documented, maintained, and updated over time. The explicit representation of problem-solving knowledge and factual knowledge can greatly enhance the role of a knowledge acquisition tool by deriving from the current knowledge base, the knowledge gaps that must be resolved [14]. Basili et al. [7] suggest a framework for collecting data in IT projects that consist of the following six schemes:

1. Establish the goals of data selection
2. Develop a list of questions of interest
3. Establish data categories
4. Design and test data collection form
5. Collect and validate data
6. Analyze the data

The configuration projects are categorized as software projects, however in configuration projects, configuration engineer utilizes information coming directly from the domain experts and internal documentation systems in order to construct the configuration model. In PCS the implementation and management are of more importance, as the obstacles are greater and data failure will have damaging consequences on the project procedure.

28.3.2 Stakeholder Analysis

In the knowledge acquisition process, there is a need for configuration engineer or system analyst to identify the different stakeholders and the sources of knowledge [11]. Nollore et al. [15] focus on the initial specification process in product development and propose a model to manage the interaction of the different stakeholders in the early stages. Forsythe et al. [16] define the importance of building the knowledge base and to gather data through face-to-face interviews between domain experts and knowledge engineers, in order to avoid practical problems in communication criteria between the knowledge engineers and the domain experts when developing the systems. Hvam et al. [17] suggest a methodology based on the representation of the product in a hierarchical structure using UML to package and present the product information for a targeted set of stakeholders (knowledge domain). In the context of configuration, at least three viewpoints are relevant: the customer, engineering, and production views, which correspond to the most important stakeholders in the PCSs projects. Furthermore, Felfernig et al. [3] introduce UML as the language that is a de facto software engineering industry standard and thus more easily accessible for the stakeholders in a development project. Mortensen et al. [18] discuss a procedure to handle the conceptual modeling, which is expected to improve the conditions to involve the relevant stakeholders early in the projects and improve conditions.

Based on IT projects, categorization of the requirements can be done according to two main aspects, which are functional and nonfunctional requirements. Nonfunctional requirements or general quality attributes, which emphasize that quality means compliance to requirements. A requirement that describes not what the software will do, but how the software will do it is called a nonfunctional requirement [19]. Jiao et al. [20] illustrate the steps for nonfunctional requirements identification, which are demonstrated in Table 28.1.

Table 28.1 Features for nonfunctional analysis [20]

Features	Explanation
Requirement elicitation	This is to extract and make an inventory of the requirements of stakeholders
Requirement analysis	This is to interpret and derive explicit requirements that can be understood by everybody
Requirement specification	Requirement specification is about the definition of concrete product specifications (FRs) in the functional domain

A functional requirement is a requirement that specifies each function that a system must be capable of performing [19]. Lim at al. [21] provide the following features, which are demonstrated in Table 28.2 used to identify and prioritize those requirements.

The MoSCoW rules are commonly used when prioritizing stakeholder needs. MoSCoW is derived from the first letters of the following criteria: Must have (Mo), Should have (S), Could have (Co), Want to have (W) [22]. To improve the quality of prioritization, the analysts can merge different statements referring to the same requirement. Future work will consider crowdsourcing the stakeholders to detect duplicates and to improve the quality of the requirements, as well as integration with existing requirements management tools to support other methods of eliciting requirements (e.g., use cases, user stories, and goal modeling) [21]. Based on the literature extensive researches have been done with regards to commercial IT or web-based tools to identify and categorize stakeholders. In Table 28.3 the current literature for stakeholder analysis, both for the IT projects and PCS projects, is summarized.

Table 28.2 Features to identify and prioritize requirements [21]

Features	Explanation
Identify requirements	The list of requirements could be elicited from interviewing an initial subset of stakeholders
Prioritize requirements	Prioritizes requirements using the stakeholders' ratings on the requirements and their influence in the project [21]
Recommend requirements of interest	Predicts a stakeholder's preference on unrated requirements using collaborative filtering techniques, and then recommends requirements with the highest predicted ratings to the stakeholder [21]
Highlight stakeholders in conflict	Highlights stakeholders with conflicting preferences for requirements [21]

Table 28.3 Stakeholders analysis literature

References for Stakeholders' analyses	IT projects	PCS projects
Forsythe et al. [16]		✓
Ebert et al. [19]	✓	
Jiao et al. [20]	✓	
Hvam et al. [17]		✓
Lim et al. [21]	✓	
Nollore et al. [15]		✓
Felfernig et al. [3]		✓
Bittner [22]	✓	
Mortensen et al. [18]		✓

28.3.3 Validation, Test, Maintenance, and Documentation of the Data

In industrial companies delivering complex and highly engineered products, it is crucial to have an efficient system for the documentation of attributes and rules implemented in the PCS, which enables communication of the product knowledge with domain experts. The documentation is also important for the configuration engineering team working with the PCSs to enable them to do future development and maintenance of the systems. Studies in companies using product PCSs have revealed that without documentation system companies are unable to develop their configurators, which can lead that they are forced to abandon or rebuild PCS [23]. It is therefore of importance to have reliable product documentation, i.e., without technical errors and mirroring exactly the customer's expectations [24]. Documentation is vital for all IT projects as it is used for sharing knowledge between people and it reduces knowledge loss when team members become unavailable [25]. The underlying product model is a "living organism" and will quickly become obsolete if not maintained [26]. Tiihonen et al. [27] reflect on the challenges of using PCSs and one of them is that many practical configuration models are poorly documented, incomplete, difficult to understand, or outdated. The maintenance of the PCSs can be divided into two general areas: maintenance of the product model and maintenance of the IT system [26].

28.4 Framework Development

To avoid the risk of failure when using data collection methods, this framework helps configuration engineers and the organizations to become more efficient in this processes. Timeliness of data collection and data validation is quite important for the accuracy of the development. In the area of PCS, the knowledge to be acquired can be both unstable and contentiously changing [11]. As the system grows and get more successful, the users expect more and have new requirements [13]. That is why an iterative framework is needed so it can be used during the project development as well as after the development and in the production phase. Based on Basili's [7] six steps approach, which was explained previously in the literature, it is possible to specify the data collection framework for a configuration project is shown in Fig. 28.1.

The suggested framework is built on a five steps that can be used iteratively. In order to accomplish the framework all the sub steps listed under each step have to be finished. Aligned with different projects in different companies and with different types of stakeholders the sub steps might have to be adjusted. This framework should enable the configuration engineers to be more in control of the level of details to be included in the project by knowing the exact outputs and thereby being able to ask for the relevant input needed for the development of the PCS.

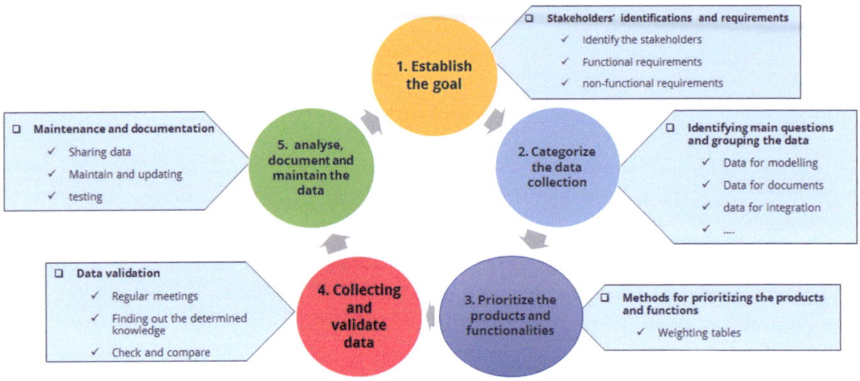

Fig. 28.1 The proposed data collection framework illustration for configuration projects

28.4.1 *Establish the Goal of the Data Collection*

Considering product configuration as a requirement for highly engineered products, the team needs to understand what kinds of outputs are needed for the project accomplishment. A goal determination is used to increase the understanding of the environment and to focus the attention on techniques that are useful at this stage [7]. Jiao et al. [20] defines the customer's requirements in general in three aspects which are: requirements elicitation, requirements analysis, and requirements specification. Based on the RUP methods, the stakeholders and their necessities can be drawn through two specific methods: the first one is by using process flowcharts (TO-BE process) [17], and the second one is by utilizing the use case diagrams from the RUP method [10]. The TO-BE flowcharts can be drawn according to different scenarios to determine the future process [28]. A use case is a pattern for a limited interaction between a system and actors in the area of application. Use case diagrams are the means of expressing the requirements and the actors involved in the project. According to the RUP rules the same use case is utilized in system analysis, design, implementation, and testing [17].

28.4.2 *Categorize the Data Collection*

The most efficient way of data categorization is to determine the most important output data according to the stakeholders' requirements and subcategorizes step by step. Such grouping permits the type of data in the PCS project with respect to the needed data. In the configuration projects, the data needed can directly come from the stakeholders' requirements determination, e.g., data for documents needed from configurator, data needed for integration and software development, and data needed for price calculations. This way the data needed can be categorized according to the desired deliverables.

28.4.3 Prioritizing the Products and Functionalities

Component-based development concerned with how to build quality systems that satisfies the business needs quickly, preferably by using parts rather than handcrafting every individual element [29]. The purpose of using a component-based structure based on RUP approach [10] is to break a large and complicated project into smaller pieces in order to make the process easier both for the users and developers. As Felfering et al. [3] describe using a component-based strategy can be very helpful in complicated and highly engineered projects. When categorizing the expected results and outputs from the configurator due to the requests from stakeholders, the expectations from the project become clearer. By giving components weight to determine their priorities and importance in the project can help in the initial assessment [28, 29]. The weight can be an index for scoring the value adding activities. The comparison between the tables related to the components weights gives a sense of the importance of the components regarding different aspects such as stakeholders' requirements, sales rates, market needs, or even the complexity of the component.

28.4.4 Collecting and Validating the Data

The accuracy and correctness of data is checked by the domain expert for correctness, consistency, and completeness during the project iteration [30] in each version of PCS. A number of methods have been used to help the engineers to do project tests iteratively [31]. In order to gather the information the best option is often to have regular meetings to ask for the knowledge, receive feedback and validation.

28.4.5 Analyzing, Maintaining, and Documenting the Data

Documentation systems are one of the vital tools during the project development, which permit the domain experts to be involved in the process from the first phases of the PCS project. The presentation of the knowledge in the PCS projects in the phenomenon model structure is one of the greatest challenges in these kinds of projects [32, 33]. The ideal situation is to have a documentation system and exchange the knowledge inside the PCS with domain experts to allow them to test, verify, and update the knowledge in the system iteratively. This can reduce costs due to preventing potential mistakes in the final stages of the project. The results indicate that having documentation system available during the system development reduces the maintenance time by approximately 20 % [34]. There are previous researchers who had been worked on representation techniques [8, 35–40]. In this step it is recommended to use Product Variant Master (PVM) associated with Class, Responsibilities, and Collaboration (CRC) Cards and class diagrams, which are built on UML. The reason for the selection of this representation technique is based on experience of the research team.

Product Variant Master. The PVM presented by Hvam [41] represents the product knowledge in a structured format from three different aspects, which are customer's view, engineering view, and production/part view. The different aspects are chosen to represent the most important stakeholders of the system. Furthermore, the relation between the different views allows identification of none value adding activates and complexity. The PVM is built of the Product Family Master Plan that is used developing "product families," based on the architecture presented by Harlou [42]. For visualizing and facilitating product knowledge, the PVM has proven to be successful in number of cases.

CRC Cards. The CRC cards were first proposed by Cunningham [43] as a way to teach object-oriented thinking. Hvam et al. [41] later presented several revised definitions of the CRC cards to be used in product configuration projects. The CRC cards are used associated with the PVM and the class diagram in order to contain more detailed information.

28.5 Case Study

The proposed framework was tested at an industrial engineering company, which is specialized in production of heterogeneous catalysts and in the design of process plants based on catalytic processes. The framework was used in the early phases in PCS project for Wet Sulphur Acid (WSA) processes plants used in industries like oil refining, coking, coal gasification, and viscose fiber use.

28.5.1 Establish the Goal of the Data Collection

Aligned with the literature, workshops where held in order to determine the goal of the data collection after determining the main stakeholders. The main tools that were used in this phase were flowcharts to determine the To-Be processes and use case diagrams for visualization and communication with the domain experts. A long list of functional and nonfunctional requirements for individuals parts of the system where identified. In Table 28.4 some of the stakeholders' requirements have been prioritized according to the MoSCoW principles.

In this case use case diagrams used for the project visualizing where the main actors involved in the configuration processes along with the functional requirements.

Table 28.4 Examples of stakeholders' requirement prioritization

List of requests	Must have	Should have	Could have	Want to have
Combining document snippets into full technical or commercial proposals (sales people and cost estimators)		✓		
Loading data from the configurator into tables in the technical and commercial (sales, cost estimators and marketing group)			✓	
Price calculation, bills of material and scope of supply (all stakeholders)	✓			
Having colors for different components in user interface				✓

Table 28.5 The categorized phases for the case study

Categorized phase	Output needed
Configuration requirements	There is a need for the products data for configuring the product according to the stakeholders' order in the execution of the system
Calculation pre-requirements	There is a need for the data are used in the calculation inside the configuration engine for constraint parts
Needed documentation requirements	There is a need for the data are used in the documentation part for Price Calculation Sheets (PCS), Bills of Materials (BOM), Scope of Supply (SOP) and…
Integration requirements	There is a need for data is used for the integration section ✓ For calculation ✓ For flow diagrams

28.5.2 Categorize the Data Collection

At this stage of the project, there is a need to determine what kinds of information are needed based on the stakeholders' requirements and the project goal. In Table 28.5 a categorization of the information is listed.

28.5.3 Prioritizing the Products and Data

Weighting tables are used to determine the importance of different components [30]. In this case, the weighting is according to the stakeholders' requirements and the complexity of the components. It means the project will start with the components that receive the highest score according to the stakeholders' needs and the lowest score for the complexity. The complexity in this case is determined by comparing number of rules and attributes across different products. Products

Table 28.6 Weighting Tables

Product 1	Importance (0–10)	Complexity (0–10)
Stakeholder 1	10	10
Stakeholder 2	9	6
Stakeholder 3	8	9
Stakeholder 4	10	10
Stakeholder 5	8	8
Stakeholder *n*	….	….
Mean value of importance	$\frac{10+9+8+10+8+\ldots}{n}$	$\frac{10+6+9+10+8+\ldots}{n}$

categorized with low complexity therefore have few rules and attributes compared to other products. In Table 28.6 an example of component weighting is shown. In this table the product is requested highly by the stakeholders but also earns high degree of complexity. The weighting table could contain other factors to make the decision making easier for the configuration group.

28.5.4 Collection and Validation of the Data

In this particular case, the close relation with domain experts was really helpful to gather and validate data for the project. In this step the following achievements are fundamental for the project success:

- Logical consistency: the attributes, variables, and constraints should be consistent when entering the PCS.
- Validate the model with domain experts: there must be an efficient communication method available between the configuration group and domain experts. Therefore, domain experts are able to check and validate all the knowledge modeled in the PCS. A communication system based on the PCS data extraction used in this case [9].

28.5.5 Analyzing, Maintaining, and Documenting the Data

The documentation system at the company illustrates the knowledge in the PCS in the form of PVM and class diagrams. The system has been developed to have a proper communication with the domain expert during the project development as well as for the documentation and maintenance of the knowledge and for the future updates and changes. An example of the PVM that was made for this project is shown in Fig. 28.2.

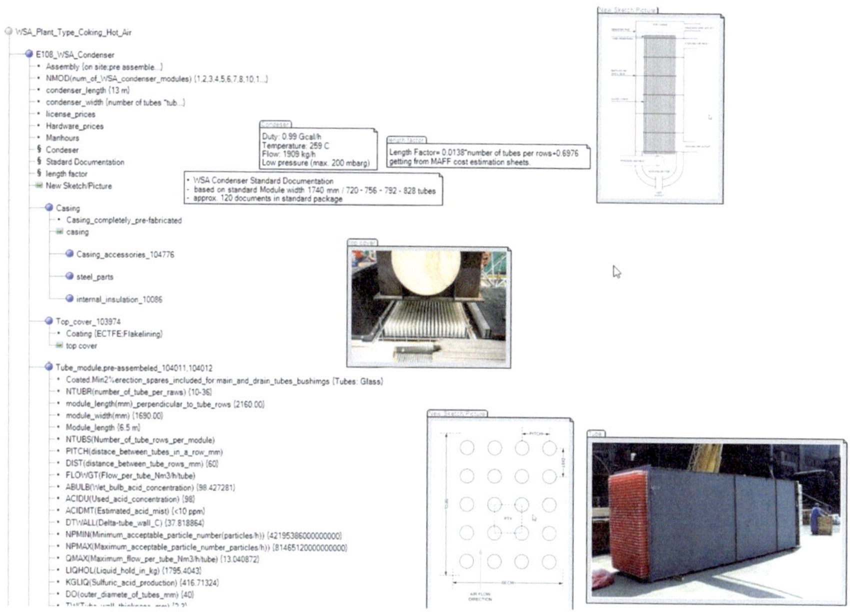

Fig. 28.2 The example for PVM from the case study

28.6 Discussions and Conclusion

The suggested framework for data collection is developed based on literature and experiences from implementing PCSs and IT projects. Companies investing in PCSs aim to use the PCS as a solution for decreasing complexity, and make the sales and engineering processes more efficient [44]. Without a clear framework for data gathering from the early stages, the PCS tends to get complicated as a result to lack of focus on the level of the data details. The framework proved to be useful for the project team by supporting early clarification of the project goal, identification of stakeholders' requirements, data categorization, products prioritization and finally for the validation of the data and maintenance and documentation. The framework helped to focus and give priority only to needed parts of the PCSs and reduce time spent in the early phase of the project. The suggested framework has been tested in an ETO company on a couple of PCS projects. In terms of future studies several areas have been identified that are listed below:

- More testing for different types of project and in different companies.
- More research on the categorization of data.
- Other available tools and methods for prioritization of the products and functionalities.

Acknowledgement This work was conducted within a Company called Haldor Topsoe A/S. The granted project and financial support is gratefully acknowledged.

References

1. Haug, A., Hvam, L., Mortensen, N.H.: Definition and evaluation of product configurators development strategies. Comput. Ind. **63**, 471–481 (2012)
2. Fleischanderl, G., Friedrich, G.E., Haselböck, A., Schreiner, H., Stumptner, M.: Configuring large systems using generative constraint satisfaction. IEEE Intell. Syst. **13**, 59–68 (1998)
3. Felfernig, A., Hotz, L., Bagley, C., Tiihonen, J.: Knowledge-Based Configuration: From Research to Business Cases. Morgan Kaufman, Waltham (2014)
4. Mildred, L., Shaw, G., Woodward, J.: Modeling expert knowledge. Knowl. Acquis. **2**, 179–206 (1990)
5. Turley, F.: An Introduction to PRINCE2®. Management Plaza (2010)
6. Taiichi, O.: Toyota Production System: Beyond Large-Scale Production, English translationth edn. Productivity Press, Portland (1988)
7. Basili, V.R., Weiss, D.M.: A methodology for collecting valid software engineering data. IEEE Trans. Software Eng. **10**, 728–738 (1984)
8. Haug, A.: Representation of industrial knowledge—as a basis for developing and maintaining product configurators. PhD thesis. Technical University of Denmark, Lyngby (2007)
9. Shafiee, S., Hvam, L., Kristjansdottir, K.: An agile documentation system for highly engineered, complex product configuration systems. In: 22nd EurOMA Conference: Operations Management in an Innovation Economy Neuchatel, Switzerland (2015)
10. Kruchten, P.: The Rational Unified Process: An Introduction. Addison-Wesley, New York (1998)
11. Friedrich, G., Jannach, D., Stumptner, M., Zanker, M.: Knowledge engineering for configuration systems. In: Felfernig, A., Hotz, L., Bagley, C., Tiihonen, J. (eds.) Knowledge-Based Configuration: From Research to Business Cases, pp. 139–155. Morgan Kaufmann, Waltham (2014)
12. Felfernig, A., Friedrich, G., Jannach, D., Stumptner, M., Zanker, M.: UML as knowledge acquisition front end for Semantic Web configuration knowledge bases. In: CEUR Workshop Proceedings (2002)
13. Barker, V.E., O'Connor, D.E., Bachant, J., Soloway, E.: Expert systems for configuration at Digital: XCON and beyond. Commun. ACM **32**, 298–318 (1989)
14. Ramachandran, S., Gil, Y.: Knowledge acquisition for configuration tasks: The EXPECT approach. In: AAAI Workshop on Configuration (1999)
15. Nellore, R., Söderquist, K., Eriksson, K.Å.: A specification model for product development. Eur. Manag. J. **17**, 50–63 (1999)
16. Forsythe, D.E., Buchanan, B.G.: Knowledge acquisition for expert systems: some pitfalls and suggestions. IEEE Syst. Man Cybernet. Soc. **19**, 435–442 (1989)
17. Hvam, L., Mortensen, N.H., Riis, J.: Product Customization. Springer, Berlin (2008)
18. Mortensen, N.H., Harlou, U., Haug, A.: Improving decision making in the early phases of configuration projects. Int. J. Ind. Eng. Theory Appl. Pract. **15**, 185–194 (2008)
19. Ebert, C.: Dealing with nonfunctional krav in large software systems. J. Software Eng. **3**, 367–395 (1997)
20. Jiao, J., Chen, C.: Customer requirement management in product development: a review of research issues. Concurr. Eng. **14**, 173–185 (2006)
21. Lim, S.L., Damian, D., Finkelstein, A.: StakeSource2.0: using social networks of stakeholders to identify and prioritise requirements. In: Proceedings of the 33rd International Conference on Software Engineering. ACM (2011)
22. Bittner, K.: Use Case Modeling. Addison-Wesley Longman, New York (2002)
23. Haug, A., Hvam, L., Mortensen, N.H.: Implementation of conceptual product models into configurators: from months to minutes. In: 5th World Conference on Mass Customization & Personalization MCP (2009)
24. Forza, C., Salvador, F.: Managing for variety in the order acquisition and fulfillment process: the contribution of product configuration systems. Int. J. Prod. Econ. **76**, 87–98 (2002)

25. Paetsch, F., Eberlein, A., Maurer, F.: Requirements engineering and agile software development. In: Proceedings of the Twelfth International Workshop on Enabling Technologies, Washington, DC (2003)
26. Hvam, L., Pape, S., Nielsen, M.K.: Improving the quotation process with product configuration. Comput. Ind. **57**, 607–621 (2006)
27. Tiihonen, J., Heiskala, M., Anderson, A., Soininen, T.: WeCoTin—a practical logic-based sales configurator. AI Commun. **26**, 99–131 (2013)
28. Hvam, L.: Mass customisation in the electronics industry: based on modular products and product configuration. Int. J. Mass Customization **1**, 410–426 (2006)
29. Tiihonen, J., Soininen, T., Männistö, T., Sulonen, R.: State of the practice in product configuration—a survey of 10 cases in the Finnish industry. Knowledge Intensive CAD **1**, 95–114 (1996)
30. Shafiee, S., Hvam, L., Bonev, M.: Scoping a product configuration project for Engineer-to-Order. Int. J Ind. Eng. Manag. **5**, 207–220 (2014)
31. Hirsch, M.: Making RUP Agile. OOPSLA 2002 Practitioner Report (2002)
32. Sabin, D., Weigel, R.: Product configuration framework—a survey. IEEE Intell. Syst. Appl. **13**, 42–49 (1998)
33. Hansen, B.: Development of Industrial Variant Specification Systems. PhD thesis. Department of Manufacturing Engineering and Management, Technical University of Denmark, Lyngby, Denmark (2003)
34. Tryggeseth, E.: Report from an experiment: impact of documentation on maintenance. J. Empir. Softw. Eng. **2**, 201–207 (1997)
35. Aldanondo, M., Rouge, S., Veron, M.: Expert configurator for concurrent engineering: cameleon software and model. J. Intell. Manuf. **11**, 127–134 (2000)
36. Chao, P., Chen, T.: Analysis if assembly through product configuration. Comput. Ind. **44**, 189–203 (2001)
37. Margo, D., Torasso, P.: Decomposition strategies for configuration problems. Artif. Intell. Eng. Design Anal. Manuf. **17**, 51–73 (2003)
38. Tseng, H., Chang, C., Chang, S.: Applying case-based reasoning for product configuration in mass customization environment. Expert Syst. Appl. **29**, 913–925 (2005)
39. Jinsong, Z., Qifu, W., Li, W., Yifang, Z.: Configuration oriented product modelling and knowledge management for made-to-order manufacturing enterprises. Int. J. Adv. Manuf. Technol. **25**, 41–52 (2005)
40. Yang, D., Miao, R., Wu, H., Zhou, Z.: Product configuration knowledge modelling using ontology web language. Expert Syst. Appl. **36**, 4399–4411 (2009)
41. Hvam, L.: A procedure for the application of product modelling. Int. J. Prod. Res. **39**, 873–885 (2001)
42. Harlou, D.: Developing Product Families Based on Architectures Contribution to a Theory of Product Families. Technical University of Denmark, Lyngby (2006)
43. Beck, K., Cunningham, W.: A laboratory for teaching object-oriented thinking, object-oriented programming systems. Languages and Applications Conference, pp. 1–6 (1989)
44. Hippel, E.: User toolkits for innovation. J. Prod. Innov. Manag. **18**, 247–257 (2001)
45. WiseGEEK. http://www.wisegeek.com/what-does-a-configuration-engineer-do.htm

Chapter 29
Minecraft and the Resource-Scarcity Advantage

F. Xavier Olleros

29.1 Introduction

Whenever we see a good design, we tend to ascribe its existence and merits to the enlightened mind that created it. In so doing, we gloss over the particular circumstances that might have influenced the design as much as the creator's mind did. Some creative constraints, in particular, can be very fertile [1]. This can be as true for successful technological and business model innovations as it is for other types of creative achievements.

Along these lines, resource scarcity has long been acknowledged by economists not only as a powerful stimulus for innovation, but also as an active selector and shaper of the type of innovation that will be favored by entrepreneurs and their clients [2]. More recently, a school of thought has emerged to underline the importance of "frugal innovation," that is, innovation that perceives resource constraints not as an impediment but as an enabler of innovation [3]. Arguments for frugal innovation often highlight the fertile ground that poorer countries may represent for the deployment of cutting-edge new technologies and business methods, such as solar power, innovative eye-surgery procedures, or affordable ambulance services for all [4, 5]. If so, the argument goes, multinational companies should abandon old assumptions about the optimality of always launching new products and services in rich countries, and consider the profitable opportunities afforded by innovations first targeted at poorer nations [6, 7].

This chapter argues that frugal innovation can be just as optimal and necessary in rich as in poor regions. My argument is simple: in a resource-abundant setting, a

F.X. Olleros (✉)
Department of Management and Technology, University of Quebec in Montreal (UQAM), Montreal, QC, Canada
e-mail: olleros.xavier@uqam.ca

J. Bellemare et al. (eds.), *Managing Complexity*, Springer Proceedings in Business and Economics, DOI 10.1007/978-3-319-29058-4_29

designer can spend her way out of sloppy choices and avoidable mistakes, and she'll therefore have little incentive to always strive for the best possible solution. Contrarily, by denying the designer ample room for error, resource scarcity forces her to be very disciplined and to always choose smartly.

Few successful entrepreneurs have acknowledged the salutary effects of resource scarcity in more eloquent terms than Pierre Omidyar, the founder of eBay. As is well known, eBay's design as a two-sided platform that would only facilitate market transactions, without having to worry about products, prices, and logistics, was critical to its scalability and rapid success [8]. In a May 2002 commencement address at Tufts University, Omidyar frankly admitted that his "enlightened design choices" were less a stroke of genius than a response to the money and time constraints that he was operating under. As he put it: "Almost every industry analyst and business reporter I talk to observes that eBay's strength is that its system is self-sustainable to adapt to user needs, without any heavy intervention from a central authority of some sort. So people often say to me, *when you built the system, you must have known that making it self-sustainable was the only way eBay could grow to serve 40 million users a day*. Well… nope. I made the system self-sustaining for one reason: back when I launched eBay on Labor Day 1995, eBay wasn't my business—it was my hobby. I had to build a system that was self-sustaining because I had a real job to go to every morning. I was working as a software engineer from 10 to 7, and I wanted to have a life on the weekends (…) If I had a blank check from a big venture capitalist and a big staff running around, things might have gone much worse. I would have probably put together a very complex, elaborate system—something that justified all the investment. But because I had to operate on a tight budget, tight in terms of money and tight in terms of time, necessity focused me on simplicity. So, I built a system simple enough to sustain itself" [9].

In the following pages, I will argue that, like eBay years ago, Minecraft—the immensely popular indie game—has also benefitted from the resource scarcity that weighed on its creator during the early stages of the game's development. The second section of this chapter highlights Markus Persson's excellent choices for his game, by comparing them to those made by another user-built, open-ended virtual environment, Second Life, several years before. The third section of the chapter describes the scarcity of time and resources that Persson experienced from May 2009 to December 2010, as the sole creator of an increasingly complex world and the sole moderator of an increasingly large community of gamers. The fourth section of the chapter compares Minecraft with LEGO Universe in order to better assess the extent to which resource scarcity may have predisposed Persson to make some of his most consequential design and business model choices. I conclude that in an innovative economy, small and frugal entrepreneurial initiatives, such as Persson's, sometimes have important advantages that need to be recognized and supported more fully.

29.2 Persson's Superb Design and Business Model (or Why Minecraft Succeeded Where Second Life Failed)[1]

Markus (aka "Notch") Persson started developing Minecraft (MC) on May 10, 2009. Barely 1 month later, he posted an alpha version of the game for free download and started accepting and charging pre-orders for the full game at preferential prices. An unfinished MC version would always be free, but users who chose to immediately pay $15 for the game would be entitled to free futures updates. Those who waited for the beta version would pay $20 and those who waited for the final version would pay $26.

Very gradually—so gradually that Persson could not quit his day job until August 2010 and could not hire suitable programming help until December 2010—MC became a runaway success. Officially released in November 2011, by September 20, 2014, MC had become the best-selling computer game of all time. By March 26, 2015, MC's computer version had reached 19 million sales [11]. To these figures, we need to add the sales of MC versions for game consoles and wireless devices. As of September 2014, there had been more than 12 million downloads of MC on the Xbox console alone. And by mid-January 2015, MC's Pocket Edition, for smartphones and tablets, had sold over 30 million copies [12].

MC's profitability has been no less astounding. During the 15 months following the founding by Persson of his Mojang startup (September 2010—December 2011), MC's revenues reached $78,722,300. Of that, more than $69 million was pure profit. Not surprisingly, in September 2014, Microsoft paid 2.5 billion US $ for Mojang and all of its assets, including MC [13, 14].

So, what exactly is so special about MC? In January 2012, the late Greg Lastowka, a professor at the Rutgers Law School tried to explain it. He first expressed the conventional wisdom amongst game developers: "Players do not have the creative skills to make games, so why should they have robust creative tools? They'll just make boring junk" [15]. He then went on to argue that Minecraft's success had turned that bit of expert wisdom on its head: "Minecraft is a hit because it was a game that took the creativity of players just as seriously as it took the creativity of its creator (…) Most games run players on rails through pre-programmed content, showing off what game developers make and giving players little freedom to shape their virtual worlds. Minecraft, intentionally or accidentally, took the opposite approach. It recognized that players appreciate artistic and creative freedom" [15].

Although correct, professor Lastowka's account neglected to acknowledge the fact that MC had not invented the open-ended approach to user-built virtual worlds that he so praised. Linden Labs did so, back in 2003, when it launched Second Life (SL). SL, however, has failed to reach mainstream success, and remains a stagnant world, kept alive by a relatively small population of hardcore fans. In 2009, on average, some 133,000 people—very few of them under 18—were responsible for about 90 % of the traffic in SL servers [16]. Since then, the number of concurrent SL logins has fallen, rather than risen [17].

[1] Unless otherwise indicated, the data reported in this section come from [10].

MC, on the other hand, is booming. How much bigger than SL is MC today? Here's a piece of comparative data, among others. Twelve years after its official launch, SL has an average of about 45,000 concurrent logins [17]. Only three-and-a-half years after its official launch, the multiplayer version of MC today has an average of about one million concurrent logins [18]. And to that figure, we need to add the millions of users—mostly very young kids—concurrently using the single-player MC apps on their computers or tablets, at any given time.

Why this sharp divergence of diffusion paths? Why has MC succeeded so massively while SL merely sputtered along? Both MC and SL are open-ended virtual worlds that depend entirely on user-generated content, impose minimal constraints and offer neither a blueprint nor preset goals. But SL has several design flaws that explain why it never became a mainstream success. They are the following:

- A steep learning curve, beyond the mere touring of the virtual space.
- An expensive merchant economy: little of interest can be done in SL without disbursing serious amounts of money. Both Linden Lab and SL's residents make money through the trading and monthly renting of "virtual real estate."
- A single, seamless space, shared by all SL visitors at all times. This may have some advantages for some people, but it has mostly disadvantages for most people. For example:
  - Users face a considerable risk of unpleasant encounters with total strangers who can disguise their true identity and intentions any way they want.
  - Given the absence of search filters and guides, finding interesting stuff within the SL world is a very time-consuming affair, at best.
  - Portability to wireless devices has been impossible, given the huge size of the SL digital file.
  - SL offers zero room for platform customization, rather than simple customization of play, as any user's platform changes would impact everybody else's play experience.
- Since SL sits in Linden Labs' servers, response times for client computers become poor on slow Internet connections. The result often is awkward interaction between gamers.
- The complexity of SL's platform and economy has resulted in a hefty stack of rules and regulations. Its "Terms of service" contain over 45,000 words, spread across some twenty documents that the user is supposed to have read, understood, and agreed to before accessing SL or its forums [19].

All of the above flaws and limitations have resulted in a virtual world that can only attract a fringe population. It is very touching to read that "the disabled can walk, run, work and dance inside this virtual country" [20, p. 267], but a platform that is ideal for people who are old or disabled is not necessarily optimal for millions of kids and teenagers eager to explore and experiment.

Minecraft is a totally different type of virtual world, largely thanks to some early design choices made by Markus Persson. While staying faithful to the idea of an open-ended, user-built world, Persson chose to develop a platform, a business model, and a user experience that are the exact opposite of SL's.

Consider first the most salient characteristics of MC's user experience.

- MC's play entails a very smooth learning curve. As a result, it offers a child-friendly experience. Many children start playing MC when they are 4 years old, as soon as they learn that their daycare friends are on it.
- MC is inexpensive to buy and free to play (i.e., no monthly fees required) in single-player mode, and even in some communitarian multiplayer platforms. Furthermore, we can surmise that thousands of poor kids are using pirated copies of the full edition, since the MC file is fairly small and Persson and Mojang have been uninterested in scaring away freeloaders.
- MC is compatible with all the major game consoles and computing platforms, wired or wireless (except Nintendo), helped by its small and self-contained digital file.
- MC is very customizable: one can choose between single-player and multi-player games, and if the latter, one can choose between family-friendly servers, run by Mojang, or third-party servers. In all cases, the player can select her game partners as well as the mode of play: spectator, creative, survival, or adventure.
- Though initially smooth, MC's learning curve need not ever reach a plateau for the keenest players, since they can transition from simple play to programming their own MC mods and plug-ins [12, 21].
- The emergent and self-directed style of play encouraged by MC means that each gamer's experience is unique. This in turn favors the YouTube *Let's Play* phenomenon that has contributed so much to MC's viral promotion and spread (more on this, below).

This attractive set of qualities in MC's user experience is not a fluke. They all have resulted from a few important choices made by 'Notch' Persson in 2009–2010, regarding both his game design and his business model [15, 20, 22–25]. For example:

- From the earliest days of the MC project, Persson kept a running blog (appropriately named "The Word of Notch"), where he expounded on various matters concerning the evolving MC universe, for the benefit of its users. In addition to this information venue, he read hundreds of fan emails and actively participated in MC-oriented forum discussions. He also maintained a Twitter feed followed by more than two million fans.
- Persson launched an alpha version (May 2009) and a beta version (June 2009) of MC and asked gamers for early feedback as to possible improvements. A perfect illustration of the importance of user feedback to MC's success was Persson's change of mind regarding what turned out to be the most widespread and impactful form of MC use: its "creative mode." Initially, he was not keen on favoring this play mode, figuring that without monsters and other dangerous creatures, the game would lose much of its playful tension and would become terribly lame. But when he saw what some people had managed to build with MC blocks and the success that their showcase videos garnered in YouTube, he changed course and made sure that in the beta version of the game the creative mode would be more fully supported.
- While those early versions of the game were free, Persson immediately started accepting and charging pre-orders of the full version (launched in November 2011) at a discount.

- Persson did not provide any game tutorials or user instructions for MC, thereby encouraging others to fill the void. Which they did in droves, usually through *Let's Play* home-made videos posted on YouTube. As of this writing (August 2015), more than 45 million videos about Minecraft have been posted on YouTube. Thus, MC's success is a stellar example of viral marketing: Mojang has never needed to spend any money on advertising; the user community has spontaneously done all the promotion in the cheapest and most effective way possible.
- Persson did not involve his startup Mojang in the setup and management of multiplayer servers until the summer of 2013—through its family-friendly Minecraft Realms service—more than 4 years after the game's development started and almost 2 years after its official launch. A mix of user-communities and third-party entrepreneurs rushed to fill this void, at considerable expense of time and money [25]. In early 2015, there were over 200,000 third-party-run MC servers [24].
- Persson also encouraged users to produce plug-ins, e.g., "skins" for new custom textures, and mod packages that add additional tools and features to the game. And again, that opening served to mobilize the initiative and energy of third-party developers and entrepreneurs who rushed to fill the void.

To summarize, between May 2009 and September 2010, Persson made several shrewd design and business model choices. Arguably, the most important of them was giving MC players unprecedented degrees of creative freedom. In addition to giving players many in-game options, he expanded their creative freedom in at least four other ways. First, by soliciting early gamers for design suggestions and implementing the best of them; second, by encouraging players to show off their own MC creations and to broadcast their own tutorials to the game, typically through the YouTube platform; third, by allowing gamers to host and manage their own servers for multiplayer games; and fourth, by allowing gamers to code their own modifications to the game.

All of these collaborative and decentralizing decisions proved to be critical to the formation of the vibrant value ecosystem that has since evolved around the MC game. What factors could explain such flawless design choices? Markus Persson is undoubtedly a very competent designer but, based on the evidence presented by his own biographers and on a detailed comparison of MC and LEGO Universe, I will proceed to argue that his penchant for delegating and externalizing critical tasks was largely due to his lack of the time and money with which to attend to such matters himself, or to hire and train a team that could do so reliably.

29.3 Persson's Creative Constraints

From May 2009 to December 2010, Markus Persson worked on Minecraft's development alone, and only part-time until August 2010. By the end of September 2010, his two Mojang cofounders, Jakob Porser and Carl Manneh, started helping him with strictly administrative matters [10, p. 132].

Detailed accounts of those early months before the core of the Mojang team came together, repeatedly mention the time and money constraints that Persson was under, all through that critical period. Moreover, when he was asked why he decided to release his game so early (in alpha version on May 17, 2009, and in beta version on June 13, 2009) and to start charging for pre-orders of the finished game, Persson's answer was concise and frank: "I would never have been able to finish it otherwise" [10, p. 99]. Understandably, back in May 2009, it was essential for Persson to turn MC into a stable source of funds as soon as possible, so as to be able to quit his day job, turn his hobby into a full-time occupation, and perhaps even have enough money to hire some help. Right after the above quote from Persson, the authors of the most authoritative book yet published about the MC story added: "Anyone looking for a more refined business logic behind what would become the most profitable gaming phenomenon of the last decade is on a fool's errand" [10, p. 99].

By the time Mojang was formed, in September 2010, Persson was already netting tens of thousands of dollars daily from MC sales. Money was never again a problem for him, but finding the right people to hire certainly was. Only in January 2011, did a second programmer, Jens Bergensten, start helping Persson with the coding of MC. And only in the summer of 2012, was Persson able to fully delegate MC's further development to a Mojang team led by Bergensten [10, pp. 130–143].

Even after that date, Persson and his team kept playing catch-up with the various needs of their exploding user community. For example, even though Persson always encouraged users to produce plug-ins and "mods" for the game, Mojang failed to develop an official MC API until February 11, 2013, and only for the Raspberry Pi. As of this writing, all other computing and game platforms are still waiting for their corresponding MC API [14]. Likewise, the frequently requested, family-friendly, Mojang-operated servers for the multiplayer version of MC only became a reality in the summer of 2013 [26].

Important as the above evidence is to my argument that resource scarcity can help rather than hinder creative thinking and implementation, I suggest that we can get a more complete and convincing picture of this argument through another pertinent comparison. Thus, in the following section, I will offer another useful point of contrast with Minecraft's success: LEGO Universe's failure.

29.4 Why Lego Could Not Possibly Have Invented Minecraft[2]

MC is often referred to as the virtual-reality version of Lego, the immensely popular construction toy for kids. And Lego executives have frequently admitted that they wish they had invented MC themselves [28]. In early June 2015, the media

[2] Unless otherwise indicated, the data reported in this section come from [27].

trumpeted the arrival of LEGO World as a belated attempt to imitate MC's successful formula [29, 30]. Few analysts seemed to remember that, back in 2010, the toy giant from Billund (Denmark) had already launched a massive 3D virtual world. It was called LEGO Universe, and its very expensive failure is as pregnant with lessons for frugal innovators as that of Second Life's tepid success.

Back in 2005, when the LEGO Universe (LU) project started, Lego seemed ideally suited to build a massive, open-ended 3D construction game for kids. Despite a financial slump in 2002–2004, it had been a leading force in the global toy industry for more than 50 years. In addition to its long industry experience, its plentiful resources and its profitable product line, Lego had an enviable brand name that extended to its online website, visited by more than 20 million unique visitors every month, already back in 2005.

Unfortunately—and perhaps inevitably, given Lego's corporate culture—LU was conceived as a very complex and expensive high-end entertainment experience. And at that it simply failed, as it could not compete against the top-of-the-line role models it was trying to emulate: World of Warcraft, EverQuest, EVE Online, and similar massively multiplayer online role-playing games.

So, unlike MC, LU was a very big bet right from the start. A 5-year, $50 million project, LU involved directly or indirectly some 350 Lego employees, in addition to the 75 employees of NetDevil, the Denver (Colorado) game studio subcontracted by Lego to develop it. Oddly enough, and although LU was designed as a massively multiplayer online game for kids, building new structures was quite secondary to the LU experience. In LU's basic plotline, Lego mini-figures operated by the various players joined the righteous Nexus Force to defeat the sinister Maelstrom and its evil architect, Baron Typhonus.

The LU development process was methodical and painstakingly slow. No expense was spared in the pursuit of a flawless product launch. The green light for the project only came after some 40 meetings of the Lego brain trust, and 51 potential development studios were screened before Lego executives selected the Denver-based studio for LU. Later, as a long series of ideas and game prototypes were proposed by the NetDevil developers, a carefully vetted group of about a 100 adult Lego players were engaged to evaluate them (why adults were selected rather than kids, I do not know).

Since it was conceived as a high-end massively multiplayer online game, in its directives to the NetDevil team, Lego always insisted on the highest possible quality regarding both image definition—down to the virtual bricks, each of them impeccably rendered and imprinted with Lego's logo—and software bugs. As the worldwide launch date approached, Lego readied its own servers at considerable expense and it established a correspondingly pricy revenue formula: $40 for the game's DVD, plus a $10 monthly subscription.

The results of the LU market launch in October 2010 were deeply disappointing: in total, only 38,000 DVDs and subscriptions were sold across the world. On January 30, 2012, only 15 months after launch, LU was abandoned and its servers were gradually shut down. In his detailed analysis of Lego Universe's commercial flop, David Robertson concludes: "Lego Universe's collapse shows how difficult it is for any big company to change its product development system so as to deliver a disruptive innovation (…) Rather than let NetDevil act like the startup that it really was and ignore what counted as prudent in Billund, the Danish company's senior management pushed the Colorado coders to develop Universe the Lego way" [27, p. 231].

By contrast, and against the backdrop of LU's failure, MC's success shines as an archetype of what a "disruptive innovation" [31] of the game industry should be like. While Lego tried to battle with the gaming industry giants for the market's high end, Markus Persson developed an initially underwhelming game primarily aimed at a much younger crowd. More recently, and in classic disruptive fashion, MC has started moving upmarket, largely driven by its own users' inventiveness, particularly in the field of child education [25]. Moreover, with the proliferation of affordable 3D printing services, MC may soon pose a direct threat to Lego's classic block toys, as kids will be able to cheaply replicate the plastic versions of their own digital creations, using Mineways or similar printing platforms [32].

29.5 Strategic Implications

In this chapter, I have contrasted the swift success of Markus Persson's Minecraft game with the disappointing performances of Second Life and LEGO Universe. I conclude that Minecraft's success is an eloquent illustration of the somewhat paradoxical role of resource scarcity as an enabler of, rather than a hindrance to, innovation. If correct, my argument has implications regarding the strategic advantages of startups and their proper funding. I will close by discussing such implications.

The management literature has identified and documented three advantages that startups often have over large incumbents. First, compared to a large incumbent, a small startup can more easily start anew and reinvent processes and value networks from the bottom up, without preconceived notions or legacy burdens [33]. I will call this a startup's "clean slate" advantage. Second, compared to a large incumbent, a small startup can envisage and develop business models that will allow it to thrive on much lower profit margins [34–36]. I will call this a startup's "lean business model" advantage. Third, unlike a large incumbent, a startup has few concerns that could keep it from giving full attention to the one project that it has selected to develop and on which it hopes to build a prosperous future [33, 37]. I will call this a startup's "laser focus" advantage.

In this chapter, I have introduced a fourth and different startup advantage. While the comparative analysis of Minecraft and Second Life has allowed me to highlight the merits of Persson's game and business model design, the comparative analysis of LEGO Universe and Minecraft illustrates the fact that smart designs may partly be the happy result of resource scarcity. If so, unlike large incumbents, frugal startups will naturally gravitate toward more efficient designs and more scalable business models. I will call this a startup's "resource scarcity" advantage.

In an age of cloud computing, learning algorithms, and mass collaboration, scalable solutions tend to be increasingly optimal, especially in software-based products and services [38, 39]. Thus, if my argument is valid, the importance of startups will only grow in the digital economy, as such firms will be able to better leverage their meager resources to transform lagging sectors with innovative solutions and business models.

Lastly, and although Mojang never needed to raise external capital, its success also has something to teach us about the proper way to finance startups. The argument has often been made that a venture-capital-funded startup needs to be frugal in order to avoid diluting its founders' ownership and control of the company [40–42]. This chapter has presented a different argument in favor of frugal innovation by startups everywhere, and regardless of their source of funding.

My argument for frugal innovation suggests that feeding money to startups at an optimal rate, neither too much nor too little, is critical to their success. If this argument matters for optimal startup growth, it also matters for public policy regarding startup funding.

Many technologically advanced countries have devised a type of organization that is supposed to favor frugal innovation: the VC-funded startup. But while financial angels and venture capitalists are, in principle, well suited to impose a frugal discipline on young firms, the chronic cyclicality of venture capital markets [43] is a serious problem. Inevitably, in times of overabundant capital, the likelihood of funding undeserving ventures increases—just as the likelihood of rejecting meritorious ventures rises in times of scarce capital. But an additional problem of investment bubbles is that the likelihood of giving too much money too soon to perfectly viable startups can bias their design in deleterious ways. Having so much money in the bank, a startup will be more likely to make costly mistakes. More specifically, and as argued in this chapter, resource overabundance may lead them to quickly settle on suboptimal designs and business models, rather than strive for the best possible ones.

In this context, the potential of equity crowdfunding—typically small amounts of money canvassed from a large crowd of small investors—as a better way to fund the first round of financing for innovative startups should be worth exploring.

References

1. Stokes, P.D.: Creativity from constraints: what can we learn from Motherwell? From Mondrian? From Klee? J. Creat. Behav. **42**, 223–236 (2008)
2. Rosenberg, N.: Technological innovation and natural resources: the niggardliness of nature reconsidered. In: Rosenberg, N. (ed.) Perspectives on Technology, pp. 229–259. Cambridge University Press, Cambridge (1975)
3. Radju, N., Pradbhu, J.: Frugal innovation: a new business paradigm. http://knowledge.insead.edu/innovation/frugal-innovation-a-new-business-paradigm-2375#tVShrCzxiJSz3QuP.99. Accessed 10 Jan 2013
4. Christensen, C., Craig, T., Hart, S.: The great disruption. Foreign Affairs **80**, 80–95 (2001)
5. Hart, S., Christensen, C.: The great leap. MIT Sloan Manag. Rev. **44**(Fall), 51–56 (2002)
6. Immelt, J.R., Govindarajan, V., Trimble, C.: How GE is disrupting itself. Harvard Bus. Rev. **87**(10), 56–65 (2009)
7. The Economist. The world turned upside down: a special report on innovation in emerging market. http://www.economist.com/node/15879369. Accessed 15 April 2010

8. Cohen, A.: The Perfect Store: Inside eBay. Little Brown, New York (2002)
9. Omidyar, P.: From self to society: citizenship to community for a world of change. Tufts University Commencement Address. http://tuftsalumni.org/images/uploads/omidyar-speech.pdf. Accessed 19 May 2002
10. Goldberg, D., Larsson, L.: Minecraft: The Unlikely Tale of Markus Persson and the Game that Changed Everything. Seven Stories Press, New York (2013)
11. Minecraft Wiki. http://minecraft.gamepedia.com/Minecraft_Wiki. Accessed 18 June 2015
12. Popper, B.: Why parents are raising their kids on minecraft, The Verge. http://www.theverge.com/2014/9/15/6152085/why-parents-love-minecraft. Accessed 15 Sept 2014
13. Stuart, K., Hern, A.: Minecraft sold: Microsoft nuys Mojang for $2.5bn. The Guardian. http://www.theguardian.com/technology/2014/sep/15/microsoft-buys-minecraft-creator-mojang--for-25bn. Accessed 15 Sept 2014
14. Wikipedia, Minecraft. http://en.wikipedia.org/wiki/Minecraft. Accessed 20 June 2015
15. Lastowka, G.: Minecraft, intellectual property, and the future of copyright. http://www.gamasutra.com/view/feature/6591/minecraft_intellectual_property_.php. Accessed 17 Jan 2012
16. Au, W. J.: Stats shock: just 133K second life users responsible for nearly 90% total monthly SL user hours! http://nwn.blogs.com/nwn/2009/08/sl-has-133k-hardcore-users.html (2009)
17. Nino, T.: Second life statistical charts. http://dwellonit.taterunino.net/sl-statistical-charts/. Accessed 20 June 2015
18. Maiberg, E.: Minecraft hits one million concurrent players on PC. http://www.gamespot.com/articles/minecraft-hits-one-million-concurrent-players-on-p/1100-6424579/. Accessed 10 Jan 2015
19. Nino, T.: The second life terms of service are a part of the problem. http://dwellonit.taterunino.net/2014/07/16/second-life-terms-service-part-problem/. Accessed 16 July 2014
20. Johnson, P.: Second Life, Media and the Other Society. Peter Lang Publishing, New York (2010)
21. Whale, D., O'Hanlon, M.: Adventures in Minecraft. Wiley, New York (2014)
22. Lastowka, G.: Minecraft as Web 2.0: Amateur Creativity & Digital Games http://papers.ssrn.com/sol3/papers.cfm?abstract_id=1939241 (2012)
23. McVeigh, T.: Minecraft: how a game with no rules changed the rules of the game forever. The Guardianhttp://www.theguardian.com/technology/2013/nov/16/minecraft-game-no-rules-changed-gaming. Accessed 16 Nov 2013
24. Landay, L.: Spawn this: Minecraft as a virtual world. https://www.academia.edu/12003027/Spawn_This_Minecraft_as_a_Virtual_World (2015)
25. Ito, M.: Why Minecraft rewrites the playbook for learning, BoingBoing. http://boingboing.net/2015/06/06/why-minecraft-rewrites-the-pla.html. Accessed 6 June 2015
26. Handrahan, M.: Mojang targets families with Minecraft realms subs service. http://www.gamesindustry.biz/articles/2013-03-13-mojang-targets-families-with-minecraft-realms-subs-service. Accessed 13 Mar 2013
27. Robertson, D.: Brick by Brick: How LEGO Rewrote the Rules of Innovation and Conquered the Global Toy Industry. Crown Business, New York (2013)
28. Collins, C.: Lego: 'We Wish We Had Invented Minecraft', Wired. http://www.wired.co.uk/news/archive/2014-08/20/lego-fusion. Accessed 20 Aug 2014
29. O'Connell, J.: Can Minecraft compete with LEGO worlds? http://www.hardcoregamer.com/2015/06/03/can-minecraft-compete-with-lego-worlds/151758/. Accessed 3 June 2015
30. Kairit, N.: LEGO launches their Minecraft-inspired game, LEGO worlds. http://www.hardcoregamer.com/2015/06/01/lego-launches-their-minecraft-inspired-game-lego-worlds/151496/. Accessed 1 June 2015
31. Christensen, C.: The Innovator's Dilemma. Harvard Business School Press, Boston (1997)
32. Humphries, M.: Mineways lets you export Minecraft models for 3D printing. http://www.geek.com/games/mineways-lets-you-export-minecraft-models-for-3d-printing-1454333/. Accessed 3 Jan 2012
33. Thiel, P.: Zero to One: Notes on Startups, or How to Build the Future. Crown Business, New York (2014)

34. Johnson, M.W., Christensen, C.M., Kagermann, H.: Reinventing your business model. Harvard Bus. Rev. **86**, 51–59 (2008)
35. Dyer, J., Gregersen, H., Christensen, C.: The Innovator's DNA: Mastering the Five Skills of Disruptive Innovators. Harvard Business Review Press, Boston (2011)
36. Ries, E.: The Lean Startup. Crown Business, New York (2011)
37. Graham, P.: Startup = Growth. http://www.paulgraham.com/growth.html. Accessed September 2012
38. Andreessen, M.: Why software is eating the world. Wall Street J. http://www.wsj.com/articles/SB10001424053111903480904576512250915629460. Accessed 20 Aug 2011
39. Chavez, M.: Paradigm shifts in software. http://www.goldmansachs.com/our-thinking/pages/paradigm-shifts-in-software.html. Accessed April 2014
40. Graham, P.: The 18 mistakes that kill startups. http://paulgraham.com/startupmistakes.html. Accessed October 2006
41. Wasserman, N.: The Founders' Dilemmas. Princeton University Press, Princeton (2012)
42. Knowledge@Wharton. The 'Too Rich to Succeed' challenge facing start-ups'. http://knowledge.wharton.upenn.edu/article/what-happens-when-start-ups-raise-too-much-capital/. Accessed 10 Nov 2014
43. Gompers, P., Lerner, J.: Short-term America revisited? Boom and bust in the venture capital industry and the impact on innovation. In: Jaffe, A.B., Lerner, J., Stern, S. (eds.) Innovation Policy and the Economy, vol. 3. The MIT Press, Cambridge (2003)

Chapter 30
Design and Innovation Beyond Methods

Caroline Gagnon and Valérie Côté

30.1 Introduction

Nowadays, society and companies are challenged by numerous issues. For instance, the coming of new technologies, the collaborative economy, the aging population, the environmental and social crises as well as the pressure on public finances. In fact, some authors suggested that design, as a strategy for innovation, is a promising way to rise up to these challenges [1, 2–4]. Moreover, the recent enthusiasm for design thinking [5–8] in management contexts tends to show that an innovation model oriented towards design would offer really interesting perspectives and tools to address the many issues that society faces presently [1, 9–12]. In that view, design thinking gathers a variety of tools and methods in order to generate innovation based on how designers reason as well as solve problems holistically and iteratively. Therefore, design thinking is considered as an innovation methodology [8, 11, 13]. However, Kimbell [13] underlines the difference between the practice of design usually focused on the doing and the design thinking centered on the thinking, the latter being increasingly generalized outside of the traditional practice of design [13]. Therefore, this article will integrate the design thinking notion, namely an approach with a set of methods for designers to tackle the different practices and knowledge of design.

A growing number of international publications are trying to demonstrate that design, as a strategic approach, could be a powerful drive for innovation [2–4, 9, 11, 14]. Furthermore, in the new economic context, Holston [15] underlines that the designer is more of a strategist whose abilities go beyond the formal aspect and

C. Gagnon (✉)
Laval University, Quebec, QC, Canada
e-mail: caroline.gagnon@design.ulaval.ca

V. Côté
University of Montreal, Montreal, QC, Canada
e-mail: valerie.cote.5@umontreal.ca

J. Bellemare et al. (eds.), *Managing Complexity*, Springer Proceedings in Business and Economics, DOI 10.1007/978-3-319-29058-4_30

manufacturing process of objects, but is also able to collaborate, to manage complex socioeconomic issues, to better understand processes linked to users, to increasingly be part of the decision making and to reveal business opportunities. Similarly, the Design Council [11] proposes that design centered innovation is based on three features: multidisciplinary work, engagement towards users/citizens as well as a holistic approach in the development of products and services. Thus, it is also mentioned that this perspective can lead to surpass silo organizational structures and favor collaboration; that it is a continuous validation process through iteration and prototyping, generating low risks; that it is centered above all on human needs, on diverse and extreme users as well as being linked to the question of consumers' heterogeneity and mass customization; and that its tools offer actual results to raised problems with tangible solutions [6, 11]. These elements are the basis of design thinking and so, a prerequisite for innovation by design.

30.2 The Design Innovation Models

In the design field, some models were elaborated to sustain this perspective. After a literature review on design innovation, relevant and thorough models were chosen to address a framework on innovation by design and our article is mainly based on Manzini [2, 16], Gardien et al. [4], and Verganti [17, 18] models. Nevertheless, it is important to keep in mind that, more often than not, publications on the subject are statements of intents rather than real empirical studies.

For Manzini [2, 16], innovation in design should be of social nature to address, firstly, challenges generated by the enduring economic crisis (mainly in Europe) and, secondly, to favor the transition towards sustainability. The notion of social innovation can be defined as a transformation approach. In other words, and very succinctly, social innovation can be understood as ''a new idea that works in meeting social goals" [19]. Furthermore, Mulgan [19] offers that "[a] more detailed definition could be the following: Social innovation is a process of change emerging from creative re-combination of existing assets (from social capital to historical heritage, from traditional craftsmanship to accessible advanced technology), the aim of which is to achieve socially recognized goals in a new way" (p. 57).

Moreover, Manzini [3] highlights that social innovation evolves with society and has a window of opportunities never explored at the moment [20–23]. Manzini [3] also proposes that design could offer a variety of initiatives that would allow for the creation of more realistic, efficient, sustainable, and reproducible social innovations. Furthermore, Manzini [3] emphasizes that design for innovation assumes a dynamic process of creative and proactive activities where the designer is often seen as a mediator between the different stakeholders and as a facilitator of ideas as well as initiatives from participants. The designer's role is then more one of conceiving and carrying out design opportunities through creativity. Specific know-how enabling designers to promote, sustain, and orient socially innovative projects. Even more so, Manzini [3] suggests that the designer can be more than a facilitator by becoming a social change agent. Thus, designers can do more than assume the traditional role of facilitators often attributed to them in co-design teams by becoming design activists who provoke socially significant initiatives [24–26]. In doing so, Manzini [3] considers that designers could take advantage of their unique skills and their empathy to create initiatives and generate new discussions in regards to current social problems of our societies.

The perspective regarding design in a transformative economy of Gardien et al. [4] introduces the different changes in design practice in line with the present socio-economic issues. The authors based their arguments on the categorization of the different economic paradigms that the history of design practice went through (industrial, experiential, knowledge-based, and transformative) and suggest that to innovate in a perpetual shifting society we have to know how to adapt to social change. However, most of the players presently adapting to the changing economic paradigms are usually start-ups rather then established companies often resulting in outdated mindsets in traditional business models. The transformative economy presented by Gardien et al. [4] suggests that design should address social needs. For instance, through living-labs that can allow contextual experimentations and data collection in order to get a greater understanding of social problems and design opportunities that could solve them.

The idea of design as an interpretation [17] is less centered on social innovation and more on innovation in the design field. Verganti [17] highlights that the unique knowledge of tools and techniques, as seen in the solely application of design thinking methods, are not enough and that design should be, first and foremost, a capacity to interpret the world and give it meaning through a product or a service. Furthermore, it should allow the transformation of negative experiences into positive ones, in other words, to go from a hostile environment to a comfortable environment [18] or at least a socially acceptable one. Norman and Verganti [18] also suggest that design projects based on design innovation research can lead to radical innovation to the meaning given to products and/or services if the goal is to get a new interpretation of what is important to people. Moreover, Norman and Verganti [18] suggested that design innovation research based on interpretative processes could lead to radical changes that would be recognizable and reproducible.

In this perspective, we think, after a broader literature review on design thinking, design and innovation as well as with the experience of teaching design research tools associated with design thinking but within design practices [1], that design if seen as an innovative approach can enable three types of business changes, in: the changing processes, the generated human experiences, and the organizational structure.

30.3 Three Types of Business Changes

30.3.1 The Changing Processes

Design thinking is increasingly adopted as a creative approach allowing innovation in businesses. The design thinking process inspired by the way designers tackle problems holistically is generally translated into four or five steps based on convergent and divergent thinking. The Design Council proposes that the approach unfolds,

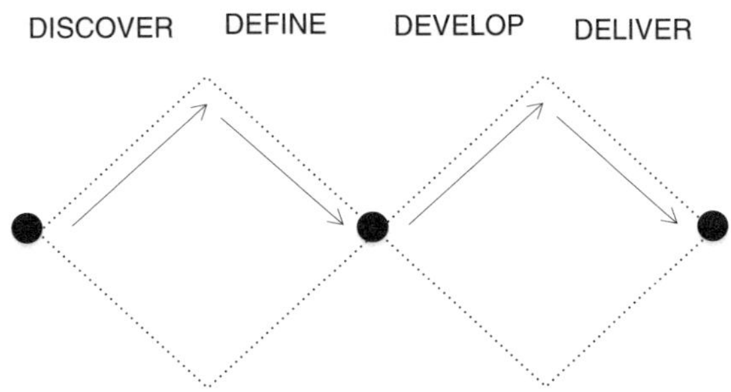

Fig. 30.1 The double diamond diagram explaining design thinking as divergent thinking and convergent thinking. *Source*: Design Council [11]

firstly, by a discovery phase where different perspectives fuse and then converge towards a problematic definition. Thus, some qualify this phase as the empathic stage aiming at collecting information to get a better understanding of the lived and felt experience of the people linked to the products and/or services. Afterwards, the concern is to develop propositions and to deliver them [11] (Fig. 30.1).

For Kimbell [27], the design process is characterized by different but connected phases going from exploration, to interpretation, to proposition, and to iteration often pulled off in a disorderly or not always in a linear way. Furthermore, the author adds that design thinking and design practice are two different perspectives and that the design thinking methods have frequently evolved outside of the traditional design practice where brainstorming is often realized implicitly and intuitively [13, 27]. However, designers adopt these tools increasingly and their use will vary with the project parameters.

In 2007, the Design Creates Value, National Agency for Enterprise developed an integration design ladder for the Danish economy (Danish Design Ladder) going from a minimal integration at the first level to a strategic integration at the fourth level [12]. Specifically, design in the first phase of the ladder is not involved in the product and service development (no design). In the second phase, design is considered as styling (design as styling) and, in the third phase, design is an integral part of the development process (design as a process). Finally, in the fourth phase, design is seen as one of the key strategic means to encourage innovation (design as strategy). In 2011, the Sharing Experience Europe (SEE) also considered a scale enabling the SEE to understand the design range of intervention in design politics and the maturity level of its integration; going from no politics to a vision for industrial design, to service design, and finally, to strategic design. Messier [28] drew his analysis from this ladder and suggested that design as innovation in Quebec is situated at the second level, meaning that design is a strict vision of industrial design for businesses. In 2013, the Design Council published a study on the role of design for the public goods where a three-level scale was developed to gain a better idea of the impact of innovation by design in the public sector. The design for discrete problems is the first level of this ladder and describes design essentially as a professional practice helping to improve particular situations inscribed in a limited project. This approach which is limited to product and service design in a constrictive way is not

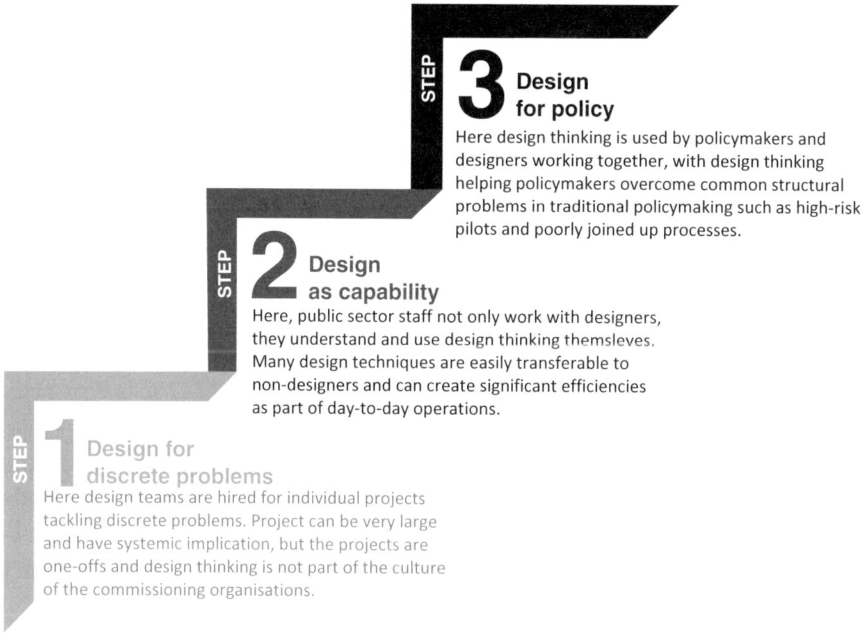

Fig. 30.2 The public sector design ladder. *Source*: Design for Public Good Report, http://www.theservicedesignprogramme.org/wp-content/uploads/2013/07/design-for-public-good.jpg

properly called design thinking because it is not inscribed in the strategic development of broader services. The design as capability is the second level of the ladder and proposes the integration of design in the public service project culture, both, in the way of operating them and in the decisional process. In this approach, the manager skills to capture the design role are present and allow design professionals to integrate a project as well as encompass problems with an overall design innovation procedure (design thinking). The final level of the ladder considers design as a strategic approach of political innovation. In this perspective, design thinking is included in the development of public policies. In a nutshell, the second and third levels of this ladder allow designers to act as facilitators of innovation processes as well as letting ideas generated by all the stakeholders involved (managers, citizens, experts, designers) materialize in tangible projects, meaning deliverable products and services (Fig. 30.2).

This changing process usually implies the integration of a higher level of sensitivity to human experiences in order to develop the desired innovative solutions. This kind of sensitivity was largely handled by the introduction of empathy in the design process as demonstrated by design thinking and illustrated by the work of the IDEO firm [29]. For the most part, these approaches are getting designers to grasp a better understanding of the complexity of a design problem, first and foremost, from a user's and/or citizen's point of view in everyday life. Furthermore, this perspective overlaps what is now called design thinking on research methods and data collection aiming at feeding the design project [13, 30]. Therefore, empathic design is an ensemble of approaches, techniques and tools insisting on the creative understanding of user/citizen experience with the objective of feeding and orienting the design project [1]. In fact, in the product design field, it is often mistaken for the user-centered approach, as we pointed out in our study on empathic design teaching: "In fact, user-centered design is primarily concerned with the functional usage of a product and little with the overall experience brought by it, like empathic design is seeking to accomplish" [1].

30.3.2 The Generated Human Experience

In the Norman and Verganti [18] article, the authors are questioning the contribution of design in leading to innovation, and particularly to radical innovation. In this line of thinking, even though design thinking may be considered as a sensitive approach to human experiences, it does not always generate innovation. In this perspective, Verganti [18] also suggests that it is the contribution of a significant experience that leads to innovation in design and that changes in product and service experiences can bring radical innovation. Moreover, the example of Philips is quoted in the article and highlights that the major transformation for the healthcare experience is not linked to a new technology, often associated to radical innovation, but rather to the lived experience itself. For instance, all the possible medical examinations were reevaluated, notably scans, to offer a reassuring and immersive experience to people in the healthcare system who often suffer from anxiety and have a certain apprehension of invasive health check-ups. In doing so, Philips seeks the experience of a real connection with people. These approaches based on experiences are also supported by innovation strategy in services. Thus, the last Design Council report showed that the global product and service experience is extremely important to value creation in businesses. These experiences linking the tangible to the intangible are leading to what many are defining as service design [13, 27, 31]. These approaches integrate right away the design thinking strategies and methods.

On this basis, the upstream introduction of empathic approaches (transversely to idea refinement techniques or in a technology) induces this experience transformation more significantly and is greatly supported by service design approaches. This understanding of the human experience could allow for the establishment of opportunities in line with perceived and lived realities. However, Postma et al. [32] as well as Köppen and Meiner [30] showed that it is still difficult to introduce the interpretation and translation of experiences into design opportunities in the different organizational structures of products and services development. In fact, Postma et al. [32] noticed this situation in design teamwork because the emotional character of the experiences gathered in the discovery and problem definition phases is often lost in the process. Thus, the authors suggest a sharable framework allowing design teams to quickly identify and structure data to prevent losing analytical sharpness. The authors also proposed to support the work with a reference framework derived from a literature and study review linked to the problematic. Moreover, an incomprehension towards the argumentative role and strategic approach of the positioning of a design project is still persistent in the traditional design practices and even more so, in the standard management structures.

30.3.3 The Organizational Structures

The findings discussed earlier brought us to question the strategic role of design in businesses. In the same way, Postma et al. [32] showed that the introduction of empathic approaches in the project processes and the emphasis on human experiences are not sufficient and that this perspective should be supported more and preserved across all the organization's departments, from the design team to engineering and marketing. Furthermore, the recognition that design is an innovation factor in businesses is not surprising to the extent that the managerial approach is design-driven and is supported by the company's management team [33–36]. Thus, it is comprehensible that if design is to be profitable it has to transversely integrate all of the organizational components (marketing, engineering, etc.). This kind of transversal integration is possible when the company's management team states its need and supports its integration, is brought by a strategic culture of design as an innovation methodology rather than a unique and punctual professional and creative expert design intervention in a project. In this way, design should become more of a strategic approach than an operational expertise. However, there is still a lack of studies on that subject matter to draw more general conclusions. Nonetheless, organizational changes are an important aspect of innovation by design even if they are difficult to set up and are often one of the major obstacles to their integration in businesses.

30.4 Discussion: Towards Value Creation Within the Business by the Creation of Sensible Products and Services?

In a very recent study [37], an analysis of design for innovation in services was conducted and it was noted that designers tend to work in the traditional logic of product delivery to give an answer to the differentiation of the market offer. However, service design is more associated with an approach implying more profound changes in organizations and in the offer configuration as a whole. Furthermore, Bason [38] in his recent work on design in the transformation of political contexts explains that design is also changing and that we can no longer consider its purpose uniquely as a way to create tangible products.

"Design as a discipline is thus undergoing a significant transformation, which perhaps places it more squarely at the heart of an organization's ability to create new valuable solutions. Variations such as participatory design and service design, which focuses on (re)designing service processes from an end-user perspective, are in rapid growth [38, 42]" (p. 4).

Moreover, service design is inscribed in an interdisciplinary perspective in design and is often confused with User Centered Design in web-based application development. Nonetheless, Kimbell [27] points out that the service design perspective adopts a broader attitude and embrace towards innovation in businesses. The author, inspired by the works of Herbert Simon [43] on design as a process to resolve problems and by Christopher Alexander [44] on design as object shaping, also suggests that design is particularly successful when generating transformations in organizations [27]. In this way, Kimbell [27] proposes a model that combines the ideas of knowledge generation and idea development between the internal company perspective and the users' world. We think that this perspective of innovation by design is interesting and should be considered because it proposes design as an approach to create value in businesses by the development of organizational processes as well as by the proposal of products and services adapted to consumers. Moreover, this perspective unites the three aspects discussed in this paper, namely the changing processes, the generated human experience, and the organizational structures. Kimbell [27] also argues that innovation should come upstream of the design process contrarily to traditional practices where design tardy intervenes. The innovation process should have more influence and should be conducted at the beginning of a project to lead to more efficient and less costly transformation approaches (Fig. 30.3).

30.5 Conclusion

To value the interpreter status of the designer it is essential to better comprehend the role of creativity and innovation in society and its underlying models [17, 45] as well as its facilitator or strategic role in the development of innovative design opportunities [10]. When the stakeholders of a project are engaged and valorized in the change processes with different tools and approaches, design can allow the visualization and the proposition of future scenarios, which are more powerful and innovative alternatives to traditional solutions in the different sectors of society and management [6]. Cope and Kalantzis [10] suggest that design acts at the interface of the knowledge society and the creative economy. Therefore, design can be considered as the capacity to act in the world where many stakeholders interweave with needs and aspirations waiting to be seized, understood, and analyzed. Finally, Best [6] underlines that design is concerned with creative economy and green economy

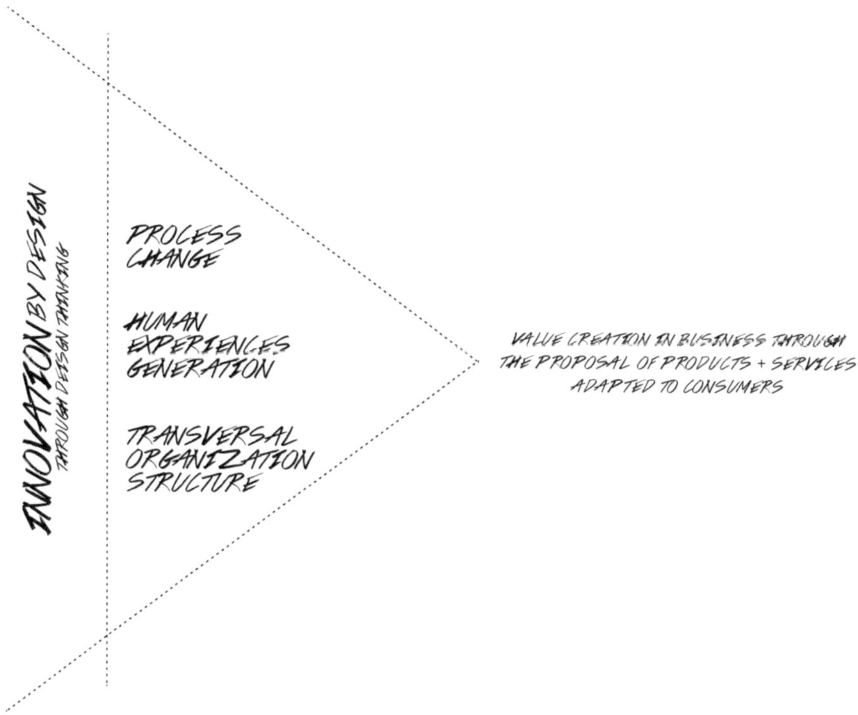

Fig. 30.3 Innovation by design framework

enabling the combined integration of the creativity benefits and the generation of new ideas; the environmental and social equity questions as a means of stimulating the economy and increasing the well-being of populations. The author also proposes that design is a transformation process centered on humans and moving beyond traditional management approaches to trigger a major shift in the way socioeconomic problems are tackled. This can also be translated in the Design Orders model of Richard Buchanan that illustrates the expanding scope of design practice [46] (Fig. 30.4).

In this way, we think that design can bring interesting perspectives to the table. In fact, in the light of this paper, we think that it is essential to act on the design processes and the development of innovative design opportunities with design thinking, to create value mainly from the product and service experience and transformation as well as, to integrate design in organizations transversely. Thus, these changes bring us to consider design as a strategic approach in businesses and as an innovative approach by service design.

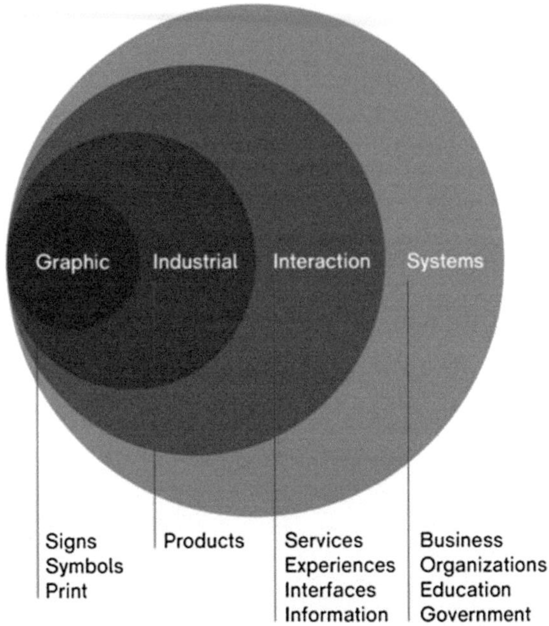

Fig. 30.4 Design orders diagram by Richard Buchanan. *Source*: Design for Europe, http://www.designforeurope.eu/news-opinion/we-need-new-ways-discuss-design

References

1. Gagnon, C., Côté, V.: Learning from others: a five years experience on teaching empathic design. In: Lim, Y.-K., Niedderer, K., Redström, J., Stolterman, E., Valtonen, A. (eds.) Proceedings of DRS 2014: Design's Big Debates. Design Research Society Biennial International Conference 16–19 June 2014, Umeå, Sweden, pp. 113–126. Publications of Umeå Institute of Design, Umeå University, Umeå (2014)
2. Manzini, E.: Making things happen: social innovation and design. Des. Issues **30**(1), 57–66 (2014)
3. Kolko, J.: Wicked Problems: Problem Worth Solving. A Handbook & a Call to Action. AC4D, Austin (2013)
4. Gardien, P., Djajadningrat, T., Hummels, C., Brombacher, A.: Changing your hammer: the implications of paradigmatic innovation for design practice. Ind. J. Des. **8**(2), 118–139 (2014)
5. Brown, T.: Design thinking. Harvard Business Review. June. 1–11 (2008)
6. Best, K.: Design as enabler of change. DMI Viewpoints. July (2012)
7. Kimbell, L.: Beyond design thinking: design-as-practice and designs-in-practice. In: CRESC Conference, September, Manchester (2009)
8. Kimbell, L.: Rethinking design thinking: Part I. Des. Cult. **3**(3), 285–306 (2011)
9. Steinberg, M.: Public-sector chief design officers, anyone? Des. Manag. Rev. **23**(2), 38–41 (2012)
10. Cope, B., Kalantzis, M.: Design in principle and practice: a reconsideration of the terms of design engagement. Des. J. **14**(1), 45–63 (2011)
11. Design Council: Design for Public Good. Design Council, London (2013)
12. DBA, A. P. D. I. G. i. a. w.: Design and the Public Good: Creativity vs The Procurement Process? A report by the Associate Parliamentary Design & Innovation Group in association with the DBA, London (2010)
13. Kimbell, L.: The turn to service design. In: Julier, G., Moor, L. (eds.) Design and Creativity: Policy, Management and Practice. Berg, Oxford (2009)

14. Brown, T., Wyatt, J.: Design thinking for social innovation. Stanford Social Innovation Review, Winter (2010)
15. Holston, D.: The Strategic Designer: Tools & Techniques for Managing the Design Process. HOW Books, Cincinnati (2011)
16. Manzini, E.: Design, When Everybody Designs: An Introduction to Design for Social Innovation. The MIT Press, Cambridge (2015)
17. Verganti, R.: Design-Driven Innovation: Changing the Rules of Competition by Radically Innovating What Things Mean. Harvard Business Press, Boston (2009)
18. Norman, D.A., Verganti, R.: Incremental and radical innovation: design research vs. technology and meaning change. Des. Issues **30**(1), 78–96 (2014)
19. Mulgan, G.: Social Innovation: What It Is, Why It Matters, How It Can Be Accelerated. Basington Press, London (2012)
20. Bauwens, M.: Peer to peer and human evolution. In: Foundation for P2P Alternatives. p2pfoundations.net (2007)
21. Tapscott, D., Williams, A.D.: How Mass Collaborations Changes Everything. Portfolio, New York (2007)
22. Leadbeater, C.: We-think. Profile Books, London. http://www.wethinkthebook.net/home.aspx (2008)
23. Murray, R.: Danger and opportunity. Crisis and the social economy. In: NESTA Provocation 09,London. http://youngfondation.org/publicatoins/reports/danger-and-opportunity-september-2009 (2009)
24. Meroni, A.: Design for services and place development. In: Cumulus Shanghai Conference Proceedings, 7–10 September, Shanghai (2010)
25. Staszowski, E.: Amplifying creative communities in NYC: a middle-up-down approach to social innovation. In: SEE Workshop Proceedings, Florence (2010)
26. Simeone, G., Corubolo, M.: Co-design tools in 'place' development project. In: Designing Pleasurable Products and Interfaces Conference Proceedings, June 22–25, Milano (2011)
27. Kimbell, L.: The Service Innovation Handbook: Action-Oriented Creative Thinking Toolkit for Service Organizations. BIS Publishers, Amsterdam (2014)
28. Messier, P.-A.: Towards a design policy for Québec. EMBA-MCGILL-HEC, Hiver 2013, Montreal (2013)
29. Kelley, T., Kelley, D.: Creative Confidence: Unleashing the Creative Potential Within Us All. Crown Business, Park Avenue (2013)
30. Köppen, E., Meinel, C.: Knowing people: the empathic designer. Des. Philis. Pap. **1**, 1–10 (2012)
31. Stickdorn, M., Schneider, J.: This is Service Design Thinking: Basics-Tools-Cases. BIS Publishers, Amsterdam (2010)
32. Postma, C., Lauche, K., Stappers, P.J.: Social theory as a thinking tool for empathic design. Des. Issues **28**(1), 30–49 (2012)
33. Szostak, B., Dhuyvetter, W., Dechamp, G.: Impact de la relation dirigeant-design dans l'intégration du design en PME. Étude exploratoire sur le territoire de La Loire. In: Colloque AIMS, Nantes, 7 au 9 juin (2011)
34. Rampino, L.: The innovation pyramid: a categorization of the innovation phenomenon in the product-design field. Int. J. Des. **5**(1), 3–16 (2011)
35. Baglieri, E, Zamboni, S., Secchi, R., Rampino, L.: Design as a strategic competence for continuous innovation. In: International DMI Education Conference. Design Thinking: New Challenges for Designers, Managers and Organizations, 14–15 Avril, EESSEC Business School, Cergy-Pointoise, France (2008)
36. Jenkins, J.: Creating the right environment for design. Des. Manag. Rev. **19**(3), 16–22 (2008)
37. Sangiorgi, D., Prendiville, A., Jung, J., Yu, E.: Design for service innovation & development final report. DeSiD (2015)
38. Bason, C. (ed.): Design for Policy. Gower Publishing, Farnham (2014)
39. Bate, P., Robert, G.: Bringing User Experience to Healthcare Improvement: The Concepts, Methods and Practices of Experience-Based Design. Radcliffe Publishing, Abingdon (2007)
40. Shove, E., Watson, M., Hand, M., Ingram, J.: The Design of Everyday Life. Berg, Oxford (2007)

41. Brown, T.: Change by Design: How Design Thinking Transforms Organizations. HarperCollins Publishers, New York (2009)
42. Cooper, R., Junginger, S.: General introduction: design management—a reflection. In: Cooper, R., Junginger, S., Lockwood, T. (eds.) The Handbook of Design Management. Berg Publishers, Oxford (2011)
43. Simon, H.: The Sciences of the Artificial, 3rd edn. MIT Press, Cambridge (1996)
44. Alexander, C.: Notes on the Synthesis of Form. Harvard University Press, Cambridge (1971)
45. Krippendorf, K.: The Semantic Turn: A New Foundation for Design. Taylor and Francis, Boca Raton (2006)
46. Sangiorgi, D.: We need new ways to discuss design. Design for Europe. http://www.designforeurope.eu/news-opinion/we-need-new-ways-discuss-design (2015)

Chapter 31
Co-creation of Experiences in Retail: Opportunity to Innovate in Retail Business

Ron Journée and Marcel Weber

31.1 Introduction

Many retailers are facing hard times. The context in which they operate is changing constantly and fast. Most retailers operate in mature and competitive markets. The advent of the Internet has also changed the retail landscape revolutionarily. Online Internet shopping is growing dramatically, as physical shopping declines, initially affecting sectors where posting or downloading of products or services is most suitable but lately spreading to other sectors, including food and fashion [1]. Products and services are nowadays available all over the world and have become virtually indistinguishable from each other, resulting in the effect that many retailers are competing on price, decreasing their profits.

Consumer habits have also changed radically because of technology with the spread of mobile, Internet, and social media networks [2]. The growth of social media and emerging channels like mobile also enhances the empowerment of customers. Retailers face the fact that customers have become more demanding, are better informed and are looking for personalized products and services. Customers are using social media channels to voice their experiences with companies. They can go wherever and whenever they want. Social media has become a primary way for individuals to share thoughts—both positive and negative, as well as likes and

R. Journée
Windesheim University of Applied Sciences, Zwolle, The Netherlands
e-mail: rja.journee@windesheim.nl; ron.journee@abnamroverzekeringen.nl

M. Weber (✉)
Windesheim University of Applied Sciences, Zwolle, The Netherlands

3CI Customer Co-Creation, Waalwijk, The Netherlands
e-mail: mea.weber@windesheim.nl; marcelweber4@gmail.com

J. Bellemare et al. (eds.), *Managing Complexity*, Springer Proceedings in Business and Economics, DOI 10.1007/978-3-319-29058-4_31

dislikes with their online network of friends and colleagues. It has opened another channel for retailers to reach out to their target audience and has provided another medium to differentiate the customers' shopping experience. Multiple channels can be used by consumers and businesses; any channel is suitable, as long as it allows the retailer to connect with the customer. Companies need to listen to the voice of the customer and provide the service they expect, or they'll go elsewhere [3].

These market revolutions, the current recession, and increasing competition to win and retain customers are driving many long established retailers out of business. This is evidenced by a steady decline in store numbers on high streets and an exodus of retail in town centers. Most large retailers have reacted by changing into multichannel firms, where the same customer visits the retailer via different channels for different purposes (e.g., obtain information online, purchase offline, and customer contact/support via telephone), expanding their focus from selling products to engaging and empowering customers, with the ultimate goal of creating a rewarding customer experience [4]. But smaller retailers seem to lag in their response, often leading to their demise [5]. In this landscape, delivering the highest standards of customer service is critical to corporate success—and even survival. Bad service isn't tolerated by customers who simply switch to competitor's websites at the touch of a mouse. Service quality and customer satisfaction may actually be declining as customers often receive service and quality that falls well below their expectations [6]. "At heart, the message is relatively simple: if you sell undifferentiated products, you compete solely on price; but if you provide experiences that consumers want, you offer a differentiated service for which a premium can be charged. The difficulty, of course, is how to create and manage these unique experiences. How to create relevant customer experiences?" [7].

In this chapter, we start the debate by systematically reviewing the concept of customer co-creation of experiences for the retail. We will particularly discuss a framework for customer co-creation, in particular, co-creation of experiences in retail SME, which we have developed in and derived from a previous study on customer experience management [8]. In the end, we will reflect on this body of knowledge, contemplating on its use for the retail in practice.

31.2 The Concept of Customer Experience and Experiential Marketing in Retail

Journée and Weber [8] conducted a systematic literature review of the concept of customer experience in B2B. They observe that the concept is applicable to all kinds of businesses or markets, including retail, since it concerns customers who are human. They posit that a perfect customer experience will aid in the attraction and retention of customers and is a highly desirable goal for organizations wishing to improve customer loyalty and enhance profitability [6]. Journée and Weber [8] define customer experience as "… a personal and subjective response that customers have on direct or indirect contact with an organization. By influencing this customer experience the organization tries to evoke several kinds of perceptions to a customer: emotional, physical, sensorial, rational, and relational, where customers and the organization co-create unique, meaningful experiences in order to achieve a profitable, durable and affective relationship that gives value to all stakeholders."

Creating superior customer experience seems to be one of the central objectives in today's retailing environments. Retailers have embraced the concept of customer experience management, ensuring a positive retail experience for customers by focusing on convenience, value and quality, or the best customer experience in the markets the firm serves [9]. Focus on product assortment is unlikely to lead to long-lasting competitive advantage for retailers. They primarily sell products manufactured by others and rarely derive sustainable benefits from exclusivity in their product assortment, because comparable products may be available elsewhere. A successful business model in the retail focuses not only on what a retailer sells but, more importantly, on how the retailer sells.

Traditionally, the "buying experience" concept dates back to 1973, when Philip Kotler [10] noted "atmospherics as a marketing tool" and suggested that store spaces and environments have to trigger certain emotional effects in the customer that influence the likelihood of making a purchase. There is now a widespread acceptance about the marketing being no longer sufficient to ensure long-term customer loyalty and, that creating a positive customer experience, leads to high levels of customer satisfaction and is an important step towards durable customer relationships [11]. Retailers recognize that greater understanding of customers can enhance customer satisfaction and retail performance [12]. To be responsive, retailers need to create memorable experiences to each customer for the purpose of generating greater economic value, instead of simply selling products and delivering services to the customers [13]. To manage a customer's experience, retailers should understand what "customer experience" actually means. They have to understand that focus on the customer's shopping experience [14] implies that customer experience is evoked at every point of contact at which the customer interacts with the retail business, its product, its service, its communication and other utterances made by the retailer, including advertising. In a retailing context, it is not only the buying moment of the customer that is important, but the internal response that this provokes. As a result, retailers have to move from a focus on selling goods and services to enhancing the customer experience [4]. Customer experience management represents a business strategy designed to manage the customer experience. It represents a strategy that results in a win–win value exchange between the retailer and its customers.

Journée and Weber [8] have developed a universally applicable, conceptual model, based on systematic review of extant literature on customer experience and customer experience management. The model has been enhanced with ongoing research and is depicted in Fig. 31.1.

Starting in the middle and glancing to the left we observe that the evoked customer experience is a result of both business and customer activities. The business actions aimed at creating a certain experience are incorporated in the organization's customer experience management. Customer activities entail search, online and offline contacting of the organization, the buying, the consumption or use, and other pre- and after-sales behavior of the customer. Customer activities and customer experience

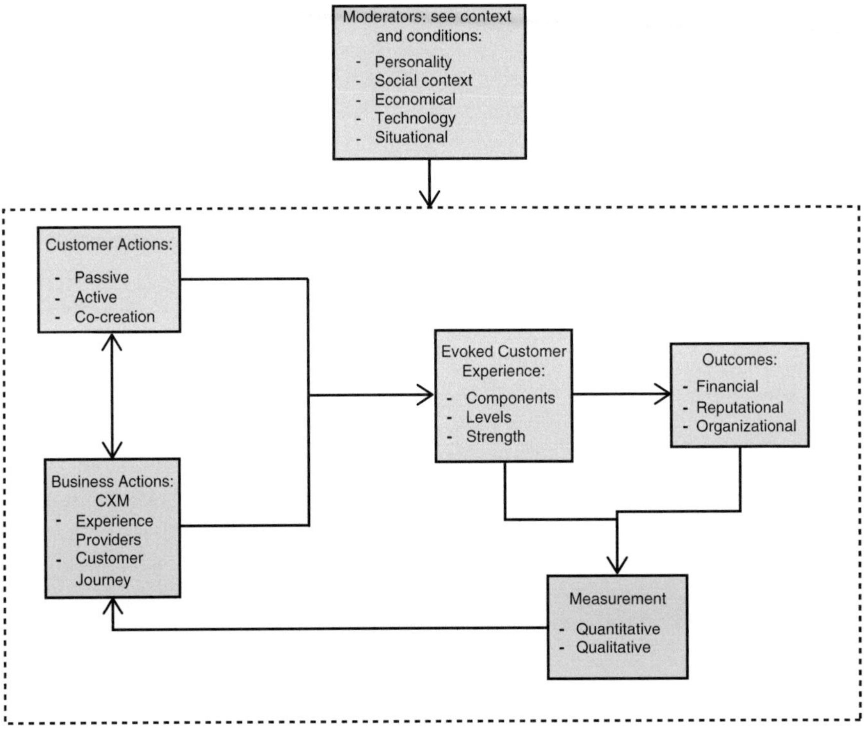

Fig. 31.1 Conceptual framework for customer experience management

management also influence each other. From an organization's perspective, the evoked experience will lead to certain outcomes, such as financial, reputational, and organizational results. However, the degree in which these outcomes are obtained is influenced by several moderating conditions, like economic condition of the market, competition, market type, and activities from other stakeholders (government, opinion groups, media, other customers), and customers' personality. For instance, customer experience in retail is not only created by elements which the retailer can control (e.g., service interface, assortment, price, promotions, loyalty programs and the supply chain/process and location/place), but also by elements that are outside of the retailers control (such as the environment, influence of others, purpose of shopping) [9]. As we have experienced from our literature review, contextual conditions that shape business' and customers' activities also act as moderating variables for the evoked experience and organization's outcomes.

Both outcomes and evoked experience can be measured by the company, which, in turn, can lead to an adaptation or even innovation of its customer experience management. We will not review all the elements of this model, but will elaborate on Business Actions and Customer Actions in retail perspective, which will be done in the next sections.

31.3 Business Actions: Customer Experience Management for Retailers

Actions that retailers have to take entail the implementation and management of the so-called "Experience Providers" [15], such as communication, visual and verbal identity, product presence, and social media. One important provider is people, personnel, or staff; organizations need to carefully select, train and motivate personnel because they are a critical in respect of customer experience, particularly front office staff. In order to determine where and when to implement these experience providers, retailers have to analyze the customer's experiential world with the Customer Journey approach [16]. The Customer Journey approach entails a description of all experiences a customer is exposed to; from long before the transaction starts to long after this moment, depicting all touch-points the customer has with the organization and other stakeholders during that time. Touch-points are the many critical moments when customers interact with the organization and its offerings on their way to purchase and afterwards [17]. Some of these touch-points are crucial and of high impact in creating a meaningful and intended customer experience [18] and are therefore called "moments of truth." The experience during the customers' journey depends on expectations, and the interactions the customer has.

The implications of the recommended actions for retailers are many-sided. Retailers should consider shopping orientations, such as the experiential or task-focused orientation [19], to guide customer segmentation and the tailoring of marketing instruments to customers' shopping orientations. But, retailers also engage in direct interactions with end customers, often with a large number of them. This underscores the importance of the customer interface: retailers have the possibility to emphasize their store's style and overall atmosphere [20]. At sensory and affective level, consumers experience retail environment in ways that involve their senses. It can evoke emotional responses like pleasure and energy, which in turn influence patronage. Research [21] suggests that the interaction of various atmospheric stimuli will strongly influence patronage behavior. A positive emotion results in a more positive evaluation of a retail experience and visitors are more likely to return and through post-purchase satisfaction, they are more likely to recommend [12].

The Internet has increased the efficiency of the shopping experience by reducing customers' search costs and by allowing them to purchase products that were previously not geographically accessible [22]. In most countries, an increase of sales by web shops is observed and projected for the coming years [5]. Retailers are facing a change in shopping habits, customer demand in shopping experience and options in services. Many retailers therefore embrace the concept of experience management and try to create positive experiences in their physical shop by distinguishing themselves from competition and web-retailers [5, 9] and increasing sales and loyalty by making shopping a pleasant experience. Although some companies may choose a single-channel strategy, many more are developing marketing strategies based on multiple channels. Studies [23] show that multiple channel retail strategies enhance the portfolio of service outputs provided to the customer, thus enhancing customer satisfaction and ultimately customer retailer loyalty. These results suggest that multiple channel retailing can be a useful strategy for building customer retailer loyalty. Implementing an online channel strategy is challenging because of its implications; managers must consider covering an extended period, influencing multiple customer touch-points, implications for face-to-face channels, and functional results. Firms need to have a clear scope and express an achievable development plan for building dynamic capabilities [22]. With most organizations today operating in a multichannel environment, the channel strategy should seek to ensure that a perfect or outstanding customer experience is created both *within* channel and *across channels* [6]. Using different channels, customers want a perfect and consistent shopping experience, and a superior customer value. This requires a total integration of all aspects of the supply chain and to integrate the activities in the different channels. The main benefit and main challenge of the new multichannel environment are to create immediate access throughout all channels to not only gain insight in the new customer-experience-driven behavior pattern, but to deliver the corresponding experiences.

However, retailers are also confronted with an increase in costs [4], demanding an emphasis on cost reductions and efficiency in the sector. In an environment with increasing competition and a growing need for operational efficiencies and customer orientation, retailers are looking beyond their organizational boundaries to develop and leverage the resources and capabilities of their supply chain partners to create superior value and competitive advantages in the marketplace, and relationship innovation [24]. Cost savings can also be realized by adopting new technologies that automate processes previously handled by employees, e.g., self-checkout technology [4] and in-store navigation [5].

In the case of implementing new hardware or software, or a combination of both, retailers have to investigate the impact of technology on the marketing mix (the 7 P's of retailing and service marketing). These P's represent the specific dimensions that retailers can use to strategically differentiate themselves from other retailers so as to produce mutually satisfying exchanges with the target market across diverse channels [25]. The challenge for retailers is to determine which technologies and innovations are likely to have the biggest impact on the retail experience and should, therefore, be adopted to remain competitive.

31.4 Customer Actions: Experience Co-creation

As concluded by Journée and Weber [8], companies ask a lot from their customers these days, even though they view customers as passive recipients of their services [26]. But, customers are expected to be more than mere passive recipients. The advent of self-service technologies, tasks previously performed by service employees, are transferred to the customer and transforming the customer role from essentially passive to active [27]. Mass customization technology has increased this active role of customers into designers to provide products which are better adapted to individual customers' aesthetic and functional preferences, contributing to unique customer experiences [28]. Customers are more and more into creating and co-producing their own experiences [29–31]. Co-production of experiences creates value for the customer, such as time-saving, ease, and the enabling a customer-specific and positive experience [32]. Co-creation of experiences has an important role in seeking to develop an outstanding or perfect customer experience by engaging the customer in a dialogue and interactions with suppliers during product design, production, delivery and subsequent consumption [6]. Co-creation of experiences seems to become an important condition for services [33] and new product development [34], where firms no longer provide experiences, but provide artifacts, content and platforms where customers can create their own, unique experiences [35]. Each customer designs his own experience in the unique context of each interaction they have with the company [8].

It would seem logical that this trend towards an increase of customers' preference for experience co-creation also applies to the retail sector. Customers are becoming increasingly accustomed to making buying decisions and shopping using a combination of shopping platforms: in the bricks and mortar shop, online, in e-mail campaigns, on mobile devices, etc. At all levels, across several retail sectors, sometimes resulting in the creation of (online) communities [36] where consumers in the role of "prosumers," take responsibilities for physical distribution, market timing, and financial risk—activities that, for customers of traditional retailers, are assumed by the retailer [37]. Stone [38] concludes that one way forward is to give customers the tools *to manage their own experience*.

Consumers will be provided with ever more sophisticated information and communications technology, e.g., through a co-creation platform [39] particularly but not solely by suppliers who use a multichannel approach. Eventually, the self-fulfilled experience will be the norm, except in most difficult and risky purchases. Consumers will increasingly use their own technology, especially smartphones, to identify the best offers, make comparisons between products and services, and receive offers from national and local suppliers. For more complex products and services or buying situations, expert third-party diagnosis of customer needs will be replaced by self-diagnosis [38].

Whether co-creation is integrated into an existing business model or has triggered the creation of a new one, if supported by appropriate format and activities, it creates customer experience and it enables a governance mechanism that can create significant value for customers, some of which can be appropriated by the retailer [4]. This means the company needs to make its products interactive, train its people for co-creative dialogue, redesign its physical places for two-way interactions, and open up the architecture of its digital sites to other processes and content that the company doesn't control.

Alternatively, opportunities for co-creation can also be extended to suppliers and other actors, like the retailer's network of partners throughout the supply chain, involved in creating and delivering customer experiences [4]. For instance, customers are coproducers in many retail environments, such as banking (e.g., Internet banking) and grocery shopping (e.g., self-scanning and self-checkout); the design of retailer interfaces is largely based on customer content (e.g., user reviews); and, in mass customization customers co-create assortment. Suppliers can also help shape retailers' assortments and interfaces to enhance the customer experience by modifying their own supply chain in response to customer needs. Such a network of retailers, customers, suppliers, and other necessary stakeholder can manifest itself as a community in which members collaborate in value co-creation that provides value for consumers [36, 37, 40]. This requires careful selection and education of co-creating customers [26, 41].

31.5 Discussion and Implications for Retail Management

In the perspective of growing globalization, fierce competition, and digitalization, retailers have to deliver the highest standards of customer service to achieve corporate success—and even survival in this continues changing landscape. Retailers face the "fight, flight, or freeze" options.

Based on extant literature review, we present a customer-centric strategy with the aim of co-creating a unique customer experience in retail. To achieve this, retailers need to adapt a customer-centric point of view and enhance the possibilities of new technologies (as mentioned in the previous section) and customer (online) communities. Retailers need to encourage the engagement and commitment of customers and employees and improve customer-listening and -connecting efforts, through a well-thought multichanneling strategy at all important touch-points in the "customer journey."

31.5.1 Customer Centricity

An important strategic goal for retailers is "customer centricity." Market-led demands and "more power to the people" lead to a shift towards consumer centricity and the creation of a more engaging experience, promoting feelings of relevance and encouraging consumer participation.

An important premise for this customer centricity is the ability to listen to the customers. Listening, as discussed here, is more than the information-gathering of traditional market research. Listening is establishing and building rapport, with the goal of creating a different, more collaborative relationship with the customer [42]. Listeners concentrate on the key distinctions that can serve as a bridge between their world and the customers. The skilled listener becomes ever more sensitive to how a customer's past has shaped his or her view of the market and the world. Most retailers know that listening to customers makes good business sense. Businesses have much to gain from actively seeking and encouraging customer participation, which we define as getting customers to provide constructive suggestions and share their ideas on how to shape service offerings. Yet while the idea that soliciting and listening carefully to customers is old, many companies only pay lip service to it [3]. Herein lies the transformational challenge that retailers face as they become customer experience co-creators.

31.5.2 Customer Engagement

Higher levels of customer role readiness, technology integration, and connectivity positively affect different co-creation experience dimensions. The impact of these dimensions on the overall co-creation experience, however, differs according to customers' expectations in terms of co-creation benefits [43]. Conceptualized in a retail setting, the customer experience entails a personal involvement at different levels. In the choice of a retail business model, retailers can design their activities in such a way that the level of customer engagement is enhanced. Customer experiences should be designed to evoke emotional involvement that goes "beyond purchase" [44], for example, by strengthening customer-brand identification [45]. Engagement goes beyond satisfaction; it represents an active, rather than passive, involvement with the product or retailer brand [4]. Many retailers adhere the idea that new technology is for rational, discount searching customers only, because these technologies do not evoke emotional experiences. However, the abundance of technology-based retail and other examples shows that online customer experiences really create affection, delight, and advocacy. Some retail examples are:

- Walmart's (http://www.walmart.com) app that helps consumers find where they can buy a product, check that it is in stock, locate it in the store itself, create shopping lists, mobile ordering with site-to-store shipping, and switch into "shopping mode" once they are in-store to access an array of tools including store maps
- Walgreens' (http://www.walgreens.com) one-stop-shop app creates an integrated experience for health and wellness for consumers that need prescription refills via barcode scanning, medication reminders, photographic orders, and loyalty card point tracking
- Home Depot's (http://www.homedepot.com) mobile app not only provides consumers with access to an "endless aisle" of more than 400,000 products, but also applies augmented reality to visualize with the phone's camera how products at home will look like

Receiving a great customer experience motivates customers to spread the word to family and friends, contributing to the retailer's reputation. Encouraging customers to take part in spreading positive word of mouth can also result in new customer acquisition for the retailer.

31.5.3 Multichannel Strategy

One of the major developments in retailing has been the emergence of the Internet as a channel for commerce (e.g., online vs. stores). A multichannel strategy integrates e-commerce with m-commerce and social commerce, without forgetting to reinvent the brick-and-mortar store to keep the customer surprised [2]. The entire shopping process from information search, communication and selection, transaction, delivery to after sales, can be conducted across different channels. Allowing customers to purchase online and pick up at a store, or access the retailer's larger online assortment, while shopping in store where they can take advantage of customer support [46]. A multichannel approach not only offers additional opportunity to push products and services but also helps in turning this challenge into an opportunity to connect with customers [47], for instance through social media [22].

A multichannel strategy should seek to ensure that an outstanding and consistent experience is created both within channel and across channels [6]. This requires a total integration of all aspects of the supply chain and all activities in the different channels, which influence the multiple customer touch-points during the customer journey. Research shows that firms with well-integrated channels are more successful than single-channel firms or multiple, but poorly integrated channels [48]. These results suggest that multiple channel retailing can be a useful strategy for building customer retailer loyalty. Channel integration can enrich customers' experiences with retailers, and can strengthen customers' overall perceptions regarding the image of a retailer [6]. Developing multiple, well-integrated channels should therefore be an important goal and challenge for retailers.

31.5.4 Co-creation

The Internet has increased the prevalence of customer co-creation [4]. The proliferation of social networks represents an extraordinary opportunity for companies wishing to increase customer participation. Co-creation involves customer engagement in the creation of offerings through ideation, design, and development [26, 43].

Minkiewicz et al. [49] assert that, in order to co-create experiences and value, the individual consumer has to actively participate in one or more activities performed in the experience (co-production), transcend into a psychological state of cognitive and emotional immersion (engagement), and tailor the experience to meet their needs through customization, interaction with service representatives, and technology (personalization). It is the experience that is co-created, with value as a derived outcome [49]. This is achieved through the provision of stimulating and engaging retail environments where consumers are inspired to create and tell their own stories around their retail experience. Frontline staff can fulfill specific consumer requests and personalize the experience to each individual consumer that would be best suited to their needs. This can lead to the origination of new, consumer-created brand communities into which retailers can tap to engage customers in co-creation. Retailers can tap into customer communities, listen to the voice of the customer, see what they suggest, comment or even complain and address these immediately, engage customers and have them get a feeling of ownership and connection with the brand [3]. Retailers should therefore observe both customer-to-business interactions and customer-to-customer interactions, e.g., reviews, and assistance [26], across all channels.

31.5.5 Aimed at Co-creating Unique Customer Experiences

The challenge also lies in co-creating optimal and consistent customer experiences across multiple channels during the whole customer journey. Connecting the different touch-points and integration between the channels are essential. Customers make their choice for a device or a channel, depending on where they are located: it depends where they are and what they expect concerning customer experience.

Yet, many small and midsized retailers are still one-way focused, centered on products and often not familiar with developing engagement platforms and using multiple channels and different technologies. Many companies find it difficult to deliver—needless to say co-create—consistent and unique experiences across the channels, leading to bad experiences which can lead to negative word of mouth [50].

In endorsing Sorescu et al. [4], we conclude that linking retail activities on co-creation with multichannels may require significant changes to a retailer's business model but it is a change worth considering, given its high potential on value creation. A retailer can engage customers by not just selling products, but co-creating an entire experience that adds an entirely new exciting layer to the retail setting. For instance, themed brand stores are exponents of a retail brand ideology meant to immerse the customer in a complex experience, which includes socialization, co-creation, and embedding of the brand into personal memories.

31.5.6 Implications for Future Research

We observe, however, that these recommendations are still of an abstract and conceptual nature, needing empirical evidence. We intend to further investigate the recommendations in practice, in order to develop adequate design propositions [51] for retail business model innovation and improvements. In that respect, we are currently involved in two pilots in the Netherlands, i.e., Project Retailpower [52] and Project "Vibrant Inner Cities" [53]. We expect to present results of this research in the future.

31.6 Conclusion

Based on a systematic review of extant research on the subject of customer experience, we have recognized that customer experience management can be a new way for retailers to create a competitive advantage, simultaneously creating more value for their customers. We exposed the conceptual framework of customer experience management (introduced by Journée and Weber [8]) to the retail business. We especially researched the possibilities for customer co-creation of experiences by retailers because of the active role that customers are claiming nowadays. We have also discussed how retailers face the options to either "throw in the towel" or survive by embracing new technologies, with the aim to enhance customer co-creation of experiences. To achieve this, retailers need to adapt a customer-centric, outside-in point of view, by improving their customer-listening efforts through a well-thought channeling strategy at all important touch-points in the "customer journey" and engaging these customers throughout the whole journey. But, because of cost implications, retailers need to seek a trade off with the cost and benefits of customer co-creation of experiences. We have indicated the possibilities with new technologies, customer co-creation, and experience building through customer (online) communities. Developing co-creation means encouraging the involvement and commitment of customers and employees, making optimum use of opportunities to expand and grow. This goes beyond the traditional marketing and leads to a perfect customer co-creation of experiences.

We also observed that current research has not reached a sufficient level of knowledge to aid retailers in achieving this optimum. We intend to further investigate the application of customer experience and customer co-creation concepts in Retail in ongoing studies.

References

1. McIntyre, C.: Physical retail space and place: the historical development of a social psychogeography of liminal consumption. In: 16th Biennial Conference on Historical Analysis and Research in Marketing (CHARM), Copenhagen, DK, pp. 144–159 (2013)
2. Nueno, J.L.: The decline of main street, the rise of multichannel retail. Expert IESE Insight **9**, 46–53 (2013)
3. Merlo, O., Eisingerich, A.B., Auh, S.: Why customer participation matters. MIT Sloan Manag. Rev. **55**(2), 93 (2013)
4. Sorescu, A., et al.: Innovations in retail business models. J. Retail. **87**(Suppl. 1), S3–S16 (2011)
5. Molenaar, C.: Red de winkel! Integreer internet in je winkel en je winkel op internet, p. 192. Academic Service, Den Haag (2013)
6. Frow, P., Payne, A.: Towards the 'perfect' customer experience. J. Brand Manag. **15**(2), 89–101 (2007)
7. Patterson, A., Hodgson, J., Shi, J.: Chronicles of 'customer experience': the downfall of Lewis's foretold. J. Mark. Manag. **24**(1–2), 29–45 (2008)
8. Journée, R.J.A., Weber, M.E.A.: A bonded experience: "value creation as the creation of an experience, within a business relationship". In: Brunoe, T.D. et al. (eds.) Proceedings of the 7th World Conference on Mass Customization, Personalization, and Co-Creation (MCPC 2014), Aalborg, Denmark, February 4–7, 2014, pp. 1–16. Springer International Publishing (2014)
9. Verhoef, P.C., et al.: Customer experience creation: determinants, dynamics and management strategies. J. Retail. **85**(1), 31–41 (2009)
10. Kotler, P.: Atmospherics as a marketing tool. J. Mark. **49**(4), 48–64 (1973)
11. Barnes, J.G., Wright, J.W.: In: Kandampully, J. (ed.) A framework for applying customer insight and context to the development of a shopping experience strategy, in service management, pp. 43–65. Springer, New York (2012)
12. Puccinelli, N.M., et al.: Customer experience management in retailing: understanding the buying process. J. Retail. **85**(1), 15–30 (2009)
13. Nasermoadeli, A., Erfan, S., Kwek Choon, L.: Exploring the relationship between social environment and customer experience. Asian Soc. Sci. **9**(1), 130–141 (2012)

14. Grewal, D., Levy, M., Kumar, V.: Customer experience management in retailing: an organizing framework. J. Retail. **85**(1), 1–14 (2009)
15. Schmitt, B.: Experiential marketing. J. Mark. Manag. **15**(1–3), 53–67 (1999)
16. Zomerdijk, L.G., Voss, C.A.: NSD processes and practices in experiential services. J. Prod. Innov. Manag. **28**, 63–80 (2011)
17. Rawson, A., Duncan, E., Jones, C.: The truth about customer experience. Harv. Bus. Rev. **91**(9), 90–98 (2013)
18. Normann, R.: Service Management: Strategy and Leadership in Service Business, 3rd edn. John Wiley & Sons, Chichester (2007)
19. Büttner, O.B., Florack, A., Göritz, A.S.: Shopping orientation as a stable consumer disposition and its influence on consumers' evaluations of retailer communication. Eur. J. Mark. **48**(5/6), 1026–1045 (2014)
20. Rayburn, S.W., Voss, K.E.: A model of consumer's retail atmosphere perceptions. J. Retail. Consum. Serv. **20**(4), 400–407 (2013)
21. Hart, C., Stachow, G., Cadogan, J.W.: Conceptualising town centre image and the customer experience. J. Mark. Manag. **29**(15–16), 1753–1781 (2013)
22. Klaus, P., Nguyen, B.: Exploring the role of the online customer experience in firms' multi-channel strategy: an empirical analysis of the retail banking services sector. J. Strateg. Mark. **21**(5), 429–442 (2013)
23. Wallace, D.W., Giese, J.L., Johnson, J.L.: Customer retailer loyalty in the context of multiple channel strategies. J. Retail. **80**(4), 249–263 (2004)
24. Ganesan, S., et al.: Supply chain management and retailer performance: emerging trends, issues, and implications for research and practice. J. Retail. **85**(1), 84–94 (2009)
25. Lee, L., Meyer, T., Smith, J.: In: Kandampully, J. (ed.) Reinventing the Customer Experience: Technology and the Service Marketing Mix, in Service Management, pp. 143–160. Springer, New York (2012)
26. Weber, M.E.A.: Customer co-creation in innovations. A protocol for innovating with end-users, p. 507. Eindhoven University of Technology, Eindhoven (2011)
27. Hilton, T., Hughes, T., Chalcraft, D.: Service co-creation and value realisation. J. Mark. Manag. **28**(13–14), 1504–1519 (2012)
28. Franke, N., Schreier, M.: Product uniqueness as a driver of customer utility in mass customization. Mark. Lett. **19**, 93–107 (2008)
29. Ford, R.C., Dickson, D.R.: Enhancing customer self-efficacy in co-producing service experiences. Bus. Horiz. **55**(2), 179–188 (2012)
30. Mascarenhas, O.A., Kesavan, R., Bernacchi, M.: Lasting customer loyalty: a total customer experience approach. J. Consum. Mark. **23**(7), 397–405 (2006)
31. Ramaswamy, V.: Co-creation of value—towards an expanded paradigm of value creation. Mark. Rev. St. Gallen **26**(6), 11–17 (2009)
32. Zeithaml, V.A., Bitner, M.J., Gremler, D.D.: Services Marketing. McGraw-Hill, Boston (2008)
33. Sfandla, C., Björk, P.: Tourism experience network: co-creation of experiences in interactive processes. Int. J. Tour. Res. **15**(5), 495–506 (2013)
34. Füller, J., Hutter, K., Faullant, R.: Why co-creation experience matters? Creative experience and its impact on the quantity and quality of creative contributions. R&D Manag. **41**, 259–273 (2011)
35. Prahalad, C.K., Ramaswamy, V.: Co-creation experiences: the next practice in value creation. J. Interact. Mark. **18**(3), 5–14 (2004)
36. Muñiz Jr., A.M., Schau Jr., H.J.: Vigilante marketing and consumer-created communications. J. Advert. **36**(3), 35–50 (2007)
37. Hunt, D.M., Geiger-Oneto, S., Varca, P.E.: Satisfaction in the context of customer co-production: a behavioral involvement perspective. J. Consum. Behav. **11**(5), 347–356 (2012)
38. Stone, M.: The death of personal service: why retailers make consumers responsible for their own customer experience. J. Database Mark. Cust. Strat. Manag. **18**(4), 233–239 (2011)
39. Ramaswamy, V., Gouillart, F.J.: Building the co-creative enterprise. Harv. Bus. Rev. **88**(10), 100–109 (2010)

40. Nambisan, S., Baron, R.A.: Virtual customer environments: testing a model of voluntary participation in value co-creation activities. J. Prod. Innov. Manag. **26**(4), 388–406 (2009)
41. Füller, J., et al.: Consumers' creative talent: which characteristics qualify consumers for open innovation projects? An exploration of asymmetrical effects. Creativity Innov. Manag. **21**(3), 247–262 (2012)
42. Flores, F.: Innovation by listening carefully to customers. Long Range Plan. **26**, 95–102 (1993)
43. Verleye, K.: The co-creation experience from the customer perspective: its measurement and determinants. J. Serv. Manag. **26**(2), 321–342 (2015)
44. van Doorn, J., et al.: Customer engagement behavior: theoretical foundations and research directions. J. Serv. Res. **13**, 253–266 (2010)
45. Borghini, S., et al.: Why are themed brandstores so powerful? Retail brand ideology at American girl place. J. Retail. **85**(3), 363–375 (2009)
46. Berman, B., Thelen, S.: A guide to developing and managing a well-integrated multi-channel retail strategy. Int. J. Retail Distrib. Manag. **32**(3), 147–156 (2004)
47. Deshpande, D. et al.: Emerging Multichannel Customer Experiences in Grocery Retail. WIPRO Technologies, Bangalore (2011)
48. Sousa, R., Voss, C.A.: Service quality in multichannel services employing virtual channels. J. Serv. Res. **8**(4), 356–371 (2006)
49. Minkiewicz, J., Evans, J., Bridson, K.: How do consumers co-create their experiences? An exploration in the heritage sector. J. Mark. Manag. **30**(1–2), 30–59 (2014)
50. Song, J. et al.: Fostering a seamless customer experience in cross-channel electronic commerce. In: 2013 SIGBPS Workshop on Business Processes and Service, Milan, Italy, pp. 21–26 (2013)
51. Weber, M.E.A.: Ontwerpstellingen en ontwerpprincipes. In: van Aken, J.E., Andriessen, D. (eds.) Handboek ontwerpgericht wetenschappelijk onderzoek. Wetenschap met effect, pp. 61–77. Boom Lemma, Den Haag (2011)
52. ter Halle, I., Galama, M., de Vries, S.: Retailpower—urban living labs and their contribution in retail contribution. In: 4th ENoLL Living Lab Summer School, Manchester, UK (2013)
53. Weber, M.E.A. (ed.) Klanten winnen en behouden met klantbeleving. Onderzoek in en naar klantbelevingsmanagement bij het MKB, p. 92. Christelijke Hogeschool Windesheim, Zwolle (2015)

Chapter 32
Proximity Marketing as an Enabler of Mass Customization and Personalization in a Customer Service Experience

Nataly Levesque and Harold Boeck

32.1 Introduction

The literature aptly proclaims that mass customization and personalization increase the consumer experience with regard to the service. In fact, a study of the American customer satisfaction index [1] and one of the European customer satisfaction index [2] demonstrates that the experience of service through personalization and mass customization has an effect on customer satisfaction. In addition, this factor increases customer loyalty [3]. Some authors such as [4] even claim that mass customization is at the heart of differentiation on the market, since it allows the customer to create a product according to his own needs [5]. Going further in this process, it is possible for companies to benefit from this client participation in order to adjust their offer. This involves the co-creation of values between the supplier and the customer. In this context, it must be emphasized that it is not the offer of the supplier itself that takes precedence, but the value perceived by the customer, namely the "value-in-use." With this in mind, it goes without saying that companies integrate this perspective of "value-in-use" and work on the co-creation of the client's experience in order to explore and implement strategies [6] geared toward the latter. Thus, personalization and mass customization prove to be the keys for companies to successfully increase the experience of personalized service. Managers must therefore be in a position to comprehend and master these two concepts in order to respond to the consumer's needs and to contribute to the company's ROI and brand image.

N. Levesque • H. Boeck (✉)
School of Management, ESG-UQAM, Montreal, QC, Canada
e-mail: boeck.harold@uqam.ca

J. Bellemare et al. (eds.), *Managing Complexity*, Springer Proceedings in Business and Economics, DOI 10.1007/978-3-319-29058-4_32

We postulate that one form of emerging marketing contributes to personalization and mass customization: the Proximity Marketing. The latter is based on the identification of the customer's geographical position using a wireless connectivity. The transmission of information is executed on the condition that the consumer is equipped with an electronic device (such as a smartphone or an electronic tag) and has given his prior consent. In this way, businesses can send the advertising content to a specific person and a specific location. This contributes to a personalization and customization of the service for an increased experience. As an example, millions of consumers in Japan have agreed to receive mobile alerts from McDonald's who offers customized messages comprising discount coupons, competitive opportunities, invitations to special events, therefore unique content that is specific to the brand [7]. Thus, Proximity Marketing adapts the marketing offer and personalizes it, in real time and in terms of the customer's location. In addition, Proximity Marketing provides tangible benefits for consumers, such as time (shopping, cash register, delivery) [8–10] and money savings (promotion) [9]. These prove to be well-known advantages desired by consumers that are leading to an enriched experience with the service.

32.1.1 Problem

The personalization and mass customization have already demonstrated their effectiveness in relation to the experience with the service. Currently, a growing number of companies are interested in the added value of these concepts in the framework of Proximity Marketing [11]. Although the concepts are fairly well documented individually, their interrelationship is not well established academically and the definitions are divergent. Researchers do not agree on a common conceptualization of mass customization and personalization [12]. This consequently leads to the following problem; there are a lot of discrepancies in the definitions of the concepts.

32.1.2 Research Question

We have established a clear research question, being: "Does Proximity Marketing contribute to mass customization and personalization of an improved service experience?"

The structure of this chapter is the following: Introduction to the literature about Proximity Marketing and definition of five concepts in connection with this type of marketing. These concepts are mass customization, personalization, product versioning, co-creation, as well as reverse marketing. Next, we will address the methodology used in order to select relevant cases. Subsequently, we will focus on Proximity Marketing and the concepts previously defined, as well as their values adding to the experience with the service. Finally, we will offer interesting managerial implications for brands wishing to employ this type of innovative marketing.

32.2 Theoretical Background

32.2.1 *Proximity Marketing: Development*

We must go back to the beginning of the 1920 to review the first research about positioning geography as core of commercialization activities [13]. However, it is only in the year 1990 that marketing agencies have actually implemented the use of systems based on the customer's geographic information [14]. Thus, Proximity Marketing, a new form of marketing which includes geo-localization had emerged. "We define Proximity Marketing as the wireless and localized distribution of advertising content related to a specific location. It involves geographic identification of consumers by means of technology such as wireless devices, GPS, radio frequencies, Wi-Fi, Bluetooth Low Energy, and Near Field Communication. Proximity Marketing implies that firms must dispatch their advertising contents to targeted geographic locations where potential customers have been identified [15]." Since this is a totally new field on which, to our knowledge, no academic reflection of great scope has been carried out [16], we believe it is interesting to focus on this emerging marketing.

We have listed three major benefits of Proximity Marketing use for customers. First, the real-time aspect [17]. As a result, Proximity Marketing allows immediate access to interesting information about liked brands. Second, the relevant and valuable content added for the consumer. In fact, if the client receives information about brands he likes, or even promotions, coupons and discounts, the content represents the information consistent with his tastes. Therefore the content is interesting to him [18]. Third, the personalization of a global offer, which is available through information obtained via loyalty programs [18] or other processes. In addition, Proximity Marketing naturally integrates the approach of Customer Relationship Management (CRM) combining the direct relationship with the customer, the geomatics, and the logistics [19]. The inclusion of CRM via Proximity Marketing is also representative of being empathic toward the consumers and understanding about their feelings regarding their customer experience [20].

We understand therefore that the combination of a customer's behavioral, geographical, spatial and sociodemographic information, gathered by means of sophisticated tools, provides patterns of reactions, habits and market analysis leading to the efficiency of Proximity Marketing.

We came to the conclusion that researchers fail to agree on the concepts surrounding Proximity Marketing. A confusion between personalization and mass customization often occurs. Yet, they each have their own meaning. It is therefore necessary to highlight their respective differences. As mentioned previously, three concepts related to this are apparent, namely product-visioning, co-creation, and reverse marketing. Subsequently, we will define the aforementioned.

32.2.2 Mass Customization

Mass customization is the term most often used when referring to made to "measure service." The concept started to occur toward the end of the 80s and can be considered as a natural consequence of processes becoming more flexible and improved in terms of quality and costs [21]. The concept does not only have one definition, but several. Davis speaks of mass customization when "the same large number of customers can be reached as in mass markets of the industrial economy, and simultaneously they can be treated individually as in the customized markets of pre-industrial economies [22]." Other authors agree to define mass customization more precisely: "They define mass customization as a system that uses information technology, flexible processes, and organizational structures to deliver a wide range of products and services that meet specific needs of individual customers (often defined by a series of options), at a cost near that of mass-produced items [23]." This permits us to understand that mass customization offers standard products, but the consumer can adjust them according to his tastes and available options. The consumer participates in the improvement of his purchase or service. In addition, mass customization represents an important differentiation approach due to its versatile organizational process [4, 5].

The fact that mass customization has a flip side cannot be disregarded either. The customer may be confused by the amount of opportunities that are available to him and thus has difficulty, due to the lack of information, to define his preferences and therefore does not benefit from a full product/service satisfaction [24, 25]. This phenomenon is called "mass confusion" [24]. To obstruct this deficiency, managers came to opt for a one-to-one personalization [26]. Unlike the mass customization, the one-to-one personalization relies on strengthening the unique interaction between the company and the client. It is a Business-To-Customer (B2C) approach pertaining to preserve the personalized value in terms of service, information, and support [5]. Hence, we came to the realization that mass customization is juxtaposed to another concept that is especially focused on the customer profile: the personalization.

32.2.3 Personalization

"Personalization can be defined as the ability to proactively tailor products and product purchasing experiences to tastes of individual consumers based upon their personal and preference information [27]." Our attention focused on a comparative definition of concepts that we considered explicit. "Personalization (or individualization which are used object's nature synonymously) in general means matching one object's nature with one subject's needs (i.e., customize products, services, content, communications to the needs of single customers or customer groups). Mass customization is the individualization of products (and services) at the cost of one-size fits all [28]." In addition, some articles cautioned us specifically not to confuse mass customization and personalization. According to Tseng and Piller, mass customization affects modification, assembly or modification of products or services according to the needs and desires of customers. As to the personalization, a communication and a sustained interaction between the customer and the supplier are implicated in the process in order to have knowledge, amongst other things, of the customer's profile [29] and to be able to propose offers accordingly.

To be in the position to distinguish between the two concepts, [30] p. 7) provide the example of the online catalog from the clothing retailer Land's End. In 1999, Land's End proposes a "virtual model" as well as online recommendations allowing to feature sets of clothing linked with the profile of the customer. Although the sets are adapted to the taste and measurements of the latter, the garments remain standard. It is therefore a question of personalization. In 2001, Land's End adds mass customization to its offer. In fact, the client can now order "made-to-measure" sweaters within a substantial number of designs. However, in this case, the personalization does not support the mass customization, because the client must know his measurements, since they are not automatically linked at the time of the order.

Finally, some authors describe the concept by mixing the two terms, resulting in "customized personalization" [31]. The customized personalization is based on the desire to assist the client in receiving the best possible form of offered service corresponding to his needs. However, mass customization combined with personalization—the advice and attention directed toward the customer—allows for an increased customer confidence since he is convinced that he has obtained the best alternative to meet his needs [31]. This therefore supports the thesis of [30] affirming that the two concepts are distinct and we particularly understand that their combination would increase the experience with the service greatly. Although the three following concepts are not yet genuinely exploited in connection with Proximity Marketing, they represent the continuity for both, personalization and mass customization. Consequently, we were not able to go beyond the extent of product versioning, co-creation, and reverse marketing.

32.2.4 Product Versioning

The first phase of the individualized service experience including personalization and mass customization is to provide an offer of product versioning. This simply means to offer different versions of a base product to serve a large portion of consumers. We often see this type of application in the software industry [5]. In other words, this represents the initial step for offering an increased individualized experience with the service by combining the two basic concepts. The literature subsequently suggests other marketing opportunities geared toward this type of unique setting.

32.2.5 Co-creation

For some time, the "buzz word" co-creation occurs. This concept allows for an interaction with the business sector starting at the first stage of product development. The client is therefore actively involved in the process "Co-creation is about joint creation of value by the company and the customer. It is not the firm trying to please the consumer [32, p. 4]." This statement seems to define co-creation fairly precise. An excellent example is that of Lego. In fact, the company offers the customer, after being registered online, the possibility to design his own custom Lego mini-figurine [33]. This may even become a viral popularity within the community of Lego followers. Co-creation operates therefore on an even more profound level than mass customization and personalization, and offers real benefits in relation to the knowledge of the customer base for businesses.

32.2.6 Reverse Marketing

Deriving from co-creation, reverse marketing is a type of promising marketing that follows into the tradition of options to increase the experience with the service. Reverse marketing represents the most sustained degree in terms of combination of mass customization and personalization. This type of marketing leaves the entirety of the product creation to the consumer. For example, the brand Swarovski allows the customer to create his entire jewelry [5]. In conclusion, "In the information-rich regime, marketers need to evolve further toward customer-configured offerings, where the customization is done by customers, and not by marketers (Sawhney and Kotler 2001 in [5])." We are convinced that it is a path that will be widely developed in the next few years, because if you want it done right, you've got to do it yourself.

In the light of this introduction to the literature, we believe that the five previously defined concepts in connection with Proximity Marketing are independent and by no means subgroups of personalization. For this purpose, Table 32.1 represents the various levels that Miceli et al. [5] shall appoint "A personalization continuum." Since to our opinion, the concepts complement each other, particularly in the case of mass customization and personalization, we would be tempted to rename this table as being the "Continuum of individualized experience with the service."

32.3 Methodological Process

We have listed some of the methodologies in relation to Proximity Marketing in order to be inspired for the establishment of our own selection. Researchers have conducted qualitative studies using discussion groups or scenarios [34]. Others

Table 32.1 Continuum of individualized experience with the service: adapted from [5]

Approach	Interactional flexibility	Example	Level of individualized experience
Product versioning	Low	Software	Basis of first efficient functioning for client
Mass customization	Low	Cars	Consideration of standard desires of client
One-to-one personalization	High	Travel	Integration of personal needs of client
Co-creation	High	Lego	Adaptation of product by client and company
Reverse marketing	High	Swarovski	Total adaptation of product by client

have prioritized a quantitative approach in form of questionnaires [8, 35] which occurs when respondents provide answers biased toward what would be more socially acceptable [36]. In addition, the results such as those of the approach by means of scenarios may be distorted, because they are not necessarily representative of "real life" [37]. These deficiencies are the reason why we opted for the case study. "Theory developed from case study research is likely to have important strengths like novelty, testability, and empirical validity, which arise from the intimate linkage with empirical evidence [38, p. 548]." As specified in the article of [39] p. 201): "If properly conducted research by these methods can provide a deep understanding [40], a fuller contextual sense of the phenomena under study [41], and an explicit provocation toward theory building that often is missing from both simple descriptive work and most cause-and-effect research [42]." Thus, the case study allows to closely analyze daily situation and to better understand the habits of consumers.

32.3.1 Case Studies Identification

We were inclined to use the RFID technology, a technology often serving as a basis for this type of marketing, since there is no existing database on Proximity Marketing. We have listed case studies in the specific database dedicated to this technology, which are IDTechEX and the industry newspaper RFID Journal. Finally, we have completed our research via Google Scholar with the following keywords: Mass Customization + Proximity Marketing, Personalization + Proximity Marketing, Consumer Experience + Proximity Marketing, Consumer Experience + RFID, Mass Customization + RFID, Personalization + RFID.

32.3.2 *Case Studies Selection*

As of June 19, 2015, the database of IDTechEX contained 4828 case studies covering 124 countries. In order to make our research more efficient and appropriate to Proximity Marketing, we concentrated on a particular sector where the Proximity Marketing implementation works, namely the "Retail." In this category, the database contained 660 case studies. Due to this fact, we have refined our observations by using the search tool. We selected the cases in which the benefits of "convenience" and "customer service" were apparent. In fact, these advantages are connected to the experience with the service and the convenience offered by Proximity Marketing. We have excluded the non-relevant cases, which were primarily those that made reference to the RFID chips affixed on pallets carrying merchandising goods and therefore rather applied to logistics. Hundred and seven Cases were found in "convenience" and 53 in "customer service" for a total of 160. For its part, the database of RFID Journal contained 44 case studies in the section "Retail" and the results of Google Scholar did not provide cases more relevant than those we had already selected.

After observation, we found the content of the IDTechEX database less comprehensive and detailed than that of the RFID Journal. Since the case list of the RFID Journal was better presented, we relied on this latter in order to read and annotate the most appropriate cases. Thus, of the 44 potential cases of the RFID Journal, our selection concluded at 11 cases. Given that we followed the principle of theoretical saturation [43], that is to say, the moment where there is no more new data that emerges from the cases, we cut two cases for a final total of nine. Then, we systematized by field of activity and by country.

During the selection, choosing the cases coherent with mass customization and/or personalization and service experience in the context of commercialization took precedence. Our research allowed us to determine that this combination is not frequent, and RFID technology is rather geared toward (practical/beneficial) convenience, for example, "cashless payments," "access control," or "data analytic" than the marketing experience as such.

Up to now, we have individually defined the various concepts surrounding Proximity Marketing and we have established links with the possible increase of the experience with the service. Now, we demonstrate the role of these concepts specifically applied to a Proximity Marketing strategy.

32.4 Results

Our nine cases were held between 2002 and 2014. They all come from the "Retail" sector, but in five different industries and six countries are represented across three continents. Table 32.2 presents a summary of the cases selected. It allows us to understand that Proximity Marketing is a facilitator of mass customization and personalization. In addition, the table illustrates the contribution of Proximity Marketing to a better experience with the service.

Table 32.2 Proximity marketing cases as an enabler of mass customization and personalization in a customer service

Case study	Company	Customer experience	Enabler of
1	Aki Choklat (United Kingdom, 2014)	The bag owner can identify the bag's location through GPS on his phone and share it on FB. This operation is comparable to a digital travel diary	Personalization
2	Osheaga (Canada, 2014)	The festival visitor personalizes his experience with the help of a RFID bracelet that permits him to synchronize the bracelet chip to his FB account. He can consequently share his experience "live"	Personalization
3	Sephora (United States, 2014)	The mobile application of Sephora proposes personalized offers/informations according to the client's physical proximity to the product in the store	Personalization
4	Helsinki Mall (Finland, 2012)	The client carries a key equipped with a RFID chip inside the shopping center. The moment he approaches a screen, the latter transmits content/offers according to the customer's habits/tastes. Also rebates are proposed	Personalization
5	Nissan Paris Motor (France, 2012)	During the show, the client, who has linked his RFID bracelet to his FB page and accepts automatic sharing on his account, receives a VIP card in order to customize his own car	Mass customization
6	Nike Action Sports (n.d., 2012)	During a snowboard event, the participant, equipped with RFID bracelet, can customize his own sweater. He can use a photo of himself in action from during the day	Mass customization
7	McDonald's (Japan, 2008)	Through RFID technology, the client receives offers/coupons based on his shopping history directly to his phone. The customer can even pay through his smartphone	Personalization
8	Curves International (United States, 2007)	The RFID bracelet is linked to client information and the system proposes personalized training objectives to her. She can also compare her performance to her previous ones and can receive personalized messages from her trainers	Personalization
9	Prada (United States, 2002)	Due to VIP RFID customers card, the boutique's client is identified upon his arrival and his preferred vendor is notified of his presence. The client can benefit from made-to-measure offers according to his preferences and can see everything appear on a screen in the changing room	Personalization

32.4.1 Case Description

32.4.1.1 Fashion Industry. Case 1: Aki Choklat

When a customer buys a luxurious bag by Aki Choklat online, he receives an email in order to personalize his purchase through linking his profile. In addition, it is suggested to download an application to identify the location of the bag via GPS on the customer's smartphone. Thus, when the customer visits a new city, the bag can be located and the customer can then share his position on his social networks [44]. The Proximity Marketing contributes to the personalization, because the customer keeps his own "digital travel dairy" in real time. This represents an ingenious strategy, because several hundred Facebook "friends" can be reached and made aware of the brand through the client's sharing of his position.

32.4.1.2 Event Industry. Case 2: Osheaga

The Music and Art Festival Osheaga (Montreal) provides a bracelet for participants fitted with an RFID chip during its weekend of activities. When the participant enrolls in the festival, he is then invited to link his bracelet to his personal information, and even age, sex, and music interests [45]. In the following, the company proceeds to the identification of the consumer and his location with the help of the bracelet that emits a wireless signal when it is presented to the various terminals located on the site of the festival. The consumer can also synchronize his bracelet to his Facebook account. In this way, the moment he scans his bracelet at a terminal, he is automatically recognized and the information then appears subsequently as "instant check-in," along with photos taken on the site including the logo of the sponsor on his page. The participant sees his experience with the event more satisfying, because it is personalized. Everything goes viral and the company draws real profits from Proximity Marketing. In addition, the "sponsor activation" is inviting for donors.

32.4.1.3 Cosmetic Industry. Case 3: Sephora

When a customer comes into the Sephora shop and selects the mode "shopping" through the shop application on her smartphone, she receives a welcome message inviting her to choose information from a menu relating to products near her physical location in the store. Further, she can use the application to scan the bar code of a product to display comments, as well as to look at her past purchases or her wish list. In addition, the application is designed to inform users of new promotions and discounts during their anniversary month. Furthermore, the application allows users to receive special offers and information triggered by tags installed inside the store [46]. The contribution of Proximity Marketing to the personalization is effective, because the client herself chooses the products she is interested in.

32.4.1.4 Fashion Industry. Case 4: Helsinki Mall

Since 2014, the Citycenter shopping Helsinki mall offers a key to its customers that is fitted with an RFID chip enabling shoppers to receive marketing messages related to their shopping behavior, as well as discount offers for the stores they visit regularly. In addition, when a customer is approaching a screen, the latter disseminates personalized promotions according to the customer's preferences. Beyond these attributes, the collected information is then stored and analyzed according to the movements of each client. Relevant to this Proximity Marketing strategy is to be able to know the time when a client modifies his journey in consequence to the presentation of the personalized content on the screen [47]. The collaboration of the marketing and the personalization of offers in real time leads to an increase of the experience due to the understanding of the customer's needs.

32.4.1.5 Event Industry. Case 5: Nissan Paris Motor

At the Paris Motor Show, visitors receive a "hands-free passive ultrahigh-frequency" device. By means of this device, visitors cannot only interact with other RFID devices, but Nissan also planned a customized experience for them. In fact, each person accepting the automatic sharing of their experience on their social networks received a custom VIP card allowing them to design their own car. In addition, they could create an audible noise for the Nissan Leaf and appear on the cover of a magazine with the Micra model [48]. This customization is an excellent opportunity to make the visitor dream. The contribution of Proximity Marketing is therefore increasing the individualized experience with the event.

32.4.1.6 Sports Industry. Case 6: Nike Action Sports

The shoe manufacturer Nike, through the company Snapsportz, implemented a system for taking photos in action at ski resorts or snowboarding events. All participants accessing the site receive a free bracelet equipped with a RFID chip. Athletes are invited to share their photos via the Nike Facebook page. Certainly, all the pictures are taken with the Nike logo "Just do it" and therefore contribute to the advertisement for the brand. The snowboarders also have the possibility to customize a sweater. They choose a "background" and a picture of themselves in action, and then have the image printed on a sweater [49]. Proximity Marketing enables an enriched experience and contributes to the customization.

32.4.1.7 Restauration Industry. Case 7: McDonald's

Since 2008, the restaurant chain McDonald's (Japan) permits its customers to use RFID technology in order to receive custom rebate coupons. In fact, these coupons are based on the customers' personal purchase history and can be obtained directly on their smartphones. In addition, they can download rebate coupons—even rebate coupons in connection with their preferences. Clients have the possibility to pay through approaching their mobile device close to a device provided for this purpose [50]. In this case, Proximity Marketing has demonstrated its impact on the personalization in order to offer an interesting experience with the service to the customer.

32.4.1.8 Sports Industry. Case 8: Curves International

During the registration at the "computer kiosk," the client of the gym receives a bracelet equipped with a RFID chip that is linked to her membership information. In addition, the results of her gym performance are entered in the database, which allows the system to establish specific objectives for the client's training session. The personalization does not end here. The electronic device, accessing the profile of the client, also provides a real-time feedback as to whether the client has reached her goal at the workout equipment. At the end of the session, she can consult the number of calories burned and compare the results with her ten most recent workouts. In addition, the client can receive personal messages from coaches, such as birthday wishes or notifications regarding her membership [51]. This is a good example where Proximity Marketing contributes to the personalization of the experience and pushes the client to succeed during her training sessions.

32.4.1.9 Fashion Industry. Case 9: Prada

Prada is an experience in itself. Incorporating a Proximity Marketing strategy ensures that the customers are treated to the height of the expenditures that incur at the luxurious boutique in New York. By using his VIP customer card upon his arrival at the boutique, the customer is identified immediately. If the latter has a preferred vendor, the vendor is immediately notified of the client's presence in the store. Besides, any associate in the store may know the client's preferences due to his registered profile and is able to advise him about items in relation to his recent purchases. If he proposes an item that is too dark for the taste of the client, the vendor may present different shades on the screen in the changing room by using his portable reader. To do so, he only has to scan the label on the garment [52]. As precursor of RFID chips integration, Prada has been able to demonstrate the effectiveness of Proximity Marketing on the customization by focusing on the customers needs. The costumer has an increased experience.

The selected cases show that the experience is enriched through the key concepts illustrated in this chapter, namely mass customization and personalization. We believe as well that Proximity Marketing contributes actively to the individualized experience of customer service.

32.5 Discussion

Through the intermediary of the introduction to the literature and the case analysis, we believe that we were able to respond to the research question, namely "Does Proximity Marketing contribute to mass customization and personalization of an improved service experience?" We claim therefore that mass customization and personalization are real concepts increasing the service experience provided to customers. In addition, we were able to respond to the research problem, that is to say we clarified the definitions of two key concepts linked to Proximity Marketing—mass customization and personalization—and their correlation with the increase of the individualized service experience.

During the analysis of the cases, we noticed that those cases where mass customization is applied in relation to Proximity Marketing (2) are much less present than those linked with personalization (7). Is this due to the fact that this type of marketing is emergent and that managers have not yet seized/understood the opportunities that customization and Proximity Marketing present? Or is it due to the fact that we analyzed only nine cases in detail? This will certainly be subject to observation in the following few years.

32.5.1 *Managerial Implications*

Organizations benefit from the advantages of mass customization and personalization via Proximity Marketing and are enabled to improve sales due to better logistics [53] and foremost are capable of offering added value to customers. In fact, a tangible potential for rapid return on investment occurs [54]. For example, the Helsinki mall confirmed having increased the movement of consumers in the Citycenter by 14.5 %. In addition, it has increased the amount of time that consumers spend at the commercial center by 21.7 % [47]. This type of marketing creates a positive effect for companies, because the consumer, who uses the proposed technology, perceives the company as being innovative and this enhances its brand image [55]. Thus, we are inclined to believe that this type of marketing is financially advantageous for the organizations in the "Retail" sector. In addition, businesses can increase their market shares through their new complementary capabilities and, thus, the improvement of the company's value chain [53]. In this place we represent two managerial implications interesting for marketers.

Through the performed analysis of cases, we realize that the aspect of the logistical efficiency is accentuated when the RFID technology is in question. By contrast, we have demonstrated that these chips are an element improving the experience. To conclude, due to this technology, Proximity Marketing is a facilitator of the two key concepts considered in this chapter. As discussed earlier, the mass customization is underused when it comes to Proximity Marketing strategies. Managers should not only incorporate mass customization, but also combine it with personalization.

Finally, as shown by means of the case analysis, the social network component wins greatly through the integration into a Proximity Marketing strategy. The objective is to create a craze in order that clients want to interact instantly on their social platforms during their experience in the store [56]. This transforms satisfied customers into real ambassadors for a brand as a result of their increased experience. It is at this moment that Proximity Marketing acquires its most effective form.

32.6 Conclusion

From the outset, we established three major advantages of Proximity Marketing, namely the real-time aspect, the relevant content, and the personalization. Through the identification and the analysis of nine cases, we were able to demonstrate that Proximity Marketing contributes to both, the customization and personalization in the "Retail" sector. Consequently, an increase in the experience with the service develops. In addition to generating many benefits for organizations, such as ROI and brand image, this type of marketing allows for an application of an effective customer relationship management (CRM) strategy, because the customers feel unique, VIP. Their personal needs are met.

As future managerial avenue, we should look at another sector that the retail to validate the effectiveness of Proximity Marketing. Plus, if personalization allows the differentiation [57], why shouldn't we think "out the box" and incorporate the co-creation or even the reverse marketing into the service? This would permit a mass customization and a total personalization of the offered service, a proof of an exchange of mutual values between the two creative parties. The objective is to elevate the individualized customer experience to the point where the customer has a feeling of total control over his purchase.

The famous essayist of the nineteenth century, Robert Waldo Emerson, could not have expressed the thought better, which savvy marketers in 2015 must possess: Do not go where the path leads. Create your own path [58].

Acknowledgment This research was supported by the Social Sciences and Humanities Research Council of Canada.

References

1. Johnson, M.D., Gustafsson, A., Andreassen, T.W., Lervik, L., Cha, J.: The evolution and future of national customer satisfaction index models. J. Econ. Psychol. **22**, 217–245 (2001)
2. Fornell, C., Johnson, M.D., Anderson, E.W., Cha, J., Everitt Bryant, B.: The American customer satisfaction index: nature, purpose, and findings. J. Mark. **60**(October), 7–18 (1996)
3. Kwon, K., Kim, C.: How to design personalization in a context of customer retention: who personalizes what and to what extent? Electron. Commer. Res. Appl. **11**(2), 101–116 (2012)
4. Pine, B. J. (1993). Mass Customization: The New Frontier in Business Competition. Boston: Harvard Business School Press.
5. Miceli, G., Ricotta, F., Costabile, M.: Customizing customization: a conceptual framework for interactive personalization. J. Interact. Mark. **21**(2), 6–25 (2007)
6. Rowley, J.: Understanding digital content marketing. J. Mark. Manag. **24**(5), 517–540 (2008)

7. Edelman, D.C.: Branding in the digital age you're spending your money in all the wrong place. Harv. Bus. Rev. December, 14 (2010)
8. Yeh, H.: Effects of RFID in retailing on customer trust. J. Serv. Sci. Manag. **2013**(June), 143–150 (2013)
9. Slettemeås, D.: RFID—the 'next step' in consumer–product relations or Orwellian nightmare? Challenges for research and policy. J. Consum. Policy **32**(3), 219–244 (2009)
10. Uhrich, F., Sandner, U., Resatsch, F., Leimeister, J.M., Krcmar, H.: RFID in retailing and customer relationship management. Commun. Assoc. Inf. Syst. **23**, 219–234 (2008)
11. Petro, G.: How proximity marketing is driving retail sales. Forbes, pp. 1–4 (2014)
12. Sunikka, A., Bragge, J.: Applying text-mining to personalization and customization research literature—who, what and where? Expert Syst. Appl. **39**(11), 10049–10058 (2012)
13. Sheth, J.N., Gardner, D.M., Garrett, D.E.: Marketing Theory: Evolution and Evaluation. Wiley, New York (1988)
14. Gallopel, K., Cliquet, G.: Géomarketing et espace publicitaire. Décisions Mark., no. 26 (2002)
15. Levesque, N., Boeck, H., Durif, F., Bilolo, A.: The impact of proximity marketing on consumer reaction and firm performance: a conceptual and integrative model. In: Twenty-First Americas Conference on Information Systems, pp. 1–8 (2015)
16. Volle, P.: Produit et information géographique: le géomerchandising. Géomarketing 1–27 (2002)
17. Sill, H.E., Fisher, S.L., Wasserman, M.E.: Consumer reactions to potential intrusiveness and benefits of RFID. Int. J. Inf. Technol. Manag. **7**(1), 76 (2008)
18. Tcuentofr: Les avantages du Marketing de proximité, TC group solutions [Online]. http://www.retail-intelligence.fr/2012/02/les-avantages-du-marketing-de-proximite/ (2012). Accessed 24 June 2015
19. Cova, B., Louyot, M.-C., Louis-Louisy, M.: Les innovations marketing en réponse è la montée de l'hédonisme: Articulations avec le CRM? In: 3e Congrès sur les tendances du marketing en Europe (2003)
20. Fournier, S., Avery, J.: Putting the 'relationship' back into CRM. MIT Sloan Manag. Rev. **52**(3), 63–72 (2011)
21. Da Silveira, G., Borenstein, D., Fogliatto, H.S.: Mass customization: literature review and research directions. Int. J. Prod. Econ. **72**(49), 1–13 (2001)
22. Davis, S.M.: Future Perfect **28**(1) (1989)
23. Joneja, A., Lee, N.K.S.: Automated configuration of parametric feeding tools for mass customization. Comput. Ind. Eng. **35**(3–4), 463–466 (1998)
24. Huffman, C., Kahn, B.E.: Variety for sale: mass customization or mass confusion? J. Retail. **74**(4), 491–513 (1998)
25. Bettman, J.R., Luce, M.F., Payne, J.W.: Constructive consumer choice processes. J. Consum. Res. **25**(3), 187–217 (1998)
26. Peppers, D., Rogers, M., Dorf, B.: Is your company for one-to-one marketing ? Harv. Bus. Rev. **77**(1), 151–160 (1999)
27. Chellappa, R.K., Sin, R.G.: Personalization versus privacy: an empirical examination of the online consumer's dilemma. Inf. Technol. Manag. **6**, 181–202 (2005)
28. Totz, C., Riemer, K.: The effect of interface quality on success—an integrative approach on mass customization design. In: 1st World Congress on Mass Customization, pp. 1–9 (2001)
29. Tseng, M.M., Piller, F.T.: The Customer Centric Enterprises Advances in Mass Customization and Personalization. Springer, Berlin (2003)
30. Tseng, M. M., & Piller, F. T. (2003). The customer centric enterprise. In The customer centric enterprise (pp. 3-16). Springer Berlin Heidelberg.
31. Surprenant, C.F., Solomon, M.R.: Predictability and personalization in the service encounter. J. Mark. **51**, 86–96 (1987)
32. Prahalad, C.K., Ramaswamy, V.: Co-creation experiences: the next practice in value creation. J. Interact. Mark. **18**(3), 5–14 (2004)
33. Bigus, P., Meister, D.: The Lego Group: Building Strategy, pp. 1–12. Ivey Publishing, London (2011)

34. Angeles, R.: An empirical study of the anticipated consumer response to RFID product item tagging. Ind. Manag. Data Syst. **107**(4), 461–483 (2007)
35. Boeck, H., Roy, J., Durif, F., Grégoire, M.: The effect of perceived intrusion on consumers' attitude towards using an RFID-based marketing program. Procedia Comput. Sci. **5**, 841–848 (2011)
36. Gagné, C., Godin, G.: Les théories sociales cognitives: Guide pour la mesure des variables et le développement de questionnaire (1999)
37. Pramatari, K., Theotokis, A.: Consumer acceptance of RFID-enabled services: a model of multiple attitudes, perceived system characteristics and individual traits. Eur. J. Inf. Syst. **18**(6), 541–552 (2009)
38. Eisenhardt, K.M.: Building theories from case study research. Acad. Manag. Rev. **14**(4), 532–550 (1989)
39. Bonoma, T. V. (1985). Case research in marketing: opportunities, problems, and a process. Journal of marketing research, 199-208.
40. Geertz, C.: The Interpretation of Cultures. Basic Books, New York (1973)
41. Miles, M.B.: Qualitative data as an attractive nuisance: the problem of analysis. Adm. Sci. Q. **24**(4), 590–601 (1979)
42. Van-Maanen, J.: Reclaiming qualitative methods for organizational research: a preface. Adm. Sci. Q. **24**(4), 520–524 (1979)
43. Yin, R.K.: The case study crisis: some answers. Adm. Sci. Q. **26**(1), 58–65 (1981)
44. Swedberg, C.: Aki Choklat brings authentication and personalization to luxury bags. RFID J. 1–2 (2014)
45. Swedberg, C.: Osheaga music festival to include new NFC-enabled features. RFID J. (2014)
46. Swedberg, C.: Sephora sees beauty in beacon-based services. RFID J. 1–2 (2014)
47. Swedberg, C.: RFID delivers location-based data at Helsinki Mall. RFID J. 1–4 (2015)
48. Parkinson, D.: Nissan social experience at Paris motor show kicks off with Google+hangout (2012)
49. Personalized t-shirts onsite using Snapsportz images. Snapsportz [Online]. http://www.snapsportz.com/customers/nike.html. Accessed 24 June 2015
50. Swedberg, C.: McDonald's Japan brings discounts to phones. RFID J. 1–2 (2008)
51. Weil, M.: RFID pumps up women's workouts. RFID J. 1–4 (2007)
52. N.d.: Learning From Prada. RFID J. 1–4 (2002)
53. Rekik, Y., Sahin, E., Dallery, Y.: Impact de l'utilisation de la technologie RFID sur la disponibilité des produits en magasin. Lab. Génie Ind., no. 1972 (2003)
54. Goel, R.: Managing RFID consumer privacy and implementation barriers. Inf. Syst. Secur. **16**(4), 217–223 (2007)
55. Bertin, L.: Marketing mobile : les objectifs. L'observatoire du commerce en réseau [Online]. http://www.distribu.fr/marketing-mobile-les-objectifs/. Accessed: 24 June 2015
56. Johnson, J.R.: Social media and RFID: a golden opportunity for retailers. RFID 24–27 (2012)
57. Tam, K.Y., Ho, S.Y.: Understanding the impact of web personalization on user information processing and decision outcomes.Management Information Systems Research Center **30**(4), 865–890 (2006)
58. Violaine, R.W.: Emerson ou l'anticonformiste! www.violaine.net [Online]. http://www.violaine.net/developpement/developpement_personnel/ralph_emerson.html. Accessed 24 June 2015

Chapter 33
Investigating the Impact of Product Volume and Variety on Production Ramp-Up

Ann-Louise Andersen, Mads Bejlegaard, Thomas D. Brunoe, and Kjeld Nielsen

33.1 Introduction

Today's global competitive environment is characterized by increased variety in product offerings and shorter product life cycles, due to fragmented customer needs, increased need for customized products, and rapid technological innovations [1, 2]. As a consequence, manufacturing companies are compelled to find solutions for realizing the competitive strategy of mass customization and delivering individually configured products at a cost near mass production [3, 4]. A key enabler of mass customization is modular product design, where end-variety is achieved through configurations of standardized product modules [5]. However, simply introducing modular product models is not enough for manufacturing companies to gain competitive advantage, as the products need to be produced and delivered to the market at the right time and cost [6]. Reconfigurable manufacturing systems are an attractive option for this, due to their ability to efficiently produce a variety of products grouped into product families and rapidly adapt production resources to varying demand and new product models [7, 8]. Thus, reconfigurable manufacturing systems reduce the traditional trade-off between efficiency and flexibility by incorporating advantages of both traditional mass production and flexible production systems, which is suitable for mass customization production [6].

One of the key premises of reconfigurable manufacturing is that reconfigurations in terms of capacity and functionality can be carried out continuously and without significant losses in productivity [6]. This means that production start-up and ramp-up are frequent events that will occur numerous times in the life of the system, which therefore must be effectively controlled and continuously reduced [9]. Moreover, with brief windows of market opportunities, the importance of short ramp-up periods

A.-L. Andersen (✉) • M. Bejlegaard • T.D. Brunoe • K. Nielsen
Department of Mechanical and Manufacturing Engineering, Aalborg University, Aalborg, Denmark
e-mail: ala@m-tech-aau.dk

J. Bellemare et al. (eds.), *Managing Complexity*, Springer Proceedings in Business and Economics, DOI 10.1007/978-3-319-29058-4_33

cannot be overstressed [10]. In particular, this applies to mass customization companies, where not only continuous improvement of existing products, but also frequent product introductions have proven to be a necessary precondition for maintaining long-term competitiveness [11, 12].

Production ramp-up is generally regarded as the period where production processes and systems are scaled to full volume from initial pilot series, where production targets related to cycle-time, capacity, and quality are reached [10, 12, 13]. Such ramp-up periods occur with the introduction of new products, where it is the last step of the product development process, but also when a new production system is started [14]. Clearly, this initial stage of production is different from mature production phases in a number of different ways. The most common differences mentioned in literature are low level of knowledge about product and processes, gradual learning, low output, low production capacity, high cycle-time, high demand, lack of planning reliability, and a high degree of disturbances in processes, in the supply chain, and in quality [10]. These differences to mature production underpin the conflicting characteristic of production ramp-up, which is pressure for meeting high product demand with high uncertainty and constrained capacity [15]. This also means that ramp-up periods usually contain many fire-fighting activities and unforeseen problems that need to be immediately addressed [10, 16].

In current research, there are several case studies investigating ramp-up problems and how they can be classified into generic problem classes [13, 17–19]. One of the widely applied categorizations is based on the work by Fjällström et al. [17] and Nyhuis and Winkler [20] and is summarized by Surbier et al. [10]. This categorization contains seven main categories of sources that ramp-up problems typically belong to: product, processes, logistics, quality, methods and tools, cooperation and communication, and personnel. From this classification, it is obvious that ramp-up periods are subject to a high degree of uncertainty, since problems and challenges occur in many different areas, from inadequate product specification to lack of information sharing in the company. Identifying and solving such different problems is one of the main activities during production ramp-up, which stresses the importance of identifying typical problems and the sources of these in order to improve performance [14]. However, it is reasonable to consider that some ramp-up problems are more or less typical in different ramp-up projects, depending on the company, the product, and the production system involved [10, 21]. Currently, ramp-up problems are mainly studied in a high-volume industrial context, such as the automotive industry [10]. Only few studies consider production start-up and ramp-up problems in low-volume industries, where products are typically more customized and order based than in high-volume industries [18, 21–23]. The current studies in low-volume industries highlight that these ramp-up projects are very different from typical high-volume ramp-up projects. Among the main differences are fewer engineering prototypes, limited pre-series production, use of existing or modified production set-up, and extensive focus on product functionality [22]. However, with the limited research in this area, an important research focus is the identification of differences in typical ramp-up problems uncounted during ramp-up in different types of industries. Therefore, the aim of the research presented in this chapter is to investigate the impact of product volume and variety on the production ramp-up following a new product introduction.

33.2 Methodology

In order to address the research aim stated above, multiple-case study is selected as the research method. According to Yin [24], a case study is an appropriate research method when a contemporary phenomenon is investigated in depth and is believed to be highly pertinent on the subject that is studied. This is the exact concern of this research where the aim of understanding ramp-up challenges encompasses important contextual conditions, such as the type of industry, product, and production setting. However, despite the fact that case studies rely on contextual factors and by nature are situationally grounded, some sense of generality should also be present, which is referred to as the duality criteria by Ketokivi and Choi [25]. Therefore, in the research presented in this chapter, two cases from Danish industry are selected: a large enterprise with high-volume standard products and an SME with low-volume customized products. These two cases represent different empirical settings that allow for cross-case comparison that potentially leads to more general insight.

33.2.1 Data Collection

The two cases included in the research represent two different ramp-up projects, here referred to as case A and case B. Case A is a ramp-up project of electronic products in a large Danish company, whereas case B is a ramp-up project in a smaller Danish company producing excavators and earth moving equipment. Case A is considered a large enterprise with more than 4000 employees, while case B is an SME with approximately 150 employees. Moreover, the electronic products produced in case A are low-variety products with a monthly volume of more than 100,000 units, whereas case B represents high variety and customized products with an annual volume of approximately 250 units.

Generally, different complexity degrees of ramp-up projects exist, depending on the degree of change in product and processes [10, 26]. If either processes or the product remain unchanged or slightly modified, as in, e.g. a face-lift of an old model, managing production ramp-up is less complex than when it concerns a completely redesigned product that requires entirely new technological processes. Therefore, in order to allow for comparison between the two cases in this research, both cases represent ramp-up projects of highest complexity, where completely new products and production processes are involved.

In each of the cases, empirical data on production ramp-up challenges has been collected through a number of semi-structured interviews performed in spring 2015. At this point in time, both ramp-up cases were in their final stages. A total of seven interviews were carried out in case A, including key employees from the NPD project and operations. In case B, one of the authors is involved in the daily operations without being directly engaged in the specific project and has therefore been a main source of information. In addition, three interviews were carried out with key

Table 33.1 Ramp-up problem categories [10]

Problem category	Types of problems
Product	Problems related to product specification, product maturity, changes in product design, and late engineering changes
Technical process	Problems related to maturity of production processes and technology, unforeseen bottlenecks, process difficulties, manufacturability of products, set-up times, and processing times
Logistics	Problems related to setting up the supply chain, availability of supplied parts, and quality of supplied parts
Quality	Problems related to the quality of the end-product, scrap, and rework
Methods and tools	Problems related to methods and tools used for controlling the ramp-up project, e.g. resource planning, knowledge management, project management, and data management
Personnel	Problems related to definition of responsibilities, qualification of employees, and training of employees
Cooperation and communication	Problems related to sharing information across functions and in the project team, information losses, and sharing information across firm boundaries

employees directly involved in the product introduction project. Furthermore, data and information on ramp-up performance were included for both cases, e.g. project reports and data on capacities, output volume, cost, and quality. This data was used in two ways: first in regard to validating the information collected in the interviews and secondly in regard to forming an overview of each case.

33.2.2 *Data Analysis*

In order to collect information in each case on the challenges that have occurred in the ramp-up period, interviews were performed based on open questions related to each problem category identified by Surbier et al. [10]. The problem categories are summarized in Table 33.1. The reason for using this as a basis for the interviews is that it forms a general and comprehensive list of problem sources that may occur in ramp-up projects.

Based on the semi-structured interviews, a large number of different problem statements could be identified. It should be noted that the term problem here is defined as in the work by Surbier et al. [18], as an unwanted gap between desired state and reality, which with difficulty is solvable in some way. In both cases, the statement collected in the interviews were analysed through conducting the following steps:

- Step 1: Transcription of interviews and identification of problems, critical events, and challenges.
- Step 2: Exclusion of statements that are vague or not sufficiently described.

- Step 3: Identification of identical statement.
- Step 4: Identification of problem type, e.g. inadequate product specification, or lack of information sharing.
- Step 5: Classification of problems of similar type.
- Step 6: Classification of problem types into main sources in accordance with the seven problem types in Table 33.1.

As a result, a total of 37 unique problem statements and a total of 17 main problem types belonging to 6 of the overall problem categories were identified in case A. In case B, a total of 31 unique statements and 10 problem types belonging to 4 of the problem categories were identified. Eventually, the classifications of problem statements were verified by managing employees in each case company.

33.3 Empirical Findings

33.3.1 Case Study A

The ramp-up period studied in case A is the final phase of the introduction of a new electronic product. This product is a completely redesigned product, which is intended to replace older versions of products in a particular market segment. A total of eight different variants of the product were offered from the beginning, with a target of 1.4 million units produced per year. The production set-up is fully automatic and requires entirely new machinery and a reorganization of the shop floor. Only a minor part of the complete production process involves existing equipment, which is modified to the new product. The product introduction project in case A consists of a traditional stage gate approach, where a project team is formed across organizational functions, involving purchasing, operations, logistics, sales, development, and internal equipment suppliers. The management of the project is highly formalized and includes numerous levels of project managers from different functions.

During the data collection, the project was in its final stage, but gaps between planned and realized performance regarding cost and capacity were still occurring. Moreover, time plans and milestones were delayed throughout the project. In Fig. 33.1, an overview of case A is presented.

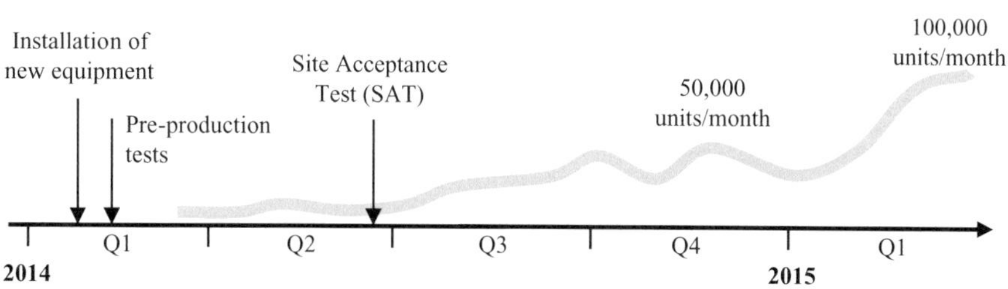

Fig. 33.1 Overview of case A

The problems that occurred in the ramp-up period are presented in Table 33.2, and cover six of the categories presented in Table 33.1. The main problems relate primarily to the product, processes, personnel, and management methods applied in the project. The problems in the product and process categories were closely interlinked, as changes in one of them typically resulted in changes in the other. This concurrent engineering of equipment and product specification happened throughout the entire project, and the number of changes in product specification was rather high and included both minor adjustments and comprehensive design changes that required significant changes in the production set-up. In fact, at the time of the interviews, a major design change in the product was still to be implemented, due to difficulties in automating the initially intended manufacturing processes. The result of this specific problem was an increase in unit cost, as some processes had to be performed manually during ramp-up. An important aspect of this relates to the maturity of the production equipment and the product design, where important milestones in both areas were delayed. Prior to the production ramp-up, product design freeze was postponed, which created great difficulties in specifying the new production equipment. Obviously, this caused great difficulties, in particular because product demand already was high and pressure for deliveries increased throughout the project. Consequently, compromises between delivering orders and improving processes had to be made, which resulted in increased cycle-time and unit cost.

Problems related to personnel were mainly related to the fact that numerous functions in the organization were involved in the project, which resulted in contradicting priorities, division-thinking, and vague borderlines of responsibilities. Moreover, as the project involved entirely new product design and production processes, employee experience was limited in some areas, which also was indicated as a main challenge during the production ramp-up. The methods and tools category contains issues related to the management of the project, e.g. control, follow-up on project goals, and feasibility of project plans.

33.3.2 Case Study B

The ramp-up project that is investigated in case B is the final stage of the introduction of a new series of excavators, which is completely redesigned from existing products offered by the company. The excavator series contain three main types of machines with a total of 30 variants, which can be further configured through an additional eight parameters with numerous different options to choose from. Thus, practically none of the produced excavators are identical, as they are configured directly to the needs of the customers. The production involves a manual final assembly set-up, and a steel processing set-up that involves both manual handling of large components and automatic welding. The annual volume of the entire production set-up is approximately 250 units, while the excavator series has a target of at least 100 units per year. The project team in case B consisted mainly of employees from development, tooling

Table 33.2 Ramp-up problems in Case A

Problem category	Problem source	Problem type
Product	Simultaneous engineering	Numerous component revisions
		Late component revisions
	Maturity of product	Revisions according to international standards
		Postponement of variants
		Change in component design
	Manufacturability of product	Change of product specification
		Late changes in product design
Process	Introduction of new technology	New technology for welding
		New technology for fitting
	Process difficulties	Problems with specific processes
	Manufacturability of product	Higher cycle-time than planned
		Increase in manual labour
	Delayed product design	Time pressure on line and equipment specification
		Uncertainty in line and equipment specification
Personnel	Definition of responsibilities	Ill-defined responsibilities
		Lack of line responsible
		Hand-over from project to production
	Employee skills and experience	Lack of employee experience
		Lack of process knowledge
		Training of operators
	Division-thinking	Contradicting goals and priorities
Cooperation and communication	Cooperation with equipment supplier	Choice of wrong supplier
		Taking home immature equipment
		Pressure for unrealistic plans
	Communication between departments	Lack of information sharing
Quality	Adaption of quality control to new product	Communication problems with testing equipment
		Failing testing equipment
Methods and tools	Evaluation and documentation	Lack of evaluation from test series
		Unclear milestones that are not kept
		Unclear goals and follow-up
	Late involvement	Involvement of internal equipment supplier
	Sophistication of methods	Trial and error approach
	Uncertainty in plans	Unrealistic time plans
		Lack of commitment
		Delayed SAT
		Delayed design freeze and subsequent milestones
		Large test orders in run-in

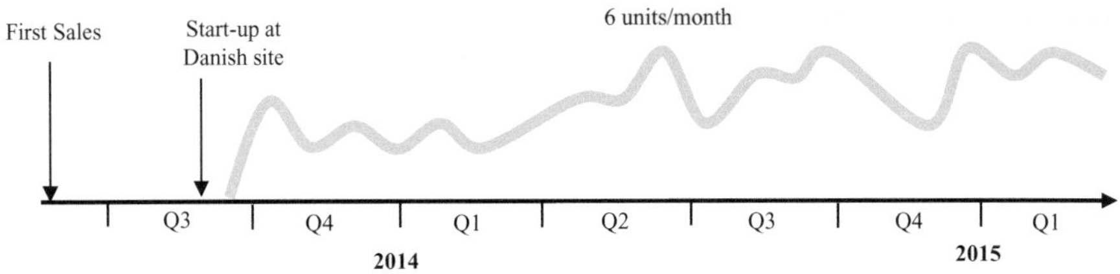

Fig. 33.2 Overview of case B

construction, purchasing, and operations. However, due to the size of the company, a less formalized approach to project management is applied involving only a few project leaders.

The investigated ramp-up project was in the final stage during data collection. Production start-up initiated in fall 2013 following a period of more than 2 years of product development, prototyping, and testing. In comparison with the initial plans, the production start-up was delayed approximately 4 months, due to difficulties with a specific main component supplied from an external vendor. Some of the main components of the excavator, e.g. motors, gears, and electronics were sourced externally, which made co-development between the case company and their main suppliers an important factor in the product introduction project. Moreover, the initial prototypes were made at another production location than the final one in Denmark, as a late decision regarding production site was made in the project. Figure 33.2 provides an overview of the ramp-up in case B.

Table 33.3 presents the ramp-up problems identified in case B. Several problems were found to be related to the processes and production set-up required for the new product, which involved entirely new tooling. However, the process of constructing these was rather long and resource intensive, which meant that the product had to be ramped-up with only few tools being ready. This increased operating cost, prolonged lead-times, and created difficulties in terms of estimating processing times, which resulted in lack of planning foundation.

In case B, problems related to the product were mainly linked to a specific main component that was sourced externally and had to be developed in cooperation with the supplier. The specific main component had to be specified by the case company, and developed by the supplier, which created great difficulty and delayed the entire project.

This was a critical issue in the project, as excavators were already sold and had to be delivered to the customers. Furthermore, the time plans for the project were tight as production start-up initially was planned at a different site than the Danish production site. At the time of the decision of changing production location, preparations and prototype production had already been ongoing for a while at the other site. Consequently, problems in regard to knowledge transfer between the two sites were experienced during ramp-up.

Table 33.3 Ramp-up problems in case B

Problem category	Problem source	Problem type
Product	Inadequate product specification of externally supplied component	Difficulties in specifying specific component
		Delayed and failing tests of component
	Manufacturability of product	Change of product design in accordance with processing requirements
Process	Construction of new tooling	Production start-up without all tooling
		Completely new tooling needed
		Time-consuming tooling construction
	Run-in of new tooling	Changes in tooling
		Time-consuming run-in of tooling
		Longer lead-times than planned
	Uncertainty in processing times	Difficulties in estimating processing times
		Lack of planning foundation in ramp-up
		Difficulties with tact-line introduction
	Assembly of new product	Numerous adjustments in assembly
		Start of assembly with temporary drawings
Cooperation and communication	Cooperation with supplier of main component	Delayed deliverables from supplier
		Shorter test periods on delivered parts
		Difficulties in implementing supplied component
Methods and tools	Uncertainty in forecasting sales	All configuration options available in ramp-up
		Difficulties in forecasting variants
		Difficulties in procuring components with long lead-time
		Forecasted mix not matching actual mix
	Uncertainty in plans	Late start of tooling construction
		Unrealistic and tight project plans
		Time pressure for delivery of sold products
	Change in decision on production locations	Late decision on production location
		Preparation done at initial location
		Prototype completed at initial location
		Transfer of know-how between locations

33.4 Discussion

The two investigated cases represent very different types of industrial settings. Therefore, the empirical findings provide valuable insight into the differences in ramp-up challenges faced by low-volume and high-volume production companies with different degrees of product variety and complexity. Table 33.4 presents a comparison of the two cases based on key product and production characteristics inspired by Javadi [22]. In the following, the findings in regard to the different ramp-up problem categories will be discussed and related to the different characteristics of the two cases. In particular, focus will be on the product, process, and method categories, as these were dominant in each of the two investigated cases.

33.4.1 Product

In both case A and B, critical problems occurred during ramp-up in regard to the product. In case A, product problems were mainly related to numerous and significant changes to the product design and specification, from which many were closely related to the design of the fully automatic production system. Many large pre-production test series were performed prior to production start-up on the new production equipment. In case B, product problems were solely related to a specific component that had to be co-developed with a supplier, which meant that the overall product design and specification was well established at the time of the production start-up. Prior to the start-up, only two prototypes had been produced, which were needed in order to launch product sales, but also to test the functionality of the product and complete its design. Thus, a key difference between the two cases is that in the high-volume standard production, the interrelation between product and process design is a source of numerous problems, as the efficient manufacturability of the product is a key issue in the fully automatic set-up. In contrary, this simultaneous adaption does not occur to the same extent in the low-volume case. This key difference may be attributed to the fact that manufacturability of products and its concurrent specification is not as critical in a highly flexible set-up, which is designed to produce customized products. These findings are to some extent similar to those of

Table 33.4 Comparison of case A and B

	Case A	Case B
Company type	LE with >250 employees	SME with < 200 employees
Product type	Electronic product	Excavator
Volume	>100,000 units/month	<200 units/year
Variety	Few products variants	Customized products
Production system	Fully automatic and dedicated set-up	Mostly manual set-up with dedicated tooling
Assembly cycle-time	App. 10 s	App. 15 h
Planning policy	Make-to-stock	Make-to-order

Javadi [22], suggesting that functionality of products have higher priority than manufacturability in the early stages of low-volume production start-up and that pilot and pre-series are completely dependent on customer demand.

33.4.2 Process

At the time of production start-up, the state of the production systems and processes were very different in the two cases. In case A, a fully automatic and dedicated production system had to be purchased and installed prior to start-up. In contrary, in case B production start-up happened at a time where only a small part of the new tools and fixtures was fabricated. Due to the flexible and highly manual set-up in case B, production start-up of the new product was possible with the existing production set-up, if increased manual labour, increased time cost, and longer processing times were accepted. More than a year after production start-up, the new tooling was still not entirely implemented in the production set-up. These significant differences between the processes and production set-up during ramp-up are to a high extent related to the volume characteristics of the two cases. Moreover, the ramp-up pattern is less gradual in case B compared to case A, due to its more sporadic low-volume sales. This conclusion is consistent with the findings of Javadi [22], stating that in low-volume industries, necessary changes to the existing production system are considered in late stages of the production introduction or even in the normal production stage.

A similar characteristic of the two cases is the complexity of the product introduction and ramp-up process. In both cases, substantial resources and time were used on developing and implementing a new production set-up for the new product. In case A, a completely new dedicated production system was introduced and in case B, entirely new dedicated tools and fixtures were introduced. Even though, case B has higher reuse of the existing production set-up due to its less gradual ramp-up, both case companies could reduce the complexity of this process by introducing higher reuse of existing production resources. Through effective reconfigurations of existing production resources rather than replacing these completely, time and resource usage could be reduced.

33.4.3 Method and Tools

In both cases, various different problem statements were identified as having the project management methods and tools as the main source. In particular, uncertainty in plans and time pressure was a critical issue in both ramp-up projects. However, the origins of the plan uncertainty are to some extent different. In case A, the tight, uncertain, and unrealistic plans were generally a result of delayed design freeze, immature production equipment, and pressure for delivering large orders for customer tests during run-in. In case B, plan uncertainty was mainly related to a tight initial schedule, due to a late change in decision on production location. However, uncertainty was also created by the lack of planning foundation from the beginning of production start-up, as tooling was not ready and processing times were highly uncertain. In addition, uncertainty in regard to forecasting sales and procuring components was evident, as the product design is highly customizable. Generally, these findings suggest that uncertainty in plans is main factor in all types of ramp-up projects, but that increased variety and customizability increases uncertainty in plans and resources during ramp-up.

In case A, late involvement, lack of evaluation between stages, and unclear milestones were widely mentioned, which was not the case in case B. This relates to the project management method and the size of the project and company, which is further described in the following section.

33.4.4 Other Categories

Cooperation and communication challenges between the case company and its main suppliers occurred in both of the investigated ramp-up projects. However, cooperation issues between internal departments, e.g. lack of knowledge sharing, were only identified in case A. Furthermore, issues in regard to the personnel category, e.g. division-thinking, lack of information-sharing, and ill-defined responsibilities were also only mentioned in case A. This finding represents a notable difference in the two ramp-up project, which is likely to be attributed to the size of the company and its organizational structure. Main structural dimensions, such as formalization of project procedures, level of specialization of jobs and departments, and span of management control [27] are very different in the two cases, which is likely to explain this difference in ramp-up challenges.

In the analysis of both case A and B, not all of the problem categories proposed by Surbier et al. [10] are represented. In neither of the cases, problems were identified as having logistics as the main problem source. Examples of such problems are availability of supplied materials, the set-up of a new supply chain, and bottlenecks in transportation [20]. Even though, none of the problem statements in either of the cases were related primarily to this group of problems, it does not mean that problems primarily were related to the activities inside the boundaries of the firm. For instance, problems in the cooperation category involve external suppliers. This implies that the ramp-up problem categories are closely interlinked, and that some problems might be argued as belonging partly to multiple categories. Similarly, problems in regard to personnel and methods are widely related to each other, as the behaviour of employees depends highly on the way the project is organized.

33.5 Conclusion

Production ramp-up is a critical phase in any production system, which needs to be effectively and efficiently managed in order for manufacturers to succeed in the global competitive market. In particular, this applies for manufacturers seeking to realize the competitive strategy of mass customization through quick reconfigurations and adaptions of production resources. Therefore, the aim of the research presented in the chapter was to identify challenges in production ramp-up and compare these across different industrial settings, represented by a large enterprise producing high-volume electronic products and an SME producing low-volume customized excavator. Both companies faced problems in the ramp-up period that affected time and cost targets. Nevertheless, some significant differences were identified between the two cases, which can be explained by different characteristics in terms of product volume, product variety, organizational structure, and size. In the low-volume high-variety case, ramp-up is sporadic and dependent on customer orders and only few prototypes are produced, while in the high-volume case, volume is gradually built up and large test series are run. The main differences in identified problem types and sources are:

- In the high-volume case, the interrelation between product and process design and the manufacturability of the product were main problem sources, while this simultaneous adaption did not occur to the same extent in the low-volume case.
- In the low-volume case, ramp-up initiated in the existing production set-up where only few changes and implementations of new tooling had been completed. In contrary, the high volume and low variety in the other case required a widely finished new production line.
- Uncertainty in plans was a main problem in both types of ramp-up projects, but was increased with higher variety and customizability.
- Personnel, cooperation, and communication related factors were a main problem source in the large enterprise, e.g. division-thinking, information-sharing, and definition of responsibilities, but were not present in the SME.

These findings represent a valuable extension of current research, which is primarily related to high-volume industrial settings. Moreover, identification of which ramp-up challenges that are typical in different industrial settings represents an important step towards improving ramp-up performance in practice. In regard to this, it should be highlighted that intensive time and resource usage was needed in both cases, in order to establish a manufacturing set-up that could accommodate the product being introduced. The difficulty of this activity could be reduced by a higher degree of reuse of existing production systems. Thus, introducing reconfigurability in the production systems could potentially lead to a less complex product introduction, resulting in reduced time-to-market and time-to-profit.

References

1. ElMaraghy, H., Schuh, G., ElMaraghy, W., et al.: Product variety management. CIRP Ann. Manuf. Technol. **62**, 629–652 (2013)
2. Westkämper, E.: New trends in production. In: Dashcenko, A.I. (ed.) Reconfigurable Manufacturing Systems and Transformable Factories, pp. 15–26. Springer (2006)

3. Gilmore, J.H., Pine, B.J.: Markets of One: Creating Customer-Unique Value Through Mass Customization. Harvard Business School, Boston (2000)
4. Pine, I., Joseph, B., Victor, B.: Making mass customization work. Harv. Bus. Rev. **71**, 108–117 (1993)
5. Ulrich, K.T., Tung, K.: Fundamentals of product modularity. Sloan School of Management, Massachusetts Institute of Technology (1991)
6. Koren, Y.: The Global Manufacturing Revolution: Product-Process-Business Integration and Reconfigurable Systems. Wiley, New York (2010)
7. ElMaraghy, H.A.: Flexible and reconfigurable manufacturing systems paradigms. Int. J. Flex. Manuf. Syst. **17**, 261–276 (2005)
8. Koren, Y., Heisel, U., Jovane, F., et al.: Reconfigurable manufacturing systems. CIRP Ann. Manuf. Technol. **48**, 527–540 (1999)
9. Koren, Y.: General RMS characteristics. Comparison with dedicated and flexible systems. In: ElMaraghy, H.A. (ed.) Reconfigurable Manufacturing Systems and Transformable Factories, pp. 27–45. Springer (2006)
10. Surbier, L., Alpan, G., Blanco, E.: A comparative study on production ramp-up: state-of-the-art and new challenges. Prod. Plan. Control **25**, 1264–1286 (2014)
11. Piller, F.T.: Mass customization: reflections on the state of the concept. Int. J. Flex. Manuf. Syst. **16**, 313–334 (2004)
12. Slamanig, M., Winkler, H.: An exploration of ramp-up strategies in the area of mass customisation. Int. J. Mass Custom. **4**, 22–43 (2011)
13. Almgren, H.: Pilot production and manufacturing start-up: the case of Volvo S80. Int. J. Prod. Res. **38**, 4577–4588 (2000)
14. Terwiesch, C., Bohn, R., Chea, K.: International product transfer and production ramp-up: a case study from the data storage industry. R&D Manag. **31**, 435–451 (2001)
15. Terwiesch, C., Bohn, R.E.: Learning and process improvement during production ramp-up. Int. J. Prod. Econ. **70**, 1–19 (2001)
16. Säfsten, K., Harlin, U., Fjällström, S., et al.: Proactive and reactive ways of managing product introduction. In: Second Swedish Production Symposium (2008)
17. Fjällström, S., Säfsten, K., Harlin, U., et al.: Information enabling production ramp-up. J. Manuf. Technol. Manag. **20**, 178–196 (2009)
18. Surbier, L., Alpan, G., Blanco, E.: Identification of problem types during production ramp-up. In: International Conference on Industrial Engineering and Systems Management (2009)
19. Winkler, H., Heins, M., Nyhuis, P.: A controlling system based on cause–effect relationships for the ramp-up of production systems. Prod. Eng. **1**, 103–111 (2007)
20. Nyhuis, P., Winkler, H.: Development of a controlling system for the ramp-up of production systems. In: International Conference on Competitive Manufacturing (2004)
21. Bellgran, M., Aresu, E.: Handling disturbances in small volume production. Robot. Comput. Integr. Manuf. **19**, 123–134 (2003)
22. Javadi, S.: Towards tailoring the product introduction process for low-volume manufacturing industries. Mälardalen University Press Licentiate Theses, No. 201 (2015)
23. Javadi, S., Bruch, J., Bellgran, M.: Challenges in the industrialization process of low-volume production systems. In: International Conference on Manufacturing Research (2013)
24. Yin, R.K.: Case Study Research: Design and Methods. Sage (2003)
25. Ketokivi, M., Choi, T.: Renaissance of case research as a scientific method. J. Oper. Manag. **32**, 232–240 (2014)
26. Almgren, H.: Pilot production and manufacturing start-up in the automotive industry: principles for improved performance. Doctoral thesis, Chalmers University of Technology, Gothenburg (1999)
27. Daft, R.: Organization Theory and Design. Cengage Learning, Mason (2012)

Chapter 34
Implementing 'Design for Do-It-Yourself' in Design Education

Jan Willem Hoftijzer

34.1 Introduction

Technological advancements, strongly influencing the practice of product designing and making, exemplified by the so-called scale-free-ness of tools [5] and the infinite availability of information, have clear consequences for the field of product design. As a consequence, the changes may have a large effect on the type of profession designers will have in the near future [1, 2]. Together with new and democratizing tools and toolkits, providing the means for non-designers to play a role in the product development process, comes the possibility for people to exert their creativity and express their identity. Historical analysis and analogies have indicated that these demands are actually elements of people's inner needs [6–9]. Obviously, mass customization and even more co-creation practices are clear examples of how technological innovation enables product design to anticipate customer heterogeneity: to offer so-called long tail products [10], as referred to by the organizers of the MCPC conference.

In Fig. 34.1 that shows both extremes of (1) mass production (referring to relatively large distance between maker and user) and (2) fully autonomous DIY activity (referring to a conjunction of both producing and consuming [11]), mass customization and co-creation are positioned in between these two extremes. This chapter, and the research project to which it contributes, focuses on the left side of the graph: facilitating Do-It-Yourself activity.

Today, many design education programs prepare students for a future in a traditional user–supplier relationship context, but it seems tenable that anticipating the recent changes in the field should be considered more seriously. Recent developments and changes require the rethinking of design vision, design strategies, design structures and design implementation. As one of the probable consequences of 'open

J.W. Hoftijzer (✉)
University of Technology Delft, Delft, The Netherlands
e-mail: j.w.hoftijzer@tudelft.nl

J. Bellemare et al. (eds.), *Managing Complexity*, Springer Proceedings in Business and Economics, DOI 10.1007/978-3-319-29058-4_34

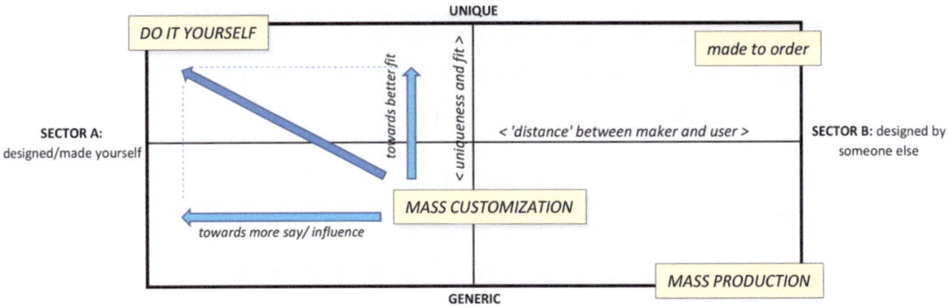

Fig. 34.1 Mass customization and Do-It-Yourself (DIY) on the axes of (**1**) uniqueness and (**2**) 'distance' or separation between maker and user

design' (design whose makers permit, e.g. modifications and derivations of it) [12], Caroline Hummels refers to educational models that need to become more open and flexible [4]. Pieter Jan Stappers and Elizabeth Sanders, referring to Ivan Illich, and Mike Press are only some of many recent authors who state that people seek after more than only passive consumption and shopping [13–15], which is a strong argument for rewriting the starting points of many design education curricula.

Considering the fact that changes in society and practice take place more rapidly than they do in educational programs, there's a chance that design education will run behind today's world. This chapter attempts to provide answers to the question of how a product design curriculum could anticipate some of the changes that take place.

34.2 Definitions

Obviously, new technologies as additive manufacturing and e.g. laser cutting techniques, are having an increasing effect on the industrial design practice. Not only because designers need (want) to anticipate the altered industrial manufacturing opportunities (operational), moreover because the process of designing, iterating, physically visualizing, simulating and testing has changed profoundly (also operational). Besides these operational adaptations, the entire meaning of product design seems to be changing: what is and will be the reason for design, if consumers themselves become designers? And of course the structure and approach of design is changing: stakeholders and the stages of the design process.

It is this context of the professional field of product design that serves as a reference for the technological advancements and parallel societal changes that occur, and for the consequences those changes might have, or even should have, to design education.

34.2.1 Industrial Product Design Definition

To be able to describe the (relatively) full spectrum of consequences to the industrial design field, it is important to enlist some of the most important characteristics.

Industrial design has its origin in the emergence of industrialization in the eighteenth century. When industrial manufacture changed, consumer products changed as well: products started being mass produced. Products entered the marketplace, and along came changing consumption behaviour, by a larger and more homogeneous population. The relationship between the person making a product and the person using it changed dramatically: 'the purpose of production shifted from use to exchange' [16, p. 40]. It marked the start of the so-called Second Wave Civilization, which stood (and still stands) for '() mass production, mass distribution, mass consumption, mass education, mass media, mass recreation, mass entertainment, ()', according to Alvin Toffler [16].

The function of 'Industrial Design' was and is largely related to this industrial context: the design of products for mass consumption. And in order to ensure profit from the manufacture of products, the industrial designer (serving either the corporation or the hired agency) has always taken a short range of second Wave principles into account, as there are for example standardization (of components), specialization of tasks (division of labour) and maximization (of turnover, profit, size) [17–19]. The Industrial Designers Society of America defines the field of Industrial Design as follows: 'Industrial design (ID) is the professional service of creating and developing concepts and specifications that optimize the function, value and appearance of products and systems for the mutual benefit of both user and manufacturer'.

34.2.2 The 'Do-It-Yourself' Scenario

As can be derived from the definition as described in the paragraph above, the industrial designer profession has always been strictly linked to the objectives of the corporation, either in a direct way (cost, efficiency or brand identity requirements) or indirectly by focusing on market trends and consumers' wishes.

Today's context (trends and developments at a Meta level), as shortly referred to in the introduction, gives room to various new situations, therefore to various future scenarios. Relevant developments in the industrial design field are, e.g. the merge of design disciplines (public, graphic, product, etc.), service design, digitization (of the product itself AND of the design and manufacture), smart products and systems, internet of things, design thinking and for example post-industrial design.

Additionally, what could be considered a major development is the democratization of product design, as a consequence of both technological advancements (e.g. digitization) and consumer behaviour. With Second Wave, top-down, business structures as reference, new business models have emerged hugely since digitization and, for example, scale-free manufacture were introduced at the end of the twentieth century. At the same time, consumer behaviour is altered from merely passive consumption to various ways of having more influence on what's offered, as referred to in the introduction.

This chapter builds on the future scenario in which the professional designer's job will be to enable people to execute design and making steps by themselves:

'Design for Do-It-Yourself'. However, in order to enable non-designers to actually participate in the design of a product, specific considerations and preparations are required. The educated designer would be the suitable person to pick up this facilitating task [14]. The 'DIY' scenario was derived from the combination of factors (1) enabling technology, and (2) people's continuous need to express their identity and employ their creativity [20, 21]. The emerge of DIY activities and even of a new DIY culture shows various striking parallels to historic DIY eras within and outside the field of product design [22].

34.2.3 Do-It-Yourself Definition

As referred to in the foregoing paragraph, Do-It-Yourself means—generally speaking—a conjunction of both producing and consuming. Instead of having someone else create the things they need or use, people—at least partially—create their own. In terms of economics that refers to a major shift from sector B to Alvin Toffler's sector A [16], see also Fig. 34.1. The specific definition of DIY depends on its context, depends on which antipode is topic of a discussion. DIY could mean the following:

34.2.3.1 DIY Versus Industrialization

The phenomenon of DIY has a significant meaning when compared to the emergence of the marketplace and since the rise of industrialization. Before products were used as trade, and before products were cheaply mass produced in the industrial age, people tended to make and mend their own or their family's tools and objects themselves. When the marketplace and—later—industrialization appeared, self-sufficient communities changed into an industrial civilization, and consumption became routine.

34.2.3.2 DIY Versus Hiring a Professional

Modern 'Do-It-Yourself' is the term used by Paul Atkinson and refers to the 1950s and 1960s, when the DIY era came of age [23]. World War II ended, causing much work to be done; in the USA there was no skilled labour available, nor money to pay for professional help [21]. Post-war DIY activity was seen as a democratization of the work process, allowing decision-making and freedom from supervision [21, 24]. At that time, many consumer (power-) tools became available, allowing amateurs to construct and repair for themselves, helped by the abundant availability of manuals, toolkits and magazines. This DIY era concerned objects with a certain size: construction projects inside and around the house, and it meant the birth of today's widely spread DIY shops. There was a clear difference between the segments of consumer products and the products created by DIY activity.

34.2.3.3 DIY Providing Options to Choose and Design

Today, this kind of 'modern' DIY project activity is obviously still there. Maybe it is less there than in the 1960s because the further commercialization of society since then, and maybe more because of the enormous availability of 'how-to' information.

But what is new today (since the computer, digitization and user generated content made their entrance) is that DIY applies to consumer products as well. Either through mass customization, co-creation, hacking or modelling and 3d-printing your own objects, technology: tools, kits and information, enable people to have early influence on the product they prefer. Again referring to Fig. 34.1, DIY activity can be seen as the most self-sufficient way of having that influence: designing and making your own product. As was the case in the 1950s, DIY in today's context also means doing parts of the product design process yourself instead of having it done by a professional. The DIY activity in this paragraph is the kind of DIY this chapter refers to.

34.2.4 Design for DIY Definition

'Design for DIY' (DfDIY) represents a special case of a product design process. 'Design for DIY' refers to the envisioned scenario in which professional designers are there to provide the means that enable amateurs to design and make for themselves. The means comprise either the platform, toolkits, templates, tools, information, inspiration, preliminary designs or i.e. the (physical) workshop environment. In terms of process-steps, DfDIY requires taking an important range of extra process steps in consideration (see Fig. 34.2 and Box 34.1).

34.3 'Design for DIY' and Industrial Design Education (Consequences for Design and Design Education)

As discussed, the Industrial Design field is in many cases strictly related to the industrial character of most of our society; the Second Wave society. Students learn how to design products suitable for mass production, and considering the fact that (mass) production and consumption of certain products will not disappear, it is defendable to teach students how to do that. It seems wise though to anticipate societal and technological changes as mentioned and add curriculum elements that incorporate scenarios of a new designer–consumer relationship [1, 4, 25]. The reason to anticipate the technological and societal changes could be founded on, among others, Victor Papanek writing that engineers should take responsibility to respect people's true demands [26], instead of creating obsolete products. On the other hand, to quote a classic statement, 'design does not only shape but also reflects society'.

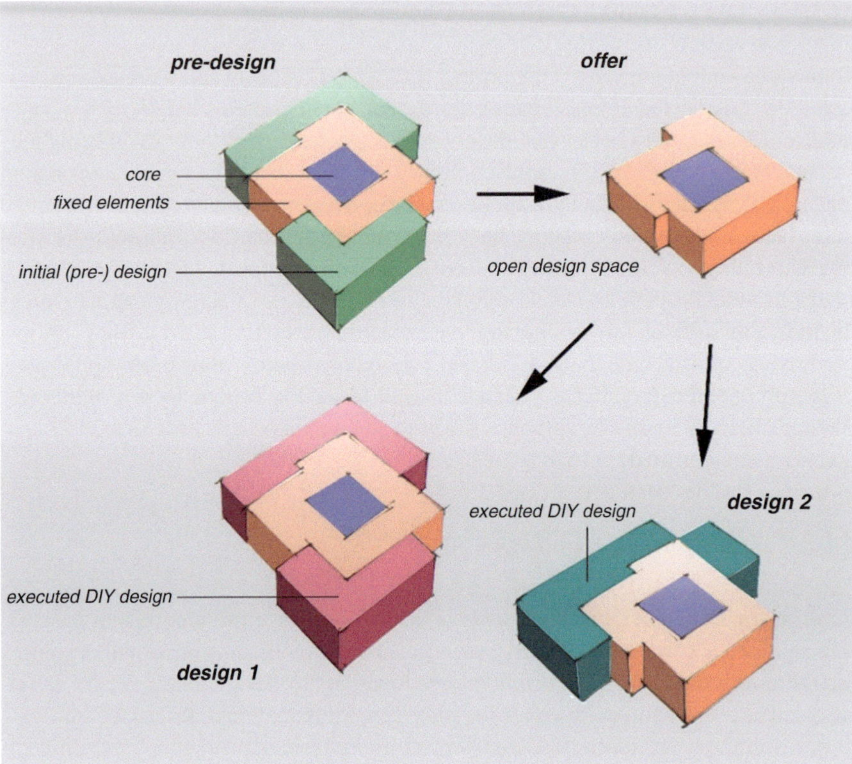

Fig. 34.2 Pre-design > the offer > DIY design

Box 34.1: Generic DfDIY Steps

Generic DfDIY steps:

1. Before a DIY design project can be offered to the amateur, there is the pre-design stage, including project goals, and the preliminary product design that distinguishes 'product core', 'fixed elements' and (preliminarily filled) design space. See Fig. 34.2.
2. DfDIY is to be seen as providing a product–service combination: the provided non-finished product design requires tools, a toolkit and additional information (design environment or extended toolkit) to let the amateur take up his or her part of the design process.
3. Parallel to and in between the amateur's DIY design steps, the facilitating professional designer guides the process, gives feedback and then brings the design to the final stages. In this phase, the amateur designer may execute his or her design steps with the help of a template in a certain format.
4. After the amateur has created his or her design, the result perhaps will have to be translated into information or a format that is producible, and maybe modified to even better answer the wishes of the amateur designer.

Such a step does not only require to implement new technological elements and tools for design operation (e.g. 3d printing, Arduinos), or to discuss new methodologies to apply with students, but it would mean that the vision upon the industrial design profession needs to be reassessed. The changing designer–user relationship, and the scenario of design democratization (for which Design for DIY stands) would require such a visionary change of perspective to be applied to design education. Design education should teach students about business models, and the designer's position therein, that are entirely different from traditional business models. Examples of some relevant and profoundly different business models can be found in practices as 123D, Quirky, Zazzle, Kickstarter, Ponoko, 3dhubs, design-2gather, Thingiverse, in which traditional structures of top-down mass manufacturing have been replaced by offering digital making tools, customization, funding platforms, one-off creation, community-design and distribution of digital content. Alvin Toffler's 'Third Wave Society' that he described in 1980, that represents a digitized society of de-massified information and production and a return of the prosumer (of the 'sector A' society [16]), has clearly entered the arena.

In a design-educational context, the topic of DfDIY means that students need to learn and practice how to facilitate the amateur, anticipating the practice situation in which professional designers and non-professionals cooperate (facilitate, respectively design for themselves). Simply put, Design for DIY implies two layers of design: first design of a product, and secondly, around that, design of the interaction between designer (facilitator), platform and amateur, depicted in both Fig. 34.2 and in Box 34.1.

34.4 A Series of 'Design for DIY' Pilot Projects

In order to explore and test the process of 'designing for DIY' (facilitating/providing the means and platform to help amateurs design for themselves), a series of five 'design for DIY' experiments were executed. Apart from being experiments, these projects have served as pilot course modules in which students researched, explored and executed 'design for DIY', constructed solution spaces. For all experiments, the initial problem statement was, generally speaking, the gap between (a) the technological tools that are increasingly available and (b) the limited ability of many non-professionals to make use of these means.

For all of these pilot studies, some generic design steps were defined, such as the elements enlisted in Box 34.1.

The changes that concern the technology of making, the so-called scale-free-ness of tools for example [5], have an effect on the type of profession designers will have in the near future. Relationships between suppliers, manufacturers, designers and users alter rapidly, causing business models to change as a consequence [1, 2], as many economic principles seem to be no longer valid [3].

While many education programs prepare students for a future in a traditional user–supplier relationship context, anticipating the recent changes should probably be considered [4].

In the subsequent paragraphs, two of these pilots will be briefly explained and illustrated. These studies help, together with the forgoing conclusions, to define what 'design for DIY' could mean in the context of design education. Learnings from these experiments apply to all three levels of how design could be managed and viewed: (1) the strategic (Meta) level, (2) the organizational level (level of tactics, the design process) and (3) the operational level (level of implementation). This categorization is commonly used in design management literature [27, 28].

In the subsequent paragraphs, the pilot studies and the outcomes will be regarded from the perspective of the various design levels, resulting in a list of recommendations for design education.

34.4.1 Project 1: 'DIY Design: Developing a Toolkit for the Layman Designer' (Enabling the Layman Designer to Design a Desk Lamp)

34.4.1.1 Introduction Project

As one of the project's starting points, Mark Sypesteyn takes Turkka Keinonen's Design Contribution Square a basis to argue for a DIY situation in which the non-professional designer (the user) should be thought in a proactive position [25]. The required pro-activity level of the (facilitating) professional designer here depends on the amount of assistance that is required. A desk- (or table-) lamp suits the goal of this project, as a lamp could offer design space regarding both aesthetics and functionality, and offers a manageable level of complexity (Fig. 34.3). The DIY desk lamp project has integrated factors of didactics and fun in a DIY interaction proposal with a rather physical nature.

34.4.1.2 Description

Based on the three general facilitation principles: guidance (design space and activities), complementation (resources, materials, tools) and expansion (varying the complexity/amount of decisions to be taken), a design kit was designed according to the steps shown in Box 34.1. The toolkit offers a specific low-quality ribbon, to be physically manipulated into a certain three-dimensional curved shape (DIY design). The resulting non-professional design result is to be documented by the amateur him or herself, according to a provided strict photo-protocol (Fig. 34.4). The professional designer, who is the creator of the pre-design and the supplier of the toolkit and the environment (webpage), takes the uploaded photographs as a starting point for the making of a corresponding CAD model and the largely 3d printed end product.

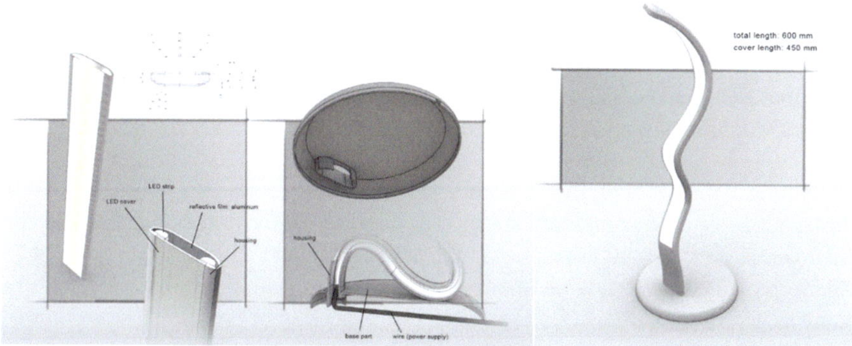

Fig. 34.3 Structure and elements and pre (-liminary) design of the Sculpture desk lamp [29]

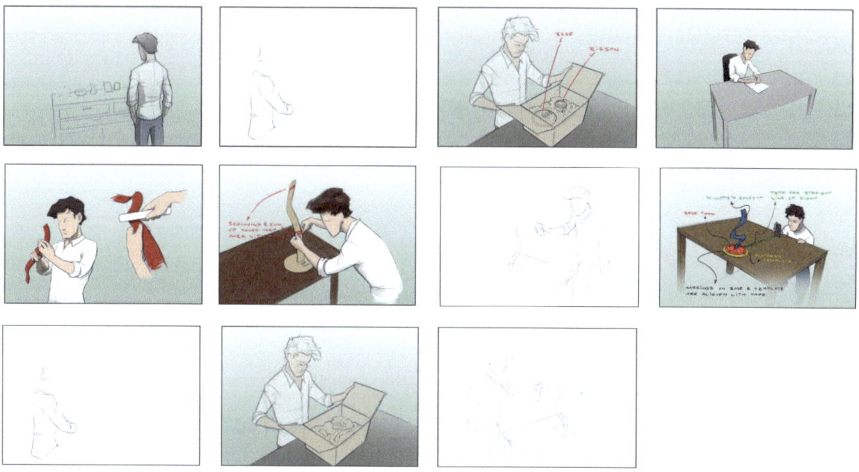

Fig. 34.4 The envisioned DIY design process: (**1**) inspiration, (**2**) ordering the kit, (**3**) receiving physical kit, (**4**) ideation, (**5**) making preliminary sculpture, (**6**) preparing documenting, (**7**) 3d scanning, (**8**) following photograph protocol, (**9**) uploading pictures, (**10**) receiving final 3d printed result, (**11**) adjusting [29]

34.4.1.3 Test Results and Conclusions

With the composition of the design kit package, the path was chosen to physically involve the non-professional designer and ask the amateur to manipulate/modify a shape with his or her hands, then make photos of it. It was a deliberate choice not to provide a software tool for this; hence to increase the level of involvement. It proved to be a very interesting approach with specific attention for the interaction between the professional and the non-professional designer, providing the 'guidance' that was aimed for.

Based on seven trial runs, a questionnaire helped evaluating. Results show that, although instructions for the use of the kit were clear, more attention is required for the explanation, to the participants, of the overall approach and theme. This indicates that the audience should be informed sufficiently regarding the DIY framework in which they perform, in order to increase involvement. Another result from the interviews was that the perceived time spent during the DIY design steps was much longer than it actually took. Sypesteyn suggests replacing long and textual explanations by video instructions on the website environment. People said the documentation part (making photographs of the sculpted DIY result, according to a strict protocol) was easy, while the actual design part: sculpting/modifying the ribbon was perceived hardest. This indicates that following of prescribed steps requiring no creativity is perceived less hard than being creative and freely defining a curve of your own choice. Another conclusion was that sufficient communication about what to expect is crucial: although the subject of colour was of no importance in the sculpting phase, some people expected that the colour combination in the 3d printed end result would be comparable to the colour combination used for the DIY sculpting kit. This could indicate that managing expectations is important, but also, nonetheless, that the facilitating professional designer should as well keep in mind people's imagination capacity.

34.4.2 Project 2: 'Designing a DIY Design Toolkit: Enabling People to Design Their Own Headphones'

34.4.2.1 Introduction Project

Headphones were chosen as DfDIY subject of Karim de Waard's project because of the fashionable and technically manageable character of the product [30]. Young trend sensitive people were the target audience. This project deals, for example, with the chance that people might prefer to buy a branded set of headphones, instead of their own creation. Referring to Nikolaus Franke and Martin Schreier, Karim de Waard aims for a high perceived value, depending on the preference fit, the process enjoyment and the process effort [31]. Another important aspect of the project was the modular architecture, required to be able to assemble both the fixed base of the product and the DIY designed parts.

34.4.2.2 Description

During the pre-design stage, the decision was made to clearly distinguish (a) the (unique) elements or areas of the product, to be modified by the amateur, and (b) the base structure of the product that was to be predefined and even presumed to be mass manufactured (Fig. 34.5).

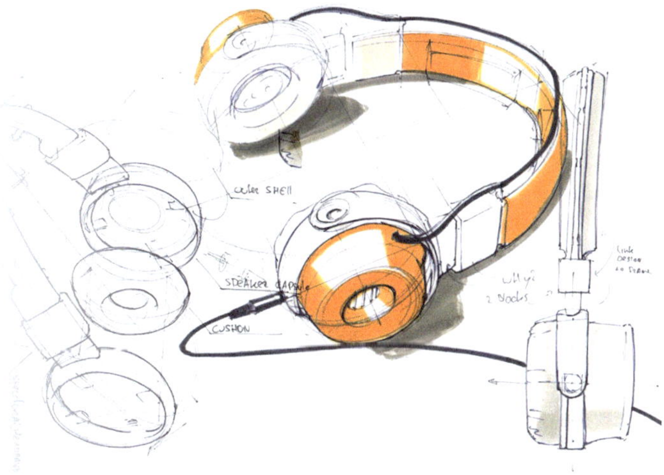

Fig. 34.5 Pre-design stage: distinguishing (**1**) fixed and (**2**) open design space

The DIY area: the design space, comes down to the headphone shells, and the headband sleeve, the most eye-catching elements. The headphones pre-design was the basis for the kit that was to be offered and distributed through an online platform. In constructing the toolkit, the opportunity was taken to not only provide preference fit and enjoyment of the process: specific attention was paid to the DIY interaction medium: to the amateur's effort of drawing his or her design, as a true designer. The drawing medium functions as a suitable learning environment, as an integral part of the kit, it provides the option for the amateur to improve his or her design drawing skills and improve each attempt. De Waard included an in between 'redesign' step to find out whether a final designer's fine-tuning step would be appreciated.

34.4.2.3 Test Results and Conclusions

A series of very helpful insights were the result of a combination of trial (test) runs and interview. To begin with, De Waard distinguished two types of outcomes of the DIY design processes: the first is a literal translation of the side view drawings (on template) into a 3d CAD model and the second an interpreted translation by the professional designer. After having done the test run, participants were asked a short list of questions. Some of the outcomes: people perceived a better fit between the resulting model and their expectations in case of the 'interpreted' design. The interpreted design was appreciated as 'looking better' than the literal design (3.3 vs. 2.8 on a scale of 5). The perceived influence people had on the design of the headphones was (obviously) higher in the literal design scenario. These results indicate that the interpretation step at the end of the DIY design cycle helps to make the design fit better, although influence is perceived less. The fact that (as an answer to

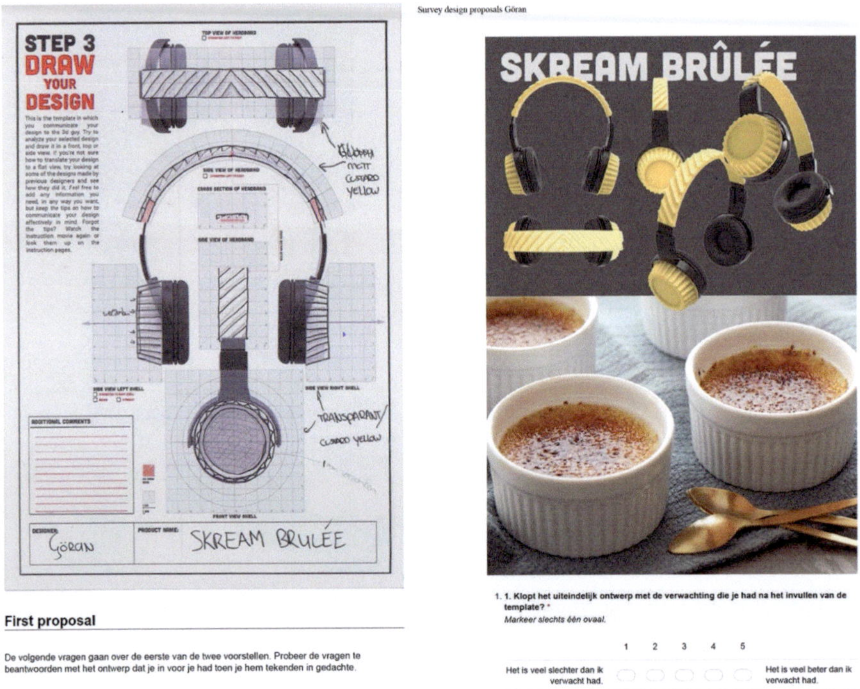

Fig. 34.6 DIY design stages: (*left*) drawing design intention in template (kit), (*right*) inspiration theme and final result

another question) people were willing to pay more for the interpreted headphones design, confirms this. The prices people were willing to pay for their self-designed headphones seemed very similar to prices they wanted to pay for existing products (Fig. 34.6).

Apart from the non-professionals (the target audience) who executed the test runs, some extra runs were done by industrial design students. Differences between the design student runs and the non-professionals were interesting, mainly regarding the freeness of the approach: design students felt free to use the template paper in front of them to also sketch and visualize some extra (even 3d) views to make sure their design intention was clear to the facilitator, while amateurs drew strictly between the lines. Design students translated the provided inspiration theme into an abstract interpretation of that theme, while non-professionals in many cases strictly copied the lines, rims and volumes provided by inspirational theme pictures. The presence of drawing skills seemed to enable participants to design further than mere copying of a theme. In general, the use of the drawing templates as design kit for the amateur seemed to work very fine. However, one should not rush when trying to help develop the drawing skills of the amateur: the template should be kept very clear and preferably 2d (asking non-designers to visualize their thoughts on a '3d' grid was a step to far).

34.5 Overall Conclusions and Steps to Take (Value of the Outcomes)

These pilot studies have illustrated ways of how to construct a DIY project for non-designers. They refer to a situation in which the professional designer is the creator, manager and even seller of the initial pre-design that is to be modified with the help of the provided design kit as part of the offered DIY environment. The execution of these projects (and of the other, not here discussed projects) turned out to be a very logical activity for the Industrial Design students. With their projects, they anticipated the changing role of a designer, the increase of information availability and designed products and the toolkit for modifying. The projects, the processes and the results from tests and interviews have raised some new questions and issues, and doing so help to establish a clearer view of what Design for DIY could do and how to approach. In terms of potential lessons to be taken for design education, the outcomes of execution and evaluations can be viewed from different perspectives. For that reason, as mentioned, it would be wise to view the conclusions from these projects from the various perspectives: different levels, as suggested in Sect. 4.

34.5.1 The Strategic Level of DfDIY: The Meta Level

When discussing the Meta level, design for DIY obviously seeks for another structure than design for mass production does. Not only provides DfDIY design space for the non-professional designer, it also specifies a new job for the professional designer, who traditionally decides on the outcome of a project himself or herself. Moreover, DfDIY does not aim for mass production or mass consumption. A broad digital distribution of the proposed templates and process steps may very well describe the suitable scenario.

Executing DfDIY requires to do the extra steps compared to a traditional design process, as discussed, but choosing for the DfDIY scenario also requires an ethical consideration, a vision: why would you do that? There are actually many reasons, for which I refer to literature in which the active participation of the end user is regarded to increase people's awareness, product attachment and as a solid way to activate people's creativity [7, 8, 20, 31–33]. From an educational viewpoint, it would be wise to acknowledge these arguments, and search for ways how to incorporate the new role of the designer.

However, many considerations should be taken into account, as can be concluded from the pilot studies. Obviously, most important is the search for the right borderline that distinguishes (1) the fixed elements to be decided and established by the professional designer and (2) the amateur's design space. The experiments tell that one always needs to evaluate the freedom provided and the assistance, facilitation and the expertise of the professional designer.

34.5.2 The Organizational Level of DfDIY

As is referred to in this chapter, DfDIY requires a different vision, different starting points, a different structure and it also means different (extra) design process steps (see Box 34.1). Therefore, from an organizational perspective, most existing process models and charts should be altered in order to include the (extra) steps to be taken by both designer and non-designer.

The phenomenon of DIY product design itself could very well be regarded with the help of the abstract 'Vision in Design' model [34], that gives room to depict both past and future context factors, in a search for the definition of a future scenario (Fig. 34.7). This model could help positioning the old (Toffler's Second Wave) situation and the new (Third Wave) society, both in one model. Referring to the same model, the transition from past to future context results in a new interaction vision (#5: human–product interaction), which is what Design for DIY is exactly about.

For design education, viewing the organizational level of design and of DfDIY means that designers and design students should even more integrate changing context factors and people's demands in their process steps.

As new business models occur one after the other, this organizational aspect of the design for DIY field is a very dynamic one. But it is necessary to include all of the new approaches of design systems in the curriculum. In the past 10 years, many new platforms and companies based on new business structures have appeared, of

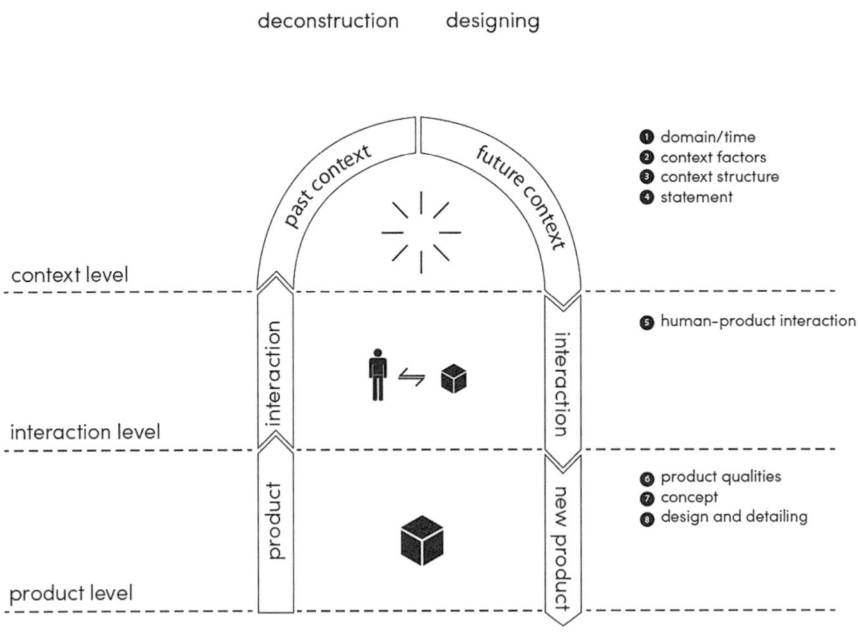

Fig. 34.7 ViP (Vision in Product Design) model [34]

which some have been mentioned in Sect. 1. These are commercial businesses but also freely accessible or even non-profit platforms and services. Design education should include these new structures and designer jobs in their curriculum, at least facilitate the discussion as soon as possible.

34.5.3 The Operational Level: How to Practice?

From the experiments, operational design guidelines can be retrieved that concern lists of requirements, process steps advised and boundaries not to cross. For design education, DfDIY seen from the operational perspective, the various tasks to be executed by the professional designer require a lot of expertise and design skills. Either designers will have to be equipped with all those knowledge and skills, or facilitator tasks should be divided. In short, the designer's job has increased in size: not only bear in mind the requirements and the product design synthesis, but also consider the design environment and design vehicle for the amateur, plus consider all related 'Design for DIY' requirements, divided under DIY product requirements, DIY process requirements, toolkit requirements and platform requirements.

A series of additional trial projects as the ones described in this chapter will be necessary in order to construct a framework for DfDIY. Such a framework will help to establish a clear basis for thinking of, envisioning and executing the new area of Design for DIY.

References

1. Winsor, J.: Beyond the Brand: Why Engaging the Right Customers Is Essential to Winning in Business. Dearborn Trade Publishing (2004)
2. Hoftijzer, J.W.: Design for DIY, beyond the fixed solution space. In: MCPC: World Conference on Mass Customization, Personalization, and Co-Creation 2011. San Francisco Airport Marriott Waterfront (2011)
3. Haegens, K.: Het economie onderwijs ligt onder vuur: voor kritische vragen is geen tijd. In: NV Weekblad De Groene Amsterdammer. De Groene Amsterdammer, Amsterdam (2015)
4. Hummels, C.: Teaching attitudes, skills, approaches, structure and tools. In: Van Abel, B., Evers, L., Klaassen, R., et al. (eds.) Open Design Now. BIS Publishers, Amsterdam (2011)
5. Anderson, C.: In the next industrial revolution, atoms are the new bits. In: Wired Magazine. The Conde Nast Publications, San Francisco (2010)
6. Maslow, A.H.: A theory of human motivation. Psychol. Rev. **50**, 370–396 (1943)
7. Max-Neef, M.: Development and human needs. In: Ekins, P., Max-Neef, M. (eds.) Real-Life Economics: Understanding Wealth Creation, p. 488. Routledge, New York (1992)
8. Ehrenfeld, J.R.: Sustainability by design: a subversive strategy for transforming our consumer culture. Yale University Press, New Haven (2008)
9. Hoftijzer, J.W.: DIY and co-creation: representatives of a democratizing tendency. Des. Princ. Pract. Int. J. **3**(6), 69–81 (2009)
10. Anderson, C.: The long tail: why the future of business is selling less of more. Hyperion Books, New York (2006)

11. Edwards, C.: Home is where the art is. J. Des. Hist. **19**(1), 11–21 (2006)
12. Abel, B.V. et al.: Open Design Now, p. 320. BIS Publishers, Amsterdam (2011)
13. Press, M.: Design management: a future vision. In: Keynote Presentation to DME Award Launch. Eindhoven (2007)
14. Hoftijzer, J.W.: Sustainability by do-it-yourself product design user design opposing mass consumption. In: DRS 2012. Chulalongkorn University, Bangkok, Thailand (2012)
15. Sanders, E.B.-N., Stappers, P.J.: Convivial Toolbox, p. 310. BIS, Amsterdam (2012)
16. Toffler, A.: The Third Wave. Bantam Books (1980)
17. Taylor, F.W.: The Principles of Scientific Management. Harper Bros, New York (1911)
18. Hudson, P.: The industrial revolution: change and opportunity in economy and society, c1750–1830. http://www.ehs.org.uk/industrialrevolution (2008)
19. Mintzberg, H.: Structures in fives, designing effective organizations. Prentice-Hall, Englewood Cliffs (1983)
20. Schreier, M.: The value increment of mass-customized products: an empirical assessment. J. Consumer Behav. **5**(4), 317–332 (2006)
21. Atkinson, P.: Do it yourself: democracy and design (editorial). J. Des. Hist. **19**(1), 1–10 (2006)
22. Hoftijzer, J.W.: Co-creation: het Nieuwe Doe-Het-Zelf? In: Product Magazine, pp. 12–14. Media Business Press, Rotterdam (2008)
23. Goldstein, C.: Do It Yourself: Home Improvement in 20th-Century America. Princeton Architectural Press (1998)
24. Nieuwenhuys, C.: Opstand van de Homo Ludens. Paul Brand, Bussum (1969)
25. Keinonen, T.: Design contribution square. Adv. Eng. Inform. **2009**(23), 7 (2009)
26. Papanek, V.: Design for the Real World, p. 394. Thames & Hudson Ltd., London (2006)
27. Borja de Mozota, B.: Design Management; Using Design to Build Brand Value and Corporate Innovation. Allworth Press, New York (2003)
28. Best, K.: Design Management; Managing Design Strategy, Process and Implementation. AVA Publishing SA, Lausanne (2006)
29. Sypesteyn, M.: DIY design: developing a toolkit for the layman designer. In: Industrial Design, p. 208. Delft University of Technology, Delft (2014)
30. Waard, K.D. Designing a DIY design toolkit: enabling people to design their own headphones. In: Industrial Design, p. 146. TUD, Delft (2014)
31. Franke, N., Schreier, M., Kaiser, U.: The 'I designed it myself' effect in mass customization. Manag. Sci. **56**(1), 15 (2010)
32. Chapman, J.: Emotionally Durable Design, Objects, Experiences and Empathy, p. 211. Earthscan, London (2005)
33. Csikszentmihalyi, M.: Creativiteit, p. 460. Boom, Amsterdam (1998)
34. Hekkert, P., Dijk, M.V.: ViP (Vision in Design). BIS (2014)

Chapter 35
Apparel Technology Integration and Development for Purchase Activated Manufacturing

Muditha Senanayake, Peter Kilduff, and Bill Grier

35.1 Introduction

Apparel manufacturing has gone through number of transitions within the continuum from standardization to customization. These were discussed by researchers under the manufacturing strategy umbrella in mass production (MP), quick response (QR), just in time (JIT), flexible manufacturing (FMS), lean production, made to order (MTO), demand activated manufacturing, mass customization (MC), made to measure (MTM), and fully custom. Most of the apparel businesses in the standardization end are driven by risk based forecasting and planned over production at every link in the supply chain and have been discussed as unsustainable. While technology has had its greatest impact in every step of the apparel pre-production area, growing emphasis has been given to integrating the pre-production, production, and post-production processes during the last decade to achieve integrated manufacturing systems. Today, computers are extensively used in integrating efforts called "seamless solutions" to coordinate activities to bring new synergism to the apparel industry. The objectives are to reduce the inventory build-ups, lead-time delays, and imbalances that are inherent in apparel's batch processing methods. Integration of technologies, especially that are working with digital technologies, such as body scanning, computer aided design, textile design, digital printing, new digital fabric coloration methods, marker making/nesting, fabric cutting, and sewing systems have created opportunities to work with digital inventories before physical products are made. This approach with digital technologies also has created the opportunity for the customer to participate in the product development process allowing customization and

M. Senanayake (✉) • P. Kilduff
California State Polytechnic University, Pomona, CA, USA
e-mail: mudithas@cpp.edu; pkilduff@cpp.edu

B. Grier
AM4U Inc., Rancho Cucamonga, CA, USA

J. Bellemare et al. (eds.), *Managing Complexity*, Springer Proceedings in Business and Economics, DOI 10.1007/978-3-319-29058-4_35

personalization. Considering these dynamics a team consisting of educators and technology companies has developed a demand driven virtual inventory manufacturing system for Purchase Activated Manufacturing business named apparel made for you (AM4U). AM4U embraces ecommerce solutions, business integration systems, a new sustainable fabric coloration technology with latest digital technologies and flexible manufacturing systems to color fabrics and produce apparel on demand. The team has developed its first Integrated Mini Factory that can demonstrate this strategy under one roof. The objective of this chapter is to discuss the purchase activated manufacturing (PAM) concept.

35.2 Background

To meet the demands of the increasingly geographically dispersed economy with production for efficiency and low costs, the system known as Mass Production (MP) was developed. MP system required standardized products because any changes or custom work will result in bottlenecks in the production process causing higher costs. With the importance of the scale of the standardized nature of products to maintain low costs, MP system was highly dependent on machines, specialization of work, and division of labor. In the MP practice, standardization of customer taste allows for standardized design, which allows for mechanized MP and thus allows for mass distribution [1]. The MP is also characterized by long runs of identical products based on the principles of economies of scale and Frederick Taylor's specialization of labor [2].

In the assemble to order (ATO) or made to order (MTO) MP practice, after receiving an order, if the raw materials are not available in house, the manufacturing entity starts purchasing fabrics, accessories, and sub-systems, waits for the goods to arrive, and then assembles them into finished or partially finished products. The lead-time will increase with any additional supply chain link, which has to go through the same process of ordering and waiting for its parts. The general MP manufacturing practice in a "batch" and "queue" environment causes the delay at every workstation in the manufacturing process. Even though low cost, high quality, and quick delivery are simply qualifiers in the purchasing process, manufacturers must personalize products to meet customer needs and simulate market demand.

Based on the drawbacks of MP systems to meet the volatile customer demand, quick response (QR) apparel manufacturing became important [3]. The demand simply is to deliver the goods on short notice, in small lots, and a broad assortment. This was a more effective competitive strategy for apparel companies demanded by the volatile apparel markets that needed large products variation. Quick response manufacturing (QRM) is identified as a company wide strategy that pursues the reduction of lead-time in all aspects of company's operations. This strategy is explained in two contexts: externally as responding to those customers' needs by rapidly designing and manufacturing products customized to the needs, and internally as reducing the lead-time for all the company's own operations and tasks resulting in improved quality, lower cost, and quick response. QRM is a practical strategy that embodies the mindset of pursuing lead-time reduction, along with detailed management principles, manufacturing methods, techniques and tools, and step-by-step methodology to achieve the desired reduction in lead-times [4].

Another strategy to overcome the drawbacks of MP was just-in-time manufacturing (JIT), which became popular and has been adopted by some companies. The strategies based on these Japanese principles were named as Lean Manufacturing [5]. According to Suri [4], QRM also finds its roots in strategy used by Japanese, later known and documented by American authors as "time-based competition" (TBC). The underlying principle of TBC is the use of speed by a company to gain competitive advantage thus delivers products or services faster than its competitors.

Flexible textile and apparel production technologies are increasingly in demand with the evolvement of new apparel business environments needing customization and personalization. The demand for apparel manufacturing flexibility is not new and has been on research agendas for the last few decades. Manufacturing flexibility in the apparel industry has ranged from made-to-measure tailored clothing shops to plants set up with an effort to manufacture a volume of one single product. The need to shift towards flexible apparel manufacturing plants, which offer multi-product flexibility thus faster market responsiveness, is inexorable. In the past, the change from one style to another even with a smaller change was considered as a disruptive event in apparel production as the factories wanted to continue production without changes over a longer time. However, the numbers of "one-product" apparel companies have reduced rapidly. The shift towards apparel manufacturing plants offering multi-product flexibility and faster market responsiveness has become the key towards success.

Demand activated apparel manufacturing is another phenomenon that has been discussed for many decades. The demand activated manufacturing architecture (DAMA) project launched in 1993 was one that studied this strategy in detail. This project discussed an inter-enterprise architecture and analysis for supply chains that enable improved collaborative business across the apparel supply chains. The goals of this architecture were to reduce new product realization cycle time that will result in significant cost savings and inventory reduction. Taking the advantage of then upcoming secured Internet communication, developing and making technologies commercially available, developing an inter-enterprise architecture for supply chain collaboration, and using modeling and simulations to analyze supply chain configurations were accomplishments expected from the DAMA project. A secure infrastructure data communication tool for collaboration named TEXNET was developed while a true collaboration beyond data sharing was emphasized. To validate the DAMA architecture, models were developed and simulated. Resulting two models addressed lead-time calculations and inventory levels at different stages of the supply chain. The results from the simulation proved that this architecture did, in fact, had a positive impact. The project concluded that the supply chain issues such as long lead times were caused by the inability to accurately forecast what consumers would buy 6–12 months in the future, the complexity of the process steps in the pipeline, and the lack of synchronization among the supply chain partners. These boundaries lead to enormous inventory requirements at every stage of the apparel pipeline [6].

More than two decades after the DAMA project, the apparel industry still deals with major issues such as forecasting inaccuracies, large inventory costs, and large proportion of markdowns leading to an unsustainable industry with countless amount of apparel products end up in landfills. Globalization once considered, as the weapon for profits, has become a recipe for disaster for some companies with above issues. While these changes were taking place, the consumer behavior with their demands and expectations has changed. The business dynamics and competitiveness have amplified. As an answer to some of these problems, a considerable push inbringing back manufacturing to the US is apparent. As a positive outcome, the technology has improved tremendously specially in the digital front over the last decade. These digital technologies have allowed the supply chain partners to work in virtual settings. Considering the issues apparent in current industry dynamics, the objective of this chapter is to discuss a purchase activated manufacturing system with virtual inventory for apparel with technology integration. Various new digital technologies and new operational capabilities arising from these technologies are focused. The purchase-activated apparel manufacturing system is expected to lead to manufacturing of apparel for individual customer demand with customization and personalization capabilities.

35.3 Purchase Activated Apparel Manufacturing

Among the recent technological advances, there is a growing interest of capturing human body measurements using the scanning technology. An accurate data set of the surface of the body is needed in order to develop consistent body measurements and thus accurate patterns. There is a very high expectation for this technology to drive towards Fit Customization [7]. The made-to-measure apparel requires the underlying technology to facilitate acquiring human body measurements and extracting appropriate critical measurements so that patterns can be altered for the customer.

The 2-D and 3-D CAD technology related to apparel has been quite successfully developed such that the technology is capable of handling number of pre-production and production functions using complex computer software and hardware systems. Most of these developments of CAD have been comparatively well discussed in the literature. These CAD technologies today have better capabilities that provide rapid garment pattern generation based on customer's body measurements to produce made-to-measure patterns. However, when such CAD technology for customization and personalization of apparel is concerned, still significant amount of research and development work is essential to provide the color and fit of each garment demanded by individual customers. This observation has also been discussed in previous research [8]. CAD systems with faster pattern making abilities and automatic pattern alteration methods provide the opportunity for fit customization for customized and personalized apparel manufacturing practices.

Innovative ink-jet inks and ink-jet printers have recently been developed, bringing textile printing one step closer to becoming completely digitized. Especially formulated water-based ink-jet inks now make digital ink-jet textile printing possible with simplified fixation equipment. The advanced Raster Image Processing (RIP) software systems provide the capability of obtaining planned and expected colors for the apparel products [9]. These new technologies are expected to revolutionize designing, sampling, and short-run textile printing by offering digital ink-jet print capability on a wide range of fabrics [10].

Sublimation printing is a versatile, digital printing method that opens up a world of opportunities to consumers needing decorated apparel goods. Sublimation is a process by which sublimation dyes are printed onto a transfer medium (donor) with especially prepared ink-jet printers [9]. Thereafter, those dyes are transferred from the donor medium to a (Receiver) fabric under the heat and pressure delivered by a commercial heat press. When the heat and pressure are applied, the dye on the transfer medium sublimates, or becomes a gas, and is then absorbed into the receiving fabric.

Active Tunnel Infusion (ATI™) is a replacement technology for traditional wet printing, chemical dye processes, and sublimation printing. This patented permanent fabric coloration process is an important technology for customization and personalization of dying, printing, and imprinting. This process uses heat and photon stimulation to generate capillary action to pump dye inside the fibers before being trapped for permanent coloring. The system needs no pre- or posttreatment processes and enables dyeing, printing, and labeling in a single pass. The technology looks similar to sublimation printing but claims to be superior. It is a pollution free, waterless technology which provides better color fastness for synthetic fiber fabrics [11].

Fabric cutting using optical scan technology or digital vector technology has been developed for flexible fabric cutting. A cutter with a scanning system can scan the garment patterns on fabric and send the contour information to the cutter, which will then be converted to a cut path to cut the garment patterns. In a digital vector cutting system, a vector pattern file, which is created during preparation stage with identification marks, will be sent to the cutter and the cutter follows the vector outlines and the registration marks to identify the cut path and cut the fabric accurately.

The flexible sewing systems such as modular and unit production systems (UPS) are popular to sew apparel products with small order quantities and can be adopted for customized or personalized apparel production.

35.3.1 Technology Integration

Customized or personalized manufacturing requires a new generation of shop floor control systems that can dynamically respond to each customer and adopt unanticipated changes in the production environment. Requirements in this regard include re-configurability, decomposability, and scalability to achieve make-to-order with a short response time. Efforts have been made to design control systems for some industries by leveraging recent progresses in computing and communication technology including new software engineering methods and control technologies such as barcodes, QR codes, smart sensors and actuators, open architecture, fast reliable networks, and cloud computing [12].

It is suggested that the old batch oriented computer systems to handle customer orders need to be replaced with the advanced computer integrated manufacturing (CIM) factory control systems that would broadcast the order requirements to every station in the production line based on the bar code or QR code of the unit at that station. Each station needs to be informed what unique operation that needs to be performed for the product at the station [13]. As Silveira et al. [14] discuss, mass customization enabling technologies that support the implementation are advanced manufacturing technologies (AMT) such as computer numerical control (CNC) and FMS, and communication and network technologies such as CAD, CAM, CIM, and electronic data interchange (EDI) [14].

With the integration of various product development and product technologies connected with customer information in a digital framework have enhanced the ability for computer integrated, flexible, customized, or personalized apparel manufacturing. The motivation to customize will lie in finding efficiencies in two key dimensions: to include each customer's specifications in the product design and to utilize a modular design to achieve manufacturing efficiencies closer to MP efficiencies [15]. The literature suggests that customer driven manufacturing originates from traditional manufacturing paradigm, "one-of-a-kind production" [12].

Therefore, the challenge is to develop more flexible apparel manufacturing facilities. The flexibility of the process needs to be covered throughout all the functions, not just cutting, sewing, and finishing. The concepts of purchase activated manufacturing and integrated mini factory are efforts to address this challenge, which is discussed below. With the idea of customized or personalized manufacturing requirements, the next section will provide an overview of the PAM concept.

35.4 Technologies for Purchase Activated Apparel Manufacturing

When a customer is interested in purchasing an apparel product, he/she will need to participate in a process to make a fit decision. Strategies such as selecting his/her size from a micro sizing system or matching his/her body to an avatar from an existing database (may be based on some responses to questions) will lead to detecting the appropriate garment patterns for the style selected. These garment patterns decorated with his/her choice of fabric, color, prints, logos, and/or logo placements will be virtually draped on his/her body and visualized with selected sewing features to make the purchase decision. The customer's style configuration decisions and fit pattern selection initiates the demand driven manufacturing process. This process will be managed using an internal virtual inventory manufacturing software system covering the existing ERP, MRP, and PLM functionalities.

The framework begins with identifying the customer demand for demand driven manufacturing as a way to reduce the forecasting error for businesses. This may also lead to more satisfied customers with fewer returns. The product in discussion can either be a complete product such as a garment, a partial product such as a fabric with a surface design on it, or an intermediate product or work-in-progress such as cut fabric panels waiting for assembly. The digital technologies such as body scanning, avatar generation, 2-D and 3-D CAD systems, fabric drape simulation and visualization systems that provide fit choices and software programs that provide configuration of product options such as design, style, fabrication, color, and feature choices provide a customer (B to C) or a company (B to B) to develop and create unlimited virtual product inventories which can be stored on a cloud server. This cloud space can be identified as the "digital virtual inventory closet space" for a consumer or a "digital virtual inventory material storage" for a company that can be accessed for future purchase decisions. This virtual inventory will provide market-testing opportunities in making forecasting decisions without producing physical inventories thus greater cost savings for apparel companies. This is one of the principal opportunities in the proposed PAM.

Unlike in the apparel mass production process where decorated fabric is ordered well in advance, spread, cut and sew based on forecasts, this process decorate the garment pattern pieces digitally based on customer choices. These decorated garment patterns will either digitally printed on fabric, digitally donor print on paper and transfer print on fabric, or sublimation printed on fabric on the fly. The novel patented sustainable on demand fabric coloration and printing technology; Active Tunnel Infusion (ATI™), provides enormous printing capabilities for demand driven manufacturing. This process reduces the fabric coloration and printing lead-time from weeks and months to minutes and hours. This process provides a demand activated digital virtual inventory fabric manufacturing process in a B to B setting.

Once the decorated garment panels are printed on the fabric, the cutting can be done using single ply contour vision cutting systems or using digital single ply vector cutting systems. This is to ensure that the garment panels are cut precisely around the decorated pattern pieces. This also can be a B to B situation where cut garment panels are delivered for sewing and finishing process. These off-loaded garment panels will be sewn using apparel production systems such as the modular or Unit Production System (UPS) before it is delivered to the customer in a B to C setting.

The individual garment panels can be tracked using a digitally printed barcode or QR code on a garment panel, which can be a part of the garment label. The code may consist of information that can be scanned to activate manufacturing processes such as the UPS. Further, this label consisting of customer information will be scanned to generate address labels to ship the product to the consumer.

References

1. Lampel, J., Mintzberg, H.: Customizing customization. Sloan Manag. Rev. **38**(1), 21–30 (1996)
2. Hayes, R.H., Wheelwright, S.C.: The dynamics of process-product life cycles. Harv. Bus. Rev. **57**(2), 15–22 (1979)
3. Pine-II, B.J.: Mass Customization: The New Frontier in Business Competition. Harvard Business School Press, Boston (1999)
4. Suri, R.: Quick response manufacturing: a competitive strategy for the 21st century. http://www.bu.edu/mfg/pdf/SuriPaperQRM.pdf (n.d). Accessed 23 Oct 2014
5. Womack, J.P., Jones, D.T.: Lean Thinking: Banish Waste and Create Wealth in Your Corporation. Simon & Schuster, New York (1996)

6. Demand Activated Manufacturing Architecture: DAMA final report. http://techexchange.com/library/DAMA%20Final%20Report.pdf (2001). Accessed 15 May 2015
7. Senanayake, M.M., Little, T.J.: Mass customization: points and extent of apparel customization. Int. J. Fash. Mark. Manag. 14(2) (2010)
8. Istook, C.L.: Enabling mass customization: computer-driven alteration methods. Int. J. Cloth. Sci. Technol. **14**(1), 61–76 (2002)
9. Watkins, P.: Innovations and developments in ink-jet printing. Text. Outlook Int. **74**, 62–76 (1997)
10. Deverakonda, S.: Ink jet printing vs screen printing: bridging the gap between ink-jet flexibility and screen-printing quality. Textile world. http://www.textileworld.com/Issues/2004/October/Dyeing_Printing_and_Finishing/Screen_Printing_Versus_Ink-Jet_Printing (2004). Accessed 15 May 2015
11. AM4U Technology: http://www.sresearch.com/am4u/technology/ (n.d.). Accessed 8 June 2015
12. Tseng, M.M., Piller, F.T. (eds.): The Customer Centric Enterprise: Advances in Mass Customization and Personalization. Springer, Heidelberg (2003)
13. Eastwood, M.A.: Implementing mass customization. Comput. Ind. **30**(3), 171–174 (1996)
14. Silveira, G.D., Borenstein, D., Fogliatto, F.S.: Mass customization: literature review and research directions. Int. J. Prod. Econ. **72**, 1–13 (2001)
15. Duray, R., Ward, P.T., Milligan, G.W., Berry, W.L.: Approaches to mass customization: configurations and empirical validation. J. Oper. Manag. **18**(6), 605–625 (2000)

Chapter 36
Co-design Visual Displays in Virtual Stores: An Exploration of Consumer Experience

Juanjuan Wu, Natasha Thoreson, Jayoung Koo, and Angella Kim

36.1 Introduction

Wu et al. [1] previously adopted a Facet Theory Approach and quantitatively categorized the various visual components of user-generated visual merchandising directives in 3D virtual store environments. This research continued the study of co-designed visual displays but focused on women's apparel and utilized pre-designed modules as inspirations to aid the consumer co-design process. Consumer co-designing visual displays using 3D technology is still a new research direction, which enables direct access to first-hand information about consumer needs and preferences. Albeit new in the field of visual display and merchandising, co-design has been a popular subject in product development in both academic research and industry practices. Co-design, that involves consumers to co-create a product Piller et al. [2], characterizes mass customization, which is thought to provide a revolutionary solution to many fundamental limits of mass production, such as uncertain consumer demand, uniform product offerings, high inventories, and markdowns [3]. Though in virtual worlds such as The Second Life consumers have already started to make changes to their retail or entertainment environments through consumption and socialization co-designing visual displays in retail stores have been largely kept as an in-house function. However, the development of virtual worlds and the advancement in 3D technology may eventually enable individualization of virtual stores online. The purpose of this research is threefold: (1) to discover consumer preferences for visual displays in virtual stores. We aim to reveal, in an ideal world, how consumers prefer merchandise to be displayed in a virtual store for them; (2) to describe consumer's co-design experience, including any frustrations

J. Wu (✉) • N. Thoreson • J. Koo • A. Kim
University of Minnesota, St. Paul, MN, USA
e-mail: jjwu@umn.edu; thor0330@umn.edu; kooxx051@umn.edu; kimx3119@umn.edu

J. Bellemare et al. (eds.), *Managing Complexity*, Springer Proceedings in Business and Economics, DOI 10.1007/978-3-319-29058-4_36

they may experience during the co-design process; (3) to investigate consumer's evaluation of the potential to make this co-design experience a retail offering.

36.2 Method

Research participants went through a co-design process to individualize a virtual store according to their unique needs or preferences. A total of 39 co-designed virtual stores were content analyzed based on a framework that was developed from the literature and used by Wu et al. [1]. And 37 written reports were also analyzed to reveal participants' co-design experiences and their evaluations of the experience as a retail offering.

36.2.1 Pre-designed Modules and Procedure

Participants were recruited from students taking retail merchandising classes, mainly in the Visual Merchandising classes at a mid-western university in the USA. All participants have acquired basic training to use the 3D virtual retailing software package Mockshop. They were given course credits and had a chance to win scholarships as incentives. Research participants were shown modules of six virtual stores that the researchers predesigned that displayed merchandise based on three grouping methods of lifestyle, brand name, and color, and varied in high or low density. The module only populated half of the store leaving the other half empty. The research participants, acting as co-designers, could pick and base their design on any one module as a point of reference. However, they were encouraged to make changes and display merchandise that reflects their own needs and preferences. Participants were asked to only use merchandise (Target's Ready to Wear) we provided to populate the fixtures. They could also only use the types of fixtures that are already in the module that they chose. However, they were allowed to use as many merchandise options or fixtures as they want to achieve their desired density and effect through a function called duplication.

36.3 Results and Discussion

36.3.1 Content Analysis of Co-designed Virtual Stores

We adopted Wu et al.'s [1] framework and analyzed three major aspects of the 39 co-designed virtual stores, including merchandise presentation, in-store environment, and in-store promotion. Participants' preferred primary product grouping methods were based on (listed in descending order): color (10) lifestyle (9), a combination of lifestyle and brand name (7), brand name (6), product category (6), and only outfitting (1). The method of outfitting is simultaneously used along with all other grouping methods except for five virtual stores that only used product category as a grouping method. This finding indicates the importance of color and lifestyle as primary product grouping methods to guide consumer interest. Outfitting has also become a nearly universally preferred displaying technique. Regarding product display density, compared to four ways and convertibles, tables and back or side walls featured significantly fewer numbers of merchandise, indicating the tendency of consumers emphasizing the "displaying" instead of "stocking" function of these fixtures. All stores used "hanging" instead of "folding" as the main manner of presentation, which again signifies the importance of showcasing the totality of a product in a store. Four ways and convertibles were generally preferred fixtures. Back or side walls were used mainly in stores that used lifestyle, color, and product category as primary grouping methods perhaps because they are suited to create a visual impact when viewed from a distance. Freeform, followed by racetrack, was a predominant layout used by the participants, which fits with the needs of the stores selling women's apparel.

On the other hand, grid layout is often perceived as more utilitarian and emphasizing speed and ease of shopping [4]. Flooring mainly featured wood in brown. Wall colors were mostly neutral with few exceptions of using patterned wallpapers or multicolors. Signage and text within the stores were generally used to enhance the merchandising directives instead of creating extra excitement. Overall, participants preferred to use product grouping and displaying techniques that are visual stimulating. Offering styling suggestions through outfitting has become a must-have when selling women's apparel.

36.3.2 Content Analysis of Written Reports

Results of a content analysis of 37 usable written reports were tabulated based on levels of fun, levels of interest, levels of difficulty, design issues, software issues, and practical application. Our findings indicated that the co-design experience was fun and interesting for most participants. Because the modules were already built, participants did not have to focus on the more technical aspects of the program. As one respondent explained, "I really only needed to do the fun part of designing: picking colors, floors, and aesthetics." While many respondents praised the co-design experience as a way to better understand their own shopping preferences, most respondents were interested in the experience from a more practical perspective. The respondents, students in an undergraduate retail merchandizing program, were excited to learn the Mockshop software as well as practice designing a store. For many, this was an eye-opening experience. For instance, several respondents noted the "attention to detail that it takes to design a store" or "how much thought is actually put into designing a store." Other respondents saw the co-design experience as a way to explore designing with "limitations" and problem solving within "constraints." This may indicate that the value of the experience came from its real-life applications; most retail merchandizers will be limited by available signage and store fixtures.

Some respondents commented on the learning curve required to best use Mockshop. Some indicated they felt uncomfortable or frustrated using the software, even after training. Others noted that the process of learning the software was time consuming and required many questions to unpack. Such critical feedback was also expressed when respondents were asked if the co-design experience could be offered to customers in a field setting. Nearly all respondents agreed that Mockshop was a difficult, complicated, and tedious software to learn, requiring an extensive amount of time-consuming training. As such, many thought customers would not enjoy the co-design experience.

Opinions did differ. Some respondents felt the co-design experience would be most beneficial to individuals with a high level of interest in shopping. Others suggested the co-design experience would offer customers the same type of engagement that they themselves experienced—it would allow customers to better understand their own shopping preferences. In addition, respondents thought customers would respond positively to the opportunity to show retailers what they wanted in a retail experience. One respondent noted that customers would have more agency in their retail experience and thus more "freedom, and say in what happens in their favorites [*sic*] shops."

References

1. Wu, J., Kim, A.J., Koo, J.: Co-design visual merchandising in 3D virtual stores: a facet theory approach. Int. J. Retail Distrib. Manag. **43**(6), 538–560 (2015)
2. Piller, F.T., Moeslein, K., Stotko, C.M.: Does mass customization pay? An economic approach to evaluate customer integration. Prod. Plan. Control **15**(4), 435–444 (2004)
3. Anderson, L.J.: Mass Customization: A Business Strategy. Apparel Research Seminar of the American Apparel Manufacturers Association Task Group, Atlanta (1999)
4. Vrechopoulos, A.P., O'Keefe, R.M., Doukidis, G.I., Siomkos, G.J.: Virtual store layout: an experimental comparison in the context of grocery retail. J. Retail. **80**, 13–22 (2004)

Chapter 37
Seven Steps Manufacturers Must Take to Begin Offering Mass Customization to Their Customers

Jocelyn Bellemare and Serge Carrier

37.1 Introduction

The goal of mass customization is to efficiently provide customers with what they want, when they want it, at an affordable price. Inala [1] contends that mass customization has become a competitive strategy for businesses that want to offer personalized products. The more a business provides opportunities to personalize its products, the more competitive it becomes [2]. Mass customization offers a new business model and growth opportunities for small manufacturing businesses and specialized industries. Indeed, from mass or large volume production, businesses in different industries can profit from this value-adding advantage. According to Zipkin [3], this type of production will become possible on a large scale as new technologies become more easily accessible. The question is how to find the "optimal" product configurator with the capacity to efficiently translate customers' desires and associate them with their personalized characteristics.

The market has evolved in the last few years and technologies, both at the consumer's and at the manufacturer's end of the supply chain, are becoming ever more important. The industrial era was characterized by a top-down approach to business and was managed by whoever controlled the financial, physical, and human production resources. The digital era has changed this paradigm to a new one where everyone with some computer savvy may start a business and contribute to the economy.

The volume and scale economies allowed by the globalization of production have made technologies available to anyone and, as Pine and Korn [2] state, have made them ever easier to understand and use. This consumer ownership of information

J. Bellemare (✉) • S. Carrier
Department of Management and Technology, University of Quebec in Montreal (UQAM), Montreal, QC, Canada
e-mail: bellemare.jocelyn@uqam.ca; carrier.serge@uqam.ca

J. Bellemare et al. (eds.), *Managing Complexity*, Springer Proceedings in Business and Economics, DOI 10.1007/978-3-319-29058-4_37

technologies (IT) is central to the theme of mass customization as it is at the root of the consumer's participation and collaboration movement (virtual fitting room, augmented reality, etc.). For the consumer, the availability of those technologies not only provides a more efficient and economical way to shop but it also adds to the pleasurability of the shopping experience.

Past research has demonstrated the importance of understanding mass customization within the context of trade globalization which has led to ever more ferocious industrial competition. Moreover, as some products now seem to have an ever shorter life cycle, a phenomenon which is exacerbated by the introduction and implementation of new business models, businesses' commercial strategies face mounting pressure. This situation forces industrial players to revise their organizational strategies in order to survive. Organizations must reinvent themselves and find new ways to satisfy their customers. In order to grow, to maintain the current level of employment and possibly increase it, producers need to develop new manufacturing strategies by orienting local production toward a flexible, quick-response system that allows for the production of various types of orders (small quantities, short deadlines, skilled labor, etc.).

It is now essential for businesses to implement new strategies that correspond to the reality of current markets in order to keep up with the rhythm of short cycle production. Businesses need to focus on flexibility, adaptability, and agility [4].

37.2 Uniqueness

Reviewing the writings on this subject tells us that paradoxically, at a time where the global key word in most industries is standardization, the focus in the apparel industry is on "uniqueness." With the recent surge in the use of new media and telecommunication, consumers are more and more demanding and informed. They are no longer satisfied with standardized products that force them to make compromises. The internet influences customers' buying habits by creating needs that have to be satisfied instantaneously.

Many businesses are currently researching technological ways to produce, adjust, sell, and deliver, in a systematic and automatized system, personalized and made-to-measure products. Nevertheless, mass customization somehow remains misunderstood or is rarely used by important actors mainly because of the widely variable human measurements, of the problems in adapting processes, of the need for flexibility, and of manufacturing delays and methods.

A mass customizer must first identify the idiosyncratic needs of its customers; specifically, those product attributes along which customer needs diverge the most [5]. As a result, it must provide a personalized product and possibly use a different method of production in what Haug et al. [6] referred to as personalized and handcrafted production. Likewise, in order to be able to meet the demands of mass customization, all of a manufacturer's operations have to be based, according to

[3], on flexible processes that allow it to respond rapidly to customers' requests. More often than not, mass customization consists in, for example, assembling basic items according to specific orders.

Mass customization therefore becomes a crucial development alternative for businesses specialized in apparel manufacturing and distribution [4]. In this industry, the demand for mass customization is growing stronger, and it is becoming possible thanks to the contribution of new technologies. Yet in this case, as with technical products, mass customization requires a very thorough understanding of the expectations and specificities of each individual; it rests mainly on a successful integration of the value chain. In some respects, businesses must accomplish a feat by performing well on two axes that are generally on opposite sides of the spectrum: maintaining short supply lead times while offering custom-made products that correspond to clients' specifications.

37.3 Mass Markets

That there are mass markets for some customized products is not questioned—the emergence of mass customization demonstrates it [3]. The main problem of mass customization is related to the conception and manufacturing of products, specifically in the apparel industry, according to the customers' requirements. Moon and Lee [7] state that because of their lack of knowledge and experience, consumers do not really know what they want. It is thus important to focus their request by offering them some guidance. Doing so not only requires knowing a customer's measurements, specifications and style, but also obtaining information that he never reveals: what the literature refers to as "sticky information."

The term "sticky information" was coined by Von Hippel [8] as information hidden by a customer that provides, in certain cases, a company with a key competitive advantage and offers significant opportunities for innovation. Consumers know their needs and tastes better than manufacturers. Yet it is difficult for a manufacturer to obtain some of this information which is either confidential or perceived to be so irrelevant that consumers will only reveal it sporadically, at best. This unknown data, like perception, fit, style, proportions, preferences, and the like, are essential to the production of custom products. Because of the importance of sticky, personal information, for some kinds of innovation and product customization, Von Hippel [8] suggests that in certain circumstances the innovation will be increasingly accomplished by end-users (user innovation) rather than an expert provider within an established innovation process. The lack of understanding (or knowledge) of this sticky information is, in the apparel industry, the source of most purchase returns. The increase in product returns, both in stores and on the Web, creates headaches for retailers as it bears consequences on their brand image, not to mention their profitability. Failing the customer's participation in the product conception/creation process, this hidden information must then be inferred, or decoded, by manufacturers.

37.4 Configuration and Product Design

Configuration processes play a crucial role to manage this task by providing customers support and navigation in co-designing their individual product or service. There is nothing simple about mass customization and it is not a simple strategy to undertake organizationally; it is not even a simple concept to comprehend. Today's market heterogeneity, increasing variety, steadily declining product life cycles, decreasing customer loyalty, and the escalating price competition in many branches of industry are the main motivators for firms going into mass customization [4].

Configuration is an essential aspect of mass customization because it creates the possibilities to guide customers as they are making choices. Pine [4] contends that the primary objective of a configurator is to facilitate the decision-making process of customers using a web-based interface. Product configuration systems play an important role in supporting the mass customization paradigm, as they help to determine the degree of personalization that a business will offer. Thus, the role of the configurator is to create a link between consumers and manufacturers [1]. Mass customization does not equate to an increase in costs. According to Piller and Blazek [5], using a configurator could significantly reduce costs since its web-based technology decreases the time required to take orders and the application of toolkits for customer co-design may be the most used approach to help customers navigate choices in a mass customization system.

In the current context, businesses use catalogs and manual production methods. Catalogs provide a predefined and limited number of combinations for a product without necessarily fulfilling all of a customer's specific needs [9]. Manual configuration, on the other hand, essentially relies on human expertise and necessitates competent and highly skilled workers [10]. However, a lack of expertise eventually requires investments in terms of time and efforts; moreover, it forces employees to keep up to date with frequent technical changes and improvements. As a result, the configuration of a product to meet a customer's requirements can become a complex task that gets more demanding as the number of components and options increases. When the configuration requires numerous variations, the possibility of errors also rises which can result in production delays. The repetition of subsequent steps may be required which can be costly. Mass customization creates various technical challenges that need to be overcome before mass customized garments can be produced.

The technological risks associated with a configurator project are essentially related to the development of a system that can share and process data and parameters (the parameter configurator) originating from various sources such as: the data entry tools (e.g., the body scanner), software, the automatic process, and the administrative and financial data. Rogoll and Piller [10] indicate that the optimal product configurator needs to create an interface between different programming languages and function entirely independently. A product configurator must be used along with a high-performance technological platform so as to allow for interaction between customer and manufacturer as the product is designed. This creates an interface between the customer and the supplier that provides opportunities for value co-creation.

The most important mass customization prerequisite is the understanding that mass customization itself is a highly customized strategy and that you cannot imitate someone else's successful mass customization strategy. If prime producers want to make the most of this project, they will have to better understand what can be done in terms of personalization and mass customization so as to formulate an appropriate strategy on how to use their configurator.

37.5 Mass Individualism

In an increasingly small and standardized world, the overconsumption of short-lived and highly available products will likely not be sustainable over the long term. But in addition to environmental and ecological considerations, consumers are better informed and thus more demanding and careful. Today, consumers still hope to obtain what they want, but not at any price, nor by acquiring merely any product. In their quest for uniqueness and authenticity, they require more sophisticated, customized, and carefully tailored products fitting their specific needs. The creation of value will therefore require the recognition of these two elements and new priorities for companies wishing to take advantage of this rapidly growing market segment.

According to our research, more than 60 % of consumers like to have guidance when making a choice in buying an apparel product and interior furnishings. A large majority (75 %) of this same group also expresses that it is difficult to trust the big brands and that the quality of their products has increasingly diminished. By contrast, when a brand offers reliability, honesty, and consistency, both in its creative approach and in terms of distribution, these customers tend to remain faithful.

Mass individualism is a philosophical, political, social, and moral concept that privileges the rights, interests, and values of the individual over those of larger groups, such as social media networks. It promotes individual autonomy over various other social and political institutions (the family, the clan, the corporation), which exercise multiple pressures on the individual. It should not be confused with mass egotism, as egotists only consider their personal interests, including through collective means, while mass individualists consider the general interests of all individuals, not just their own. At the same time, it should not be considered a form of political autocracy or total independence; being part of an organization, exchanging and communicating are not incompatible with the principle of mass individualism. This phenomenon reminds us that everyone desires to be different, but with the goal of maintaining individuality for all, using many different means, including collective ones. Therefore, manufacturers must strive to understand how to target this growing segment of consumers, not to seduce them into buying products, but also to give them what they truly want.

As always, consumers continue to dream, but today, they want different, unique products at better prices. As consumers strive for individuality and authenticity, they want more elaborate, personalized, adapted, and specific garments. The creation of value will thus only occur as manufacturers recognize these changes and define new priorities to respond to these new demands.

It is important to understand mass customization and personalization within the context of trade globalization, which has led to ever more ferocious market competition. But one question remains: why is it that some manufacturing industries are failing to provide this? It seems that one of the greatest obstacles for manufacturers is proving that they have the ability to adapt in order to create custom products of relatively high quality in a short time period that respond to the exact needs of customers; they must also offer them at a relatively low cost and at a volume that will respond to and fulfill mass demand. The principal cause of this obstacle is the lack of integration between technologies currently in place and those offered by suppliers that do not adequately respond to the needs of manufacturers and distributors.

The literature demonstrates the many limits to understand the different levels of customization. For those companies that wish to offer mass customization, it is necessary to gauge what impact this would have on their organization; they must have the capacity to respond to the demand. The goal of manufacturers today is to be as competitive as possible by developing a strategic flexibility in terms of their production processes. But more importantly, the comprehension of this process must be communicated, so that the consumers have access to the right information on which to base their consumption choices.

37.6 What Steps the Industry Must Take to Correct the Problem

Our research leads us to identify seven steps that manufacturers must take to begin offering mass customization to their customers:

1. Define a manufacturing strategy

   For manufacturers and retailers, customization must allow for the improvement of sales and profits, an increase in customer satisfaction, and the creation of a lasting competitive advantage.
2. Identify areas where customer needs diverge the most

   Mass customization begins with figuring out how shoppers' desires differ and deciding upon the most useful product features to customize, unlike mass production, which tries to fit the needs of universal customers. Customers will gravitate toward products and experiences that offer individual focus, interaction, and involvement in the entire value chain process, and they will only be attracted to products and experiences they perceive as meeting their unique needs.
3. Establish high levels of quality and reliability

   One of the challenges of mass customization is ensuring it does not impede upon supply chain operations. For this, manufacturers need a solid design and the ability to reengineer the organizational process and value chain resources to deliver customized solutions with near mass-production efficiency and reliability.

4. Generate involvement and genuine personalized communication

   Consumers will want the opportunity to interact at an individual level with retailers and suppliers. Taken to the extreme, they will seek out opportunities for involvement in the entire chain of activities that brings a product to market—from conception, design, and creation, to marketing and retailing, even to funding and true rewarding. For example, manufacturers and retailers need to understand they will be awash in consumer data. This will give rise to new customer-driven metrics designed to improve the shopper's journey and point-of-purchase experience by slowing down information overload for consumers.
5. Reduce the complexity of the product offering

   If mass customization offers too many choices, customers may find the selections too overwhelming and spend their money elsewhere. However, companies can minimize this risk by implementing a system that offers options based on personal information provided by the customer. While the authors concede that it is difficult to make the organizational changes to implement mass customization, they claim that doing so will lead to long-lasting competitive advantages.
6. Develop a strong sense of differentiation

   For manufacturers, it is necessary to strongly differentiate their brand, products and services from the competition, in terms of brand authenticity, product quality, and corporate identity. As well, a collaborative approach must be generated between companies and customers (e.g., co-fitting, *fitthinking*) because consumers as co-creators enhance their ability to get what they want by participating in the value chain as creators, adapters, re-mixers, repackagers, and re-merchandisers. Unprecedented levels of customer connectivity will actively engage consumers in the development and customization of their own products and shopping experience. We will see more customer-driven R&D, more mass customization, more personalization, and more onsite "manufacturing." Personalization will thrive in the digital world, unhampered by time and materials costs; and at the same time, more and more brick-and-mortar enterprises will benefit by incorporating personalization options into the mix, as well.
7. Create a unique value chain

   Out of mass merchandising, the next value chain will need to support niche merchandising, down to the specific location and customized individual unit. It will be defined by connectivity, the early capture of true demand signals, total visibility, shared data, real-time information, real-time response, decentralization, and integrated shared logistics. It will enable much clearer insight into true demand via the proliferation of interactive "choice-boards" designed to help consumers see and select from the full extent of product options available. We will see a transition to true demand (what the customer wants vs. what the customer was forced to buy) and even a transition to leaner consumption (minimizing waste by producing on demand and giving consumers what they want from the beginning). And it is in this difference that value is added. Manufacturers must know the difference in order to adequately respond to this demand for mass individualism and the negotiation between uniqueness and conformity.

37.7 Conclusion

These findings should encourage the actors that make up this industry to readjust. The above list should also trigger a wake-up call to the new reality: more demanding consumers, increasingly globalized markets, new technologies, etc. Manufacturers must be proactive, adapt, and adopt new mind-sets and management tools to take full advantage of information technologies. To successfully implement mass customization, it is of the utmost importance that they emphasize analysis, decision-making, performance evaluation, and added value. Most of all, flexibility is a must, as the market increasingly expects it; and in this time of mass individualization, the new economic realities require it.

References

1. Inala, K.: Assessing product configurator capabilities for successful mass customization. Theses, University of Kentucky (2007)
2. Pine, J., Korn, K.C.: Infinite Possibility: Creating Customer Value on the Digital Frontier. Berrett-Koehler, San Francisco (2011)
3. Zipkin, P.: The limits of mass customization. Sloan Manag. Rev. **42**, 81–87 (2001)
4. Pine, B.J.: Mass Customization: New Frontier in Business Competition. Harvard Business School Press, Boston (1993)
5. Piller, F., Blazek, D.: Core capabilities of sustainable mass customization. In: Knowledge Based Configuration (Chap. 9), pp. 107–119 (2014)
6. Haug, A., Hvam, L., Mortensen, N.H.: Definition and evaluation of product configurator development strategies. Comput. Ind. **63**(5), 471–481 (2012)
7. Moon, H., Lee, H.-H.: Consumers' preference fit and ability to express preferences in the use of online mass customization. J. Res. Interact. Market. **8**(2), 124–143 (2014)
8. Von Hippel, E.: "Sticky information" and the locus of problem solving: implications for innovation. Manag. Sci. **40**(4), 429–439 (1994)
9. Qin, S., Yang, L.: Fuzzy optimisation modelling for apparel fit from body scanning data mining. In: Proceedings of the Sixth International Conference on Fuzzy Systems and Knowledge Discovery, Tianjin, China, vol. 7, pp. 255–259 (2009)
10. Rogoll, T., Piller, F.: Product configuration from the customer's perspective: a comparison of configuration systems in the apparel industry. In: Proceedings of International Conference on Economic, Technical and Organisational Aspects of Product Configuration Systems, Lyngby, June, pp. 28–29 (2004)

Correction to: Does the Size of a Fashion Model on a Retailer's Website Impact the Customer Perceived Attractiveness of the Model and Purchase Intention? The Role of Gender, Body Satisfaction and Congruence

Anik St-Onge, Aurelie Merles, Florian Pichonneau, and Sylvain Sénécal

Correction to:
Chapter 22 in: J. Bellemare et al. (eds.), *Managing Complexity*, Springer Proceedings in Business and Economics,
https://doi.org/10.1007/978-3-319-29058-4_22

The book was inadvertently published with the following error and it has been corrected now:

On Page 284, Section 22.4 has been removed and the next section is renumbered consecutively as Section 22.4.

The updated online version of this chapter can be found at
https://doi.org/10.1007/978-3-319-29058-4_22

J. Bellemare et al. (eds.), *Managing Complexity*, Springer Proceedings in Business and Economics, DOI 10.1007/978-3-319-29058-4_38

Correction to: Equity Crowdfunding and the Online Investors' Risk Perception: A Co-created List of Web Design Guidelines for Optimizing the User Experience

Sandrine Prom Tep, Sylvain Sénécal, François Courtemanche, and Valerie Gohier

Correction to: Chapter 24 in: J. Bellemare et al. (eds.), *Managing Complexity*, Springer Proceedings in Business and Economics, https://doi.org/10.1007/978-3-319-29058-4_24

Chapter 24 was inadvertently published with incorrect family name of one of the authors. This has now been updated as

Given name: Sandrine

Family name: Prom Tep

The updated original version of this chapter can be found at https://doi.org/10.1007/978-3-319-29058-4_24

J. Bellemare et al. (eds.), *Managing Complexity*, Springer Proceedings in Business and Economics, DOI 10.1007/978-3-319-29058-4_39